# Economic Issues for Consumers

## TENTH EDITION

**Roger LeRoy Miller**
Institute for University Studies
Arlington, Texas

**Alan D. Stafford**
Niagara County Community College
Sanborn, New York

THOMSON

WADSWORTH

Australia · Canada · Mexico · Singapore · Spain · United Kingdom · United States

**THOMSON**

**WADSWORTH**

Editor: Robert Jucha
Assistant Editor: Stephanie Monzon
Editorial Assistant: Melissa Walter
Technology Project Manager: Dee Dee Zobian
Marketing Manager: Matthew Wright
Marketing Assistant: Tara Pierson
Project Manager, Editorial Production: Ann Borman
Print/Media Buyer: Doreen Suruki
Permissions Editor: Joohee Lee

Text Designer: Ann Borman
Copy Editor: Pat Lewis
Proofreader: Suzie Franklin DeFazio
Cover Designer: Bill Stryker
Cover Images: Photodisc, Frank Herholdt, Getty Images
Cover Printer: Lehigh Press
Compositor: Parkwood Composition
Printer: Edwards Brother, Ann Arbor, MI
Indexer: Bob Marsh

For more information about our products, contact us at:
**Thomson Learning Academic Resource Center**
**1-800-423-0563**
For permission to use material from this text, contact us by:
**Phone:** 1-800-730-2214
**Fax:** 1-800-730-2215
**Web:** http://www.thomsonrights.com

Library of Congress Control Number: 2003-106863
ISBN 0-534-62852-4

**Wadsworth/Thomson Learning**
**10 Davis Drive**
**Belmont, CA 94002-3098**
**USA**

ASIA
Thomson Learning
5 Shenton Way #01-01
UIC Building
Singapore 068808

AUSTRALIA
Nelson Thomson Learning
102 Dodds Street
South Melbourne, Victoria 3205
Australia

CANADA
Nelson Thomson Learning
1120 Birchmount Road
Toronto, Ontario M1K 5G4
Canada

EUROPE/MIDDLE EAST/AFRICA
Thomson Learning
High Holborn House
50/51 Bedford Row
London WC1R 4LR
United Kingdom

LATIN AMERICA
Thomson Learning
Seneca, 53
Colonia Polanco
11560 Mexico D.F.
Mexico

SPAIN
Paraninfo Thomson Learning
Calle/Magallanes, 25
28015 Madrid, Spain

# Contents

**Preface**

## Unit One:
## THE CONSUMER IN TODAY'S WORLD

### Chapter 1
### An Economic Foundation
### for Consumer Decisions 1

## Chapter 2
# Making Rational Consumer Choices 25

## Chapter 3
# A Flood of Advertising 48

**Chapter 4
The Many Faces of Fraud  70**

# Chapter 5
# Protection for the Consumer 94

# Unit Two: BUDGETING

# Unit Three:
# MAKING CONSUMPTION EXPENDITURES

## Chapter 10
## Purchasing Household Products 215

## Chapter 12
## Choosing a Place to Live 268

# Unit Four: FINANCIAL MANAGEMENT

# Chapter 14
# Using Credit Responsibly 328

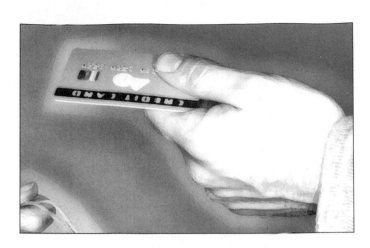

**The Ethical Consumer:**
Getting in over Your Head on Purpose? 337

**Cyber Consumer:**
Bankruptcy Information Online 346

## Unit Five:  RISK MANAGEMENT

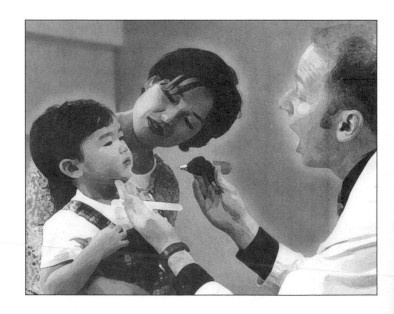

## Chapter 17
# Insuring Your Home and Your Automobile 416

## Unit Six:
## LESSONS YOU HAVE LEARNED

# Preface

As we pass through the first decade of the twenty-first century, we can look back over the past hundred years and realize how much the lives of American consumers have changed. One hundred years ago our society was primarily rural in nature. Most Americans were still riding on horses or in buggies. Women had not yet obtained the vote in federal elections. Heavier-than-air flight had just become a reality instead of a dream in the minds of the Wright brothers. Taking a spaceship to the moon was the topic of science fiction novels. Although the telephone existed, there were fewer than twenty telephones for every one thousand people. Electricity had found its way into less than 5 percent of all houses. Radio, television, DVD players, tape recorders, microwave ovens, computers, the Internet, and even movies had yet to be invented. Today, we take these products, and many more, for granted. It is safe to say that no other hundred-year period in the history of civilization has encompassed such profound and socially transforming technological change as took place in the twentieth century.

What does this mean for our nation's consumers? It means we face a more complex world and more difficult consumer decisions every day. In at least one important aspect, however, consumer decision making is the same today as it was at the beginning of the twentieth century: all consumers, everywhere on earth and at all times, are, have been, and will be faced with limited resources. Regardless of the point in time, consumers must make choices about how to spend their finite income. We assume throughout this text that all consumers have the common goal of achieving a higher quality of life through the choices they make.

In this, the tenth edition of *Economic Issues for Consumers,* many topics have been added or expanded because they are more important today than in the past. Probably the greatest force for change in our economy today is the explosive growth of the Internet. Although consumers derive many important benefits from the World Wide Web, they also bear costs in terms of their need to learn about emerging technologies and through the possible erosion of their personal privacy. Accordingly, significant portions of the first three chapters of this text are devoted to learning about the Internet and how it may be used to make better consumer decisions. All subsequent chapters make reference to the Internet in a variety of ways and include activities that link its capabilities and applications to consumer issues. Among the Internet-related topics covered are the electronic transformation of our banking and investment systems; online marketing of food, furniture, and pharmaceuticals; searching for employment on the Web; finding the best deals for insurance online; and many, many others. Other important consumer issues that are discussed include our expanding need for affordable medical care, the creation and marketing of genetically engineered food, and inflation-indexed government bonds. Actually, the list of new areas of concern that are examined in this text would go on for many pages. It is fair to say that the most important consumer issues of our day are covered.

# The Format of the Book: Chapters Highlight Consumer Issues

You will find that the format of this book lends itself to easy use and understanding. Major topics of consumer economics are presented at the beginning of each chapter. Each chapter then concludes with a **"Confronting Consumer Issues"** section that demonstrates relevant consumer skills students may apply in their own lives. For example, Chapter 4 discusses fraud in the marketplace, and the Confronting Consumer Issues section at the end of that chapter outlines measures individuals can take to protect themselves from identity theft. As another example, the chapter on health care concludes with a Confronting Consumer Issues section that describes how consumers can choose rationally among doctors and other health-care providers.

# A Central Focus

Professional reviews of the ninth edition of this text indicated a continued need to encourage student use of the Internet to find information and make rational decisions. At the same time, it was clear that many instructors wanted most Internet activities to be drawn together in a central location in each chapter. This is why Internet features have been relocated in this edition. **"The Cyber Consumer"** remains as a boxed feature within each chapter. This activity asks students to visit and evaluate a particular Web site. Other Internet exercises have been moved to the end-of-chapter activities in a section titled **"Internet Resources."** Three activities are included here. They are **"Finding Consumer Information on the Internet,"** which asks students to investigate designated Web sites that provide consumer information students could use to make better personal decisions; **"Shopping on the Internet,"** which asks students to review e-commerce Web sites that offer consumers goods or services for sale; and **"InfoTrac Exercises,"** which illustrate how the periodical research engine InfoTrac may be used to investigate consumer issues. Access to InfoTrac is provided with each new copy of this text. An emphasis on the importance of the Internet pervades the tenth edition of *Economic Issues for Consumers*. This focus makes this edition useful and interesting to students.

# Key Changes for the Tenth Edition

So much has happened since the publication of the ninth edition that a large number of changes were necessary to provide the most up-to-date information possible in a usable format for students who want to be rational consumers.

## A New Presentation for Consumer Issues

In past editions of *Economic Issues for Consumers,* the end of each chapter was set off from the body of the chapter to accommodate a Consumer Issue feature. Although these issues were related to the central topic of each chapter, instructor reviews indicated that some students did not appreciate this relationship and chose not to read or fully consider the importance of the information and skills included in these pages. In the tenth edition, the issues applications of consumer knowledge and skills have been integrated into each chapter under the major head **"Confronting Consumer Issues."** These sections still provide students with important information and consumer skills, but they are no longer set off from the rest of the chapter. This design will encourage more students to read and learn from the entire chapter.

## A New Location for "Questions for Thought & Discussion"

In the tenth edition of *Economic Issues for Consumers,* the **"Questions for Thought & Discussion"** are located in both the end-of-chapter activities and the margins of the chapter pages. This placement puts questions in the same general area where the topics to which they relate are presented in the text. Placing these questions in the margins will allow students to consider topics they reference more easily.

## Significantly Revised and Reordered Chapters

To provide students with the most relevant information in the clearest possible sequence, a number of chapters have been revised and their order adjusted.

- **Chapter 4, "The Many Faces of Fraud,"** includes a broad presentation of the dangers of identity theft. This form of fraud is different from others because it does not require the active involvement of the victim. Taking steps to protect yourself from identity theft is perhaps the most important consumer effort an individual can make.

- **Chapter 10, "Purchasing Household Products,"** concludes with an extensive discussion of how consumers may save money by saving energy. In the twenty-first century, increasing energy costs are certain to be an important factor in every consumer's budget and decision-making process. This section shows students how simple steps anyone can make will lead to major energy-cost savings. Consumers who follow these recommendations will find they have more income to spend on other goals.

- **Chapter 13, "Banks Help Consumers Save and Spend,"** consolidates Chapters 13 and 15 from the ninth edition. Professional reviews of the previous edition indicated that many instructors taught these chapters (Banks and the Banking System, and Saving) together. The most important information from these two chapters has been integrated into this new chapter, and the "Confronting Consumer Issues" section has been expanded to accommodate the growth of electronic banking alternatives.

- **Chapter 15, "Investing,"** now includes a discussion of the stock market "meltdown" of the early twenty-first century as well as the importance of investigating 401(k) plans and other investment programs. The Enron collapse and similar financial debacles of 2001–2002 are addressed, and students are provided with information to help them evaluate their investment opportunities.

- **Chapter 19, "Looking to the Future,"** maintains the future perspective of the final chapter in the ninth edition, but has been expanded to include topics from Chapters 20 and 21 of the previous edition. By combining the most important topics and consumer issues from these three chapters, this edition has saved a significant amount of space that has been used to expand on other topics. Topics addressed in Chapter 19 now include the impact of new technologies on tomorrow's consumers, the globalization of the world's economy, the potential loss of personal privacy due to technological advances, why consumers will be forced to make ethical decisions about the allocation of our medical resources, and our need to be environmentally responsible consumers. All of these topics lead students to realize that they will need to use the consumer skills they have learned to make rational choices throughout their lives.

## Additional Changes in Chapters and Consumer Issues

Other chapters and consumer issues have been updated to make sure that the text, tables, graphs, charts, and references are as current as possible. Important changes and additions include the following:

- **Chapter 3, "A Flood of Advertising,"** now includes a discussion of the Children's Online Privacy Protection Act (COPPA) of 2000.
- **Chapter 6, "The Consumer as a Wage Earner,"** discusses recent decisions regarding Title IX of the Equal Opportunity Act, which requires colleges to provide equal athletic opportunities for women and men.
- **Chapter 8, "Paying for the Government,"** explains recent and proposed reforms in federal income taxes.
- **Chapter 11, "Satisfying Transportation Needs,"** identifies and evaluates alternative means of transportation consumers should consider before choosing to purchase an automobile.
- **Chapter 12, "Choosing a Place to Live,"** discusses the costs and benefits of choosing to live in manufactured housing.
- **Chapter 14, "Using Credit Responsibly,"** explains differences between subprime and predatory lending and indicates how consumers may protect themselves from credit fraud.
- **Chapter 16, "The Health-Care Dilemma,"** discusses costs and benefits of proposals that are intended to assure the financial viability of the federal Medicare program and other medical-insurance providers.

## Revised and New Features

Several of the special features of the ninth edition of *Economic Issues for Consumers* have been maintained in this edition, and others have been revised. "The Global Consumer" is now **"The Diverse Consumer."** This change allows a greater variety of examples to be used to demonstrate how consumers with various genders, ethnicities, or nationalities approach consumer decisions. **"The Ethical Consumer," "The Cyber Consumer,"** and **"Consumer Close-Ups"** have been maintained. New end-of-chapter activities include three features under the heading **"Internet Resources": "Finding Consumer Information on the Internet," "Shopping on the Internet,"** and **"InfoTrac Exercises."**

## Access to InfoTrac

Each student who purchases a new copy of this text is provided with access to *InfoTrac* online services for his or her own personal computer. Students may be encouraged to use InfoTrac to investigate consumer issues and to update information as the economy and government laws or regulations change.

## Pedagogical Aids

Students will find a number of pedagogical aids in all chapters of this text. Each chapter begins with a **"Preview,"** a set of questions that indicate to the reader the topics that will be covered. To introduce new terminology and to allow the reader to follow the text more meaningfully, **"Key Terms"** appear in boldface type where they are first used in the text, and they are defined in the margin of that page. These key terms are also listed in the end-of-chapter materials, along with the page where they are first used. **"Questions for Thought and Discussion"** appear in the margins of each chapter. These questions can be a basis for class discussion, for individual thought, or for group work without the direct aid of the instructor. At the end of each chapter is a point-by-point **"Chapter Summary"** that can serve for review. **"Things to Do"** lists projects a class can do as a group or that individuals can do on their own at the request of the instructor. These activities demonstrate practical applications for knowledge gained in this course. Finally, **"Selected Readings"** identifies additional sources of reading for those who wish further information on subjects covered in the chapter.

## Other Useful Changes or Additions to the Tenth Edition

You will notice a wide use of illustrative materials—photographs, charts, tables, and Internet screen captures. Visualization of ideas not only aids students in understanding the material, but also makes the task of reading the text more enjoyable. All illustrations are either referred to directly in the text or have captions that include critical-thinking questions to relate them to the topics being discussed in the text.

# Supplementary Materials

A **Web site** with ongoing updates has been established to provide online support and study assistance to students who use *Economic Issues for Consumers,* 10th edition. A selection of self-help activities can be completed at this Web site to reinforce learning that has taken place in class and while reading this text. These include a **Reading Outline, Key Terms** exercise, **Using Consumer Skills** activity, **Trial Test, Short Answer Questions,** and **Suggestions for Further Learning,** for each chapter. The Web site also contains current links to other sites of interest, including many of those mentioned in this text.

A **test bank** that provides test items of varying levels of difficulty that accommodate students' individual learning styles and academic abilities is provided to instructors.

# Acknowledgements

A number of professionals in the field reviewed previous editions of *Economic Issues for Consumers*. We remain grateful to them for their thoughtful suggestions on how to create a text that best suits the needs of today's students and faculty.

Judy L. Allen
Southwest Texas State University

Howard Alsey
Arkansas State University

Anne Bailey
Miami University, Ohio

Joseph E. Barr
Framingham State College

Phillis B. Basile
Orange Coast College

Carolyn Bednar
Peru State College

Margarita Blackwell
Eastern Kentucky University

Harold R. Boadway
Moraine Valley Community College

Jean S. Bowers
Ohio State University

Margaret Jan Brennan
Western Michigan University

Mary L. Carsky
University of Hartford

Patricia A. Daly
Framington State College

Kay P. Edwards
Brigham Young University

Judy Farris
South Dakota State University

James A. Fetters
Naugatuck Valley Community-Technical College

Barbara Follosco
Los Angeles Valley College

David G. Garraty
Thomas Nelson Community College

Linda Graham
Wichita State University

Joyce S. Harrison
Middle Tennessee State University

Mary B. Harris
Sam Houston State University

Joyce S. Harrison
Middle Tennessee State University

Ron Hartje
Souk Valley College

Ann R. Hiaat
University of North Carolina at Greensboro

James O. Hill
Vincennes University

Hilda Jo Hennings
Northern Arizona University

Thomas A. Johnson
William Rainey Harper College

William L. Johnston
Oklahoma State University

Jane Buckwald Kerr
Cameron University

Hazel Kirk
Brevard Community College, Cocoa Campus

Ann Lawson
Thomas Nelson Community College

John R. Lindbeck
Western Michigan University

Dr. Merlene Lyman
Fort Hays State University

Michael Magura
University of Toledo

Allen Martin
California State University, Northridge

Esther McCabe
University of Connecticut

Roger Moore
Arkansas State University

Michael L. Oliphant
Southwest Virginia Community

Geraldine Olson
Oregon State University

Teresa M. Palmer
Illinois State University

Claudia J. Peck
Oklahoma State University

James Poley
City College of San Francisco

Thomas J. Porebski
Triton College

Rose Reha
St. Cloud State University

Shirley Schecter
Queens College

Jolene Scriven
Northern Illinois University

Candy Sebert
University of Central Oklahoma

Eugene Silberberg
University of Washington

Reuben Slesinger
University of Pittsburgh

Alden W. Smith
Anne Arundel Community College

Lisa J. Snyder
Northern Arizona University

Nancy Z. Spillman
Los Angeles Trade-Technical College

Barbara L. Stewart
University of Houston

Patti Wooten Swanson
Bethel College

Faye Taylor
University of Utah

Merle E. Taylor
Santa Barbara City College

Margil Vanderhoff
Indiana University

Mary Ann Van Slyke
North Central Technical Institute

Frank A. Viggiano, Jr.
Indiana University of Pennsylvania

Roberta A. Walsh
University of Vermont

Louise Wesswick
University of Wyoming

Mari S. Wilhelm
University of Arizona

Joseph Wurmli
Hillsborough Community College

Betty Young
Minnesota State University at Mankato

Prudence Zalewski
California State Polytechnic University
San Luis Obispo

In preparing the tenth edition of this book, we benefited from the comments of a number of users and reviewers of the ninth edition. We would like to thank the following professionals for their conscientious work:

Anne Bailey
Miami University, Ohio

Barbara Clauss
Indiana State University

John Grable
Kansas State University

Deborah Haynes
Montana State University

Roger L. Moore
Arkansas State University, Beebe

Karen Schmid
Indiana State University

Pamela Turner
Turner Consulting and Training Services

We would like to thank our editor at Thomson/Wadsworth, Bob Jucha, for his guidance and support, as well as Melissa Walter, the editorial assistant, who supervised all aspects of the supplements and many details of the text. We also thank Pat Lewis and Suzie DeFazio for their excellent copyediting and proofreading assistance. Ann Borman oversaw the production and design of the text and made updates to the art program. Ann Hoffman assisted with photo selection and research. Dee Dee Zobian worked behind the scenes to create the interactive Web site.

# To the Student

*Economic Issues for Consumers,* Tenth Edition, provides you with a foundation of information that will help you gain valuable and useful knowledge from the course you are taking. Although this text contains substantial factual data and describes many specific laws and consumer protection regulations, its greatest value lies in the practical advice it offers to help you make your own consumer choices. Much of what you learn from this text will have immediate and beneficial applications in your own life.

To use this text to best advantage, you should read the "Preview" questions that are found at the beginning of each chapter. These provide a general idea of the content of the chapter. You should then read the assigned material, taking notes on topics about which you would like more information, that you do not understand, or that have particular significance for you. Use these notes as study aids and as the basis for participating in classroom discussions. Your instructor is well qualified to provide additional information, but he or she needs you to identify the areas you find most interesting or challenging. *Economic Issues for Consumers* is a tool that will help you gain useful knowledge and skills, but it is only a tool. The value you receive from this text, and from your class, to a large extent depends on you and the amount of effort you put into your study of consumer economics.

# CHAPTER 1

# An Economic Foundation for Consumer Decisions

After reading this chapter, you should be able to answer the following questions:

- What is scarcity, and why does it necessitate choices?
- What are trade-offs and opportunity costs, and how are they involved in consumer decisions?
- How do the laws of demand and supply affect the consumer?
- What are some of the roles governments play in different types of economic systems?
- How has the U.S. economy been affected by the growth of the Internet?
- What new consumer responsibilities have been created by new technologies?
- What is ethical behavior, and how is it part of the consumer decisions Americans make?

**Consumers** Individuals who purchase (or are given), use, maintain, and dispose of products and services in their final form in an attempt to achieve the highest level of satisfaction possible with their income limitation.

**Scarcity** The condition in which we are unable to provide enough products to satisfy all people's needs and wants because of our limited resources.

**Goods** Tangible objects that have the ability to satisfy human wants.

**Services** Intangible actions that have the ability to satisfy human wants.

This is a book about consumer economics. That means it has to do both with economics and with you, the **consumer.** Its goal is to help you learn to apply economic principles when you make consumer decisions—such as whether to own or rent a house, how to choose the clothing you wear, what kind of insurance to purchase, whether a new or a used car is a better choice for you, the types of checking and savings accounts to use, and so on. The list of consumer decisions you will make in your life is virtually endless. Since we are all consumers for our entire lives, we should strive to make decisions that bring us the greatest satisfaction by analyzing our alternatives according to our individual values, goals, and objectives. An understanding of economics can help us gain greater satisfaction from being more rational consumers.

Unlike professionals with special training, U.S. consumers often make buying decisions without having had any formal instruction in the basic principles of economics that operate in the world around them. The task of this chapter is to put the horse back before the cart, so to speak, by presenting some of the fundamental economic concepts that you will encounter as a consumer. The finer points of economic theory do not concern us here. We leave that kind of instruction for a course in either microeconomics, the study of individual and business decision-making behavior, or macroeconomics, the study of economy-wide problems such as inflation and unemployment.

# Scarcity

We begin our discussion of economics with the problem of **scarcity** because it is the heart of economic analysis. Would you like to have more time to study and still have time for all the other things you want to do, such as participating on an athletic team or going to the student center with your friends? Would a larger house or apartment or a bigger room in your dormitory please you? How much value would you receive from owning better clothes or a new car? Indeed, why can't we all have more of everything? The answer is that individually and collectively we face the problem of scarcity.

## Why Scarcity Exists

Scarcity exists because we have unlimited wants and only a limited supply of resources to produce **goods** and **services** that satisfy our wants. Scarcity is a relative term. For someone who has no food, a crust of bread is a scarce product, while others may experience scarcity in their inability to afford steak or lobster every night. As long as we are not able to have everything we want, scarcity exists. This means that scarcity and poverty are not the same thing. Scarcity can exist in an environment of affluence and abundance.

Imagine that you were the richest person in the world. You still could not have everything you might want. The most obvious example of scarcity in this case would be the scarcity of time. With only one life to lead, and only twenty-four hours in a day, you would be unable to enjoy all the things you could buy. Given our finite life spans (we have only so many years to live) the scarcity of time is a problem we all face.

## How We Experience Scarcity

Usually, people experience scarcity in terms of having a limited amount of funds to spend. Although, from their personal point of view, they are correct, from the perspective of the entire economy, scarcity has a different cause.

**Question for Thought & Discussion #1:** If you won $100 million in the power ball lottery, why would you still face the problem of scarcity?

Remember, scarcity is the result of our limited ability to produce goods and services to satisfy our wants. If our government printed enough money to give every person $1 million in cash, the amount of food, clothing, houses, swimming pools, yachts, and other products we would wish to buy would not change significantly. Regardless of how much extra money people have, there will not be enough products to satisfy their unlimited wants. Scarcity would still exist because there would be no more oil in the ground, factories, workers, or other **productive resources** than previously existed. It would not be possible to create enough additional products to satisfy everyone's wants. We would all have more money, but we would not have overcome the problem of scarcity. Thus, scarcity is an unavoidable fact of everyone's life.

## Universal Scarcity

The problem of scarcity is not unique to consumers; it is faced by businesses, the government, and all nations in the world. Successful businesses in the United States earn a **profit** when they receive more income from sales than the costs they pay to produce and offer products for sale. When a firm uses its profits to buy resources to produce a television set, it cannot use those same resources to produce a DVD player instead. A business that buys machines to automatically assemble electric mixers cannot use that money to hire more workers. Businesses must make these types of choices because the resources they use to produce products to offer consumers are scarce regardless of how much profit they earn.

## Scarcity and the Government

We all receive valuable services from our government. The roads we drive on, the schools we attend, and the fire and police protection on which we rely are only a few of the many government services we all need but that most individuals could not afford to purchase for themselves. Our government pays for the resources it uses to provide these services by taxing or borrowing money. When our government buys resources, it faces the problem of scarcity because it cannot use the same money to accomplish more than one objective. Money paid for road repairs, for example, may not go to pay schoolteachers. The government may also be seen as contributing to the problem of scarcity for other parts of society. The resources it uses are not available to people and businesses to buy and use for other purposes. Within our country, consumers, businesses, and the government compete for the scarce resources we have.

## Scarcity in Foreign Lands

Just as Americans must deal with scarcity, all nations in the world face this problem. Resources that go to build automobiles in Japan cannot be used to increase the supply of affordable housing in that country. Land used to grow sugar cane in Cuba may not be used to raise other crops. Oil that is pumped from the ground in Saudi Arabia today will not be available in the future. Although the way in which economic decisions are made varies from one nation to another, all nations face this problem. Differences in the way resources are distributed throughout the world form the basis for international trade that affects all our lives. More will be said about the global nature of our economy later in this book.

**Productive Resources** Raw materials, tools, and labor that may be used to produce other goods or services that have the ability to satisfy human wants.

**Profit** The difference between the total amount of money income received from selling a good or a service and the total cost of providing that good or service.

**Opportunity Cost**   The value of a second-best choice that is given up when a first choice is taken.

**Trade-Off**   A term relating to opportunity cost. To get a desired economic good, it is necessary to trade off some other desired economic good whenever we are in a world of scarcity. A trade-off, then, involves a sacrifice that must be made to obtain something.

> **❝Every choice you make requires you to give something up.❞**

**Question for Thought & Discussion #2:**
What is your opportunity cost for taking this course?

# The Necessity of Choice

Scarcity forces us, as consumers, to make choices all the time. We must choose how we spend our time, how we spend our labor power (that is, what kind of work we do), and how we spend our income (our purchasing power). Life would be simple without scarcity. You and I would not have to bother much about consumer economics. In a world without scarcity, choices would not have to be made because all consumers would possess every good or service they want. But we never will achieve such total satisfaction. That is why a knowledge of consumer economics is essential for maximizing the value we can derive from the consumer decisions we make.

When making a budget, for example, you need to decide whether you would receive more satisfaction from purchasing new clothing or from taking a relaxing vacation. Or you may decide to save your money now so that you can buy a new car in the future. You can think of consumer decision making as a rational way of determining how to allocate your scarce time and money. Later in this chapter we will consider how consumer decisions in our country help determine the way our productive resources are allocated.

## Choice and Opportunity Cost

Every choice you make requires you to give something up. When you sat down to read this book, you chose not to do at least a thousand other things with your time. You could have read your English text, you could have watched television, you could have slept, you could have gone to the movies, and so on. Thus, scarcity of time has led you to choose to read this book rather than do something else that is presumably of less value. The something else that you chose not to do is the cost associated with spending time reading this book. Economists call it **opportunity cost.**

Let's assume that of all the other things you could have done instead of reading this book, the thing you most wanted to do, but didn't do, was watch television. If that's the case, then the value of the enjoyment you would have received from watching television is your opportunity cost of reading this book. Opportunity cost is defined as the value of a second-best choice that is given up when a first choice is taken. Opportunity cost is an unavoidable part of all decisions consumers make. It helps us place a value on the scarce resources that go into producing products that satisfy our wants.

## The Trade-Offs Facing You

Whatever you do, you are "trading off" one use of a resource for one or more alternative uses. The value of a **trade-off** is represented by the opportunity cost just discussed. Let's go back to the opportunity cost of reading this book. Assume that you have a maximum of ten hours per week to spend studying just two subjects—consumer economics and accounting. You believe that the more time you spend studying consumer economics, the higher your grade will be in this subject. The corresponding is true for

Digital Vision/Getty Images

What trade-offs do consumers make when they buy prepared foods?

accounting. There is a trade-off, then, between spending an hour reading this book and spending that time studying accounting. A better grade in consumer economics must be purchased at the expense of a lower grade in accounting.

Similar trade-offs occur for every choice you make. If you decide to join your school's basketball team, you must trade the value of other uses of your time for the value of the enjoyment received from participating in that sport. Whenever businesses, the government, or people in other nations make choices, they also make trade-offs in which an opportunity cost is paid for the choice that is made.

**Question for Thought & Discussion #3:**
What trade-offs have you made to be able to attend college?

## An Implicit Assumption

Economic analysis rests on an assumption that we should make clear: most people make choices that are intended to make themselves better off. Making oneself better off can take many forms, depending on each person's individual values. For the purpose of economic analysis, however, we may assume that people attempt to make themselves better off in terms of their ability to buy goods and services that satisfy their wants.

## Markets

In economics, a **market** is the sum of all transactions that take place between buyers and sellers of a particular type of product. Therefore, if we talk about the "used-car market," we are referring to the transactions between people who buy and sell used cars. The "labor market" is made up of the agreements between workers and their employers. There is even an "education market" in which students pay tuition to receive instruction and academic guidance.

Markets exist between and among people, businesses, governments, and other nations in the world. To demonstrate the importance of international markets, consider that in 2003 roughly $1 out of every $9 spent in this country went to buy a product that was manufactured in another nation. At the same time, millions of Americans relied for their employment on our ability to sell U.S.–made products in other nations. Despite its importance, the U.S. economy is only a part of the larger world economy. To understand markets in that international economy, as well as local and national markets, we need to learn more about the way transactions are carried out.

## Voluntary Exchange in the U.S. Economy

In the United States, consumers and producers are generally free to use their resources to buy and sell products as they choose. This means that we enjoy the right of **voluntary exchange** in most situations. You can spend your income collecting stamps, and I can buy an expensive boat. Other people are free to start businesses that offer rare stamps or expensive boats for sale. Although limitations do exist on the types of products Americans may legally buy or offer for sale (drugs, for example), individuals in the United States are free to make most economic choices.

Whenever a voluntary exchange takes place, it is reasonable to assume that both the buyer and the seller will benefit. Imagine that you paid $20 for gasoline for your car. You must have felt that the gasoline had greater value than the $20 you spent. The owner of the filling station, in contrast, must have valued

**Market** The sum of all transactions that take place between buyers and sellers of a particular type of product.

**Voluntary Exchange** Transactions completed through the free will of those involved.

**Demand**    The quantity of a product that will be purchased at each possible price.

**Law of Demand**    A basic economic principle that states that as the price of goods or services rises, the quantity of those goods and services demanded will fall. Conversely, as the price falls, the quantity demanded will rise.

**Supply**    The quantity of a product businesses are willing to offer for sale at each possible price.

**Law of Supply**    A basic economic principle that states that as the price of goods or services rises, the quantity of those goods and services supplied will increase. Conversely, as the price falls, the quantity supplied will also decline.

**Question for Thought & Discussion #4:**
What would happen to the quantity of courses demanded at your school if tuition rates doubled?

the $20 more than the gasoline. When the transaction was completed, both parties gained, because they valued the money and the gasoline differently. The same must be true of all other voluntary exchanges, because people would not choose to complete a transaction unless they felt that it would benefit them.

# Demand and Supply Analysis— A Brief Introduction

In the U.S. economy, the forces of demand and supply most often work together to determine which products will be produced and the quantity of each type of product that will be offered for sale. Later in this text you will discover how these same forces help consumers make many other economic decisions.

## Law of Demand

The term **demand** refers to the quantity of a product consumers are willing and able to buy at each possible price. The **law of demand** states that if nothing else changes, consumers will buy a greater quantity of a product at a lower price than at a higher price. Therefore, if the price of a product falls, the quantity demanded will rise; and if the price increases, the quantity demanded will fall. Imagine that you and several of your friends have made a habit of going to the movies together every Friday night after work. If the price of a movie ticket increases by $2, some of the people in your group might not be able to afford it, while others might prefer to spend their money on some other product. In either case, the quantity of movie tickets demanded would be less because of the increase in price.

The quantity of a product consumers demand at any price depends on many factors other than price. These factors include tastes, income, the price of other related products, and expectations of what may happen in the future. If your friends all choose to wear a particular brand of clothing, for example, your demand for this brand of clothing may increase even if the price remains unchanged. If you accept a new job that pays you more than you earn now, you can afford to demand a greater quantity of products such as new cars or vacations. An increase in the price of gasoline, however, could cause you to drive fewer miles. This would reduce your demand for replacement tires for your car. As a final example, suppose your boss tells you that you will be laid off in two weeks. At present you still have a good income, but you know you won't in just fourteen days. Your demand for many products will be less because you expect your income will be smaller in the future. Each of these examples demonstrates one of many possible reasons why consumers might demand more or less of a product when there is no change in its price.

## Law of Supply

**Supply** refers to the quantity of a product businesses are willing to offer for sale at each possible price. Businesses in the United States intend to earn a profit. The more consumers are willing to pay for a product, the more likely firms are to earn a profit by supplying it. Therefore, consumer acceptance of higher prices often causes existing firms to increase their production, as well as encouraging the creation of new firms that produce this type of good.

The **law of supply** states that if nothing else changes, a greater quantity of a product will be supplied at a higher price than at a lower price. Therefore, if the price of a product increases, the quantity of the product that is supplied should also increase, and if the price falls, the quantity supplied should decline. You may see the law of supply in your own life. Ask yourself how many hours you would be willing to work at $2 an hour? $8 an hour? $100 an hour? Wouldn't you be willing to supply more of your labor as the wage (price) you receive grows?

Other factors that affect the profit a firm earns will also change its willingness to supply products. For example, a firm that produces bread might choose to go out of business if a 50 percent increase in the cost of flour eliminates its profit. A firm that finds a way to produce carpet with fewer workers will be encouraged to supply more of its product because of its reduced labor costs and increased profits. Generally, an increase in a firm's costs of production will reduce the quantity of products it will supply at each possible price, while a decrease in its costs will cause it to supply greater quantities.

> **Equilibrium Price** A price at which the quantity of a good or service demanded is exactly equal to the quantity that is supplied.

## Equilibrium Price

The price you pay for most goods and services you buy is determined by the forces of demand and supply. Every product has a price at which the quantity demanded is exactly equal to the quantity supplied. Natural forces operating within markets tend to force prices toward this **equilibrium price.**

Suppose that lettuce farmers are selling their products for $3 a head. This high price will encourage many farmers to grow more lettuce, but it will also

---

### THE diverse CONSUMER

## New Demographics Equal New Demand and New Supply

On January 21, 2003, the U.S. Bureau of the Census announced that for the first time in U.S. history people who identify themselves as Latino or Hispanic outnumber all other minority groups in this country's population. According to government estimates, the number of Latinos living in the United States has reached 37 million, or 13 percent of the total population. The U.S. Latino population has been growing rapidly in recent years. In 1980, Latinos accounted for only 6.4 percent of the U.S. population. By 1990, this proportion had reached 8.9 percent. At the current rate of growth, the Latino population will account for roughly 16 percent of the U.S. population by 2010. What will this growth mean for the U.S. economy?

One obvious result of the growth of the U.S. Latino population is an increase in the production of goods and services that are intended to appeal to people in this ethnic group. If you examine any big-city newspaper or surf the Internet, you will find many advertisements printed in Spanish. Latino spending in the U.S. economy is estimated to total nearly $800 billion per year. This market is attractive to many businesses, both large and small. Demographic (population) trends in recent years have caused both the demand for, and the supply of, products designed to appeal to Latino consumers to grow.

### Figure 1.1
**Demand and Supply for Products Intended to Appeal to the U.S. Latino Population, 1980 and 2003**

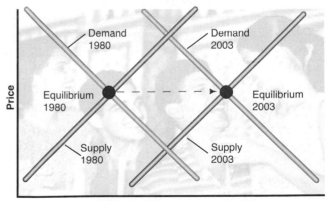

Quantities Demanded and Supplied

Consider the graph of demand and supply shown here. There are two sets of demand and supply curves, each of which is dated. What factors do you believe caused these curves to shift to the right over time? Why hasn't the equilibrium price for these products changed very much? How would you expect the location of these graphs to change in the future?

discourage consumers from buying the product. As a result, there will be a surplus of lettuce, which will force the price down until it reaches the equilibrium price (let us say $1.50) where the quantity supplied and the quantity demanded are the same.

In a similar fashion, prices below the equilibrium price tend to be forced up by demand and supply. If farmers can get only $1 for a head of lettuce, they will be discouraged from supplying many units of this product. The low price, however, will cause consumers to demand more lettuce. The result will be a shortage of lettuce that will encourage farmers to increase their price to the equilibrium price ($1.50) where the quantity they are willing to supply is equal to the quantity consumers are willing to buy.

Although in our economy prices may be set that are not equilibrium prices, the forces of demand and supply still tend to push prices to this level. Prices may remain at other levels for extended periods of time only when there is imperfect competition. This condition will be discussed later in this chapter.

To see how economists frequently represent demand, supply, and the equilibrium price, see Figure 1.2.

## Relative Price

The **relative price** of any product is its price compared with the prices of other goods in the economy. The price we pay in dollars and cents for a product at any point in time is called its **money price** (also known as *absolute, nominal,* or *current price*). Consumer buying decisions, however, depend on relative, not money prices. Consider the hypothetical example of prices of pizzas and submarine sandwiches in Table 1.1. Note that the money prices of pizzas and subs have risen during the year. That means consumers have to pay more for both of them in today's dollars and cents. If we look at the relative prices, however, we find that last year pizzas were twice as expensive as subs, whereas this year they are one and four-fifths as expensive. Conversely, subs cost only half as much as pizzas last year, whereas today they cost 56 percent as much. In the one-year period, the prices of both products have gone up in money terms, but the price of subs has gone up more rapidly. Therefore, the relative price of pizzas has fallen while the relative price of subs has risen. If the law of demand holds true, then over this one-year period a relatively larger quantity of pizzas will have been demanded while a relatively smaller quantity of subs will have been sold, other things being equal.

**To find the relative change in price of two products, you can use the following formulas. If the result of the top equation is larger than that of the bottom equation, product B's price has gone up more rapidly. If the result of the bottom equation is larger, then the price of product A has increased relatively more.**

$$\frac{\text{Original price of product A}}{\text{Original price of product B}} = \underline{\quad\quad}$$

$$\frac{\text{Current price of product A}}{\text{Current price of product B}} = \underline{\quad\quad}$$

## An Important Distinction

Once the distinction between money prices and relative prices is made, there is less chance of confusion about the effect of price increases on the quantities of different products that are demanded during a period of time when all money prices are increasing. Products whose money prices increase more rapidly than most will be demanded less, while those whose money prices increase less rapidly will be demanded more.

## The Law of Demand Holds True

Someone not familiar with this distinction might believe that the increased demand for pizzas, despite their higher price, violates the law of demand. But, as we can see, the price of pizzas must be considered in relation to the prices of other products, in this case the price of subs. Although this example involves

### Figure 1.2
### Demand, Supply, and Equilibrium Price

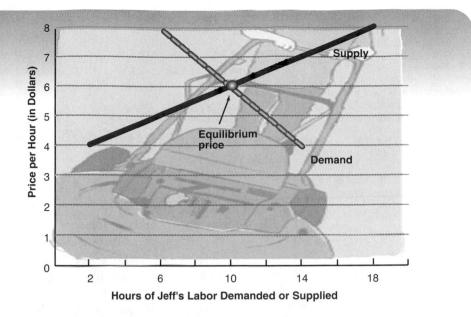

Economists often demonstrate demand, supply, and equilibrium price through tables or graphs like those shown here. Suppose Jeff, a high school student, wants to earn spending money by doing yardwork for his neighbors. He discovers that the number of hours he will be hired depends on the hourly price he charges for his labor. As you would expect from the law of demand, the higher the price he charges, the fewer the hours he works. This is shown in the table in column (1), "Price Charged," and column (2), "Quantity Demanded."

Jeff, like most teenagers, has more things he would like to do than he has time to do them. When he works, he can't be meeting with friends, playing sports, watching television, or doing homework. The number of hours he is willing to work depends on the price he is able to charge his customers. As you would expect from the law of supply, the higher the price, the more hours he is willing to work. This is shown in column (1), "Price Charged," and column (3), "Quantity Supplied."

There is a price for Jeff's labor at which the number of hours he is willing to work is exactly the same as the number of hours that will be demanded by his customers. This equilibrium price is $6 per hour and can be seen in the boldface row C in the table.

The same information can be presented in graphic form. The demand curve on the graph has been plotted from the values in columns (1) and (2) in the table. In a similar way, the supply curve has been plotted from the values in columns (1) and (3). The curves intersect at the equilibrium price of $6, where ten hours of Jeff's labor is demanded and supplied each week.

#### Demand and Supply Schedule for Jeff's Labor

|   | (1) Price Charged | (2) Quantity Demanded | (3) Quantity Supplied |
|---|---|---|---|
| A | $8 | 6 hours | 18 hours |
| B | $7 | 8 hours | 14 hours |
| **C** | **$6** | **10 hours** | **10 hours** |
| D | $5 | 12 hours | 6 hours |
| E | $4 | 14 hours | 2 hours |

### TABLE 1.1    Money Price versus Relative Price

The money price of both pizzas and submarine sandwiches has risen. But the relative price of pizzas has fallen (or conversely, the relative price of subs has risen).

|  | Money Price | | Relative Price | |
|---|---|---|---|---|
|  | Price Last Year | Price This Year | Price Last Year | Price This Year |
| Pizzas | $8 | $9 | $\frac{\$8}{\$4} = 2$ | $\frac{\$9}{\$5} = 1.80$ |
| Submarine sandwiches | $4 | $5 | $\frac{\$4}{\$8} = 0.5$ | $\frac{\$5}{\$9} = 0.56$ |

**Market Economy**  An economy that is characterized by exchanges in markets that are controlled by the forces of demand and supply.

**Perfect Competition**  A market condition in which many businesses offer the same product for sale to many customers at the same price.

**Consumer Sovereignty**  A situation in which consumers ultimately decide which products and styles will survive in the marketplace; that is, producers do not dictate consumer tastes.

**Imperfect Competition**  A market condition in which individual businesses have some power to set the price and quality of their products.

**Monopoly**  The only producer of a product that has no substitutes.

only two products, in our economy the prices of all products must be considered. The demand should grow for any product that experiences lower rates of price increase than the average for most goods, while it should fall for goods that experience price increases greater than the average.

# Consumer Sovereignty in a Market Economy

The U.S. economy is a type of **market economy** because it is characterized by transactions that are free exchanges based on the laws of demand and supply. In a market economy, individuals are free to use resources as they see fit to produce goods and services that are then offered for sale to consumers, who are free to buy or not to buy them. In an ideal market economy, there would be **perfect competition** and **consumer sovereignty.** In such a world, no firm would be large or powerful enough to set prices for its products higher than those charged by other firms that offered similar goods or services for sale. Consumers would buy products they desired at the lowest possible prices. The most profitable firms would be those that produced goods and services most efficiently. Their success would allow them to expand their own production and encourage others to open similar businesses. Money and resources would flow to the types of production that consumers demanded. Therefore, consumers would determine how scarce resources would be used to produce goods and services that were best able to satisfy their desires. If perfect competition existed in our economy, consumers would control production through the way they spend their money. They would be sovereign.

# The Real World of Imperfect Competition

In the real world, **imperfect competition** prevents consumers from always controlling production decisions and the allocation of scarce resources. There may be a limited number of producers, or other suppliers may be prevented from offering a similar product for sale by various barriers. In this event, a high price and large profits will not necessarily increase the quantity of the product that is supplied in the market or of the resources allocated to that type of production.

## Examples of Monopoly Power

Suppose you produce a medical device that doctors have found very useful in surgery. Suppose also that you have patented your device and have not sold the patent rights to any other producer. You have restricted entry into the market for this device because no other firm may legally produce it or offer it for sale while you hold your patent. Although you charge a high price for your device and earn a large profit, other individuals cannot produce the same product to compete with your product. Even if other firms could make the product more efficiently and offer better quality, they would not be allowed to do so. In this situation you would have a **monopoly.** This is a simplistic example, of course, but it illustrates how the principle of consumer sovereignty can be invalidated by firms that hold some degree of monopoly-like power.

Businesses that hold monopoly-like powers have market power, which means they do not need to respond only to the demands of consumers. Not many are sovereign, however. Most small firms are quite competitive, and there is a limit to what even the largest firms are able to do with their economic power. Consider the difficult times experienced by large firms that failed to quickly adopt Internet marketing. Some lost much of their monopoly-like powers to competition from smaller e-commerce marketers. Even when large firms use sophisticated marketing techniques, they may lose the battle for control of the market to consumers. It has been estimated that nine out of every ten new products offered to U.S. consumers fail within one year. Essentially, such failures are due to the unwillingness of consumers to demand these products at a price that would allow the producing firms to earn a profit.

Even where consumer choice exists, individuals may be forced by law to buy some products. For example, it is illegal to drive a car without also purchasing automobile liability insurance. To be sure, some people ignore the law, but if they are apprehended, they may pay a heavy penalty.

**Question for Thought & Discussion #6:**
Do you believe that there can be such a thing as too much competition? Explain your answer.

## A Range of Monopoly Power

As shown in Figure 1.3, consumers confront a range of purchasing situations. In some situations they have almost total control over production through their purchasing decisions. In others they may be faced with government regulations or with the power of firms that possess monopoly-like power. Even advertising may reduce their control over the types of products offered in the market. It is safe to say that we generally are in the middle of the range of purchasing decisions, somewhere between having only one purchase alternative and being able to make independent choices.

Given our imperfectly competitive economy, neither complete producer sovereignty nor complete consumer sovereignty can exist. As a result, from an early time in our nation's history the government has imposed regulations on various markets to protect both business and consumer interests.

## Role of Government

Everyone in the world is a consumer of goods and services when he or she uses products to fill basic needs for food, shelter, and clothing. The role and economic power of consumers vary from country to country, because each

**Figure 1.3** The Range of Consumer Choice and Sovereignty

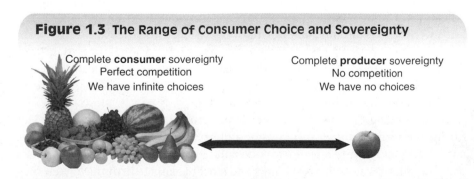

Complete **consumer** sovereignty
Perfect competition
We have infinite choices

Complete **producer** sovereignty
No competition
We have no choices

At one extreme, no one forces us to buy anything; at the other, we are required to purchase an item whether we like it or not. Generally, depending on the situation, we are somewhere in between.

Photodisc

**Economic System**  A set of understandings that governs the production and distribution of goods and services that satisfy human wants.

**Socialist Economic System**  An economic system in which there is group (most often government) ownership of productive resources and control over the distribution of goods and services.

**Capitalism**  An economic system based on private ownership of the means of production and on a demand-and-supply market. This system emphasizes the absence of government restraints on ownership, production, and trade.

nation has its own unique **economic system.** In some countries resources have been owned, controlled, and allocated by an agency of the government. Individual consumers were given little power to determine what products would be produced or how resources would be used. Many nations with these **socialist economic systems,** after experiencing great economic difficulties and individual hardship, have changed their economic systems to rely more on the forces of demand and supply.

In nations such as the United States, an economic system known as **capitalism** has dominated production. Capitalism is an economic system in which the ownership and control of resources and businesses are held largely by private individuals, and the forces of demand and supply are relied on to control the production of goods and services and the allocation of resources.

A strong theme in the U.S. economic experience has always been "the less government, the better." This view holds that we should not hamper the functioning of the laws of demand and supply in our markets with unnecessary government regulation and intervention. Individuals willing to take the risk of establishing businesses to reap profits or suffer losses should be allowed to do so. During the 1800s and early 1900s, this "hands-off" attitude prevailed. But beginning around the turn of the twentieth century, and accelerating during the Great Depression of the 1930s, some economists began to advocate government intervention in the marketplace to aid economic stability and to prevent the pain of recurring recessions. The New Deal legislation of President Franklin Roosevelt during the Great Depression was an example of this deepening involvement of government in our economic system.

This does not mean that the government had nothing to do with the economic life of the nation before the twentieth century. From the beginning of our nation's history, the government has been active in many economic areas—building harbors, constructing canals, establishing tariffs to promote domestic industrial growth, and, to a limited extent, regulating business activities to protect both businesses and consumers. Although most people think of government involvement in our economic system as a relatively recent development, it actually has a long history. Much more will be said of this involvement later in this book.

**Question for Thought & Discussion #7:**
What is a recent national event that has demonstrated the need for government involvement in the economy?

# Americans in an Electronic Market

In recent years the growth of the Internet has changed so many aspects of our lives that grasping the true dimensions of all that has taken place can be difficult. Consider your own life. As a college student, you almost certainly use the Internet to search a variety of Web sites, check electronic indexes of periodicals (access to InfoTrac®, for example, is provided with this text), or carry out interlibrary searches for information you need to complete class work. Imagine that you have been assigned to research national sales data and demographic trends for a marketing class. You can access the Internet to search government databases for the information you need, and you won't have to travel any farther than your computer.

Beyond using the Internet for academic research, you can search the Net for information about almost any good or service you might want to own or use. Once you identify a product you want to buy, you may place your order over the Internet as well. Most banks and stockbrokerage firms have Web sites that enable consumers to investigate saving accounts, investment opportunities, or sources of credit that fit their individual needs. The Internet may be used to make travel reservations or earn a college degree. And, as long as this list may seem, it is incomplete. You can surely think of other uses of the Internet that could be added.

When you want to relax, a world of entertainment is available over the Internet. You can listen to music, read the complete works of William Shakespeare, or play interactive games with people you have never met. Or, you could write a letter to one of your friends or relatives. Electronic mail has become so pervasive that the U.S. Postal Service believes it may be forced to reduce its services because of a decline in the use of first-class "snail mail."

Some people believe that they have not been affected by the Internet, but they are wrong. Even consumers who do not personally surf the Internet are affected by it because the businesses where they shop or work do access the Net. The success of our economic system and therefore the quality of our lives have become tied to the Internet. We truly live in an electronic economy.

A woman makes a purchase over the Internet. If she had made her purchase in a store, how might her shopping experience have been affected?

## Growth of the Internet

The Internet as we know it began in the 1960s as a project funded by the U.S. Department of Defense. Its original purpose was to link computers and create a nuclear-attack–proof communications system. Soon other researchers at government agencies saw the value of the Internet and began to use it. Sharing information made research more efficient and less costly. Before long the Internet grew beyond the needs of the government and its researchers. At that point the Internet's funding and administration were turned over to a variety of private organizations.

One early limitation of the Internet was that specialized training was necessary to use it. Only a relatively few highly skilled technicians knew how. In 1990, this situation changed with the development of a new computer coding system called Hyper Text Markup Language (HTML) that permitted the use of graphic forms of communication on the Internet. From this innovation came the **World Wide Web (WWW).** The WWW is an information retrieval system that organizes the Internet's resources in a graphic fashion. Through the Internet computers can link with other computers anywhere in the world.

**Question for Thought & Discussion #8:**
What are several ways in which the Internet affects your consumer decisions?

## Consumer Use of the Internet

In 1993, the creation of Mosaic, the first **Web browser,** gave ordinary people the ability to search the Web. Users no longer needed special technical knowledge to link with computers that held information they wished to access. They could identify what they wanted to find by using key words and move among Web pages to search for this information with a click of a mouse.

Internet technologies have continued to evolve since the early 1990s. New Web browsers have made surfing the Net easier and more enjoyable. Computer languages have been improved to expand the capabilities of HTML and other coding systems to transfer data. Today, the Internet is a worldwide structure that links millions of computer networks. People all over

**World Wide Web (WWW)** An information retrieval system that organizes the Internet's resources in a graphic fashion to facilitate the transfer of information between computers.

**Web Browser** A program that gives users the ability to search the Internet for specific types of information.

the world share information through computers linked on what was once called the *information superhighway.*

Technological advances have improved the efficiency and speed of the Internet, but it has retained its basic character. It is still the most common method of linking computers to share information. In Chapter 2 you will study a method you may use to identify and choose a computer and related components that can link you to the Internet. In the future new technologies will expand the capacity and speed of the Internet, but more important changes are likely to result from people finding new ways to use its power and potential. Many of these developments will affect transactions between buyers and sellers in our economic system.

## The Cyber Economy

Consumers benefit in many ways from the growth of the Internet. Two important benefits are the greater variety of choices that the Internet has made available and the increased competition that exists because of online shopping. There was a time (not so long ago) when many consumers were limited to local stores and a few mail-order catalogues when they searched for products they wanted to buy. If Moe's Appliance Store was the only business in town where you could buy a microwave oven, Moe effectively had a monopoly. Moe could charge $200 for his microwave, and people would buy from him, even if the same oven could be purchased for $150 from a store in a different community.

When consumers had no efficient way to identify alternatives, there was less meaningful competition. Today, with ready access to the Internet, this has changed. No matter where you live, all you need to do to find a better deal is to carry out a search on the Net. Using "microwave oven" or "home appliances" as key words, you will find yourself presented with a list of businesses that are happy to take your order, charge your credit card, and send you an oven via UPS today. Further, because of competition and improved business efficiency, the price you pay will be lower and the variety of microwave ovens from which you may choose will be greater. The Internet, of course, isn't just for appliances—it's there to help you find just about any product you might want to own.

# New Technology—New Responsibilities

Regardless of the growing electronic shopping (sometimes called *e-commerce*) opportunities available on the Internet, there are some advantages to patronizing a local store. When you shop at Moe's Appliance Store, you can examine the microwave you want and get a sense of how it works. You may know Moe personally, and he may know you. With a personal relationship, you may feel confident that you will not be cheated. If the microwave you buy is defective, you know you can return it for repairs, a refund, or another microwave. Maybe you do pay more than the lowest possible price when you buy from Moe, but you are also buying security, service, and peace of mind.

This type of personal relationship with the owner of a business is impossible when consumers shop the Net. A DVD player purchased from A-1 Electronics.com may have a low price, but you certainly will not get to know the firm's owner. It is likely that you will complete the entire transaction without talking to a single person. You will place your order by filling out an electronic form. If the player is defective when it is delivered, having it repaired or returning it for a refund may be a difficult and frustrating process. You will probably come to understand that price isn't the only factor worth considering when you make a purchase decision. The *Confronting Consumer Issues* feature in Chapter 3 provides many suggestions to help you make rational decisions that will bring you the greatest satisfaction when you shop on the Internet.

Take this situation a step further. Suppose the DVD player you received from your Internet order was not just defective, but was not even the product you ordered. Suppose you think you have been cheated by a dishonest business. There are few effective restrictions on what may be placed on the Internet. Just because an offer is made over the Web does not mean that the offer is honest or fair. The Internet has opened up many new ways for dishonest people to take advantage of unwary consumers. In Chapter 4, you will learn about common Internet scams and how consumers may recognize dishonest offers to protect themselves from being cheated by "cybercrooks."

# Consumer Economics in an Electronic Economy

Over the past decade, the Internet has become a pervasive force in the lives of most Americans. It must, therefore, have a central role in the study of consumer economics. With this in mind, every chapter in this text contains a wealth of Internet applications that will help you make better consumer decisions. Many Web addresses are provided to give you guidance when you investigate topics presented in each chapter. A feature titled *The Cyber Consumer* appears in every chapter. These features suggest ways in which you may use the Internet to find specific information that will enhance your knowledge and understanding of consumer economics issues. At the end of each chapter, you will find three activities under the heading *Internet Resources:*

- *Finding Consumer Information on the Internet* asks you to investigate important terms or topics from each chapter by visiting specific Web sites.
- *Shopping on the Internet* asks you to investigate the purchase of various consumer goods or services at identified Web sites of Internet marketers.
- *InfoTrac Exercises* asks you to use the InfoTrac access provided with this text to investigate recent publications that discuss issues related to consumer economics.

There is a strong possibility that Web addresses identified in this text will change over time. If any of the addresses you try to visit no longer functions, you should complete a search using any of the search engines available to you to find other similar sites that can provide you with the information you seek.

The Internet and the rules that govern its use are certain to change in the future. Technological advances will increase Internet access speed and capabilities. Nevertheless, the basic concept of the Internet will remain much as it is today. Learning to use the Internet is an important part of becoming an educated consumer. The Internet will be a central part of your life and affect the consumer decisions you will make. The more you learn about using the Internet, the more satisfied you are likely to be with your life.

# Confronting Consumer Issues: Being an Ethical Consumer

When Americans make consumer decisions, they usually consider more than the price and quality of products offered for sale. Other factors that may influence their choices include where and how the items were produced, whether they were made by union labor, and whether they were made in an environmentally responsible way. For example, you might choose not to shop at a store you believe discriminates against women or ethnic minorities. Many people are willing to pay higher prices for products manufactured or offered for sale in what they believe are socially responsible ways.

In a similar way, consumers often make choices that reflect values they have set for their personal behavior. How often have you seen a store clerk make a mistake that would have caused you to pay less than the actual price for what you were buying? Did you tell the clerk that he or she had made a mistake, or did you take advantage of the situation to pay a lower price? If you broke a dish by accident in a china store, would you feel responsible to pay for it? If you were harmed by a defective product you had purchased, would you sue for an amount of money that was greater than the value of your loss? Although some people do these things, many U.S. consumers choose to behave in what they regard as a socially/morally responsible way.

All consumers have duties and responsibilities that can be summarized in one sentence: The consumer has a duty to act honestly and ethically when purchasing products and services. An age-old aphorism holds that "what goes around comes around." In the context of consumer dealings, this means that if enough consumers act dishonestly, prices of consumer products and services will rise, harming all other consumers.

> ## Question for Thought & Discussion #9:
> What is your definition of ethical consumer behavior?

## WHAT IS ETHICAL BEHAVIOR?

**Ethical behavior** essentially means acting in accordance with one's moral convictions as to what is right and what is wrong. Many commonly held ethical convictions are written into our laws. But ethical behavior sometimes requires us to do more than just comply with laws. In some circumstances, one can break the law and be fairly certain no one will ever find out about it. But getting away with something you do is not ethical behavior.

**Being an ethical consumer means paying** attention **during large and small transactions. Did** the clerk **give you too much change? Is this a rest**aurant that **treats its employees fairly? What issues** are of **concern to you?**

Imagine, for example, that you purchased a number of items from your local Wal-Mart and discovered when you got home that the cashier had failed to charge you for a $16 CD you had chosen. What is your obligation? Obviously, the CD does not legally belong to you—you have not paid for it. But why should you have to take the time and trouble to return to the store to pay for the item? After all, it wasn't your fault that the cashier did not ring up the item. What would your decision be? Would it make a difference if the item had a much higher or a much lower price?

Consider another possible situation where you know something is wrong before you actually get a product home. Suppose you regularly buy a particular brand of spaghetti sauce that you know costs $2.69 per jar. One day while shopping at a convenience store, you see several jars that have been mispriced at $1.19. What would you do? Would you buy all the jars at the lower price, buy only one or two of them, or tell the clerk that a mistake had been made? This time you cannot rationalize your decision by saying, "I didn't know." But then it isn't your fault a mistake was made. What do you believe the ethical consumer choice would be?

**It's Not Always Easy**   These examples illustrate that it's not always easy to do the "right thing"—or even to know what the right thing to do is in a given situation. You could be pretty certain that in the case of the CD no one would ever know you had obtained the product without paying for it. Wal-Mart officials surely wouldn't report you to the police or come knocking at your door. They would have no idea where their merchandise went. The case of the mispriced spaghetti sauce is only a little different. If anyone asked you about the price, you could simply say that you paid the marked price. In situations like these, there is a temptation to take the "gift" offered to you by fate. In addition to having more money left over to buy other products, your sterling reputation in the community as an ethical person would not be marred in the slightest, because no one else would ever know what you had done. But you would. And this is what ethics is all about. At the heart of ethical decision making is determining whether you personally feel that a given action is right or wrong, and acting accordingly. After all, you're the one who has to live with your conscience.

**Somebody Has to Pay** When trying to determine the rightness or wrongness of a given action, it is helpful to consider the consequences of each alternative. Keep in mind that if you don't pay for benefits you receive, someone else will. As economists are prone to emphasize, there is no such thing as a "free lunch." In other words, somebody, somewhere, has to pay for all that is produced and consumed. And that somebody is another consumer—or, rather, other consumers. This is because sellers who absorb these added costs will pass them on eventually to all purchasers in the form of higher prices.

## EXAMPLES OF UNETHICAL CONSUMER BEHAVIOR

Although most consumers act responsibly in their purchase transactions, they are obviously not saints any more than businesspersons are. And examples are plentiful of consumers who give in to the temptation to evade the letter of the law in order to get "something for nothing."

**A Camera That Doesn't Work** Consider, for example, the following scenario: Jeannie orders by mail a new digital Nikon camera from Flash Electronics, a discount house in a distant city. The camera arrives by mail, and Jeannie immediately uses it to take photographs to be included in the book she is writing. A few days later, she drops the camera and breaks the casing. She decides to "pass the buck" to the seller and returns the camera to Flash Electronics, claiming that the camera was broken when it arrived and demanding a replacement. Jeannie, who eventually receives the replacement camera, has just saved herself the cost of repairing the broken camera—at the expense of the discount firm, of course, or the manufacturer. But she suffers few pangs of conscience about her dishonesty. After all, she reasons, Flash Electronics and Nikon are huge and profitable businesses. The repair bill would be only a drop in the bucket for Flash Electronics or Nikon, but it would represent Jeannie's entire budget for a week.

What Jeannie overlooks in her reasoning is the long-run consequences of her behavior. In the short run, yes, the dis-

count house or the manufacturer will pay for the repairs. But, ultimately, who pays? Other consumers, like Jeannie, who buy cameras or other products from the discount house and Nikon will have to pay more because of Jeannie's dishonesty. But Jeannie might still rationalize that the cost of the camera repair—when spread out over thousands of consumers—will represent no real burden to each individual consumer, which is true. But, if all—or even a substantial number of—consumers acted similarly to Jeannie, what then?

It takes little effort to imagine dozens of other ways in which consumers have behaved dishonestly or unethically to gain a personal benefit at the expense or inconvenience of others. We look here at just a few variations of this theme.

**So Sue Me** Most consumers periodically receive in the mail invitations to subscribe to certain magazines or to "sign here" and receive a product to try out for thirty days. A typical offer is to sign up for membership in a book club. All you have to do is sign and send a card to receive, say, four books for which you will be billed $1 at some future date. Of course, having signed up for membership in the club, you are obligated to purchase a given number of books per year—or at least notify the club each month if you don't want a particular book or books. You receive the four books and, after a few months, have received several more. You haven't had time to read the books, don't really want them, and don't really want to be a member of the club. But you're busy and fail to do anything about it. Eventually, the book club begins to send stern demands for payment—you owe $109. You are a struggling student, short of money, and you ignore the bill. It certainly does not take priority in your budget. If the club wants to sue you for collection, fine. You are not worried about it because you know that the amount is too trivial to justify any legal action against you by the book club. Eventually, to your relief, the club stops sending you any bills at all—your account has been written off as a "bad debt"— along with hundreds of others. And you have acquired six "free" books.

**Dishonest Returns** Toni buys an expensive new dress for a special party she has been invited to attend. She wears the dress to the party, receives many compliments on it, but decides it was really far too expensive a purchase. She returns it to the store for a refund. The sales clerk does not inspect the dress closely and fails to notice the food stains on the front. Toni gets her refund. The result? Either the next purchaser gets a slightly soiled dress instead of the brand-new garment she paid for, or the store must discount the price of the dress heavily to sell it if the stain is discovered.

*(Continued on next page)*

**Confronting Consumer Issues (Continued)**

### Make the Manufacturer Pay
In the past two decades, U.S. courts and consumer protection statutes have increasingly sought to protect the "little person" against the powerful corporate entity or business firm. This has been a boon to consumers who are injured by faulty products they have purchased. It allows them to sue sellers and manufacturers for compensation, in the form of money damages, for injuries caused by carelessness in product design or production. But now and then a consumer takes advantage of these laws and of the court system to seek damages from the product manufacturer or retailer. Assume, for example, that John, a minibike enthusiast, purchases minibikes for his two sons, aged nine and eleven. In the instruction manual, and clearly indicated in large letters on the bikes themselves, are instructions not to use the bikes on city streets and always to wear a helmet while riding them. Nonetheless, John allows his sons to ride on the city streets without helmets. One day, while racing with another friend on a minibike, the oldest son, Chad, carelessly runs three stop signs and then enters a fourth intersection while looking backward toward his friend. Chad is hit by a truck and injured. John sues the manufacturer of the minibike for damages, claiming that the minibike is a dangerous product and should not have been placed on the market.

### The Nuisance Suit
Sellers are also often faced with so-called nuisance lawsuits. A typical one might involve the following series of events: Jerry, in a daze about his latest girlfriend, walks through a hardware store, carelessly trips over a stepladder displayed very close to a wall (and definitely not a hazard), falls, and falsely claims that he injured his back. Alleging that the owner was negligent by displaying the stepladder in that way, he sues the owner for damages. Similarly, Jane sues the owner of a national chain store for $10,000, alleging that a can of paint displayed on a shelf in one of the owner's stores fell on her toe and injured it, and on and on. The store owners often settle such suits out of court, because it would cost them more to defend themselves in court than to settle. Even though most store owners carry liability insurance, out of which such claims are paid, the insurance is not free to the store—and the premiums will rise (and they have risen dramatically in recent years) as more claims have to be paid by insurance firms.

## ETHICS IN AN IMPERSONAL MARKETPLACE
In the increasingly impersonal and mechanized marketplace of today, it is much easier to lose sight of our responsibilities toward others than it once was. In today's consumer world, the "others" are usually abstract entities and not people we know personally. In the past, when stores were smaller and most transactions were conducted face to face, consumers were more motivated to act honestly and ethically because they also confronted the consequences of their actions directly. Imagine, for example, that Jeannie in the camera example had lived in 1900 instead of the 2000s. After breaking her camera, she returned it to her local camera store, claiming that it was already broken when she purchased it from the seller. Very likely, the seller would remember the transaction, would know that the camera had been in good condition, and would be aware that Jeannie was acting dishonestly—regardless of whether he could prove it. Jeannie might be deprived of—or at least face a reduced quality in—the services of that store, and her reputation in the community could be affected. Because of these possible negative consequences, it might not even occur to Jeannie to try to defraud the seller. Moreover, if she knew the merchant quite well, she might have some strong ethical reservations about requiring the merchant to pay for the broken casing for which she alone was responsible.

Now let's return to the present and to a much different marketplace. When Jeannie returned the camera to the discount house, she knew that she was being dishonest, but she did not lose sleep at night over the "victim" of her dishonesty—who was not a real person but an X quantity of "others." Moreover, and perhaps more significantly, Jeannie was quite sure that she would never be "caught." No one would ever know of her dishonesty, and she would face no negative consequences. The worst that could happen is that the discount store would refuse to repair or replace her camera. In short, Jeannie felt little incentive to be ethical.

Because there are fewer *external* constraints to guide us toward ethical consumer behavior, an understanding of our responsibilities in the marketplace is even more important today than it was in the past. Huge chain-store operations and computerized networks are increasingly hiding the identities—and the behavior—of individual buyers and sellers in the marketplace. And if we are slightly dishonest or violate our own ethical standards occasionally, who will know?

### Question for Thought & Discussion #10:
Do you consider the social, environmental, or ethical record of businesses when you shop? Why or why not?

## ETHICAL SHOPPING

Although consumers may not be sovereign, they can still affect the financial well-being of businesses. If you have doubts about the ethical behavior of a corporation, you can cast your "ethical vote" by not purchasing that firm's products. The major difficulty of **ethical shopping** is, of course, that time is required to investigate businesses to determine whether their business practices are ethical or unethical.

To assist consumers who wish to engage in ethical shopping, the Corporate Social Responsibility (CSR) News Wire Service maintains a Web site, at **http://www.csrwire.com**, that evaluates thousands of large and medium-size corporations on a variety of issues including the environment, women's and minority rights, and community outreach. Using this Web site to learn how a firm is assessed takes only a few seconds. Just click on the *reports* icon and enter the name of the business you wish to investigate. In seconds you will see a report that describes the firm's policies in relation to the environment and other social issues. You may also subscribe to a free weekly news alerts magazine that is published by CSR online. By using the key words *social responsible* and *corporation,* you can search for other Web sites that evaluate U.S. businesses.

## ETHICAL USE OF GOVERNMENT SERVICES

Our government provides services that are intended to benefit specific groups of citizens who are disadvantaged or have special needs. Some people make unethical use of such benefits. An obvious example is the able-bodied person who parks in a handicapped parking space. Other people take advantage of social programs such as welfare, food stamps, or Medicaid by providing false information to government administrators. Some health-care providers have been convicted of billing the government for Medicaid or Medicare services that were unnecessary or never performed. These practices add to the cost taxpayers bear.

On a simpler level, people may abuse or purposefully destroy public property—breaking up picnic tables in parks to build fires, stealing road signs to be sold as scrap aluminum, or charging government accounts for private travel. Such behavior adds to the cost of government and may also reduce the quantity of goods and services that government is able to provide to those who truly need them.

> **Question for Thought & Discussion #11:**
> Would you be more likely to invest your money in a firm that offers a high return on your funds or a firm that promises to use your funds in a socially responsible way?

# Cyber consumer
## The Open Directory Project

In the organization's own words, "The *Open Directory Project* is the largest, most comprehensive human-edited directory of the Web. It is constructed and maintained by a vast, global community of volunteer editors." This organization's Web site at **http://dmoz.org/about.html** provides links to hundreds of other Web sites that report on various aspects of socially responsible business and consumer behavior. Find and review a site that interests you. Write an essay that identifies the Web site you chose and assesses its value. Remember, many Web sites are created and maintained with a specific agenda. You should include an appraisal of the site's balance and accuracy in your assessment.

## ETHICAL INVESTING

**Ethical investing** is accomplished by making investments in companies that engage in socially responsible behavior. For people who are unwilling to take the time to investigate specific firms for themselves, there are many socially responsible mutual funds that will invest their money for them. The managers of these funds review the thousands of publicly held corporations to determine which ones meet their ethical standards and investment goals. The specific objectives and types of investments these funds seek vary widely. Some funds avoid investing in businesses that sell to the military, whereas others look for firms that help women or minorities advance in their careers. Still others seek firms that produce environmentally friendly products or manufacture products in an environmentally responsible way. Each fund prints a prospectus that explains its goals and objectives and how it will work to achieve them.

The performance of ethical funds varies. Studies have shown that on average their returns have been slightly lower than the average return by all mutual funds. Still, some have done very well, but others have lost significant proportions of their investors' money. Table 1.2 on the next page shows

*(Continued on next page)*

## Confronting Consumer Issues (Continued)

the one- and three-year returns (losses) of four typical socially responsible mutual funds. Although each of these funds lost money between 2000 and 2003, you should remember that the value of most stocks declined in these years. The value of the Dow Jones Industrial Average (an often-used indicator of stock values) fell 7.65 percent in 2002 and declined by an average rate of 7.2 percent each year during the three years between January 1, 2000, and January 1, 2003.

If you are interested in investing ethically, Co-op America publishes the *Socially Responsible Financial Planning Guide*, which lists numerous brokers, financial planners, insurers, bankers, and credit unions that engage in ethical investing. You may order this guide online at **http://www.co-opamerica.org**.

### ONLY YOU CAN DECIDE

Obviously, there is no exact formula for ethical behavior. Every individual has her or his own set of values and moral principles, and every situation is different. But it is impor-

tant to remember that, although moral and ethical convictions are necessarily very personal qualities, our individual behavior always, in one way or another, affects others around us. This is true in all of our activities—as family members, as citizens, as employers or employees, and as consumers. Ethical decision making involves becoming aware of how our behavior affects others and evaluating whether these consequences are desirable—and this is something only you can decide.

### Key Terms:

**Ethical Behavior**   Behavior that is directed by moral principles and values; determining what is "right" in a given situation and acting in accordance with that determination.

**Ethical Shopping**   Purchasing products manufactured by socially responsible business firms and refusing to purchase products manufactured by firms whose ethical behavior is perceived to be reprehensible.

**Ethical Investing**   Investing in corporations that are deemed to be socially responsible according to a given set of ethical criteria.

| TABLE 1.2 | Examples of Ethical Mutual Funds—One- and Three-Year Returns—January 2003 | | | |
|---|---|---|---|---|
| Fund Name, Telephone Number, and Internet Address | Minimum Investment | One-Year Return as of January 2003 | Three-Year Average Return as of January 2003 | Objective Screens |
| Neuberger Berman Socially Responsible Fund (800) 877-9700 http://www.nbfunds.com | $1,000 | −14.45% | −6.03% | environment, diversity, workplace, community |
| Pax World Balanced Fund (800) 767-1729 http://www.paxfund.com | $250 | −8.86% | −4.34% | fair hiring practices, environment |
| Green Century Funds (800) 934-7336 http://www.greencentury.com | $2,000 | −21.09% | −16.70% | environment |
| Domini Social Equity Fund (800) 762-6814 http://www.domini.com | $1,000 | −20.69% | −16.24% | social, environment, community |

SOURCE: http://www.socialfunds.com.

# Chapter Summary

1. Consumer economics is the study of how consumers can apply economic principles in their decision making. Some consumers make buying decisions without understanding the basic principles that guide economic behavior in the world around them.

2. Scarcity is a problem faced by consumers, businesses, and the government in all nations. Because of scarcity, choices and trade-offs must be made. As consumers, we must choose how we spend our time and effort, as well as how we spend our income. A knowledge of consumer economics can help us maximize the satisfaction we obtain from the choices we make.

3. Because time, money, and productive resources are scarce, whenever we use any of these resources, we must forgo some other option to which we could have devoted the resources. The value of the second-most-valuable forgone option is the opportunity cost of our first choice. A trade-off is the process of giving up one alternative for another. Trade-offs must be made because scarcity exists. An opportunity cost is involved in every trade-off.

4. A basic assumption in consumer economics is that people try to make themselves better off in their standard of living—that is, in their command over leisure time and their ability to buy goods and services that will satisfy their wants.

5. In economics, a market is the sum of all transactions that take place between the buyers and sellers of a particular type of product. Markets exist between and among people, businesses, governments, and other nations in the world. In the U.S. economy, most transactions in markets are voluntary exchanges that benefit both buyers and sellers, although not necessarily in equal proportion.

6. The law of demand states that if nothing else changes, consumers will buy a greater quantity of a product at a lower price than at a higher price. The quantity of a product that consumers will buy at any particular price may change as the result of a change in consumer tastes, income, the price of related products, or consumers' expectations of the future.

7. The law of supply states that if nothing else changes, a greater quantity of a product will be offered for sale at a higher price than at a lower price. The quantity of a product that is supplied at any particular price may change as the result of a change in the cost of supplying that product.

8. Prices are determined in competitive markets by the forces of demand and supply. The price that we pay for a good or service at any point in time is its money price. The price of an item relative to other items in the market is its relative price. Consumer buying decisions depend on relative, not money, prices.

9. In a world of perfect competition, many businesses would make the same products and offer them for sale to many customers. Consumers would control what is produced through their spending decisions. The result would be complete consumer sovereignty. To earn a profit, businesses would allocate more resources to the production of products that sell well and fewer to the production of goods or services that are less in demand.

10. In the real world, there are many cases of imperfect competition in which businesses have more control than consumers over the products they produce and the prices they are able to charge. In such situations the businesses have a degree of sovereignty over production. The most extreme case of imperfect competition is a monopoly, where there is no competition.

11. Our government has often intervened in our economic system in an attempt to make it work better or to protect the rights of consumers or business owners. Although the amount of government intervention has increased in recent years, it has been manifested to some degree since the United States was first created in the 1700s.

12. The growth of the Internet has changed many aspects of consumers' lives. These changes can be seen in how we learn, work, play, and shop. It has become possible to purchase virtually every type of consumer good or service over the Internet.

13. The Internet grew out of a 1960s project funded by the U.S. Department of Defense to link computers and create a nuclear-attack–proof communications system. At first only skilled technicians knew how to use the Internet, but the development of a new computer coding system called HTML and a Web browser named Mosaic made it possible for ordinary people to use the Web.

14. The growth of the Internet has expanded the choices consumers may make when they decide how to spend their income. It has also increased competition and therefore led to lower prices. Consumers can use the Internet to quickly compare prices and quality of similar products offered by businesses throughout the nation and the world.

15. Although the growth of the Internet has created new shopping opportunities for consumers, it has also increased their need to act responsibly. The Internet is impersonal. Consumers need to check out Internet offers before they make a purchase. Many dishonest people use the Internet to try to cheat consumers.

16. Ethical behavior means acting in accordance with one's moral convictions as to what is right and what is wrong. Consumers make choices based on their ethical values as

well as on price and quality. It is often difficult for consumers to determine what their ethical choice should be.

17. Many consumers do not behave in ethical ways. They return products that they have broken, refuse to abide by contracts they have signed, demand more or better service than they have paid for, or threaten to sue businesses in the hope of obtaining a settlement they don't deserve. The rapid growth and impersonal nature of consumer markets have made it easier for consumers to be unethical.

18. Consumers can find many sources of advice and guidance to help them carry out ethical shopping. The Corporate Social Responsibility (CSR) News Wire Service maintains a Web site that contains vast amounts of information about companies' social records and policies and is easy to use. Consumers may also choose to invest in socially responsible corporations.

## Key Terms

capitalism **12**
consumers **2**
consumer sovereignty **10**
demand **6**
economic system **12**
equilibrium price **7**
ethical behavior **20**
ethical investing **20**
ethical shopping **20**

goods **2**
imperfect competition **10**
law of demand **6**
law of supply **6**
market **5**
market economy **10**
money price **8**
monopoly **10**
opportunity cost **4**
perfect competition **10**
productive resources **3**

profit **3**
relative price **8**
scarcity **2**
services **2**
socialist economic system **12**
supply **6**
trade-off **4**
voluntary exchange **5**
Web browser **13**
World Wide Web **13**

## Questions for Thought & Discussion

1. If you won $100 million in the power ball lottery, why would you still face the problem of scarcity?

2. What is your opportunity cost for taking this course?

3. What trade-offs have you made to be able to attend college?

4. What would happen to the quantity of courses demanded at your school if tuition rates doubled?

5. What would happen to the quantity of Pepsi that would be demanded if its price increased by 10 percent while other prices increased by 20 percent?

6. Do you believe that there can be such a thing as too much competition? Explain your answer.

7. What is a recent national event that has demonstrated the need for government involvement in the economy?

8. What are several ways in which the Internet affects your consumer decisions?

9. What is your definition of ethical consumer behavior?

10. Do you consider the social, environmental, or ethical record of businesses when you shop? Why or why not?

11. Would you be more likely to invest your money in a firm that offers a high return on your funds or a firm that promises to use your funds in a socially responsible way?

## Things to Do

1. Analyze a recent consumer decision you made. Explain how scarcity forced you to make this decision. Identify the opportunity cost of your choice. Describe any ways in which government regulations may have affected your choice.

2. Describe and compare the characteristics of two markets in which you buy products. One should be essentially competitive while the other should demonstrate a degree of imperfect competition. Which market do you believe better serves your consumer interests? Explain the reasons for your choice.

3. Investigate a specific law that influences the way U.S. consumers make decisions. Write an essay that explains why the law was passed, what it was intended to accomplish, and how well it has worked.

4. Identify a product that you, or members of your family, buy regularly. Compare local retailers of this product with online marketers. What differences in price, quality of service, and convenience can you find? Which of these sources of the product would you use?

5. Suppose that in 2003 a massive earthquake struck a rural area on the western coast of Mexico. Thousands of buildings were destroyed, roads were shattered, and railroad tracks were pushed out of line. Transportation became very difficult at a time when there was great need to move both people and products. Draw a demand and supply graph for building materials that shows what happened to them as a result of the earthquake. Explain your graph.

6. Imagine that you are asked to choose between two business that will supply electricity to your home. One of these businesses burns coal in a facility that was built in 1963. It hasn't been retrofitted with pollution-control equipment, but it offers electricity at a relatively low price. The other firm generates electricity from dams, windmills, and solar panels. Its price for electricity is almost 50 percent higher than the other firm's. Which source of power would you choose? Explain the reasons for your choice.

# Internet Resources

## Finding Consumer Information on the Internet

The following Web sites have been selected for their relevance to topics discussed in this chapter. Search these sites to locate information that can add to your knowledge of competition and monopolies in the U.S. economy. Remember, Web addresses change frequently. If any of these addresses no longer function, find similar sites to investigate using any of the search engines available to you.

1. The Federal Trade Commission offers a booklet entitled "Promoting Competition, Protecting Consumers: A Plain English Guide to Antitrust Laws" at its Web site. You can find it at **http://www.ftc.gov/bc/compguide**.

2. Documents and other materials relating to antitrust cases are available at the FTC's Bureau of Competition. You can find it online at **http://www.ftc.gov/ftc/antitrust.htm**.

3. The Citizens Action Coalition discusses questions related to monopoly or competition, such as whether some utilities abuse their monopoly power. You can find the discussion online at **http://www.citact.org/competition.html**.

## Shopping on the Internet

The following Web sites have been selected because they offer consumers services similar to those described in this chapter. These are commercial sites that are designed to market products. They do not represent a comprehensive or balanced description of all environmentally responsible products available online. Remember, Web addresses change frequently. If any of these addresses no longer function, find similar sites to investigate using any of the search engines available to you.

1. Equator Appliances is a manufacturer of innovative space saving laundry appliances, stainless steel and compact dishwashers, and laundry detergent. Check it out at **http://www.equatorappl.com**.

2. Environment Friendly Products says that it is "working for peace, human rights, equality, education and the environment." It offers Feeling Goods Organic clothing, linens, personal-care products, and household products. You can find it online at **http://www.mainstonline.com/shop/environment/eco_friendly.htm**.

3. Carrier Corporation is a manufacturer of air conditioning, heating, and refrigeration equipment. Its Web site is at **http://www.carrier.com**.

## InfoTrac Exercises

Purchasers of new copies of this text are provided with access to the InfoTrac Web site. This Web site links students to thousands of recent articles published in hundreds of periodicals. Use the key words **economic demand, economic supply, economic equilibrium,** or other terms from this chapter to conduct a key-word search. Choose one article that is of particular interest to you and write a brief essay describing what you have learned from the article. Be sure to cite the author and title of the article and the name and date of the publication in which it appeared.

## Selected Readings

Brown, Alden. "The Invisible Fist." *The Futurist*, November 1999, p. 26.

Brown, Stephan P., and Daniel Wolk. "Natural Resource Scarcity and Technological Change." *Economic and Financial Review*, January 2000, pp. 2–4.

Cotterill, William P., and Ravi Dhar. "Assessing the Competitive Interaction between Private Labels and National Brands." *Journal of Business*, January 2000, pp. 109, 110.

Garratt, Rod. "A Free Entry and Exit Experiment." *Journal of Economic Education*, Summer 2000, pp. 237–240.

Greenhouse, Steven, "Update on Capitalism." *The New York Times*, September 4, 2002, p. D8.

Hansell, Saul. "War of the Browsers Resumes with More Players This Autumn." *The New York Times*, September 30, 2002, p. E1.

Miller, Roger LeRoy. *Economics Today*, 12th ed. Chapter 3. Reading, MA: Addison-Wesley, 2004.

Pearson, Steve. "Too Much Supply, Too Little Demand." *The Washington Post*, August 25, 2002, p. A1.

Sirgy, M. Joseph, and Sue Chenting. "The Ethics of Consumer Sovereignty in an Age of High Tech." *Journal of Business Ethics*, November 1, 2000, pp. 1–14.

Smith, David. "Demand Puts Supply in the Shade." *The Sunday Times* (New York), January 27, 2002, p. NR6.

"Staking Small Claims." *Business Week*, January 31, 2000, p. F28.

Yoffie, David B. "What Now for Microsoft?" *The Wall Street Journal*, November 5, 2001, p. A20.

# CHAPTER 2

# Making Rational Consumer Choices

**After reading this chapter, you should be able to answer the following questions:**

- What is rational consumer decision making?

- Why is values clarification important in decision making?

- How do your values and goals affect your consumer choices?

- How can you use a decision-making process to make better choices?

- What are some common pitfalls in consumer decision making?

- How can you manage your time to make the best use of the money and resources you have?

- How can the rational consumer decision-making process be used to choose technology-based products?

ou consume; I consume; we all consume—in one way or another. Thus, we are all consumers. And the dollar value of what we consume is staggering. The estimate for the year 2004 is close to $7.0 trillion. What do we buy as consumers? In a word, everything. Our purchases include goods as varied as 18-karat-gold wristwatches, toothpicks, racehorses, cellular telephones, four-bedroom houses, televisions, memberships in athletic clubs, shrimp cocktails, and hamster cages. Table 2.1 lists some of the broad categories of goods and services on which we spend our money each year.

## Decisions, Decisions

Behind these consumption figures lies a process of continual decision making on the part of individuals and households. For every dollar spent, a decision has been made to spend it. On one level many decisions are more or less automatic—such as the decision to buy food. But the questions of what kind of food to buy, how much, and where to buy it involve a whole host of other decisions. Should we buy nationally advertised or generic foods? Should we shop for fresh or frozen broccoli? Should we cook at home or go out for dinner? Should we order a Big Mac or a low-fat tuna salad? We make hundreds of these kinds of decisions each day, implicitly or explicitly.

When we shop, we are faced with a growing number of products to choose from and with increasingly high-tech goods and services about which to learn. Each new product that comes to our attention requires us to decide whether to buy it. How many of us, for example, have had the experience of walking by a store window, seeing something new that we find attractive, and deciding whether we like it enough to buy it without further consideration? Ten minutes earlier, there was no decision to be made—because we didn't know the product existed. Advertisements on television, in newspapers or magazines, and on the Internet acquaint us with new, sometimes very tempting products—and, of course, with still more decisions to make.

We face more consumer decisions today—and more complexity in decision making—than our parents or grandparents did in the past. We can reap

Each new product that comes to our attention requires us to decide whether to buy it.

| TABLE 2.1 | Personal Consumption Expenditures by Major Type, 2002 | |
| --- | --- | --- |
| Type of Expenditure | Total Spent (in billions) | Percentage of Total Spent |
| Total spending | $6,513 | 100.0% |
| Automobiles and parts | 370 | 5.7 |
| Furniture and household equipment | 428 | 6.6 |
| Food | 901 | 13.8 |
| Clothing and shoes | 356 | 5.5 |
| Gasoline and oil | 145 | 2.2 |
| Housing services | 874 | 13.4 |
| Medical care | 963 | 14.8 |
| All others | 2,476 | 38.0 |

SOURCE: *Economic Indicators*, September 2002, p. 21.

the greatest rewards from the dollars we have to spend by making rational consumer decisions.

# Rational Consumer Decision Making

Consumer decision making happens every time we buy or use a consumer good or service. **Rational consumer decision making** requires allocating time, money, and other resources in the way that will bring us the greatest satisfaction. Imagine that you have decided to replace your old television with a new large-screen model that will allow you to watch DVDs just as if you were in a movie theater. To accomplish this, you will want to gather and evaluate information about available large-screen televisions and their prices. But information can be costly to obtain. You will have to spend your time and possibly a portion of your income to find an *optimum* (as opposed to an *infinite*) amount of information about alternatives; the optimum amount of information is just the ideal amount—neither too much nor too little. You will learn how to determine this amount later in this chapter.

## Consider Transaction Costs

When allocating their scarce money and time, consumers should consider other costs as well. For example, you might be able to drive to a store on the other side of town to buy a new computer monitor for $10 less than you would pay at a store down the street. If you do, you should weigh the value of your time and the costs of driving your car against the value of the $10 you could save. Or, if you decide to buy from a firm that markets its products over the Internet or through a mail-order catalogue, you could pay $50 less *before* you consider the $35 shipping fee. And what if the monitor arrives broken? You might have to pay to return it to the manufacturer. Costs of obtaining information and completing exchanges that are additional to the price of the product are called **transaction costs** and are important considerations in rational decision making.

## Do You Really Need It?

Perhaps the most important part of rational consumer decision making is the initial decision of whether to even consider buying a particular type of good or service. You might ask yourself, "Do I really need a larger TV at all?" or "Should I continue to watch programs and movies on my old 19-inch model? Maybe I should save my money to buy a better car I know I will need soon." The "cost" you are considering here is an example of an opportunity cost. You are comparing the value you believe you would receive from the best possible deal you could find for a new large-screen television with the value you believe you would receive from your second-best choice—buying a better car in the future.

Remember, cost and value are not measured just in terms of money. When you decided to read this book, you gave up the value of other possible uses of your time. How do you quantify the benefits of reading a book for one more hour? Those benefits could be the value you place on the knowledge you expect to gain or on the higher grade you hope to receive. To make the best possible choice, you weigh the value of these benefits against the value of alternative uses of your time.

**Rational Consumer Decision Making** Making consumer decisions that maximize the satisfaction you can obtain from your time and money resources and that assist you in attaining lifelong, as well as short-term, goals.

**Transaction Costs** All the costs associated with completing an exchange beyond the price of the product that is purchased.

**Question for Thought & Discussion #1:**
What are several transaction costs that you must pay to purchase a college education?

**Cost-Benefit Analysis** A way to reach decisions in which all the costs are added up, as well as all the benefits. If benefits minus costs are greater than zero, then a *net benefit* exists and the decision should be positive. Alternatively, if benefits minus costs are less than zero, then a *net cost* exists and the decision should be negative.

## Cost-Benefit Analyses

Rational decision making, whether it involves time or money, is completed through a series of **cost-benefit analyses** in which choices are made that yield the highest net benefits within the constraints of limited time and income. A *net benefit* is the value of all the benefits of a decision totaled together minus all of its costs totaled together.

Using cost-benefit analysis to determine your most beneficial choice may seem difficult. When your personal values or feelings enter into your decision making, you may have a difficult time quantifying them. Can you place a value on the pride you feel in owning a new car or stylish clothing? How do you weigh the value of this pride against the value of other, less expensive products you could choose to buy? Nonetheless, you are always making choices on the basis of what you feel is best for you, given the alternatives and the opportunity costs of each possible choice. Such an analysis requires that you have a clear understanding of your priorities and of the importance you attach to different wants you have. By clarifying your priorities, which are determined by the values you hold, you can make choices that support your overall goals in life.

# Values and Rational Decision Making

When you were a child, people probably asked you what you wanted to be when you grew up. You may have said a musician, artist, scientist, doctor, lawyer, flight attendant, firefighter, or any of a number of other occupations that you were aware of at the time. For the most part, these early career goals were based on limited knowledge and experience. Later on, you probably considered more carefully occupational choices that were consistent with your individual aptitudes and resources—that is, occupations that were realistic, given your abilities, financial situation, and opportunities for education or training. Ultimately, personal values are the factors people are most likely to consider when they choose the careers they will pursue for much of their lives.

Just as people rely on their values to help them make career choices, they use values when they decide what goods and services to buy. As a matter of fact, career and spending choices are often related. You may, for example, have decided to buy a car to drive to school instead of taking a vacation or buying new clothes. A decision to purchase a new computer may help you with your studies though it prevents you from buying new furniture for your apartment. When you make decisions such as these, you are either consciously or unconsciously deciding which goods and services you believe will best satisfy you. And what satisfies you best depends on what you value most.

With all these choices, it is likely that at some time you will make a wrong decision or fail to make a decision when you should. We can all think of decisions we have made that we later came to realize were mistakes. Poor decisions can happen as the result of a failure to clarify our personal values and goals, to gather enough information, or to evaluate our alternatives adequately. One of the greatest advantages of obtaining an education is that it may help us better understand and evaluate our alternatives and therefore help us make smarter decisions in our lives.

# What Are Values?

**Values** can be defined as high-level preferences that regulate or influence our behavior. For example, a person who places a high value on being aware of world events may choose to subscribe to and read several newspapers each day. Another person who values simplicity in life may choose to become a production-line worker where few choices are required. Those who have strong religious or humanitarian values may spend time working to help others who are less fortunate.

Values are formed in a variety of ways. Probably, people most commonly base their values on the customs and beliefs of the family and society in which they were raised. If you examine your own values, you should not be surprised to find that they are similar to those of your family or of other members of your community. People who have different ethnic, religious, or cultural backgrounds are also likely to hold different values. Many people in the United States do not often think about their ancestral background, but it is likely that many of their values are similar to those held by their ancestors.

The values held by Americans are also influenced by their teachers, coaches, political leaders, and often the media. To what extent have your values been changed by what you have seen on television or viewed at the movies? Ultimately, we all have a personal set of values. To make choices that provide us the greatest satisfaction, we need to understand our personal values and consider them as we make decisions.

**Values** Fundamental concepts or high-level preferences that regulate our behavior. High-level values determine lower-level tastes and preferences that affect our everyday lives.

**Question for Thought & Discussion #2:**
What values that you hold do you consider when you make consumer decisions?

# Values Clarification

Many people are unaware of some of the values they hold, even though these values affect decisions they make. You might say to yourself, for example, that you are buying a car because you need transportation to your job. On closer examination, however, it may turn out that you really don't "need" the car to get to work—you could easily take a bus or ride with a friend who lives near you and works for the same business. When you study your motives closely, you may discover that you really want more than transportation. Perhaps you want to avoid imposing on your friend's generosity, or you could be seeking the prestige that having a car of your own could bring. Such motives are indicative of values you hold—values relating to freedom, independence, prestige, or respect for the interests of others.

**KEEP YOUR VALUES IN MIND**    There is nothing wrong with any of the values demonstrated in the preceding example. The point is that, when making the decision to purchase an automobile, you may not consciously take these values into consideration. You therefore will not weigh them against other values you may hold to determine their relative importance. A failure to consider all your values may lead to a bad decision. Suppose that saving money to attend law school is your highest priority. A decision made now to purchase an automobile could prevent you from saving the funds necessary to achieve your long-term career goal. You may be able to prevent this type of choice through a process called values clarification. This entails learning what your values are and then ranking them in terms of their relative priority.

How do you learn what your values are? One way is to ask yourself "why" whenever you make a decision. Don't stop with just one "why." Continue the

**Life-Span Goal** A central goal that you wish to achieve within your life span.

Digital Vision/Getty Images

Knowing what you want out of life—and what you don't—is central to making rational consumer decisions.

**Question for Thought & Discussion #3:**
What life-span goals have you set for yourself? How are they related to your personal values?

process, querying each answer you give until you have related the decision to your high-level values and priorities.

Another way to clarify your values is to compare your beliefs, tastes, and preferences with those of your friends and closest associates. Be candid in your appraisal. Remember that values are highly personal in nature; each individual has his or her own set of beliefs. You may be surprised at what you learn about yourself from these comparisons.

**REVISIT YOUR VALUES** Clarifying values is an ongoing process. We all change throughout our lives, and the situations in which we live change as well. You are not the same person you were five years ago, and you will be someone different five years in the future. As we evolve, it is not surprising to find that some of our values evolve at the same time. This means that we should periodically review our values to help us make more satisfactory decisions. Sometimes, discussions with your family or friends, teachers, or even a career guidance counselor can help clarify your values as you mature. When you make decisions, you are weighing values. As you consider buying a car, for example, you might ask yourself, "What will buying this car do for me that I value now?" and "How will buying the car affect my ability to achieve other goals that I will value in the future?"

Knowing what you want out of life—and what you don't—is central to making rational consumer decisions. Once you have a fairly clear idea of what you wish to achieve, you have a framework for establishing concrete goals and for evaluating alternatives that are implicit in decision making.

## Goals and Values

Personal goals are inevitably linked to values. Your values lead you to set goals that you expect will bring you satisfaction. If you place a high value on academic success, your goal may be to graduate from college with honors. People who place a high value on family relationships may choose a career that allows them to spend more time with their spouses and children.

Everyone has goals, whether or not they are well defined. To attain the ultimate goal of happiness in our lives, we set many subgoals. Yours may be to complete a degree in pharmacy and get a job in a hospital, or to learn to play the guitar and become the leader of a rock band. Your goals determine many of your consumption decisions. If one of your **life-span goals** is to be financially secure when you reach retirement age, you may choose to spend less of your income now so you can invest more for your future. You might decide to camp in a state park instead of taking a costly vacation at a golfing resort. A small home with few luxuries could be a better choice for you than a large house with many special amenities. If your goal is to lead a luxurious life, however, you may choose to take expensive vacations and live in a palatial home, even though you know that you will pay for them later in life when you have accumulated less savings.

Goals and planning go hand in hand. Consumer decisions are often based on plans that are themselves based on goals. Planning can be a difficult procedure, particularly when family members are involved in numerous activities based on a variety of individual goals. Diverse interests must be considered. Compromise is always part of making and following plans. People who "wish for the stars" invariably have to compromise and face the reality of scarcity.

Consumers who plan in a rational manner may appear to lack spontaneity, and that certainly is one possible cost of planning. But one of the benefits is that life-span goals can more often be met—on schedule and to the satisfaction of the planner. If you become a consumer who plans, then you may reasonably expect to be able to satisfy more of your desires in life. But, if spontaneity is important to you, you'll probably chafe at the prospect of making plans and following them. Just remember, spontaneity may force you to pay a price in terms of your ability to achieve your life-span goals.

# The Decision-Making Process

Up to this point, we have been looking at three of the essential requirements for rational consumer decision making: cost-benefit analysis, values clarification, and goal setting. Now let's examine the decision-making process itself. Basically, the process includes the following five steps:

1. Deciding to act.
2. Identifying alternatives.
3. Evaluating alternatives.
4. Committing to a decision.
5. Evaluating the results.

## Deciding to Act

The first step in the decision-making process is simply recognizing when a decision needs to be made. This may sound obvious, but how many times have we all let the force of inertia, or habit, make decisions for us? Have you ever delayed making an important decision until it was too late to act? If, for example, you are thinking about taking a vacation during spring break but fail to make any travel reservations, you won't go. Later you may regret your inaction, but by then it will be too late. In this case, it wasn't that you made a wrong decision, but that you made no decision at all. This type of choice may be called *decision by default*.

Consider a possible future in which you have recently moved into an apartment of your own. In the past you lived with your family, where you had access to a computer and the Internet. Now this access is lost. If you want to continue to carry out word processing or spreadsheet applications, sending and receiving e-mail, shopping online, or just surfing the Net, you must buy a computer system and subscribe to an **Internet service provider (ISP)**. Making these consumer decisions takes time and effort. You may be tempted to just let the situation "slide" and do nothing at all. Such inaction involves a cost. Until you choose to act, you will lose the value you could have derived from having a computer and access to the Internet.

Maximizing satisfaction in life—the purpose of rational consumer decision making—means controlling decisions to the greatest extent possible. One way to control decisions is to be aware of what is, or is not, happening in our lives, and realizing when choices need to be made.

## Identifying Alternatives

The next step in decision making—and a very important one—is to identify possible alternatives. Again, assume that you have realized that you want, and are willing to pay for, a computer system and access to the Internet. Your decision to act is important but not clearly defined. Before you can make any

**Internet Service Provider (ISP)** A business that provides subscribers with access to the Internet through local telephone or cable lines.

**Question for Thought & Discussion #4:**
How is it possible for two people to make different decisions in the same situation and have both decisions be rational?

choices, you have to identify the alternative products from which you may choose. This can be a challenging process. There are hundreds of combinations of components that can be configured in a variety of ways to create a computer system. You could buy a total package that includes everything you need, or you could purchase each component separately and assemble your own system. How do you find packages or components that you could buy? You could read articles in magazines such as *ComputerWorld* or *PC Computing* to identify the latest models available. Or you could talk to a college instructor or other person who you believe is knowledgeable about computers. Sales associates at computer stores are always happy to tell you about the products they offer for sale. The Internet provides an almost endless list of alternatives. Your problem is more likely to be how to narrow your choices to a reasonable number than how to find enough alternatives to consider.

Many people make mistakes in this step of the decision-making process. They may identify only one or two alternatives and then later find that these were not their best choices. Or they may spend so much time searching for every possible alternative that they never make a choice at all.

## Evaluating Alternatives

In this step of the decision-making process, you weigh the costs and benefits associated with each alternative you have identified. Remember to consider costs not only in terms of money, but also in terms of the opportunity cost you will pay when you give up other uses for your funds. If, for example, you choose to pay a high price to buy the fastest, most powerful computer system available, you will give up other goods or services you could have purchased with some of these funds. Consumers often find it helpful to write down the costs and benefits of each alternative because, as Benjamin Franklin once said, "it is hard to keep all the pros and cons in mind at one time." Toward the end of this chapter, a data grid will be introduced that can be used to organize and evaluate alternatives when you make complex consumer decisions.

**ASK YOURSELF LOTS OF QUESTIONS**  To make a rational decision when you choose a computer system, or any other consumer product, you need to ask and answer many questions. What brand is best known for reliability? How much speed and storage capacity should the system have? How large should the monitor be? Which computer retailer has the most convenient help line? Should the system be purchased from a local store, a mail-order firm, or online over the Internet? A similar set of questions needs to be asked and answered when you choose your ISP. And, just when you think you have identified all the alternatives available and asked and answered all the necessary questions, a new development is likely to occur that places a different "spin" on your situation. Consumers need to realize when they have asked and answered enough questions that they can move on to the next step in the decision-making process.

**ASSESS THE ANSWERS TO YOUR QUESTIONS**  Finding answers to consumer questions is not the same as making a decision. It won't do you any good to find the best deal in the world unless you act to accept the offer. In evaluating alternatives, you should keep the life-span goals you have set in mind. Owning a powerful computer system that provides instant access to the Internet and is able to download and print images in only seconds may be nice, but is it worth the cost of not being able to pay your tuition for the night college classes you had wanted to take? Suppose these night classes could lead to a better job and a more secure future. Is the cost of the computer then just the thousands of dollars you would pay to buy it? Would you be better off limiting your spending for a computer to only a few hundred dollars or perhaps buying a refurbished used computer such as the ones offered by Dell in Figure 2.1 instead?

**SEEK ADVICE FROM EXPERTS** When you need to make an important decision, it is often worthwhile to discuss your alternatives with another person whose opinion and knowledge you trust and respect. Take the time to identify and consult with experts who are knowledgeable about choices you are considering. You are sure to know someone, for example, who is better informed than you about computer systems and the Internet. Ask this person what products she or he believes would satisfy your needs. Don't stop after talking to only one person. Find out what a variety of people think. Ask them about the computer systems they own and the ISPs they use. How satisfied are they with the decisions they have made? Would they make a different choice if they had it to do over? In general, the more expensive a purchase or important a decision, the more time you should devote to evaluating your alternatives.

Consumers should recognize, though, that it is impossible to identify and evaluate every alternative they could choose. Trying to do so is a waste of time. Further, an effort to do this can take so long that the consumer may make no decision at all. An important part of step three in the decision-making process is to know when to stop evaluating and make a decision.

> **"In general, the more expensive a purchase or important a decision, the more time you should devote to evaluating your alternatives."**

## Figure 2.1 Reconditioned Dell Computers at Online Outlet

## Committing to a Decision

The next step in making a rational decision is choosing an alternative and committing yourself to it. Sometimes, this commitment can be an ongoing process, as when you decide to save a certain amount of your income each week. Or it may require you to complete a series of steps, such as making all the different reservations and plans necessary to take a trip. Part of making a good decision may involve persevering in its implementation. Suppose you bought your computer and subscribed to an ISP so that you could access Web sites in Mexico. Your plan is to improve your ability to read Spanish and to stay current on events in that country. Paying for your computer and ISP will not accomplish your goal. It can be achieved only through a continuing effort to carry out your plan. Many people find this type of ongoing commitment difficult to maintain. If this happens to you, try to keep your life-span goals in mind. Learning to read Spanish and knowing about events in Mexico could be important steps toward achieving your goal of becoming a marketing manager for a firm that trades with businesses in Mexico.

## Evaluating the Results

The final step in the decision-making process is to evaluate the results of your choice. There are at least two good reasons to do this. In many cases it is possible to change your mind after a choice has been made. If the ISP you choose does not provide reliable service, you can always cancel your subscription and choose another. In other cases you may not be able to change your decision, but you can still learn from your mistake. Suppose the computer system you purchased has a monitor that is not large or clear enough for you to read easily. Too late, you realize that you did not take sufficient time to identify and evaluate your alternatives. You would have been better served to purchase a higher-quality product even if it cost a little more. Now you may have to buy a new monitor before you had expected to need one. But at least you have learned from your mistake. All consumers make mistakes from time to time. Mistakes can be valuable if they are used to make better choices in the future.

# Pitfalls in Rational Consumer Decision Making

Wise consumers can avoid making a number of mistakes, or pitfalls, that frequently occur in decision making. A few of the most common ones involve impulse buying, habit buying, failing to read the fine print, using credit unwisely, and engaging in conspicuous consumption.

## Impulse Buying

To buy impulsively is to walk into a store, see something we like, and purchase it without taking time for cost-benefit analysis, values clarification, or the steps in the rational decision-making process. Of course, impulse buying does not occur only in stores. It can just as easily happen when shopping from a mail-order catalogue or on the Internet. Merchants exploit impulse buying by displaying products in a whimsical or seductive way. Inexpensive items are often placed at check-out counters where customers wait to pay their bills. After all, a candy bar or a magazine isn't very expensive. But, when

**Question for Thought & Discussion #5:**
What is a consumer decision you have made that you later realized was a bad choice? What did you learn from your mistake?

hundreds of consumers buy these items each day, they can make a big difference in a merchant's profits, and an apparent hole in consumers' wallets as well.

Sometimes, impulse buying is a wise choice—when the purchase is truly a bargain and will help increase your total satisfaction without compromising your ability to achieve your high-value goals. When you see a product you frequently use offered at a low price, it makes sense to buy, even if you had not expected to make this purchase. Nonetheless, there is a limit—usually set by our limited incomes—to the amount of impulse buying in which we can safely indulge ourselves. Most consumer economists argue against impulse buying because it can undermine a budget and may lead to financial difficulties. The best way to control impulse buying is to ask yourself five questions: Why am I buying this? Do I need it? Will I use it? Can I afford it? Will it keep me from buying something I want more later? Remember, for every purchase you make, you forgo some other alternative use of your limited funds.

## Habit Buying

Many purchases are made as a result of habits acquired through the years. No plans are involved and no impulses either, just the force of habit. A person might stop at a tavern on the way home from work every Friday night. An individual may continue to subscribe to a particular magazine he or she once enjoyed but no longer reads. Habit buying has advantages. It is easy and may be convenient—it frees us from the need to think and choose. But it often involves an important cost. When buying from habit, we may miss the opportunity to find and purchase products that offer better quality or lower prices. There is value in periodically evaluating your spending patterns. Take time to identify and consider purchases you make from habit. You may find that using the decision-making process will cause you to change your habits in a way that brings you greater satisfaction.

What problems may result from buying clothing on impulse?

## The Fine Print

If there is a single most important mistake that leads to bad choices, it might be signing contracts or purchase agreements without carefully checking their contents. One has only to spend a little time reviewing court records to realize how much trouble consumers could avoid by reading the *fine print* in documents they sign. Some sales agreements, for example, require borrowers to pay off a debt completely and immediately, if they miss as few as one or two regular payments. Other sales agreements require consumers to have products serviced at specific times to keep a warranty in force. These may, or may not, be reasonable requirements, but you certainly would want to know that such clauses were included in any contract you signed.

At one time contracts were commonly written in language that most consumers could not understand. Today, this is less of a problem, but you still have a duty to yourself to protect your own consumer interests by reading contracts before you sign. Even a clear and understandable contract cannot help you if you don't take the time to assure yourself that you understand and

## CONSUMER CLOSE-UP
### Making Impulse Purchases on Purpose

**B**eing a rational consumer does not mean that you should never make impulse purchases. As a matter of fact, many people deliberately set out to make impulse purchases every year. Nationally, these purchases amount to many billions of dollars. The best example of this phenomenon takes place in December during the holiday shopping season. As the holidays approach, millions of Americans hit the malls or log on to the Internet in an effort to find the perfect gift for Aunt Millie or Uncle Herbert. Most often they don't even care very much what they buy as long as they find something to put under the tree. Consumers typically decide how much they can afford to spend and then go looking for something their friends or relatives are unlikely to own already. Retailers have built successful businesses that cater to consumers' need to make quick purchases of unique products.

Of these firms, Brookstone is probably the best known. This business operates hundreds of stores in shopping malls and a Web site at http://www.brookstone.com that stock a wide selection of goods you are not likely to find at your local Wal-Mart. Brookstone's offerings for December 2002 included the following products:

- The Roomba floor vacuum—an automatic vacuum that would clean "any room and then shut itself off," for $200.
- Confortemp ear warmers—that would store "body heat as you get colder" while you are outside in the winter, for $20 plus shipping and taxes.
- A bed-rest massager—to "invigorate tired muscles" as you rest after a hard day's work, for $125.
- A clean-park mat—to place in your garage under your car to catch "mud, slush, and grime" that fall from your car after you drive it in the winter, for $139.95.
- A pet safety ramp—to "avoid back strain" you might suffer from lifting your pet from your car, for $189.95.

**Regardless of your preferences, thousands of these items and similar products are sold each year by Brookstone. Do you think that these purchases interfere with consumers' ability to reach their life-span goals? Are any of these products ones you would be likely to purchase? How do you go about your holiday shopping?**

agree with what it says. Consumers should also be sure that any verbal agreements they have made with a seller are included in a contract they sign. They should check to see that what the seller has promised is included and that obligations to which they have not agreed have been left out.

## The Seductiveness of Credit

We all know how attractive credit can be. The possibility of using credit is often one factor consumers consider when they evaluate alternatives. It can be tempting to buy something we want now and agree to pay later, especially if we have a steady income and feel we can afford to take on debt. But even a steady income can be interrupted. When you agree to pay $399 each month for three years to buy a new car, you don't know whether you will lose your job or, worse, be injured in an accident and have medical expenses to pay. Further, paying for one product with credit may lead you to buy other products with borrowed funds as well. You may decide that your new car needs a security system or CD player that you also finance over time. When you keep using credit this way, it can destroy your budget. Unless you have set funds aside for emergencies, you should be very careful when you buy on credit. Be sure that your borrowing today will not catch up with you and cause problems in the future. The best way to do this is to use the decision-making process.

## Conspicuous Consumption

**Conspicuous Consumption**
Consumption of goods more for their ability to impress others than for the inherent satisfaction they yield.

Many people buy products more to impress other people than because they expect to receive much satisfaction from the products' use. Such **conspicuous consumption** demonstrates a person's wealth and economic power, but also often wastes scarce resources and may cause people to receive less satisfaction than they could from their limited time and income. A person who has an expensive and powerful computer system may be able to impress his friends by accessing the Internet instantly. The choice to purchase such a system could be a mistake, however, if it means that he is unable to pay expenses such as rent, insurance, or college tuition. Conspicuous consumption is a pitfall that consumers can easily avoid if they evaluate the true value of their alternative choices.

**Question for Thought & Discussion #6:**
Have you ever decided to purchase something just because you wanted to "show off"? What were the costs and benefits of your decision?

# Time Management and Rational Decision Making

In everyone's life, at least one element is constant: we all have only a limited amount of time to do the things we want to do. Scarcity rears its head again. We must each answer the question of how to spend our time to best satisfy our individual needs and wants. Again, the decision-making process offers us a way to make the best, most rational use of our time.

Consider two people who work in the same business—a sales organization that offers Western-style clothing to consumers over the Internet. One of these people reviews orders and makes sure they are filled promptly and correctly. He works a forty-hour week and forgets his job when he goes home at the end of each day. The other person owns the firm. She arrives at 6:00 A.M. each morning and often stays until 10:00 P.M. at night. She also comes in many weekends to update her Web site and make sure her business offers the latest styles at the best prices possible.

Obviously, these two people place different values on their time. The owner probably earns more income, but she bears responsibility for the business and risks a financial loss if it fails. Even when she is at home or on vacation, business problems may occupy her mind. The employee, however, may never think about his job when he isn't working. Even when he is at the office, he may have his mind on other matters. He is able to use his free time to pursue different interests.

## Different Values, Different Opportunity Costs

Both of these individuals may be making a rational choice in the use of their time. This is possible because they almost certainly have different values and life-span goals. They have made choices in how they use their time that are intended to maximize their personal satisfaction.

To manage time, people need to consider time as if it were a commodity like food or clothing, or as if it were money that could be used to buy something of value. Suppose your employer asks you to work extra hours next weekend. You know that if you accept, you will be paid overtime and may earn several hundred dollars in additional income. Working these hours, however, would prevent you from spending time with your friends or family. You might not be able to complete household tasks or finish necessary repairs on your car. To make a rational choice, you should weigh the value of the uses you could make of the extra income against the value of the alternative uses of your time you would give up if you accept the extra hours of work. Remember, earning extra income is pointless if you have no time left to enjoy the products it may buy. It is possible that a rational choice for you would be to forgo the extra work and income so that you can spend time doing something else that has greater value to you.

## Time and Family Values

**Question for Thought & Discussion #7:**
How have you experienced the shortage of time?

People who live together, whether as friends or a family, need to manage their time jointly. It is important for them to divide household responsibilities and work in a way that best satisfies their collective needs. It may make sense for all members of a family to share household chores equitably. Or, if one member of the family has a job that pays a high hourly wage, it may be a better choice for this person to work more hours outside the home while others complete household tasks. Making decisions that involve time management is easiest when you have a clear idea of your goals and values. When a decision involves members of a family, the interests of all members should be considered, and each individual should have a voice in the decision-making process.

## Breaking Parkinson's Law

Whatever our values may be and whatever our lifestyles become, most of us share a certain psychological trait commonly known as **Parkinson's Law.** C. Northcote Parkinson, a management expert, made an observation that unfortunately seems to have universal validity: *Work expands to fit the time allotted for it.* If you're aware of Parkinson's Law, you can fight it. If you're not, it may overwhelm you. Suppose you have allotted yourself four hours to finish an assignment. You had better believe that the assignment will take you at least four hours. Had you allotted yourself, say, three hours instead, you would probably have completed the work in only three hours, and its quality would probably have been no worse. In fact, the quality might have been better because you would have started in earnest immediately instead of twiddling your thumbs, looking up unnecessary source material, and getting comfortable at your computer desk.

**THE VALUE OF MAKING LISTS**    One way to avoid falling victim to Parkinson's Law is to plan and draw up lists of things to do. The more specialized and specific the lists are, the more helpful they are likely to be. For many people, list making is beneficial in accomplishing all sorts of goals. For example, instead of just telling yourself that you really ought to read more books, you can schedule an hour of reading time every day when you make a list of your day's activities. Don't set an unrealistic goal, such as five hours of reading at a sitting, but shoot for something you know you can manage. You will be surprised at how many books you can read this way. If you really want to keep up correspondence with friends who live in other states, stop putting it off—set a specific time in your day's or week's schedule to send and respond to letters or e-mails.

**GIVE YOURSELF A REWARD**    Another way to make sure you accomplish tasks on time is to reward yourself. One writer used a system of rewards to increase his output. Because he was an avid stamp collector, he decided to reward himself with one new stamp after finishing each chapter in a book. Once his plan was in place, he never missed a deadline. When he completed an entire book, he rewarded himself with a rare stamp that he had wanted for a long time.

**Parkinson's Law**  Work will expand to fit the time allotted for it.

Techniques like these can be extended into a lifetime plan. You can set goals for each day, week, month, or even five- or ten-year period of time. Some people set more goals than they believe they can possibly achieve. But, at times, they surprise themselves—they accomplish more than they thought they could because listing their goals provided an incentive to work to achieve them.

> **"The difference between rational and irrational decision making is often knowing when to stop looking for a better deal."**

## Buying and Searching

Most consumer decision making depends on information that must be acquired through a searching procedure. The best search procedure is, of course, different for each person, but a general rule for rational consumer decision making can be made: *The larger the expected payoff from searching for information, the greater the cost that should be accepted to acquire the information.* In plain language, this means that we should spend less time trying to get the best deal on a tube of toothpaste than we spend looking for the best deal for a new computer system. The expected gain from finding the best deal on a tube of toothpaste would probably be only a few pennies, but finding the best price for a computer system could save hundreds of dollars. Remember, the total cost of any product you buy is likely to exceed the price you pay. If you spend hours of your time searching the Internet to save $1 on the price of a pair of shoes, the value of your time was much greater than your saving. The difference between rational and irrational decision making is often knowing when to stop looking for a better deal.

Some people, such as doctors, lawyers, or CEOs, consider their time to be so valuable that they seek out very little information. They may purchase products from a few relatively expensive stores just to save time. They are willing to pay higher prices because they have found that the stores carry high-quality products, and they are unwilling to spend additional time searching for alternative sources for the products they want to buy. For them, the transaction cost, or the value of the time they would spend looking for a better deal, is greater than the value they place on finding alternative choices.

Every consumer reaches a point where the cost of finding additional information is greater than any benefit he or she could reasonably expect to receive from having this information. At some time we have to stop searching and make our choice based on the knowledge we have acquired. We may make a mistake, but we can learn from it to make better choices in the future.

# Confronting Consumer Issues:
# How to Buy Technology-Based Products

Perhaps the value of the rational decision-making process can best be seen when you choose among products that are based on new technologies that you don't completely understand. More and more today, advances in technology bring products to market that not only did not exist in the past, but that most of us never even imagined a few years ago. In 1980, for example, who would have predicted how our lives would be changed by the Internet? What would your parents have said at that time if they had known that consumers would soon be able to communicate instantly with people on the other side of the earth by using computers in their homes? Could they have pictured themselves shopping, taking college courses, or completing research through electronic communications? These and other uses of the Internet would have seemed like science fiction in the past. Today, we take them for granted.

Advances in technology, of course, can be seen in many products we buy besides those directly connected to the Internet. Technologies that are changing our lives include new types of telephones, stoves, heating systems, home entertainment units, and exercise equipment, to name only a few.

In the next decades, the speed of technological change can only be expected to accelerate. As a result, some observers believe that the increasingly complex nature of the products we use will cause some consumers to suffer from *decision overload.* This occurs when people feel overwhelmed by the decisions they must make. As a result, they stop trying to make rational choices, saying, in effect, "I can't figure out what I should do no matter how hard I try, so what's the point in trying? I'll just do what feels right at the time and let the chips fall where they may." Although such an attitude may be understandable, it will almost certainly bring harm to the people who hold it. Although our lives are more complex today, and many of the products we buy are harder to understand and evaluate, we should never give up trying to make the best decisions possible to protect and enhance the quality of our lives.

The ability you will develop to make rational decisions is an important benefit you may expect to receive from your study of consumer economics. This skill can help increase the satisfaction you receive from the limited funds you have to spend, particularly when you are choosing among technology-based products. When all else changes, the ability to think logically and to make rational choices will remain the best way to protect yourself as a consumer.

## TECHNOLOGICAL OBSOLESCENCE

Have you ever felt a sense of loss when an old appliance, electronic device, or piece of furniture you have owned and used

Using cost-benefit analysis, a consumer compares brands of computers at an electronics store.

for years finally "dies" and needs to be replaced? Suppose, for example, you once owned a vintage 1994 PC that operated with a 486 chip and used DOS–based WordPerfect 5.0 as a word processing program. For years, you used your computer to write reports and printed them on a nine-pin dot-matrix printer. You realized the system was old, but it worked, and you didn't need to think about how to use it. Then one day when you turned on your machine, nothing happened, but a suspicious smell of burnt electrical insulation permeated the room. You bundled the computer up and took it to your computer doctor, who pronounced it DOA. So, did you think you could find a new computer like your old one? Of course not! Suddenly, you were thrown into the world of Pentium IV processors, CD-ROMs, Windows 2000, laser printers, and Internet access. You had to investigate your consumer choices from "ground zero." Whatever you thought you knew about computer systems in the past was out of date. What's more, in a few more years, when you need to buy your next system, you will be faced with the same situation again, thanks to changing technology.

**Outdated Products**  Consumers are often faced with **technological obsolescence,** which can make products outdated almost as soon as they are purchased. Although personal computers may be the best example of technological obsolescence, it also applies to CD players, large-screen televisions, portable telephones, and virtually every other electronic device you could buy. Consider the hand-held video cameras that were first introduced to the market in the late 1970s at a price of $2,000 to $3,000. They were heavy, typically weighing from eight to twelve pounds, and required a battery pack that weighed another ten pounds. The added cables, tape measures, special lights, and light meters that were required to operate the cameras made anyone using one look a little like a walking junkyard. Thirty years later, a video camera can be purchased that weighs less than two pounds (including its batteries), adjusts automatically to a wide range of lighting conditions, and can be operated by a three-year-old if necessary. Further, these products typically cost $400 or less—about 20 percent of the price of the 1970s models.

## Question for Thought & Discussion #8:

Do you own a technology-based product that has become obsolete even though it works perfectly? How will you decide when to replace it?

### The Downside of New Technologies

You might wonder how anyone could see anything negative in these changes. Although technological advances may seem wonderful on the surface, consumers should realize that the improvements also complicate their choices. If you know that a product you are interested in owning, say, a portable telephone, is likely to be improved in the near future, should you buy it now or put off your purchase for a few months? If you do put off buying it, how should you value the lost use of the telephone at the present? What is the value of the time you will spend trying to keep up with technological change so that you can know when to buy? These are questions that may disturb your sleep as you try to make the most rational choice for yourself.

## THE DEPRECIATION PROBLEM

Rapid advances in technology create equally rapid declines in the value of older technology-based products. Although this type of depreciation is real, it does not mean that the products are worn out or have become useless. If you bought a graphing calculator for your Calculus I class last year, you would not be surprised if your Calculus II professor required you to buy a different, more advanced calculator this year. Your old calculator might work perfectly, but it could have little value if no one else wants it. Remember, the best measure of anything's value is the amount someone else is willing to pay for it.

The more rapidly technology advances, the greater the problem of depreciation is likely to be. This type of problem can be seen in the context of decisions that are made by businesses or other organizations as well as individuals. The college you attend, for example, must consider technological depreciation as it allocates funds within its budget. Do you believe your college should spend thousands of dollars buying paper copies of books and periodicals for its library when many of these publications are available on CDs or over the Internet? What about the cost of storing all the books and magazines that have been purchased in the past? Should many, most, or all of them be discarded so that library space can be used for different purposes? Indeed, are printed textbooks, such as the one you are reading, likely to become an endangered species in the future?

## FINDING, SORTING, AND EVALUATING INFORMATION

As rapidly changing technology brings new products to market, it can become a challenge to keep up with information you need to make rational decisions. There is no shortage of information; most often the opposite is true. So much information is available that your first task is to decide how long you should search and how much you need to know. There are many publications, user groups, and experts you may consult to learn about almost any product you might want to own. Consumers may search the Internet to find reports on virtually every product offered for sale. At the end of 2003, a search of the Internet for *digital cameras* using Yahoo.com identified more than 1.9 million Web sites with thousands of pages devoted to marketing or providing information about this type of product. It would have taken many weeks, perhaps years, to examine and evaluate all the information these Web sites had to offer. To be sure, this search could have been narrowed by using additional key words, but even that would have taken time. If you spend more than a few weeks to evaluate information you have found, it is likely that new advances will take place that will make the information you gathered obsolete. What is a consumer to do?

When it is difficult to keep up with rapidly changing technology, experts suggest that "service is everything." In other words, when you buy a product you don't understand 100 percent, you had better be sure that you buy it from an established business that has a record of standing behind the products it sells. These firms work to protect their names and reputations. They are more likely to help you obtain the greatest value from your purchase than a "fly-by-night" business that is here today and gone tomorrow.

*(Continued on next page)*

**Confronting Consumer Issues (Continued)**

## TRADE-OFFS IN BUYING NEW PRODUCTS

Although many consumers feel a special form of satisfaction when they buy the newest products on the market, you should recognize that there is also a danger in buying a product that is brand new and therefore untested through consumer use. No amount of manufacturer testing is likely to uncover every "bug" in a new product. Further, the more complex the product, the greater the chance that the manufacturer missed a problem. When Microsoft's Office 2000 was first introduced, many users experienced a number of "glitches." These problems were quickly solved, but they still caused difficulties for individuals who purchased this software when it first became available. Consumers who waited a few months to upgrade to Office 2000 avoided these problems.

Many experts suggest a general rule of thumb for consumers to follow when they consider products that are new to the market: *When in doubt, wait.* Products that have just been introduced are almost always the most expensive they will ever be, and they almost always have the most defects they will ever have. It is generally better to wait a while before buying products that are new to the market.

Notice we used the word *almost.* The fact that a product is new and expensive should not deter consumers from buying it if it provides a benefit they really need. For example, if you had a dreaded disease and a new medicine that promised a cure was created, you would probably want it even if it was expensive and its effectiveness not totally proved. The same was true of professional accountants who paid more than $100 for the first hand-held electronic calculators that became available in the early 1970s. These calculators were expensive and had limited capabilities, but they were so much better than the mechanical calculators they replaced that they typically sold out in stores as quickly as they could be stocked.

### Question for Thought & Discussion #9:

If you own a computer system, describe how you chose the system. If you do not own a computer, explain why you have chosen not to purchase one. In either case, what are the costs and benefits of the decision you made?

When making a decision about a new product, you should study the trade-off between the benefits and costs of buying now versus later. If you "just have to have it now" and can afford the product, go ahead and buy it. But realize

# Cyber consumer
## Safe Shopping Online

Before you consider online shopping for components for a computer system, or any other product for that matter, you should visit a Web site created by the American Bar Association at **http://www.safeshopping.org**. According to the site's home page, it will provide you with information that will "help you order safely when shopping online. You'll find cyber-shopping is fast, convenient, and opens up a whole new world of merchandise for you and your family. Just click on any of the topics to find a smarter, safer way to shop online." Visit and investigate this Web site. Then write an essay that evaluates the information it provides.

that you could almost certainly pay less for a better model if you choose to wait.

## BUYING A COMPUTER SYSTEM: A CASE IN POINT

In recent years, U.S. consumers have seen a steady stream of new electronic goods and services brought to market. At the same time, existing electronic products, such as computers and portable telephones, have been improved at an ever-increasing pace. Today, roughly 60 percent of all Americans have access to a computer in their homes, and about 80 percent of these families are connected to the Internet. This situation did not come about by accident. It came to pass because many Americans made the choice to buy a computer and subscribe to an Internet service provider. Considering why and how so many people made these choices may help you learn to make better choices for yourself at the present and in the future.

## One Choice Is Rarely Enough

When you think about it, you will realize that no one decides to buy just a computer and stops there. Ownership of a computer leads you to buy other products such as a monitor, printer, modem, CD-ROM, subscription to an ISP, computer desk, ergonomic chair, lamp—the list goes on and on. The $800 you decide to spend for a computer is only the first of many expenses you are likely to incur. When you use the decision-making process to choose the computer that is right for you, keep the costs of associated products in mind. You are not buying one product—you are taking the first step in purchasing a computer system that will include the ongoing cost of a subscription to an ISP to access the Internet as well.

## What Do You Want? What Do You Need?

Imagine that you have taken the first step in the decision-making process—you are committed to becoming connected to the Internet from your home. For most consumers, the first step toward tapping into the Internet is buying a personal computer. As you identify and evaluate your alternative choices, remember that you will use your computer for purposes that extend beyond accessing the Internet. You may want to keep records on a spreadsheet, play interactive games, send and receive images, or burn CDs. Be sure the computer you choose has the capacity, or can be upgraded, to carry out other tasks you want it to do. A challenging but important consideration is predicting how you are likely to use your computer in the future. New uses of your computer system are likely to require greater computing power than your current uses. Investigate how difficult it would be to increase your computer's storage capacity, RAM, the megahertz of its CPU's microprocessor, or the speed of its modem. Finding the current cost of upgrades may not be particularly useful because these costs are likely to change in the future.

No one can anticipate every new use that will be found for computers, but we can choose ones that are relatively easy to upgrade. Alternatively, you might consider buying a used or discontinued model that is inexpensive now, so you won't mind discarding it when you want a more powerful one in the future.

## CREATING AND USING A DATA GRID

The decision to buy a computer system is so complicated and involves so many steps that experts suggest consumers create a data grid to help them organize, compare, and evaluate the alternatives they consider. Study the sample data grid in Figure 2.2 on the following page. The first column indicates in bold headings different components of a computer system a consumer might wish to purchase. Each heading is followed by desired characteristics of the components ranked in order of their importance to the consumer. The top squares of each of the remaining columns have spaces for models or brands of components that the consumer might consider buying with their expected costs. By placing an X in the appropriate boxes, consumers can identify components that have the desired features and project the cost of each selected component as well as the total for the entire system. If this total is greater than the amount budgeted for the computer system, the data grid can help identify where cuts can be made while keeping the most desired features. This type of data grid can be used to help consumers choose an ISP as well.

Although data grids are useful when making any type of consumer decision, they are probably most helpful when you are choosing among complex, technology-based products. Try creating a data grid to help you evaluate a different type of technology-based product, such as a digital camera or telephone. What features would you look for, and how much would you be willing to pay to own one of these products? The information you enter in your grid will be useful for only a limited time, but your ability to construct and use such a grid can help you make rational consumer decisions throughout your life.

## Key Term:

**Technological Obsolescence** When products lose value because they are technologically out of date rather than because they are worn out.

*(Continued on next page)*

**Confronting Consumer Issues (Continued)**

**Figure 2.2**

| Sample Data Grid for Choosing a Computer System | | | | |
|---|---|---|---|---|
| **Component, COMPUTER** | **Brand:** Price: | **Brand:** Price: | **Brand:** Price: | **Brand:** Price: |
| Memory: at least | | | | |
| Modem speed: at least | | | | |
| RAM: at least | | | | |
| Ergonomic keyboard | | | | |
| Price of desired component $_____ | | | | |
| **Component, MONITOR** | **Brand:** Price: | **Brand:** Price: | **Brand:** Price: | **Brand:** Price: |
| Size of screen: at least | | | | |
| Flat screen | | | | |
| Ease of adjustment | | | | |
| Number of colors: at least | | | | |
| Price of desired component $_____ | | | | |
| **Component, PRINTER** | **Brand:** Price: | **Brand:** Price: | **Brand:** Price: | **Brand:** Price: |
| Quality resolution | | | | |
| Color | | | | |
| Speed | | | | |
| Price of desired component $_____ | | | | |
| **Component, SCANNER** | **Brand:** Price: | **Brand:** Price: | **Brand:** Price: | **Brand:** Price: |
| Speed: at least | | | | |
| Quality resolution | | | | |
| Color | | | | |
| Can it work as a copier too? | | | | |
| Price of desired component $_____ | | | | |
| Total price of desired components $_____ | | | | |

# Chapter Summary

1. We are all consumers who purchase a variety of goods and services.

2. Rational consumer decision making requires allocating our time and money resources in such a way as to maximize our satisfaction.

3. A cost-benefit analysis helps us look more objectively at the potential consequences of our decisions and can therefore assist us in making more rational decisions.

4. Values are fundamental concepts or preferences that regulate or in some way affect our behavior. High-level values give rise to tastes and preferences that affect our everyday lives.

5. Through the process of values clarification, we not only learn what our values are but also establish the relative priority in our lives of various values that we hold. The resulting hierarchy of values provides a framework for directed, rational decisions that are harmonious with our life-span goals.

6. Goals are closely linked to our values. Once we are aware of what we value most, we can establish goals and sub-goals that are consistent with our values. If we place a high value on education, for example, we may establish the life-span goal of going to graduate school. A subgoal might be getting straight A's next term to ensure our acceptance at the graduate school of our choice.

7. The decision-making process involves five steps: (1) knowing when a decision is called for, (2) determining the possible alternatives, (3) evaluating each alternative through cost-benefit analyses, (4) committing oneself to a particular alternative, and (5) evaluating the results of the decision.

8. There are numerous potential pitfalls in consumer decision making. Some of the most common are (1) impulse buying, (2) habit buying, (3) failing to read the fine print of purchase agreements, (4) using credit unwisely on the assumption that one's spendable income will remain the same in the future, and (5) engaging in conspicuous consumption.

9. The amount of time that one has available is constant and unchangeable. A major choice for every consumer is how to spend it.

10. Even if you have decided to accomplish certain goals because they fit in with your values, you may not succeed if you fall prey to Parkinson's Law, which says that work will expand to fit the time allotted for it. It is your responsibility to manage your personal time most efficiently.

11. Information must be acquired before purchases can be made. It is, however, useful to acquire information only up to the point at which the expected costs of searching for more information are greater than the expected payoff from having the additional information. The larger the purchase contemplated, the more time you should spend seeking information on the best product and the best financial deal.

12. Consumers who are considering buying technology-based products may benefit from using the decision-making process. High-tech products present special problems for consumers, including technological obsolescence, rapid depreciation, product complexity, and a relatively greater chance of buying defective products.

13. The increasing complexity of products consumers buy may result in what has been called consumer overload, in which people stop trying to make rational choices. This condition is harmful to consumer interests. It can be overcome by creating and using a data grid to identify and evaluate alternatives from which to choose.

# Key Terms

conspicuous consumption **36**
cost-benefit analysis **28**
Internet service provider (ISP) **31**

life-span goal **30**
Parkinson's Law **38**
rational consumer decision making **27**
technological obsolescence **43**

transaction costs **27**
values **29**

# Questions for Thought & Discussion

1. What are several transaction costs that you must pay to purchase a college education?

2. What values that you hold do you consider when you make consumer decisions?

3. What life-span goals have you set for yourself? How are they related to your personal values?

4. How is it possible for two people to make different decisions in the same situation and have both decisions be rational?

5. What is a consumer decision you have made that you later realized was a bad choice? What did you learn from your mistake?

6. Have you ever decided to purchase something just because you wanted to "show off"? What were the costs and benefits of your decision?

7. How have you experienced the shortage of time?

8. Do you own a technology-based product that has become obsolete even though it works perfectly? How will you decide when to replace it?

9. If you own a computer system, describe how you chose the system. If you do not own a computer, explain why you have chosen not to purchase one. In either case, what are the costs and benefits of the decision you made?

## Things to Do

1. Identify several basic values that are important to you. Do you believe that these values will last throughout your life? Explain why many people's values change over time. Why should people periodically reassess their life-span goals?

2. Identify one of your most important life-span goals. Describe the steps you are taking at present that you believe will help you achieve this goal. How do these steps influence the consumer decisions you make?

3. Think back to an important consumer decision you recently made. Carefully consider the steps you took to make this decision. Compare these steps with those in the decision-making process described in this chapter. How closely did you follow this process? Might your decision have been different if you had followed the process more closely? Will you be more likely to follow these steps in the future when you make other consumer decisions? Write an essay that describes your decision and answers these questions.

4. At one time or another, almost everyone makes a choice that he or she later regrets. Describe one such choice that you have made. What steps could you have taken to make a better choice? What have you learned from this choice that will help you make other consumer decisions in the future?

5. Construct a data grid similar to the one in Figure 2.2 that could be used to help you choose an apartment to rent. Instead of listing components in the first column, list factors such as location, number of rooms, age and condition, and other facilities (such as a swimming pool or weight room). Complete the grid by evaluating apartments rented by different people you know. Which of these apartments would best fit your desires? How does this exercise demonstrate the usefulness of a data grid when making any complex consumer decision?

## Internet Resources

### Finding Consumer Information on the Internet

The following Web sites have been selected for their relevance to topics discussed in this chapter. Search these sites to locate information that can add to your knowledge of time management. Remember, Web addresses change frequently. If any of these addresses no longer function, find similar sites to investigate using any of the search engines available to you.

1. You can obtain personal time-management tips and techniques and goal-setting guidelines at the "Personal Time Management Guide." Find it online at **http://www.time-management-guide.com**.

2. More time-management tips and a personal time survey are available at **http://www.gmu.edu/gmu/personal/time.html**.

3. You can find a workbook to help you develop a personal time-management plan at **http://www.hrtoolbook.com/subsites/time/timeplan.html**.

## Shopping on the Internet

The following Web sites have been selected because they offer consumers services similar to those described in this chapter. These are commercial sites that are designed to market products. They do not represent a comprehensive or balanced description of all used-computer retailers available online. Remember, Web addresses change frequently. If any of these addresses no longer function, find similar sites to investigate using any of the search engines available to you.

1. The Used Computer Mall offers free classified ads and an indexed list of dealers who buy and sell all types and brands of used-computer equipment. You can find it online at **http://www.usedcomputer.com**.

2. 123Compute sells used PC and Mac desktops and laptops, as well as reconditioned systems. Its Web site is at **http://www.123compute.com**.

3. Abacus Used Computers offers used and refurbished computers, laptops, LCD projectors, printers, monitors, and Macs. Its Web site is at **http://www.usedcomputers-abacus.com**.

## InfoTrac Exercises

Purchasers of new copies of this text are provided with access to the InfoTrac Web site. This Web site links students to thousands of recent articles published in hundreds of periodicals. Use the key words **conspicuous consumption** or other terms from this chapter to conduct a key-word search. Choose one article that is of particular interest to you and write a brief essay describing what you have learned from the article. Be sure to cite the author and title of the article and the name and date of the publication in which it appeared.

## Selected Readings

"Are Cookies Hazardous to Your Privacy?" *Information Management Journal,* May–June 2002, pp. 33–39.

"Choosing an ISP: 10 Questions." *Network World,* December 16, 2002, p. 48.

Galotti, Kathleen M. "Making a Major Real-Life Decision, College Students Choosing an Academic Major." *Journal of Educational Psychology,* June 1999, pp. 379–388.

"Internet Service Providers." *Interactive Week,* June 4, 2001, p. S-28.

Kasser Barbara, and Robert Hansen. *Internet Shopping Yellow Pages.* New York: McGraw Hill, 2000.

Kucznski, Alex. "Lifestyles of the Rich and Red Faced." *The New York Times,* September 22, 2002, p. ST1.

Lica, Lorraine. "Getting Consumers to Buy One More Thing." *Drug Topics,* August 5, 2002, p. 78.

Lipowski, Melissa. "Generating More Impulse Buying." *Food Management,* July 1999, pp. 50–51.

McDonald, Marci. "A History of Shopping Binges." *U.S. News & World Report,* May 24, 1999, p. 92.

Morton, David. *The Art of Decision-Making.* New York: Springer Verlag, 1986.

Rushin, Steve. "Fear and Clothing in Atlanta." *Sports Illustrated,* February 17, 2003, p. 17.

Selingo, Jeffery. "Does Anybody Really Use This Stuff?" *The New York Times,* November 16, 2000, p. G1.

"Useful Marketing Web Sites." *Accounting Today,* May 21, 2001, p. 1.

# CHAPTER 3

# A Flood of Advertising

After reading this chapter, you should be able to answer the following questions:

- What are the benefits and costs of advertising to consumers?
- Who pays for advertising?
- What are the characteristics of different types of advertising?
- What is deceptive advertising?
- What is the role of the Federal Trade Commission in regulating advertising practices?
- How can consumers obtain and use privately prepared information about advertised products?
- How may consumers benefit from shopping on the Internet?
- What steps can consumers take to protect their rights while shopping on the Internet?

Whether you like it or not, this year more than 240 billion dollars' worth of producer-generated information about products and services will be aimed at you and other consumers. We call this advertising. When you turn on a commercial radio station, the sounds of advertising strike your ears at least every five minutes. When you watch a commercial television station, you are usually treated to a series of ads at least every ten minutes. Whenever you open your local newspaper, leaf through a magazine, or visit a commercial Web site on the Internet, advertisements cross your field of vision. When you receive your daily mail, a good proportion of it is likely to consist of eye-catching ads, mail-order catalogues, or product samples. When you answer your phone or door, you might hear about even more products that may or may not interest you.

And if that's not enough, you can purchase more information about every good or service you might want to obtain. You can buy books on how to invest money in the stock market, how to shop for a house or a car or a computer, how to keep fit and trim, and how to avoid being cheated. Information is all around us, bombarding us every second of every waking hour—or so it seems.

Obviously, some of this information is useful and some of it is not. Some of it can even deceive you into making a purchase you may regret. This chapter first examines the world of advertising and how it affects us as consumers. It then looks at sources of consumer information that are available for free or purchase and discusses how we, as consumers, can benefit from the abundance of information at our fingertips. Finally, it discusses how you can be a successful Internet shopper.

## Costs and Benefits of Advertising

Information in the form of advertisements relating to products in our economy has been on the upswing, as is shown in Table 3.1, which details the expansion of U.S. advertising in its various forms. On average, advertisers pay almost $860 per person each year in the United States.

There must be a fairly good reason why we are subjected to so much advertising and why it is increasing each year. Let's look at it from the advertiser's

**Question for Thought & Discussion #1:**
What are several ways you use advertisements to help you make consumer decisions?

| TABLE 3.1 | National Spending for Advertising in 1990, 1995, and 2002 | | |
|---|---|---|---|
| | **Billions of Dollars** | | |
| **Media** | **1990** | **1995** | **2002*** |
| Television | $ 29.1 | $ 37.8 | $ 52.7 |
| Newspapers | 32.3 | 36.3 | 44.8 |
| Direct mail | 23.4 | 32.9 | 45.9 |
| Radio | 8.7 | 22.3 | 18.6 |
| Yellow Pages | 8.9 | 10.2 | 13.8 |
| Magazines | 6.8 | 8.6 | 11.0 |
| Internet | 0 | 0 | 5.5 |
| Other | 20.6 | 25.8 | 44.1 |
| Total | $129.8 | $173.9 | $236.4 |

*Values for 2002 are estimates.
SOURCE: Robert J. Coen, *Universal Almanac*, July 2002, at **http://www.universalalmanaccann.com**.

> "Since businesses are out to make money, they are not going to lose money by advertising, at least not intentionally."

point of view. Most businesspeople are in business for one reason and one reason only—to make money. Obviously, they wouldn't advertise if they didn't think that advertising could help increase their profits or at least help maintain their current sales and level of profits. Thus, we can assume that businesspeople believe the additional sales they will make through advertising will cover the costs of that advertising. So the advertising explosion can be attributed partly to the realization by businesses that increased advertising yields more than enough additional sales to justify the expenditure. Of course, when you look at it this way, you also realize who ultimately pays for advertising.

## Who Pays?

Since businesses are out to make money for themselves, we can be certain that they are not going to lose money by advertising, at least not intentionally. Only when they make a mistake about the profitability of a particular advertising campaign do they pay for it themselves—by taking a loss on that particular expenditure. In general, the cost of advertising is built into the prices of the products we buy. After all, the cost of labor is built into the prices of the products we acquire and so are the costs of buildings and machines. There is very little we can do about it, except to purchase nonadvertised items that are similar in quality to advertised ones and lower in price. Consumers may do this by shopping at supermarkets that sell house brands of food and other products at lower prices.

In all fairness, it should be pointed out that the advertising industry claims that the increased sales due to advertising can motivate some companies to use more efficient mass-production techniques, which lower per-unit costs. Hence, the consumer benefits when these lower unit costs are passed on.

# Brand-Name Advertising

Many consumers look for particular brands of products to buy when they shop. Producers, therefore, often market goods or services by promoting their brand names because they believe that doing so will increase their profits. And it does, if consumers buy more of, or are willing to pay higher prices for, a brand-name product they recognize and trust. Consumers do this because they are convinced the brand-name product is superior to alternative goods or services. They may feel this way even when there is little evidence for that conviction. The most common way producers create brand-name recognition among consumers is advertising.

## Promoting Brand-Name Loyalty

A common way to develop consumer recognition for a particular brand-name product is **persuasive advertising.** The objective of this type of advertising is to associate a specific brand of product with a certain lifestyle or image in the minds of consumers. Persuasive advertisements imply that if you use a particular brand-name product, you're "cool," or you'll look young, seductive, competent, or upwardly mobile. If you drink a particular brand of beer, gorgeous women in skimpy swimsuits will suddenly find you irresistible. Or by choosing to wear a particular brand of perfume, you will find your dull, mundane life filled with mysterious men helpless to resist your

**Persuasive Advertising** Advertising intended to associate a specific product with a certain lifestyle or image in the minds of consumers.

charms. Persuasive advertising provides virtually no factual information. It is intended to change consumer tastes by appealing to their psychological needs. Persuasive advertising is sometimes referred to as *puffery* because it often is little more than hot air.

Although brand-name recognition is fostered by advertising, consumer loyalty to specific products has suffered a large dent in recent years because of the proliferation of similar but unbranded products that are offered to consumers at substantially lower prices. Informed consumers have learned to save money by purchasing these unbranded goods. To combat this trend, businesses have resorted to new techniques of advertising.

One such technique is *market segmentation*—dividing the market according to age group, lifestyle, ethnic background, or other criteria and advertising accordingly. Brand-name jeans, for example, are advertised during TV programs that predictably attract teen-agers and young adults, who most often purchase such products. Selective audiences for advertising materials are also targeted through direct-mail lists that advertisers purchase from mail-marketing firms and on the Internet at carefully selected Web sites.

## The Value of Brand Names

It is often possible to purchase unbranded products at lower prices than similar brand-name products. Why, then, do many consumers choose to buy more expensive brand-name goods and services? There are at least three answers to this question:

> **"Persuasive advertising provides virtually no factual information. It is intended to change consumer tastes by appealing to their psychological needs."**

**Question for Thought & Discussion #2:**
What brand-name products does your family purchase? Why do you believe your family buys these products instead of less expensive alternatives?

---

### THE **diverse** CONSUMER

## McDonald's Segments Its Message

Large advertisers, such as McDonald's, Nike, and General Motors, have included appeals to ethnic and minority shoppers in their advertising for many years. But, in the past, these advertisements were more often than not placed in the mainstream media, appearing in publications such as *Newsweek* and *USA Today* or on major television networks during sporting events. Furthermore, except for including black, Asian, or Latino actors, the ads made little attempt to appeal specifically to African Americans and other minorities. Aside from the actors, the ads looked like the advertisements directed to the general population. In recent years, this situation has begun to change.

In 2000, McDonald's launched an annual promotion directed at African Americans. The campaign, which was coordinated by a single office, was kept separate from the company's other marketing programs. It began with a salute to Martin Luther King, Jr., and a sweepstakes that offered to send four families on vacation to Kenya. Later, McDonald's sponsored a celebration of black music and publicized corporate contributions to African American heritage educational programs.

To be sure that its message reached the target audience, McDonald's began its advertising blitz with spots placed on cable channel BET's celebrity talk shows on four consecutive nights. One episode of BET's popular show *Hits from the Street* was actually broadcast from a McDonald's restaurant in an urban African American neighborhood. This was followed by repeated messages broadcast on local cable stations that served the African American community as well as on BET. The firm's director of African American marketing explained that McDonald's had decided to concentrate its minority advertising on cable television because this medium allows it to target a "specific" audience in a way that is not afforded by most other media.

You might wonder why McDonald's would create a special advertising program for any minority group. The answer, of course, is to earn a profit. In 2000, 15 percent of the company's sales were made to African American customers. Further, while other sectors of its market were stable or in decline, the African American sector was growing. Targeting advertisements to enhance growth makes good business sense for McDonald's or any other corporation.

SOURCE: "Segmenting the Message," *Adweek*, April 17, 2000, p. 20.

1. *Brand names may mean less variance in quality.* If we have learned that brand-name products generally vary less in quality than other products, then the brand name has value, for it tells us to anticipate fewer problems. This may be true, for example, with electronic equipment. You may decide to purchase, say, Sony stereo components because you have found out, or have heard from your friends, that the brand name Sony means less likelihood of breakage and repair. Therefore, if you go to a stereo shop and see two amplifiers next to each other, one by an unknown company and the other by Sony, you may be willing to pay a higher price for the nationally advertised brand item.

    The same is true for other products and services. People sometimes prefer to pay slightly more for nationally advertised products because they have more confidence in them. Whether national brand names and quality are always related is a moot point. Advertisers have been selling brand-name reliability for years, but only recently has such reliability been questioned. Although many consumers have concluded that brand name and reliability are not associated, many others are still willing to pay more for brand-name products.

2. *Brand-name products may offer better warranties.* When something goes wrong with the national-brand product you bought, it may have a superior warranty, and obtaining service for it may be quicker and easier than for a nonnational brand. For that reason, you may decide to pay more for the national brand.

3. *National-brand products may be repaired at a larger number of facilities.* It may be more difficult to get nonnational-brand products repaired. This is especially true for automobiles. How many gas stations can help out when a Maserati refuses to start? And even if a mechanic knows how to fix it, how quickly would he or she be able to get the parts?

## Benefits of Brand Names to Producers

Brand-name loyalty is designed specifically to benefit producers. When brand-name loyalty occurs on a wide scale, producers gain **market power,** the ability to change price and/or quality without substantially losing sales. Every producer has as its goal some degree of market power. In a perfectly competitive environment, no producer has any market power; that is, each producer must take the price of its product as given in the marketplace. In a perfectly competitive world, any producer that raises its prices will lose virtually all sales. But, as was pointed out in Chapter 1, for the most part we do not live in a perfectly competitive economy. Rather, producers have various degrees of market power. Brand-name loyalty is just one attempt by those producers to obtain more market power and to make higher profits.

# Other Types of Advertising

## Informative Advertising

**Market Power** The ability of producers to change price and/or quality without substantially losing sales.

**Informative Advertising** Advertising that simply informs.

**Informative advertising** is self-explanatory: it simply informs. Consumers see a tremendous amount of informative advertising: supermarkets advertise their prices; computer shops advertise the brands they sell and their costs; producers advertise new products that were not previously available. In other words, you, the consumer, are constantly being informed about the prices, quality, and availability of products. You can take that information for what it is worth and use it any way you want. You are not asked to believe that one product is better or worse than another or that a company does a good or a bad job. Rather, you are given the relevant information about the key aspects—price, quality, and availability—of a good or service.

In the United States, informative advertising is particularly heavy. There are some very unusual products that you would never expect to be advertised because the market is so specialized. Did you know, for example, that the producers of multimillion-dollar gas turbine generators send salespeople around to various electric utilities to inform them about the availability and the costs of different types of generators? Did you know that hundreds of specialized trade magazines are targeted to narrow fields of interest and that companies in those fields subscribe to the magazines just to find out what is happening and what products are available? Journals are published exclusively for the fields of printing, publishing, electric utilities, leatherworks, paper production, flour milling, and so on. In fact, most industries have several trade magazines in which very specialized informative ads can be found.

Since the mid-1990s, it has become common for businesses to "rent" advertising space on Web sites. From a business point of view, this makes good sense because Web advertising is relatively inexpensive and can be directed to people who are more likely to buy. Firms that offer medications for sale, for example, know that advertisements placed on Web sites that provide information about health are likely to be seen by many potential customers. Hotel chains promote their establishments on travel Web sites. Investment services advertise on sites that report on the stock market, and so on. Much Web advertising is informative and can be easily found and utilized by consumers. They are able to obtain useful information from advertisements for goods or services they want to buy because businesses know where to place their ads on the Internet to achieve the best return for their advertising dollars.

As more and more traditionally unadvertised services—such as health care, accounting and legal services, and financial services—enter the promotional and marketing fields, more informative advertising about these areas is becoming available as well.

## Comparative and Defensive Advertising

**Comparative advertising** is advertising that actually names competing brands (not merely Brand X) when comparing them with the advertised brand. Although many people once thought that such advertising was illegal, it never actually was; but in the past, radio and television stations were either reluctant to broadcast such ads or refused them altogether. When properly and honestly done, comparative advertising is obviously beneficial to consumers because it saves us the time of doing the comparisons ourselves. Consumers must be careful, however, because comparative advertising is bound to be selective and to show only what the advertiser wants us to know.

Another kind of advertising is called **defensive advertising,** and, again, it is just what the name implies. Defensive advertising fosters—and is fostered by—brand-name competition. Cold remedies are a good example. Each of the large companies manufacturing these products advertises extensively, but no single company gets an edge on the others through this advertising. If any

**Question for Thought & Discussion #3:**
Why do beauty and grooming products seldom use informative advertising?

**Comparative Advertising** Advertising that makes comparisons between a product and specific competing products.

**Defensive Advertising** Advertising intended to rebut claims made by competing firms about a firm's product or business practices.

one company stopped advertising, it would lose sales, but it gains no more by advertising than it would if *no* company advertised at all. This is what defensive advertising is all about.

# The Ugly Side of Advertising

Advertising, particularly informative advertising, is an important source of information for consumers. But advertising has its ugly side, too. One of the major criticisms of the advertising industry from the consumer's point of view is that some ads are designed to appeal not to our reason but rather to our emotions and psychological needs—our guilts or fears or our need to be accepted socially. Obviously, sellers want us to spend our money, and they are not particularly concerned whether we spend it rationally or foolishly.

In addition to this general concern, consumers face some specific problems relating to advertising. Let's turn now to two such problems: (1) deceptive advertising and (2) the intrusion of certain forms of advertising on our privacy.

## Deceptive Advertising

Numerous government agencies, both federal and state, are empowered to protect consumers from deceptive advertising. At the federal level, the most important agency regulating advertising is the Federal Trade Commission (FTC). A 1938 amendment to the 1914 Federal Trade Commission Act authorizes the FTC to prohibit "unfair or deceptive acts or practices" in the marketplace. Under this authority, one of the important functions of the FTC is to ensure that consumers are not misled by deceptive or fraudulent advertising techniques.

## What Kinds of Ads Are Deceptive?

The terms *unfair* and *deceptive* are very general, and the FTC has a difficult task in defining exactly what types of advertising are illegal under these broad guidelines. Advertising may be deemed deceptive if it is scientifically untrue. In an early case, for example, a claim that a cosmetic cream would "rejuvenate" skin was held to be misleading because there was no scientific basis for the claim. More recently, in 2002, the FTC charged "Miss Cleo," the "renowned psychic," with deceptive advertising. Specifically, two Florida corporations that manage "Miss Cleo" services, Psychic Readers Network and Access Resource Services, were charged with lying to customers and using illegal billing and collection practices. The complaint alleged that the defendants "misrepresent the cost of services both in advertising and during the provision of the services, bill for services that were never purchased, and engage in deceptive collection practices." The FTC also accused the firms of responding to consumer complaints with abusive, threatening, and vulgar language. A U.S. district court in Florida issued a restraining order requiring the firms to stop their activities or face fines of up to $10,000 for each subsequent ad.[1]

**Cyber consumer**

**Online Advice about Deceptive Advertising**

A seemingly endless number of Internet sites offer advice and information about deceptive advertising. One of the more useful sites can be found at http://www.consumervoiceusa. com/DeceptSalesPrac.html. This Web site was created and is maintained by Consumer Voice USA, a service "provided by a staff of investigators, journalists, consumer advocates, and attorneys." Its purpose is to "give the consumer a voice that will be heard."

---

1. "FTC Charges 'Miss Cleo' Promoters with Deceptive Advertising, Billing, and Collection Practices," December 2, 2002, FTC Web site, **http://www.ftc.gov**.

An ad may be deceptive even though it is literally true. For example, you see an ad for "Teak Dining Tables" at only $299 each. It sounds like an incredible bargain, so you go to the store to purchase one. At the store you learn that the tables are in fact plastic, but the manufacturer is a company called "Teak." In all likelihood, this ad would be considered deceptive because most consumers would be led to assume that the ad referred to teakwood. As a general rule, the test for whether an ad is deceptive is *if a reasonable consumer would be deceived by the ad.*

## Bait and Switch: The Case of the Obsolete Digital Camera

Let's assume that you read an ad placed in your newspaper by Joe's Camera Emporium. It says Joe's is having a tremendous sale on digital cameras for only $129. You have been thinking about buying a digital camera so you decide to pay Joe's a visit. As you walk into the store, you are greeted by a friendly salesperson named Hank. Holding a copy of the ad, you ask to see the digital camera that is on sale. Hank shows you a camera, packed in a box that is yellow with age and covered with dust. The camera is large and heavy and requires a memory card that looks like nothing you've ever seen before. The camera appears to have been manufactured at least ten years ago. It is scratched and makes a strange grating sound when it is turned on.

Hank says he is sure that a person as obviously intelligent as you would never buy such an inferior product. He leads you on a tour of the store and shows you other digital cameras priced from $399 to $599 each. Hank eventually convinces you to purchase a new model that will take high-resolution photos and has a telephoto lens. You agree to pay $499. When you get the camera home, you realize that you had not intended to buy such an expensive model and that you really can't afford to pay for it.

You have been the victim of a **bait-and-switch** ploy by the camera store. The bait was the unrealistically low-priced camera. The switch was to the much higher-priced model. The key to avoid being tricked by this type of ploy is to learn how to recognize the pattern of a bait-and-switch scheme and to walk out of a store when you think you're being duped. No one forces you to take the bait or accept the switch. You may also report stores that make these offers to your state's legal authorities, because bait-and-switch ploys are against the law.

## Other Deceptions

The FTC puts bait and switch at the top of the list of common fraudulent advertising practices or deceptions. Other deceptions that often involve advertising include the following:

● *Contest winner.* You are told you have won a contest that you didn't enter, but it turns out you must buy something to receive your prize.

● *Free goods.* You presumably will get something free if you buy something else, but you may be paying a higher price for that "something else" than you would have otherwise. For example, in one case, a paint retailer advertised that it would sell two cans of paint for the price of one but then set a very high price for a single can of paint. This was held to be deceptive advertising.

**Question for Thought & Discussion #4:**
Why is it often difficult to decide whether an advertisement is deceptive?

**Question for Thought & Discussion #5:**
Why do many consumers who "should know better" fall victim to bait-and-switch ploys?

**Bait and Switch**   A selling technique that involves advertising a product at a very attractive price (the "bait"); then informing the consumer, once he or she is in the store, that the advertised product either is not available, is of poor quality, or is not what the consumer "really wants"; and finally, promoting a more expensive item (the "switch").

## THE ETHICAL CONSUMER

# Protecting Children (and Their Parents) from Online Abuse

Thousands of commercial Web sites cater to young children. Many of these sites offer games that children can play, either by themselves or interactively with other children. In the past, these Web sites typically required children to supply information about themselves or their families. Some of this information was necessary to play the games, but much of it was not.

An FTC survey in 1998 showed that while almost all "kids' Web sites" collected information from children, only 24 percent had privacy policies and virtually none of them allowed parents to control access to the sites. A number of sites required children to provide information about their parents that clearly had nothing to do with the games or activities offered.

On April 21, 2000, the Children's Online Privacy Protection Act (COPPA) went into effect. This law requires Web sites to receive parental permission before collecting information from children under thirteen years of age online. Web sites that cater to young children must allow parents to limit their children's access. The sites may not require children to provide more information than is reasonably necessary to participate in an activity. They must also maintain the confidentiality, security, and integrity of information collected from children. The FTC

plans to enforce the act though routine monitoring carried out by its Internet lab. FTC employees will periodically search the Internet using key words such as *interactive games* or *action games*. The FTC will also investigate complaints it receives from individual consumers. America Online, Microsoft, and other firms that host commercial Web sites are taking steps to inform Internet businesses about the COPPA's requirements.

In 2000, the Better Business Bureau (BBB) applied for "safe-harbor treatment" under the COPPA for member businesses that submitted to its Children's Advertising Review Unit. This means that firms that agree to abide by the BBB's advertising rules, promise to adhere to its dispute-resolution process, and undertake a self-assessment of their site's compliance with the COPPA would automatically be certified as meeting COPPA standards. They could then advertise this certification on their Web sites.

**Do you believe the COPPA can effectively protect children and their parents from unscrupulous Web sites? Should the BBB be allowed to facilitate the certification of its member businesses? Why do some people argue that the COPPA is an assault on our freedom of speech? Where do you stand on this issue?**

- *Merchandise substitution.* In place of what you thought you were buying, the seller substitutes an item of a different variety, make, model, or quality.
- *Rebates.* By taking advantage of an advertised "manufacturer's rebate," a consumer supposedly will pay a lower price. Often, however, it is inconvenient for the consumer to collect the rebate because of (unadvertised) time-consuming or costly requirements.

## What the FTC Can Do

The FTC receives letters and other communications complaining of violations from many sources, including competitors of alleged violators, consumers, consumer organizations, trade associations, Better Business Bureaus, government organizations, and state and local officials. If enough consumers complain and the complaints are widespread, the FTC will investigate the problem and perhaps take action.

**CEASE-AND-DESIST ORDERS**  If, after its own investigations, the FTC believes that a given advertisement is unfair or deceptive, it can conduct a hearing—which is similar to a trial—in which the company that has allegedly violated FTC rules on advertising can present its defense. If the FTC succeeds in proving that an advertisement is unfair or deceptive, it usually issues a cease-and-desist order requiring that the challenged advertising be stopped. A company that fails to obey such an order may be fined $10,000 for each subsequent illegal advertisement.

**COUNTERADVERTISING**   A controversial type of sanction imposed by the FTC is known as **counteradvertising** (also called *corrective advertising*). When a firm has been found liable for deceptive advertising, a counteradvertising order by the FTC requires the company to advertise anew—in print, on radio, and on television—giving information to correct the earlier misinformation. For example, Listerine advertised that its mouthwash could prevent or cure colds and sore throats. This claim was found to be unsupported, and the FTC required Listerine to state in future advertisements that its product did not prevent colds or sore throats. In another case, the maker of Profile breads advertised the weight-reducing quality of its product by stating that Profile bread had fewer calories per slice than other breads. It turned out that this was true—but only because Profile bread had thinner slices than other breads. The FTC required the corporation selling Profile bread to spend a specific amount of money to explain, via advertising, that it had presented misleading information. Interestingly, although counteradvertising has been carried out for a number of years, there is little evidence that it has had a significant impact on consumer buying habits.

> **Counteradvertising**  New advertising that is undertaken pursuant to a Federal Trade Commission order for the purpose of correcting earlier false claims made about a product.

## What You Can Do

If you feel that you have been victimized by fraudulent or deceptive advertising, you can take several steps. If you have bought a product as a result of false advertising and want to get your money back, follow the procedures described in the *Confronting Consumer Issues* feature in Chapter 5. In any case, you should let government authorities know about the problem. You can notify the FTC by writing directly to the national headquarters (6th St. and Pennsylvania Ave., N.W., Washington, DC 20580) or contacting one of its regional branches. You may also wish to notify your state attorney general's office about your difficulty. In recent years, state attorneys general have been among the most aggressive forces in curbing deceptive advertising practices. Other government agencies that may help you are the Food and Drug Administration (FDA), the U.S. Postal Service, the Federal Communications Commission (FCC), and the Securities and Exchange Commission (SEC). Depending on the nature of your complaint, you may wish to contact one of these agencies.

> **Question for Thought & Discussion #6:**
> Would you place restrictions on how businesses can advertise their products? If so, what would these restrictions be? How would you enforce them?

## Advertising and Your Privacy

If you don't want to hear television ads, you have several choices: you can "zap" the ad by muting the sound or turning to a different channel, you can watch noncommercial and pay channels, or you can simply turn off your TV. Similarly, you can turn off your radio or listen to a public broadcasting station if you want to avoid commercials. And you can skip over the ad sections in newspapers and magazines. But some forms of advertising are more difficult to control—namely, direct-mail advertising and telephone promotions.

Direct-mail marketing now accounts for a large share of the total advertising expenditures in the nation; only TV ads attract more dollars. It is estimated that the postal service processed about 200 billion pieces of this type of mail in 2003. Commercial supporters of direct mail contend that it is more informative than radio or television ads because written ads can present more detail than can ads broadcast for thirty seconds. With the high cost of TV commercials (currently about $300,000 for a national commercial ad), it's not surprising that advertisers turn to the less expensive alternative.

Once you subscribe to any magazine or request a catalogue, your name will soon be on numerous lists that are sold or exchanged for other lists. If you apply for a credit card, your name will appear on yet other lists. In a short time, your mailbox may be stuffed with unrequested information from direct-mail promoters. Even if you do none of these things, your name will still be on numerous lists that are compiled by marketing agencies from U.S. government data, such as that available from the Census Bureau.

Many consumers feel that direct-mail and telephone advertising invades their privacy. Even those with unlisted telephone numbers cannot completely avoid the sales tactics of energetic marketers who dial every number within a certain area code.

If you want to avoid unsolicited mail or wish to have your name removed from a current mailing or telephone list, you can do the following:

- Whenever you subscribe to a magazine or request a catalogue, request in writing that your name not be exchanged or sold to others as part of a direct-mail list.

- Write to any organization that uses your name and ask that it be removed from the mailing list.

- Fill out Form 2150 ("Notice for Prohibitory Order Against Sender of Pandering Advertisement in the Mail") at the post office; by law, any sender listed on that form must drop your name from its mailing list. Incidentally, there is no universally accepted definition of exactly what *pandering* materials are. The dictionary defines *pander* as providing gratification for others and exploiting their weaknesses. To many, the word also has a sexual connotation, but, strictly speaking, mail that panders is not necessarily salacious.

- Write to Consumer Services, Direct Marketing Association, Inc., 6 East 43rd St., New York, NY 10017, and request a Mail Preference Service form. After you have completed and returned it, the Direct Marketing Association will notify its member companies of your wish to be removed from all direct-mail lists. The DMA also has a telephone preference service and will remove your name from telephone soliciters' lists if you so request. You may also have your name deleted online at **http://www.the-dma.org**.

- Go to **www.donotcall.gov**, the Web site for the National Do Not Call Registry, which is managed by the Federal Trade Commission. Most telemarketers cannot call your telephone number if it is listed in the National Do Not Call Registry. You can register your home and mobile phone numbers for free, and your registration will be effective for five years.

## Your Telemarketing Rights

In recent years, the number of firms promoting their products through telemarketing (unsolicited telephone calls) has grown to over 350,000. Although you might expect this type of communication to be protected as First Amendment free speech, in 1995 the FTC created telemarketing rules to control deceptive or abusive telephone sales practices. Under these rules, telemarketers:

- can call only between 8:00 A.M. and 9:00 P.M.,
- must state that they are making a sales call,
- must identify themselves and who they work for,
- cannot legally seek payment before promised products or services are delivered, and
- must explain any credit arrangements they offer, including the annual percentage rate and finance charge.[2]

Since 1995, the FTC has placed a number of controls on telemarketers, from what information they must disclose to when they can call.

The need for these rules is clear from an estimate by the U.S. Office of Consumer Affairs that $150 billion is spent each year through unsolicited telemarketing, with the average sale per successful call approaching $400. Although many telemarketing firms are honest and offer consumers good value for their money, a significant number do not. If you believe you have been solicited improperly, you can contact your state's attorney general's office or write the FTC. If a telemarketer does not follow the FTC's rules, you should refuse to buy and hang up.

In 1995, it became illegal for a telemarketer to call consumers who formally requested that they not be called by a particular firm or organization. If a telemarketing firm continues to make telephone solicitations after being asked not to do so, consumers may register a complaint with the FTC at 202-326-3128. An FTC official has said, "If we get a pattern of complaints about telemarketers from consumers . . . it could be the basis for law enforcement action from us." If the FTC issues a cease-and-desist order against a firm, it can lead to a fine of $10,000 for each subsequent call the company makes to consumers who have asked that they not be called.[3] In recent years many states and the federal government have established do-not-call lists and other curbs on telemarketers.

## Privately Prepared Information

If you want to buy a product but are not sure which brand to purchase, you need not rely solely on manufacturers' advertisements. You can easily find information about advertised products that has been gathered and prepared by private organizations. Although some of this information must be purchased, a large amount can be accessed for free. You should begin your search on the Internet. Suppose that you are considering buying a small sports utility vehicle (SUV), but are confused by the seemingly contradictory safety claims made by different manufacturers in their advertising. Using *sports, utility,* and *vehicle* as key words, your Internet browser will take you to hundreds of sites that offer evaluations, comparisons, and road test reports for every SUV you might consider purchasing. But be careful: many of these Web sites will be operated by manufacturers or distributors of specific brands of

> **Question for Thought & Discussion #7:**
> Think of the last important purchase you made. How did you gather the information you used when you made this decision?

---

2. "U.S. Sets Guidelines to Curtail Telephone Sales Abuse," *Buffalo News,* February 9, 1995, p. A4.
3. "Don't Call Me Again!" *Kiplinger's Personal Finance Magazine,* October 1995, p. 114.

vehicles. They may be little different from advertisements that appear on television, on the radio, or in print media. Such ads may provide useful information about a vehicle's safety equipment, but they are no more likely to offer a balanced report on competing products than are advertisements in other media. For this reason, it is important to visit many Web sites or to use ones that are not tied to any firm that has a financial interest in your purchase decision. These neutral Web sites are likely to charge a fee for the information they offer.

## Information Online

One way to access free information about advertised products is to search an electronic periodical listing service such as InfoTrac, which is provided with this book. These services identify and often provide the text of articles that report on products you might want to buy (a small SUV, for example). Even if you do not have a link to one of these services from your home, they are available at almost all public and college libraries for free. One important advantage of this type of research is the listings of related topics that they identify. By reviewing related articles, you may discover alternative choices or find information about aspects of a product that you would not otherwise have considered.

## *Consumer Reports* and *Consumers' Research Magazine*

Although various sources offer free consumer information, it can be worthwhile to pay for privately prepared information to be sent to your home on a regular basis. This can help you stay up to date on consumer issues and make you aware of problems in consumer protection. Two important private sources of information consumers may purchase are *Consumer Reports* and *Consumers' Research Magazine.* Consumers Union and Consumers' Research, Inc., the respective publishers of these magazines, report on many types and brands of consumer products to help Americans make better purchase decisions.

If you decide to rely on the recommendations of these consumer groups, you should realize that their researchers may not always present purely objective results. This is not to say that you will get misinformation, but the researchers may sometimes emphasize certain aspects of products that are consistent with their interests and preferences but not with your own. For example, recommendations about cars may give more weight to safety, gas mileage, or comfort than you personally want to give. You may opt for a different car because you place more emphasis on styling or low cost than on safety. Even though the occupants of subcompact cars face a higher probability of serious injury in an accident than do occupants of bigger cars, people continue to buy small cars because they are more economical. Obviously, you will face this problem of evaluating any information you obtain, either free or at a price.

In the last analysis, only you can make your own decisions, and they must be based in part on your personal value judgments. If you are not a tireless shopper, you may be content simply to look at *Consumer Reports* for whatever you want to buy, pick either the "best buy" or the top of the line, call your local dealer, and have your selection delivered. You may get some products you dislike, but, on average, if your tastes correspond with those of the product testers, you will save considerable time and probably will avoid seriously defective products.

*Consumer Reports* publishes an annual *Buying Guide,* in December; the *Consumers' Research* annual buying guide appears in October. Both contain a wealth of information on such things as food and nutrition, energy-saving ideas, and the like. Unfortunately, they both suffer from a problem that is impossible to avoid in a dynamic economy: certain models that are listed may no longer be available by the time you decide to make a purchase.

## Other Printed Sources of Information

At any one time, you can choose from at least a half-dozen other buying guides, such as the *Consumers' Handbook,* edited by Paul Fargas; *Better Times,* edited by Francis Cerra; and the U.S. Department of Agriculture's *Shoppers' Guide.* All are uneven in coverage, and none can be recommended without reservation.

In addition to these publications, many other privately produced information sources are available. For example, *Money,* a monthly magazine published by AOL Time Warner, is aimed at families of middle income and above. This publication emphasizes financial management—stocks, bonds, retirement, real estate investments, and commodities. It also provides valuable information about making better consumer choices in other areas as well.

*Kiplinger's Personal Finance Magazine,* published monthly, provides reports on such items as weight-reduction gimmicks, insurance plans, new tax rulings that might affect you, tips on how to get interest on your checking account, or warnings against long-term car loans.

*Newsweek* magazine carries a regular series of articles on consumer issues by Jane Bryant Quinn. Similar features appear in other news magazines such as *Time* and *U.S. News & World Report. Better Homes and Gardens, Sunset, Family Circle, Woman's Day,* and other homemaker-oriented magazines give helpful consumer information. Even less traditional publications, such as *Mother Earth,* may help consumers investigate many products offered for sale. A final source of information that should never be ignored is the firsthand experience of your friends and relatives. Knowledge of products gained through personal use may be the most valuable information a consumer can find, although it should be evaluated in light of the values and knowledge of the consumer making the recommendation.

## How Much Information Should We Acquire?

What we want is reliable information at the "right" price. In our daily lives, we *acquire information up to the point where the cost of acquiring any more would outweigh the benefits of that additional information.* In other words, we engage in rational decision making, which we defined and discussed in the previous chapter. When we decide to go shopping for goods, we may look at advertisements for only a few supermarkets instead of trying to find out the price of specials at all forty-six stores in the city. Why do we look at only a few? The reason is that we have found that it does not pay to look at any more than those few pieces of information. When we go shopping for a new car, we

**Question for Thought & Discussion #8:**
Have you ever made an important decision without first gathering enough information? What were the results of your choice?

**Comparison Shopping** Acquiring and comparing information about different sellers and different products in order to find the best price for products of substantially the same quality.

may go to only a few dealers in the immediate area. Why go to only a few and not all? Because, again, we have found that it does not pay to go to all of them.

This is because the *time* that we spend in **comparison shopping** has a cost, too—the opportunity cost of our time, which was discussed in Chapter 1. This cost varies from consumer to consumer. Whereas one individual might be content to spend five hours comparing prices of toasters to save $5, another person would not spend five hours in such comparison shopping even if $50 might be saved on the item being purchased.

Indeed, some of us may not even bother to read advertisements or to seek any additional information about the goods and services we wish to purchase. Instead, we may decide to shop at a store where only the most expensive brands are carried, on the assumption that high price means high quality or because we are "status seeking." If we shop this way, we may have decided that it is not worth our while to acquire quantities of information, and we are, therefore, essentially nonshoppers.

The most that can be said here is that an abundance of information is available to you, both through advertising and through private consumer publications. By making use of these information sources and comparison shopping, it is possible to save many of your consumer dollars. How much you save depends on the value you place on your available time, and that's a decision only you can make.

# Confronting Consumer Issues: Rational Internet Shopping

I t is almost impossible to surf the Internet without being bombarded with advertising for goods and services offered by a wide variety of Web retailers. A small sample of their many products includes vacation holidays, investment services, automobile insurance, low-cost credit, and the latest video releases. In 2003, U.S. businesses spent an estimated $6 billion to advertise on the Internet. This amount is projected to grow by as much as 30 percent per year through the next decade. Businesses are willing to pay for Internet advertising because they believe it will help them earn greater profits. Although advertising can be irritating when you surf the Internet, it provides you with a number of benefits. Web advertising, for example, often contains information about the newest goods or services available on the market. These ads may cause you to investigate new products and lead you to buy goods or services that will enhance the quality of your life.

Just as advertising on commercial television pays the cost of producing your favorite programs, Web advertising provides funding for the Internet. The Internet service provider that connects you to the Internet would probably charge much higher rates if it did not earn revenue from advertising.

Consumers should evaluate Web advertising using the same criteria that they use for advertising in other media. You have learned that all advertisements, including those that appear on the Internet, provide information or images that are intended to convince consumers to buy a particular good or service. They are not likely to present a balanced or objective description of the products they are promoting. By sorting through what is displayed in an advertisement and disregarding what is neither factual nor useful, consumers may obtain information that is helpful in completing the decision-making process.

## THE GROWTH OF E-COMMERCE

Connecting to the Internet is not the same as visiting a shopping mall. Still, the apparent differences between these alternatives are becoming smaller every day. Today, virtually all national retail chains maintain Web sites where they offer products for sale. The growth of e-commerce provided consumers with many opportunities. At the same time, it presents challenges to the consumer's ability to make rational decisions. Shopping the Net requires consumers to develop and use different skills and methods to evaluate their alternatives.

Like all retail businesses, Web retailers offer many choices that consumers should consider before they decide to buy. Consumers should always remember that they are not

**How is shopping on the Internet different from shopping from a catalog, using a telephone?**

required to shop in any way that makes them uncomfortable or does not provide what they believe is the best deal. Virtually every good or service offered on the Internet can also be purchased from local stores or by telephone from a mail-order catalogue. You should shop in the way that you believe provides you the greatest benefits at the least cost. This may or may not be shopping on the Internet. To make this decision, you must identify and evaluate the benefits and costs of shopping on the Web.

## SHOPPING BENEFITS OFFERED ON THE WEB

Web shopping eliminates many problems typically associated with shopping in a store. Unlike retail stores, Web sites are

*(Continued on next page)*

## Confronting Consumer Issues (Continued)

open for business 24 hours a day, 365 days a year. No matter what time it is, if you have a free moment or feel the urge, you can go shopping. If you don't like the styles or prices at one Web site, you can leave and, in a matter of seconds, visit a different site that offers other products, perhaps at better prices. You never have to leave the comfort of your home or drive through miles of traffic only to find that the product you wanted is not available in your size or the color you prefer. You don't have to push your way through hordes of customers in cramped aisles. You don't have to wait in line for surly clerks who ask, "Whadyawant?" In short, Web shopping is less likely to raise your blood pressure, give you a headache, or cause you to have a disagreeable disposition.

**Finding the Best Deal**   Most Internet shoppers rate convenience as the most important benefit of shopping over the Net. Running a close second is the ability to comparison shop. It takes only a few clicks of your mouse to visit retail Web sites. You can view pictures of products you might want to buy with printed descriptions of their styles and qualities. It is easy to compare prices and shipping charges. In less than an hour, you can review competing products offered at different Web sites and make the selection that best meets your needs. Accomplishing the same task by visiting stores could take you many hours or even several days.

### Question for Thought & Discussion #9:

Do you choose to shop on the Internet? What benefits and costs have resulted from your choice?

**Variety of Choice**   Retail stores are limited by the space they occupy and the inventory they are able to maintain. A store cannot possibly keep every color, style, or size in stock. When a particular product sells out, a store may have to wait weeks for its stock to be replaced, if it ever is. Although there is no guarantee that the product you want will always be in stock when you visit an Internet retailer's Web site, items are much more likely to be available than when you frequent a retail store.

When you click through the offerings of a Web merchant, you will see a wide array of products offered in many styles, sizes, and colors. This might lead you to believe that the business maintains an enormous inventory, but nothing could be further from the truth. Many Web merchants keep no inventory at all. When they receive an order, they send it

on electronically to the product's manufacturer. This firm or one of its distributors then fills the order directly. This system allows the Web merchant to market a wide range of products without investing in inventory. The cost saving this provides may be passed on to consumers in lower prices. This system also means that there is a much better chance that the business will be able to fill your order.

**Special Products**   People who need special products are likely to benefit from shopping on the Web. Some individuals require clothing in a unique size or style. Some people, for example, have very wide feet. These consumers once found buying comfortable shoes difficult or impossible. They had to shop at specialty stores where the selection was limited and prices were high. Now they can easily order the style and size they want on a variety of Web sites, usually at prices no higher than those charged for shoes in more common sizes.

**Special Sales**   Many firms maintain retail stores as well as Web sites. Some of these businesses use their Web sites to sell overstocked goods or products that are no longer in style. They have found that it is quicker and less expensive to sell these products over the Internet than to ship them to retail stores where they may or may not sell. The bottom line on this for consumers is that retail Web sites often offer special sales that are unavailable at retail stores.

**Paying Your Own Price**   Products are also marketed on the Net through electronic auctions. The sellers may be ordinary stores that wish to eliminate excess inventory quickly or individuals who use special Web sites to sell products they own for the highest bid made within a specific period of time. eBay, for example, is a Web business that offers this type of service to consumers. Suppose someone you do not know owns a complete set of *Sports Illustrated* magazines from the 1970s that you would like to buy (if you knew you could). The owner agrees to sell them through eBay to the person who makes the highest bid within the next week. He also sets a minimum or *reserve* price, which is the smallest amount he is willing to accept.

You submit a $500 bid for the magazines through eBay's Web site. If your bid is more than anyone else's offer and more than the seller's reserve price, the magazines will be yours. There is, of course, no guarantee that the owner will receive as much as he hoped to get for his magazines, and he will be charged for eBay's service, but your $500 offer might be much more than he could get by advertising his collection in a local newspaper.

Another way to set your own price for a product you want is to use Priceline.com. This Web site allows consumers to state a price they would be willing to pay for a specific good or service (often tickets) and then tries to obtain the specified product at the identified price. If the product is found at the appropriate price, the consumer is required to buy. Suppose you offer to pay $100 for a round-trip airline ticket to Chicago over next weekend. If Priceline can find such a ticket at your price, it will automatically be charged to your credit card, and you will be able to pick up your ticket at the airport.

These methods of purchasing goods or services over the Internet require consumers to be sure they understand what they are offering to buy and how much the products are worth. A consumer who fails to evaluate a bid before it is made may end up paying a heavy price.

## THE DOWNSIDE OF SHOPPING ON THE WEB

Few things in life are 100 percent good and have no risk at all. This is as true of shopping on the Net as of anything else. Some offers made by Web retailers are not good deals even though they seem to be on the surface. For this reason, Web purchases, like all important spending decisions, should be made with a clear understanding of the costs that may result from any choice that is made.

Perhaps the most important problem of shopping on the Internet is that you can't touch, feel, test, or try on products before you agree to buy them and they are delivered to your home. Clothing purchased over the Internet may not fit. A dress or suit that looks wonderful on a model who appears on your computer's monitor may not look so good when you put it on. It may be a slightly different color, be poorly constructed, or have a rough texture that irritates your skin. In a store you would immediately recognize these problems and avoid the garment. But now you own it and must decide what to do about your problem.

## RETURN POLICIES

Every reputable retail Web site has at least one page devoted to its return policies. Never shop at a site that lacks such a page. As a careful consumer, it is *your* responsibility to read this information and become confident that you know what the firm promises to do if you are dissatisfied with a purchase. Most retail Web sites have a toll-free number that you can call to obtain clarification of any part of a return policy you don't understand. At a minimum, you should find the answers to these questions:

1. Are you allowed to return a purchase simply because you don't like it, or must the product be defective?

2. Must returns be made within a limited time?

3. Will the firm charge a "restocking fee" (typically 10 percent) for returned purchases?

4. Who pays the cost of return shipping?

5. Will a refund be credited to your credit-card account, or will you receive a *store credit* that can be spent only for other products from the same firm?

6. If you receive a store credit, must it be spent within a limited time?

7. If you have a problem, can you call a toll-free number and talk to a company representative?

**Other Questions to Ask**   Additional questions should be asked when you're considering specific goods or services. Vacations sold over the Internet can present a special challenge. For these you need to know what cancellation policies apply. Can you obtain a refund if you become ill and cannot travel? Does the company guarantee that your room will be in the hotel of your choice and have a balcony overlooking the ocean? When you buy any product, you know what benefits you hope to receive from your purchase. Whether you are shopping at a store, through a mail-order catalogue, or over the Internet, always be sure to investigate what will be done to satisfy you if you do not receive the value you expect.

**Obtaining Service**   Many of the products you own require periodic service to keep them in good working order, regardless of whether they were purchased from a store in your community, through a mail-order catalogue, or on the Internet. Local stores often maintain their own service departments or contract with other firms to provide service for their customers. Obviously, an Internet retail business will not maintain a local service department for the products it sells. Some larger Internet or telephone-order businesses, such as Gateway, do maintain contracts with independent service businesses and refer customers who have problems with products they have purchased over the telephone or Internet to them. These businesses, however, generally provide service only during the time that a product is covered by a manufacturer's warranty. After the warranty expires, consumers typically must find and pay technicians to service their products.

When technology-based goods need service or repair, they must often be returned to the manufacturer or to a centralized service center. This requires owners to box and ship their products to some other location in the country. Not only do the consumers have to pay the cost of shipping and

*(Continued on next page)*

## Confronting Consumer Issues (Continued)

repair, but they will not have the use of the product during this often lengthy process. Checking out a Web merchant's service policy can be just as important as investigating its return policy.

## CHOOSING A RETAIL WEB SITE

There are hundreds, possibly thousands, of retail Web sites offering to sell virtually any consumer product you could imagine. Suppose that you are thinking of buying a DVD player. There are so many Web sites where you could place your order that just identifying the site where you want to shop can be a problem. You could do a global search using the key words *DVD player*, but they would probably identify several thousand Web sites—more than you have time to visit. Internet service providers, such as America Online and Microsoft's MSNetwork, maintain relationships with Web businesses that are easy to reach from their home pages. Most Internet browsers, such as Yahoo.com and Netscape, do the same. Figure 3.1 shows a Web page from *Yahoo! Shopping*. It quotes prices for various DVD players that could be purchased through this service in early 2003. Finding this page required only three mouse clicks after reaching the Yahoo.com home page.

But does locating a product easily on the Internet mean that this is the best deal available? You can't be sure unless you investigate your alternatives. Yahoo.com and other businesses that maintain retail Web sites are not likely to tell you about a better deal for the product at a different location on the Internet. It is worth devoting a reasonable amount of time and effort to find and evaluate alternatives. One way to do this is to visit a library and examine periodicals that report on products you are considering buying. *Popular Electronics*, for example, reports on DVD players. More significantly, it contains advertisements from firms that offer DVD players for sale over the Internet. You can find the Internet addresses of these firms' Web sites and compare the prices they charge to find the lowest one. Remember, price is

not the only factor you should consider. Saving $10 on the cost of the DVD player isn't a good deal if you can't easily return it or obtain service if it is defective.

Ultimately, Web shopping is best carried out using the rational decision-making process. The methods you employ to identify and evaluate your alternatives are a little different, and they may require you to develop some technical skills, but in the end you should rely on your ability to use reason and logic. Never rush into an important decision. Take your time and do your homework, and you'll make a good choice.

**Figure 3.1** A Web Page Listing DVD Players

# Chapter Summary

1. Advertising expenditures in the United States have climbed to more than $240 billion a year.

2. Sellers advertise to make more profits. Ultimately, it is the consumer who normally pays for advertising, because producers include the cost of advertising in the prices of their products.

3. Individuals often associate brand names with (1) reliable quality, (2) better warranties, and (3) a larger number of repair facilities.

4. In recent years, a large dent has been made in brand-name loyalty by the proliferation of similar competing products that are priced substantially lower. Brand-name producers have responded to this development by inventing effective new marketing strategies, including market segmentation and psychologically linking a brand-name product to a given image or lifestyle.

5. There are several basic types of advertising. These include informative advertising, which provides specific information about products; persuasive advertising, which is intended to change consumer tastes by appealing to psychological needs; comparative advertising, which makes a direct comparison between competing products; and defensive advertising, which is intended to rebut claims made by competing firms in comparative advertisements.

6. Advertising has its ugly side, too. A major complaint from the consumer's point of view is that advertising often appeals not to our reason but to our emotions and may cause us to bypass the rational decision-making process.

7. Advertisers occasionally engage in unfair or deceptive advertising practices, such as bait-and-switch advertising. Generally, any ad that could mislead a reasonable consumer will be considered deceptive.

8. Both state and federal agencies are empowered to protect consumers from deceptive advertising practices. At the federal level, the Federal Trade Commission can issue a cease-and-desist order to prevent a company from continuing a specific ad that has been deemed deceptive. The FTC can also direct the company to engage in counteradvertising ("corrective" advertising) to correct the misinformation given to consumers.

9. Some consumers consider telephone sales and direct-mail advertising to be an invasion of their privacy. You can request that your name be removed from direct-mail lists and telephone lists.

10. You can obtain privately produced information about products and services from *Consumer Reports, Consumers' Research Magazine, Money, Kiplinger's Personal Finance Magazine,* and other consumer publications.

11. Comparison shopping is acquiring information about alternative sources for a particular product. Because comparison shopping requires time and other resources, there is a limit to how much you will want to do. Consumers will have different limits depending on the opportunity cost of their time.

12. Consumers should evaluate advertising on the Internet by using the same criteria used for advertising in other media. Although Web advertising is unlikely to present a balanced or objective description of the goods or services it promotes, it may still contain useful information.

13. Internet shopping offers consumers many benefits including longer hours, a wider selection, easier comparison shopping, and the ability to find a better deal for quality products than conventional shopping. Some Web sites are run as auctions that allow consumers to bid for products they want to purchase.

14. Web shopping also entails some disadvantages that can be important to consumers. Products purchased on the Internet cannot be examined or tried on before they are purchased and shipped to your home. Return policies vary among Internet marketers and often require consumers to pay a restocking fee and shipping costs. It can be difficult to obtain service for defective products purchased over the Internet.

15. When consumers do Internet shopping, they should be careful to use the decision-making process to evaluate their alternatives, and they should not forget to consider non-Internet shopping alternatives that can offer better choices.

# Key Terms

bait and switch **55**
comparative advertising **53**

comparison shopping **61**
counteradvertising **57**
defensive advertising **53**
informative advertising **52**

market power **52**
persuasive advertising **50**

## Questions for Thought & Discussion

1. What are several ways you use advertisements to help you make consumer decisions?

2. What brand-name products does your family purchase? Why do you believe your family buys these products instead of less expensive alternatives?

3. Why do beauty and grooming products seldom use informative advertising?

4. Why is it often difficult to decide whether an advertisement is deceptive?

5. Why do many consumers who "should know better" fall victim to bait-and-switch ploys?

6. Would you place restrictions on how businesses can advertise their products? If so, what would these restrictions be? How would you enforce them?

7. Think of the last important purchase you made. How did you gather the information you used when you made this decision?

8. Have you ever made an important decision without first gathering enough information? What were the results of your choice?

9. Do you choose to shop on the Internet? What benefits and costs have resulted from your choice?

## Things to Do

1. Copy or cut out advertisements from your local newspaper or a national magazine that demonstrate (a) informative advertising, (b) persuasive advertising, (c) comparative advertising, and (d) defensive advertising. Identify and explain the characteristics of each ad that enabled you to identify the type of advertising it demonstrates.

2. Watch one hour of a televised national sports event and one hour of an afternoon "soap." List the products and types of advertisements that appear on each program. Write an essay that describes and explains the differences between the ads that appear on these different types of programming.

3. Search the Internet to find two businesses that offer a product that you use. Compare the information provided on these sites. Would the information be useful to you in making your purchase decision? What additional information would you want before making a purchase choice? Where might you find this information?

4. Find a print advertisement for a nationally advertised product you would like to own. Search the Internet to find the Web site maintained by the manufacturer of this product. Compare the amount of information provided on the Internet with the information in the print ad. Why do producers generally provide much more information online than in their print ads?

## Internet Resources

### Finding Consumer Information on the Internet

The following Web sites have been selected for their relevance to topics discussed in this chapter. Search these sites to locate information that can add to your knowledge of deceptive advertising on the Internet. Remember, Web addresses change frequently. If any of these addresses no longer function, find similar sites to investigate using any of the search engines available to you.

1. The Federal Trade Commission provides guides to help consumers avoid becoming victims of bait advertising and deceptive pricing. You can find these guides online at http://www.ftc.gov/bcp/menu-ads.htm.

2. Information about Internet legal issues including the problem of spam and fraudulent and deceptive advertising is available at the following Web site: http://www.publaw.com/spam.html.

3. Information about false and deceptive advertising on the Internet can also be found at http://www.transparencynow.com/internet.htm.

## Shopping on the Internet

The following Web sites have been selected because they offer consumers services similar to those described in this chapter. These are commercial sites that are designed to market products. They do not represent a comprehensive or balanced description of all brand-name products available online. Find the price of the indicated product at the Web site. How do the prices charged at these Web sites compare with the prices of these products charged by your local stores? Remember, Web addresses change frequently. If any of these addresses no longer function, find similar sites to investigate using any of the search engines available to you.

1. Go to Bettymills Cleaning/Janitorial Supply at http://www.bettymills.com and look for the price of Tide detergent.

2. Check out the price of Bayer aspirin at Eckerd.com. You can find it at http://www.eckerd.com/content. asp?content=healthcare%2Fcallcenter.

3. Shop for Cargo Pants at Orvis. Visit the Orvis Web site at http://www.orvis.com.

## InfoTrac Exercises

Purchasers of new copies of this text are provided with access to the InfoTrac Web site. This Web site links students to thousands of recent articles published in hundreds of periodicals. Use the key words **bait and switch** or other terms from this chapter to conduct a key-word search. Choose one article that is of particular interest to you and write a brief essay describing what you have learned from the article. Be sure to cite the author and title of the article and the name and date of the publication in which it appeared.

---

# Selected Readings

"Bring Back Brand X." *Advertising Age,* November 8, 1999, p. 60.

Cohen, Stanley E. "The Battle over Truth and Fairness in Advertising." *Advertising Age,* March 29, 1999, p. 120.

"Deceptive Advertising Laws." *The Kiplinger Letter,* June 15, 2001, p. 24.

Fisher, Jerry. "Dare to Compare." *Entrepreneur,* December 1998, pp. 100–102.

"Google Starts New Shopping Web Site 'Froogle.'" *Knight Ridder/Tribune Business News Service,* December 13, 2002, p. ITEM02347019.

Hertz, Lawrence M. "Advertising Regulations on the Internet." *The Computer and Internet Lawyer,* June 2002, pp. 18–27.

Hill, Daniel D. *Advertising to the American Woman.* Columbus, OH: Ohio State University Press, 2002.

Packard, Vance. *The Hidden Persuaders.* New York: McKay, 1957.

Sevetz, Kevin, and Peace Gardiner. "Resources for Web Buying." *Computer Shopper,* March 2000, p. 203.

Sinrod, Eric J. "Court Enjoins B and S Spam That Offers Gift." *Daily Business Review,* May 8, 2002, pp. A8–A9.

*The World in Your Mailbox.* Available from the Consumer Services Department, Direct Marketing Association, 6 East 43rd St., New York, NY 10017.

Tolson, Jay. "What's in a Name?" *U.S. News & World Report,* October 9, 2000, p. 52.

# CHAPTER 4

# The Many Faces of Fraud

**After reading this chapter, you should be able to answer the following questions:**

- How can you avoid being victimized by fraudulent practices in the marketplace?
- What fraudulent practices are commonly perpetrated on consumers?
- What characteristics of mail-order or Internet offers are indicators of probable fraud?
- What assistance is available from the federal and state governments to help consumers protect themselves from fraud?
- What is identity theft, and how does it differ from most other types of fraud?
- How can consumers reduce the possibility that they will become victims of identity theft?

The formal, legal definition of **fraud** is:

Making a false statement of a past or existing fact with knowledge of its falsity, or with reckless indifference as to its truth, with the intent to cause someone to rely on such a statement and therefore give up property or a right that has value.

**Fraud** Making a false statement with knowledge of its falsity, or with reckless disregard as to its truth, with the intent to cause someone to rely on the statement and therefore give up something of value.

Quite a mouthful, isn't it? The essence of this legal definition is that fraud occurs when an individual or a company knowingly misrepresents or fails to reveal an important fact to the consumer, with the ultimate result that the consumer is somehow cheated. In our judicial system, fraud is limited to deliberate deceit. In other words, it must be proved that the seller intended deceit, not just that the customer was deceived. Such proof is often difficult to obtain. Consequently, the best consumer protection is to be aware of, and guard against, the numerous fraudulent schemes that some sellers in the marketplace have devised.

Many businesspeople contend that fraud must necessarily be short lived because enterprises rely on repeat customers to stay in business. According to this reasoning, a company that engages in fraud will eventually lose money as consumers realize that it is run in a dishonest way. In our complex and multidimensional economy, however, information is not always reliable or easily found. Furthermore, with the huge population being serviced, sellers can use unscrupulous marketing techniques on many different people throughout the country. In other words, it is possible for businesses to survive for many months, or even years, without serving repeat customers. In fact, the perpetrator of fraud may intend to stay in business for only a short time and then move on to "greener pastures." This is particularly true of firms that sell products through the mail, over the phone, or on the Internet.

Deceptive advertising was discussed in the preceding chapter. Here we examine a variety of other fraudulent activities that consumers should guard against.

# Consumer Products Fraud

The increasing range and complexity of consumer goods available to U.S. buyers have created opportunities for dishonest businesses to take advantage of people by offering products for sale that are not what they appear to be. Many of these schemes are directed toward individuals who are unsure of themselves, are dissatisfied with their lives, or fail to use good judgment. When fraud takes place, more often than not the victim has overlooked a central rule for protecting yourself in any consumer transaction: "When an offer seems too good to be true, it probably is." Most, but not all, attempts to defraud the public could not succeed without the active participation of the intended victims. What follows is only a partial listing of the many different types of fraud that are perpetrated on unsuspecting U.S. consumers each year.

## Personal and Health-Care Frauds

Many consumers are dissatisfied with their appearance, concerned about their health, or disappointed in their physical condition. Such people are vulnerable to offers promising a way to become as healthy, robust, and energetic as they would like to be. Personal and health-care frauds come in many sizes and shapes, but they all take advantage of people who "ought to know better."

**FOOD PRODUCT "CURES"**    High on the list of fraudulent practices is the marketing of food products as "nutritives" that are promoted as being able to cure various physical ailments. Legally, any product making curative claims should fall into the "drug" category and be subject to regulation by the Food and Drug Administration (FDA). To avoid the FDA's drug-testing and labeling requirements, however, some manufacturers of these products market them as "foods" with instructions on the label to "take one or two a day as a dietary supplement." No curative claims are placed on the packages themselves; instead, the claims accompany the product in brochures or other literature given to the retail dealer to distribute to customers.

Many shrewd entrepreneurs, seeing fertile ground for a fast dollar, have become millionaires in this industry at the consumer's expense (and sometimes to the detriment of his or her health). Firms that produce goods labeled as food are required to meet sanitary standards, but they face fewer regulations than firms that produce drugs. Some producers of food products—but, unfortunately only a very few—have been served with injunctions that ordered them to stop a specific business practice and have had their products seized by the FDA.

The FDA has created a list of claims that are often made by marketers of fraudulent "health-care" food products:

● *One product does it all.* "This product is extremely beneficial in the treatment of rheumatism, arthritis, infections, . . . prostate problems, ulcers, . . . cancer, heart trouble, diabetes, and more."

● *Personal testimonials.* "Alzheimer's Disease!!! My husband has Alzheimer's. Now he mows the grass, cleans the garage, and weeds the flower beds, and we take our morning walk again. He is more like himself again!!!"

● *Quick fixes.* "This product eliminates skin cancer in days!"

● *Time-tested or newfound treatments.* "This revolutionary innovation is formulated by using proven principles of natural health based upon two hundred years of medical science."

● *Satisfaction guaranteed.* "Guarantee: If after thirty days, you have not lost at least four pounds each week, your uncashed check will be returned to you."

● *Paranoid accusations.* "These billion-dollar drug giants have one relentless competitor in common that they all constantly fear—natural remedies."

● *Meaningless medical jargon.* "Neutralize your Hunger Stimulation Point (HSP)." "One of the many natural ingredients is inolitol hexanicontinate."

The FDA has also published a list of steps you should take if you ever consider using a product that makes health claims that you doubt.

● Talk to a doctor or other health professional.
● Check with the Better Business Bureau or an attorney.
● Check out the product with the appropriate health-professional group; for example, check with the National Arthritis Foundation for products that claim to cure arthritis.

Photodisc/Getty Images

How can consumers check to find out whether dietary supplements are safe and effective?

**Question for Thought & Discussion #1:**

Have you, or a member of your family, ever purchased a dietary supplement in the hope that it would help with a health problem? What led to the choice to purchase the product? Did use of the product have the expected result? Would you ever buy such a product in the future?

● Contact a local office of the FDA or check out the product by visiting the FDA's Web site, **http://www.fda.gov**.

**BEAUTY PRODUCT SCAMS**  Most consumers have at least one aspect of their physical appearance that they would like to change. It might be a mole, ears that stick out, a wart, hair that is too curly, or blemished skin—the list goes on and on. Unfortunately, most of these shortcomings can be remedied only by a physician or other specialist who might prescribe expensive treatments. Still, every year, thousands of consumers are taken in by promises of a quick fix that will make them beautiful or handsome.

Although botox has been approved by the FDA to reduce wrinkling, there is no guarantee that botox injections will make you beautiful—particularly if your "doctor" isn't a doctor and doesn't even use botox. In 2002, Dorothy Chin Brandt was sentenced to jail for impersonating a plastic surgeon and disfiguring patients by injecting them with what she said was botox but was really hyacell (a compound made from rooster combs that is used as a folk remedy in South America). The hyacell had been smuggled into the United States and was contaminated with bacteria that infected patients' faces. Although no patients died, many were severely disfigured.

Cosmetics fall under the jurisdiction of the FDA, but—unlike drugs—cosmetics only have to be proved to be safe, not safe and effective. If a label on a cosmetic product reads "antiaging," who can tell if it's mislabeled? Such a description is so ambiguous and difficult to measure that legally it is not considered fraudulent. As a result, creams and lotions that boast of a "newly discovered" ingredient that will "reduce wrinkling" are marketed freely, often at exorbitant prices.

**WEIGHT LOSS PRODUCTS**  Consumers spend billions of dollars every year in an effort to lose unwanted weight. They buy pills and potions that promise to magically "melt away fat while you sleep." Many of these products do achieve temporary weight reductions through water loss. At present, however, the only effective drugs for appetite suppression are sold solely by prescription, and then only for persons who are truly obese. Nevertheless, U.S. consumers continue to buy a variety of products that are at best ineffective and at worst dangerous to their health. Although the Federal Trade Commission (FTC) has taken steps to limit the sale of these products, there are too many of them for the FTC to be totally successful in its effort.

In November 2002, the FTC issued a comprehensive report on false weight loss claims for products and services. The report included the following list of eight claims, called the "eight deadly phrases," that are signs of weight loss fraud:

1. The product will lead to *substantial weight loss for all users.*
2. The product will lead to *permanent weight loss.*
3. Users of the product can *eat as much as they want and still lose weight.*
4. Users of the product *can lose fat from specific body parts.*
5. The product *blocks or absorbs fat or calories.*
6. The product works when *applied to the body or skin.*
7. The product leads to substantial weight loss *without the need for exercise or calorie reduction.*
8. Users of the product can lose *more than three pounds per week for four or more weeks.*

## Cyber consumer
**Online Advice about Health-Care Fraud**

Suppose you believe you have been defrauded in a consumer transaction and want to receive advice and assistance from the federal government. The Web sites of the Federal Trade Commission, the Food and Drug Administration, the Internal Revenue Service, the Federal Bureau of Investigation, the Drug Enforcement Administration, and more offer help with various problems. How can you determine which agency's site might help with your particular problem and which sites would be a waste of your time? The answer is likely to be found by first visiting the U.S. Consumer Gateway, a one-stop link to a broad range of federal information resources for consumers at http://www. consumer.gov. Visit this Web site to investigate what it has to offer.

### Question for Thought & Discussion #2:
Choose a female (possibly yourself) you know well. What types of cosmetics does this person buy? What is this person trying to accomplish with the cosmetics she uses? Do you believe the cosmetics she buys are worth the amount she pays?

## THE diverse CONSUMER

### Peddling Weight Loss In Multiple Languages

In the United States, the problem of being overweight is an equal opportunity difficulty. There are people of every race and ethnic group who would like to be a little thinner. This means that scam artists are also equal opportunity crooks. Consider the case against the manufacturer of Body Solutions Evening Weight Loss Formula.

Late in 2002 the Federal Trade Commission filed suit against the officers of Mark Nutritionals, Inc., of San Antonio, Texas, the manufacturer of Body Solutions Evening Weight Loss Formula. The FTC charged them with making false and unsubstantiated claims for their product. According to the suit, the defendants promoted their product to both English- and Spanish-speaking consumers by broadcasting "testimonial endorsements" from popular radio disc jockeys on more than 650 radio stations located predominantly in the South and along the Mexican border. The defendants claimed that their product offered consumers the unique opportunity to lose substantial weight permanently without dieting or exercise. They also claimed that users would lose weight regardless of what or how much they ate.

One advertisement that was broadcast in both English and Spanish went like this:

It helped me lose thirty-six pounds and it helps me maintain through the holidays. I mean, I ate so much over Thanksgiving, I still have turkey burps. But thanks to Body Solutions, I keep the weight off and now I'm ready for Christmas.

Since 1999 sales of Body Solutions Evening Weight Loss Formula totaled more than $190 million. According to the FTC, the only thing consumers who purchased this product lost was money from their wallets.

What types of people do you believe are likely to be taken in by claims like those made by Body Solutions? What should the government do to protect them? Is it possible for the government to go too far?

**Question for Thought & Discussion #3:**
Have you ever considered using a weight loss product? What factors caused you to decide to purchase, or not purchase, such a product? Why do so many U.S. consumers spend their money to purchase weight loss products?

SOURCE: "Body Solutions Products Come under Fire by the FTC," December 5, 2002, at **http://www.ftc.gov**.

Photodisc/Getty Images

Have you used weight loss drinks or supplements? Were you satisfied with the results?

The FTC has determined that none of these claims can be fulfilled without the use of prescription drugs. If any of these claims are made by a product you encounter, someone is trying to scam you.

## Health-Insurance Fraud

At least two major types of health-insurance fraud are perpetrated on U.S. consumers each year. Both of these scams are expensive and damaging to people, and one of them has the potential to be life threatening as well.

### PAYING FOR INSURANCE THAT ISN'T THERE

In recent years the cost of medical insurance has increased to intimidating levels. In 2002, the cost of basic family medical insurance ranged from $6,000 and $14,000 depending on coverage and deductibles. Many people simply cannot afford to pay these prices. The same is true for many employers who wish to provide medical insurance for their workers.

***Real and Made-Up Insurance*** This situation has opened the door for people who sell medical-insurance policies that don't exist. Some people have scammed consumers by claiming to represent established, honest insurance companies. They tell consumers that they can get the consumers a deal for some special reason and accept premiums that they pocket for themselves. Others make up nonexistent firms that are "trying to break into the market by offering low introductory prices," but the result is the same. Consumers believe they are purchasing medical coverage, but they will never be able to collect on it.

*A Case in Point* Consider the example of Lisa and Dennis Huffstutler, Florida residents who paid $4,800 in 2000 to the Vanguard Asset Group of Lake Success, New York, for medical insurance to replace their much more expensive Blue Cross–Blue Shield coverage. Unfortunately, when their son was hospitalized, Vanguard failed to pay the claim. When the Huffstutlers called, the company told them, "The check's in the mail," but the check never came. In fact, the Florida Insurance Department closed the firm's offices, and its CEO, Dwayne Samuels, pleaded guilty to being an unlicensed operator in U.S. district court in 2001. Although Samuels went to jail, that did not help the Huffstutlers, who were still responsible for their medical expenses.

The greatest danger of nonexistent insurance is that people may be refused needed treatments when hospitals or doctors discover that they are really uninsured. If you ever consider purchasing low-cost medical insurance, be sure to carefully check out the firm and the person offering it for sale.

## PAYING FOR UNNECESSARY OR IMAGINARY MEDICAL SERVICES

A large part of spending for medical care is paid for services that were either unnecessary or never performed. Each year the Office of the Inspector General of the U.S. Department of Health and Human Services (HHS-OIG) conducts an audit of the Medicare program's service claim payments. In February 2002, the HHS-OIG reported that 6.3 percent, or $12.1 billion of the payments made by Medicare in 2000 were "erroneous." In May 2002, the National Health Care Anti-Fraud Association (NHCAA) estimated that at least $39 billion in medical-care payments made by its member insurance companies in 2001 were lost to fraud. Other estimates by government law enforcement agencies place the total cost of medical-care fraud in the range of $130 billion per year and growing.

The perpetrators of this type of health-insurance fraud are doctors and hospitals that inflate their bills to insurance providers by including unneeded services they have performed or by charging for services that were never provided at all. In some instances, this type of fraud takes place with the patient's knowledge. The doctor may forgo billing the patient in exchange for the patient's not reporting that the doctor filed insurance claims for services that were not provided. Some doctors have even given patients kickbacks from the fraudulent payments they receive.

Who is the victim when this type of fraud takes place? First, the insurance company or the government is forced to pay out higher dollar amounts in claims. But the ultimate victim is the consumer, who has to pay higher insurance premiums or taxes to cover the cost of the fraudulent claims. If you ever find that your insurer was charged for treatments you did not receive, you should investigate. An honest mistake may have been made, or you may be looking at a case of fraud.

## Fraudulent Repair Services

New technologies inevitably create opportunities for the dishonest to defraud consumers in new and creative ways. When consumers don't understand how a product works, it is relatively easy to cheat them when the product needs repair or service.

**"Most consumer groups recommend against buying service contracts because they generally cost far more than they are worth."**

**HOME COMPUTERS**   Consider home computers as an example. Roughly four out of every five households own one or more home computers. Home computers may open worlds of opportunity to their owners, but when they cease to function, most of us are forced to take them to an expert for repair.

Most computer-repair services are run by entrepreneurs who offer quality work at fair prices. Unfortunately, some are not. Some technicians use *low-ball* techniques to get consumers in the door. They may advertise a "Free Diagnosis" or an "Instant Analysis for $25." But once they have your computer, they can do almost anything. The unit may only need to have the fan that cools its CPU cleaned. But if the technician says, "Its mother board is blown. I can replace it for $450," how would you know that you are not being told the truth?

When choosing any repair service, it is a good idea to select one that has been established in your community for a number of years. Dishonest businesses tend to disappear as consumers learn not to trust them. It is also a good idea to solicit recommendations from people you know who have had their own technology-based products repaired. Another rule is to avoid businesses that advertise unrealistically low prices. The labor of a skilled computer-repair technician will cost $50 or more per hour. You should avoid any firm that charges significantly less than this amount.

**AUTOMOBILE-REPAIR SCAMS**   The days when consumers could repair their cars with a few simple tools in their own garage are long gone. Automobiles now have high-tech components that seem to require a degree in electronic engineering to repair. Count the number of electronic diagnostic devices you see the next time you visit an automobile-repair facility. Do you think you could operate any of them? If you could, would you know what the information they provide means or how to make the repairs that they indicate are needed? With few exceptions, consumers must rely on expert mechanics when they have their vehicles serviced or repaired.

Just as new technologies open the door for computer-repair scams, they create opportunities for dishonest automobile mechanics and their employers. Again, the best advice is to patronize well-established firms that are recommended by people you know and trust. Try to develop a special relationship with the business by letting its owners know that you will always bring your car to them for repairs and service if they treat you well. Avoid any garage that offers exceptionally low prices. You are likely to get exactly what you pay for.

**SHOULD YOU BUY A SERVICE CONTRACT?**   Retail establishments often claim that you can avoid costly repairs for any technology-based product by purchasing a *service contract*. Then, if something goes wrong, the product will be repaired at no extra cost. Although some of these agreements may be useful, most consumer groups recommend against them because they generally cost far more than they are worth. Today, most computers, television sets, and other types of technology-based products have relatively few problems. Furthermore, they are generally sold with a ninety-day, six-month, or one-year warranty for parts and labor. If they break, they are likely to do so within the time covered by the warranty. Paying for a service contract to extend the warranty for another year or two may not provide you with much value.

Service contracts are not fraudulent unless the promised services are not provided when they are needed. Unfortunately, this has been known to happen. You may buy a three-year service contract after being told that it covers every possible problem. But later, what will you do when your computer's power supply "cooks" itself and the "small print" in your contract says it isn't covered? If you ever do agree to purchase a service contract, be very sure you understand what you are buying. Surprises are not always pleasant.

## Land Sales Scams

The largest purchase most consumers will ever make is the home they buy. And for some, their second largest purchase is a second home. Every year thousands of consumers purchase real estate they have never seen—or have seen only in a photograph or at a Web site on the Internet. This is particularly true of older people who are looking for retirement homes.

**LAND SALES SCAMS ARE EASY TO CARRY OUT** A person does not need to be particularly clever or creative to carry out a land sales scam. A devious crook might purchase 1,000 acres of empty land in the middle of nowhere. The land could be located in a swamp in Louisiana or in an arid area of Arizona nearly one hundred miles from any large community. The owner might then build a house or two on the land and take photographs intended to show the great beauty of the surrounding countryside. There might be no electricity, water service, or sewage system and no nearby hospitals, stores, or schools. The house may look good from the outside but be completely uninhabitable. The photographs won't show what the house doesn't have, and the literature about the "Great Sky Estates" development certainly won't mention these shortcomings. The photographs will be circulated in print or on the Internet to unsuspecting consumers who may think they have found their "dream house." Unfortunately, they are not likely to learn of their mistake until after they have paid for a worthless piece of land.

**THE GOVERNMENT ACTS—WELL, MAYBE** In 1968, Congress passed the Interstate Land Sales Full Disclosure Act, which requires anyone engaged in interstate selling or leasing of land to register all offerings with the Department of Housing and Urban Development (HUD). Unfortunately, this law was amended in 1979 in a way that made it largely ineffective. Today, only limited amounts of information on interstate land sales must be made available to HUD. And little is done to publicize the information that is collected.

To combat fraudulent sales of land, many states have acted on their own. Florida, for example, has created and publicized a list of ways to recognize dishonest land offers that includes the following:

- Be wary of any exceptional offer that requires you to act quickly. Legitimate offers will be available for extended periods of time.

- Before purchasing land, always view the property, investigate any taxes or fees that will be charged to buy or own it, compare the asking price with other prices for similar properties in the area, check to see what is built on or is planned for the land that surrounds the property, and find out whether the land is served by roads and other types of infrastructure.

- Investigate the builder or business that is making the offer. Is it a reputable organization? How long has it been in business? What other developments has it built?

- Do not send money until you have actually seen the property and have had a lawyer review the proposed contract.

**Question for Thought & Discussion #4:**
Why do most land sale scams target older consumers? Why wouldn't you be a good prospect for such a scam?

**Pyramid Scheme** An illegal sales plan through which people collect fees and a share of income earned from sales made by other individuals they recruit into the program.

# Pyramid Schemes

One scam, which seems to come in cycles, rather like locusts, is the **pyramid scheme.** Often pyramid schemes use multilevel-marketing endeavors. Those who join are responsible for sharing income they generate with those who recruited them into the organization. They, in turn, are promised that they will receive a share of the income earned by the people they recruit. The organization looks rather like a pyramid—hence the name.

Glenn W. Turner, for example, had a firm called "Dare to Be Great" that sold motivational self-improvement materials. When a person sold one of Turner's courses, she or he kept some of the money and forwarded the rest up the pyramid. The seller would work up the pyramid by recruiting others into the scheme. Money was sent to the top of the pyramid, where people no longer had to recruit or sell the courses through their own effort. They could sit back and collect income from other people's work.

More recently, the Fortuna Alliance of Bellingham, Washington, conned consumers by offering a guaranteed income of $5,000 per month at its Web site. To join, people were required to pay $250 for the right to recruit other people into the organization. No product was sold to anyone. The only transactions were those between current members and the victims they recruited. When the FTC began to investigate the group, its leaders disappeared with the organization's cash. Most of the members became victims, receiving nothing for their recruitment work or investments.

The people who get in at the beginning of a pyramid scheme can make lots of money. But the majority of those involved don't make much at all. In fact, the average participant makes absolutely nothing. The problem with pyramid schemes is that the only way participants can receive a return on their investment is to find others who are willing to participate in the effort. Pyramid schemes are based on the "greater fool theory": that is, you can always find a greater fool than you out there to help you get rich. Pyramid schemes are illegal. If you are convicted of organizing or participating in one, you could be fined or even sent to jail.

# Tax Service Frauds

There are few things U.S. consumers like less than paying taxes. For this reason fraudulent tax services are able to take advantage of thousands of taxpayers every year by promising to legally reduce the amount of federal and state income tax they must pay. Many of these services are little more than scams that charge taxpayers large fees and leave them liable to fines and federal prosecution for tax evasion.

The Internal Revenue Service (IRS) has created the following list of the most common tax scams, which it calls the "Dirty Dozen":

1. *African Americans get a special tax refund to make up for the history of slavery.* It isn't true.
2. *If you avoid having taxes withheld from your wages, the IRS won't know you earned any income.* Employers report income to the IRS regardless of whether any taxes are withheld.
3. *I don't pay taxes, so why should you?* If con artists don't pay taxes, they are breaking the law. You will break the law, too, if you don't pay.

4. *Pay the tax to win your prize.* Consumers are told that they have won a cash prize but must send the tax they owe on the prize to the scammer to receive the money. Of course, there is no prize.

5. *Paying taxes is voluntary.* Consumers are told that they are not required to pay taxes if they register their opposition with the IRS, which can be done for a fee of $150.

6. *Avoid paying Social Security taxes.* Consumers are told that they can opt out of Social Security by completing a form that the scammer will prepare for a fee of $100.

7. *I can get you a big refund for an up-front fee.* Consumers are told that they will receive a large refund if they immediately send their financial records and a fee of $200. Not only are the tax refunds not filed, but the records are often used to perpetrate other frauds on the victims.

8. *Share children to receive earned income credits.* Consumers are told that if they pay a fee, they will learn how multiple taxpayers can share children to receive credits from the government. It isn't true.

9. *IRS agents make house calls to collect taxes.* IRS agents do not make house calls in any but the most exceptional situations and never to collect payments.

10. *Put your savings in a trust to keep them from being taxed.* This is an attempt to convince unsuspecting consumers to hand over their savings to a crook.

11. *Claim your home as a business deduction.* Consumers are told that they can be shown (for a fee) how to write off the cost of their home as a business expense even when they don't operate a business. It isn't legal.

12. *Claim to be disabled and get a tax deduction.* Consumers are told that for a fee they will learn how to file as a disabled person and receive a tax deduction. Only people who are blind qualify and no special forms are required.

# Steps to Take to Protect Yourself

You may have noticed that most of the suggestions for protecting yourself from fraud are lists of offers to avoid. This is not an accident. So many scams are being carried out that the government and private organizations cannot possibly protect consumers from all of them. Consequently, the best defense against fraud is the individual consumer's ability to recognize and avoid fraudulent offers.

## Be an Assertive Consumer

Assertive consumers rarely fall prey to the high-pressure pitches of some salespeople. Being assertive, of course, does not mean being rude. If you don't want to deal with a salesperson, simply refuse to do so. Salespeople are humans, too, so remember your manners, as well as your rights. If a salesperson fails to respond courteously and persists in pushing you to buy something that you do not want, you always have the right to leave the store, hang up your telephone, or go offline.

**Question for Thought & Discussion #5:**
What types of people are most likely to fall victim to an income tax scam? Why is the government able to investigate almost all of these cases and warn consumers about them?

**Cooling-Off Period**  A specific amount of time in which a consumer has the right to reconsider and back out of a transaction.

Any time you intend to deal with a salesperson, it is better to do your homework first. Before you negotiate a final purchase, you should decide what you want and—just as important—what you don't want. There is little value in paying for extra product features that you will rarely use. You should also have a good idea of the prevailing market price. A salesperson may show you a wonderful food processor that happens to be on sale at 20 percent off today. But, if the store's regular price is 30 percent more than the competition's, the processor will still be more expensive even after the discount.

When you visit a store, call toll free to make a catalogue order, or shop online, be prepared. Make a list of questions that you want answered before you complete the purchase. Find out about the product's warranty, whether there are delivery or installation costs for larger items, what the firm's return policies are, and whether you will receive a refund or a store credit if you return products you purchase. When you know the facts, you are unlikely to be pushed into a consumer decision that you will later regret.

## Cooling-Off Periods

The federal government and some states have laws that require businesses to provide a **cooling-off period** for certain consumer purchases. A cooling-off period is a set period of time (most often three business days) after a covered sale during which consumers may terminate a contract that they have signed. Typically, the transaction must have taken place in the consumer's home, and the consumer must have pledged the home to guarantee payment.

This may sound like a rare occurrence, but it isn't. Many contracts are signed in people's homes each year. Consider door-to-door salespeople who offer to install new siding or a new roof. Some of these purveyors can be very forceful. Many people have signed contracts obligating them to pay a huge amount of money to have work done that they neither wanted nor could afford. Cooling-off periods allow consumers to terminate such a contract after they have had time to think about it.

In 1973, the FTC adopted a cooling-off rule that requires door-to-door salespeople to inform consumers when they have the right to terminate a contract within a set period of time. Any such contract that is signed without this information being provided is not enforceable in a court of law. When both the state and the FTC require a cooling-off period, the consumer is given the benefit of whichever law provides the greater protection. Figure 4.1 is a typical "Notice of Cancellation" that should be available from any door-to-door salesperson.

## Know the Federal Trade Commission's Mail-Order Rules

The FTC has established the following rules to protect you when you shop by catalogue or online:

1. If a catalogue or online advertisement indicates that a product will be sent within a certain period of time, it must be shipped (not necessarily received) within that time. When no date is mentioned, the item must be shipped within thirty days.

**Figure 4.1** Typical Notice of Cancellation for Door-to-Door Solicitation Sales

(enter date of transaction)

_____

(date)

You may cancel this transaction, without any penalty or obligation, within 3 business days from the above date.

If you cancel, any property traded in, any payments made by you under the contract or sale, and any negotiable instrument executed by you will be returned within 10 business days following receipt by the seller of your cancellation notice, and any security interest arising out of the transaction will be canceled.

If you cancel, you must make available to the seller at your residence, in substantially as good condition as when received, any goods delivered to you under this contract or sale; or you may, if you wish, comply with the instructions of the seller regarding the return shipment of the goods at the seller's expense and risk.

If you do make the goods available to the seller and the seller does not pick them up within 40 days of the date of your notice of cancellation, you may retain or dispose of the goods without any further obligation. If you fail to make the goods available to the seller, or if you agree to return the goods to the seller and fail to do so, then you remain liable for performance of all obligations under the contract.

To cancel this transaction, mail or deliver a signed and dated copy of this cancellation notice or any other written notice, or send a telegram to

_____

(name of seller)

at _____ not later than midnight of _____

(address of seller's place of business)                                (date)

I hereby cancel this transaction.

_____                      _____

(date)                                (buyer's signature)

2. If the item can't be shipped within the specified time or within thirty days, the customer must be notified and given a free means of stating what should be done about the delay; the supplier must provide a toll-free telephone number or a postage-paid envelope to the customer.

3. The customer may cancel an order or agree to a new shipment date. If the order is canceled, the refund must be received or credited to a credit card within seven business days. For delays that last more than thirty days, the business must offer the customer a refund.

## Hints for Mail-Order and Online Shoppers

In addition to the FTC's rules, these hints can benefit consumers when they shop by mail order or online:

1. Keep a record of your purchase by making a copy of the catalogue or stock numbers of the items you order. Most orders taken over the telephone or online are given a number that you should keep. If a problem arises with the order, having that number will make it easier for the firm to give you the service you deserve.

2. Use a credit card to pay for your order so that you will have a record of the payment and can stop payment if the product is defective or never delivered.

3. Print a copy of the product if you order online or keep a copy of the catalogue until your order arrives. This will help you make sure that you receive the product you ordered.

4. Keep track of time. If an order is supposed to arrive in ten days, be sure you know when ten days have passed so that you can investigate what has happened.

5. Be sure you provide your order information accurately. If you order by telephone, ask the person who takes your order to read back the information you provide.

6. If you suspect fraud, take action immediately and don't procrastinate. Contact your state attorney general's office or the Better Business Bureau.

# Internet Fraud

No one likes to be cheated, and relatively few people are ready to admit their loss even when they are. Still, every year millions of U.S. consumers are defrauded of money or property through a variety of Internet scams. In 2001, the National Fraud Information Center (NFIC) reported that cases of Internet fraud filed with the NFIC cost U.S. consumers many millions of dollars and that the average cost of each individual instance of fraud exceeded $500. In 2001, this organization received an average of 230 complaints of Internet fraud per week at its Web site, **http://www.fraud.org**. The NFIC believes that only a small proportion of the people who fall victim to Internet fraud report their loss. The actual total of all consumer Internet fraud losses could reach into the many billions of dollars each year. If Internet fraud is not already the most common type of fraud perpetrated on U.S. consumers, it may soon be.

## Advantages of Internet Fraud (from the Scammer's Point of View)

The growth of Internet fraud should come as no surprise. From the scammer's point of view, the Internet is an almost perfect medium. It offers a way to get people to send money while crooks maintain almost total anonymity. Anyone with even limited technological knowledge can create a Web site that gives every appearance of being operated by a large and reputable firm. In fact, some scams have been run with the unauthorized use of names of real firms. When consumers visit these Web sites, they have no easy way to tell with whom or with what organization they are dealing.

The cost of running an Internet scam is low because only people who are interested will respond to fraudulent Internet offers. Scammers don't need to waste time making "cold calls" to people who may not be willing to listen to their spiels. Furthermore, maintaining a Web site on the Internet is less expensive than making thousands of long-distance telephone calls.

Although Internet scams are carried out in a variety of ways to take advantage of different types of consumers, they all have some characteristics in common. There are, of course, no guarantees, but learning to recognize these "flags" can help you avoid being taken for a "ride" on the Internet.

**Question for Thought & Discussion #6:**
Do you shop online? If you do, how do you keep track of goods or services you have ordered? Have you ever been dissatisfied with a purchase? Do you believe you were a victim of fraud?

**CHARACTERISTICS OF INTERNET FRAUD**   You can be reasonably sure that someone is trying to scam you if an Internet offer:

- Promises benefits that are far better than you could reasonably expect to receive for what you will be required to pay.
- Tells you that you have won a valuable prize but can receive it only if you pay for another product.
- Asks you to provide your Social Security, bank account, or credit-card number to verify your identity.
- Sets unreasonable time limits in which a deal must be accepted.
- Promises that you will earn a large income if you purchase special business plans or equipment.

**SCAMMERS SHARE LEADS**   A significant amount of Internet fraud is carried out through e-mail that is directed to specific individuals. You may wonder how scammers know where to direct their e-mails. Like other marketers, scammers buy or trade lists of e-mail addresses for people (called "mooches") who are likely candidates for their fraudulent offers. A number of ploys are used to obtain the names that appear on these lists.

Scammers often create fake Web sites that offer "free" gifts or discount coupons to people who are willing to provide "consumer profiles" about themselves. They are told that the information they provide will be used by the sponsoring corporation to design marketing plans. Nothing could be further from the truth. The information is used by the scammer to determine what types of fraudulent offers to send to which consumers. The real corporation that has been touted as sponsoring the study knows nothing about it, and no gift or discount coupons are ever sent.

Probably the surest way to appear on a scammer's mooch list is to have been successfully defrauded in the past. Once scammers find a victim, they will share that person's name and address (e-mail too) with other perpetrators of crime. People who have fallen victim to an Internet scam need to be particularly careful. They can be sure that they will be sent more fraudulent offers in the future.

## Common Examples of Internet Fraud

The National Fraud Information Center's Web site provides descriptions of many types of Internet fraud that are commonly used by scammers (see Figure 4.2 on the next page). These include Nigerian money offers, low-cost Internet access services, computer equipment sales, adult services (pornography), business franchise propositions, and credit-card offers for people with bad credit histories. According to the NFIC, the following scams are among the most common types of Internet fraud. This list is not comprehensive, but it does provide some insight into the wily ways of Internet scammers.

**INTERNET AUCTION FRAUDS**   A good way to earn a profit quickly is to sell something you don't own and never intend to deliver. For example, you might agree to auction airline tickets or passes to the Super Bowl. It's easy for a crook to claim online that he has two tickets for a Hawaiian vacation that are transferable. The scammer may declare that he purchased the nonrefundable tickets for his honeymoon, but the wedding was called off. Now he'd

**Figure 4.2** The National Fraud Information Center's home page

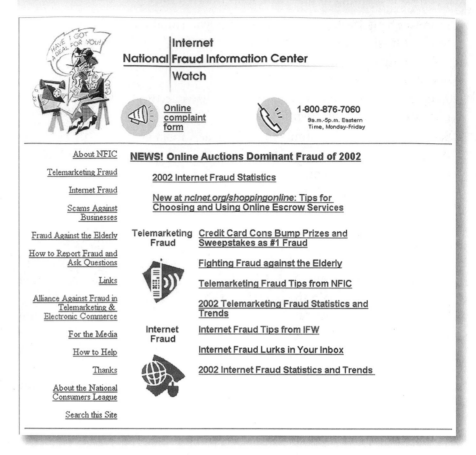

like to see someone enjoy the vacation and at least this way he'll get some of his money back. So, he's willing to sell the tickets to the highest bidder, who turns out to be you. All you have to do is send $250 for each ticket via a money order to a post office box in Boise, Idaho. The problem, of course, is that there are no tickets, and you'll never see Hawaii. Other items often offered through fake auctions include frequent-flier miles (these are not transferable), passes for Disneyland, works of art, and family heirlooms. The common denominators of all such offers are their unreasonably low price and the requirement that you send money before you receive the good or service. Greedy people are most likely to be taken in by this type of scam.

**WORK-AT-HOME OFFERS**    At any given time, millions of Americans are out of work. Other people would like to earn "a little something extra" while they stay home with their children or enjoy their retirement. People who are looking for work-at-home opportunities are often those who can least afford to be taken advantage of by Internet fraud. Unfortunately, they are the people who are most likely to be exploited by work-at-home scams.

In a typical scam, the scammer establishes a Web site that claims that consumers can earn hundreds of dollars every week by completing some sort of low-skill task in their homes. All they have to do is buy the appropriate equipment or materials from the Web site's operators. A common scheme involves clipping and sorting coupons. Victims are told that by purchasing packets of thousands of coupon circulars (for $50 each) they can earn big profits. All they have to do is clip and sort the coupons into packets to sell to their friends, charities, or even grocery store chains that will turn them into cash. In fact, not only is there no market for clipped coupons, but many of the methods for selling coupons suggested at the Web site are illegal and could result in the victim being prosecuted for breaking the law.

Consumers should be suspicious of any online offer that promises a large return for working at home. They should be particularly careful to avoid sending money to any organization they have not investigated. They should begin by contacting the office of their state's attorney general or the Better Business Bureau. Another good idea is to visit the NFIC's Web site, which lists many of the frauds currently being perpetrated on unsuspecting consumers.

**ADVANCE-FEE LOANS**  Thousands of U.S. consumers, for whatever reason, have difficulty obtaining credit but would like to borrow funds to pay for goods or services they want or need. Many of them fall victim to advance-fee loan scams offered on the Internet.

The process is really quite simple. Consumers just type key words such as *credit* or *loans* into their Web browsers. Soon they are provided with a list of hundreds of firms that offer loans or assistance in obtaining credit. Most of these firms are honest businesses. A few, however, are run by scammers and are designed to separate people from their money while giving them nothing in return. In a typical scheme, the scammer offers to find credit for people who have bad credit histories for an up-front fee. The victim is told to send a check for, say, $250, and the business will find a bank that will extend a $10,000 loan at a reasonable interest rate. In just a few days, the money will arrive and easy payments can be stretched out over the next five years. The money, of course, never arrives, and the advance fee is lost forever. The business was not a business at all. It was a scam set up to take advantage of people who are in a difficult financial situation and fail to use good judgment.

**Question for Thought & Discussion #7:**
Why are people who are having financial difficulties often easy prey for scammers?

**CREDIT-REPAIR SCAMS**  A relatively new type of Internet scam is being carried out by individuals who claim to be able to repair credit histories and eliminate debts at a fraction of the amount owed without the debtor having to resort to expensive bankruptcy proceedings. Finding such a Web site is easy— just use the key words *credit repair* or *debt reduction* to complete a search.

At these Web sites, consumers are told to provide information about their debts, assets, ready cash, and income. Armed with this information, the scammer will supposedly approach each creditor and negotiate a reduction in the amount owed. The site claims that the creditors will agree to the reductions because they will receive nothing if the consumer files for bankruptcy. The victims are typically instructed to send a check or money order for most of their cash (remember, they have provided this information) to the "lawyer" who will cash it and use the funds to satisfy their debts. The checks are cashed, but the creditors are never approached. The "lawyer" just disappears with the money.

## Protect Yourself

Consumers need to recognize that no organization or government agency can protect them in all situations. They need to use their common sense and consumer skills to protect themselves. If a deal doesn't sound logical, it probably isn't. Consumers should recognize that no one has ever been victimized by a scam offered on the Internet unless that person first responded to an offer.

Leaving a Web site is just like hanging up your telephone. If you don't provide your credit-card or Social Security number, supply your name and address, or respond in any way to a dishonest offer, you are not likely to be victimized. To be successful, a scam artist needs your cooperation. Consumers should always remember that the person or organization that creates a commercial Web site is out to earn a profit. If a deal is truly wonderful, you are probably not the only person to hear of it. There are sure to be reports about the offer in the news. Always investigate any offer, Internet or otherwise. If you are suspicious, call your state attorney general's office and the Better Business Bureau, or contact the FTC or the National Fraud Information Center. This will take some time and effort, but it can reduce the chance that you will be victimized by fraud.

# Confronting Consumer Issues:
# The Dangers of Identity Theft

Probably the most serious type of consumer fraud today is **identity theft.** Unlike other types of fraud, identity theft does not require the active participation of the victim. It occurs when crooks use information about individuals to take over their identities as consumers, borrowers, savers, and investors. At the beginning of 2003, it was estimated that as many as 500,000 U.S. consumers had been the victims of identity theft.

## HOW IT WORKS

The most important piece of information for perpetrators of identity theft is the victim's Social Security number. Having the person's date and place of birth, driver's license number, credit-card numbers, and address is helpful, but not nearly as important. Armed with this information the thief is able to either open up new credit-card accounts under the victim's name or use existing accounts without the victim's immediate knowledge. Either of these events presents grave dangers for the victim.

### Dangers of Scammers Opening New Accounts
Suppose a crook somehow obtains your Social Security number, date of birth, and other personal information and uses it to open a new charge account. She then charges thousands of dollars in purchases to the account. You know nothing of these events because a false address was given to the credit-card issuer. Bills and monthly statements are sent to a post office box, which is better from the thief's point of view than a nonexistent address because the bills are delivered and just sit in the box. They are not returned to the sender, so the credit-card company does not immediately realize there is a problem. After a few months, the company starts to look for you because you haven't paid the bills. Before that happens, though, delinquent notices appear in your file at the credit reporting agencies listed in Table 4.1.

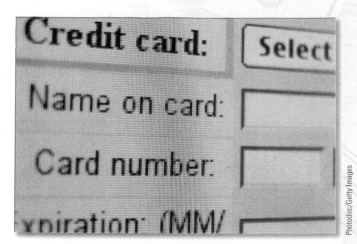

Consumers need to be careful when providing personal information.

These firms distribute the false information about you to your creditors. You might apply for a loan only to be turned down. Your credit-card accounts could be canceled. In extreme cases, people have been arrested and charged with fraud, or their wages have been taken by court orders to pay debts they did not owe. Once false information is listed in your credit file, it can take months or years to correct the situation, and the cost in terms of your time and the money you must spend can be significant.

### Dangers of Scammers Using Your Accounts
You might think that having identity thieves use your personal accounts would be less serious than having them open new accounts. At least, when you receive your monthly statement, you will discover that your account has been exploited and can act quickly to end the scam. Right? Well, maybe not.

A number of victims of identity theft report that the thieves used their existing account numbers to withdraw all

| TABLE 4.1 | Credit-Reporting Agencies | | |
|---|---|---|---|
| | **Equifax** | **Experian** | **Trans Union** |
| **Address** | P.O. Box 740241 Atlanta, GA 30374-0241 | P.O. Box 2104 Allen, TX 75013 | 760 Sproul Road P.O. Box 390 Springfield, PA 19064-0390 |
| **To order credit report** | 1-800-685-1111 | 1-888-EXPERIAN (397-3742) | 1-800-916-8800 |
| **To report fraud** | 1-800-525-6285 | 1-888-EXPERIAN | 1-800-680-7289 |

*(Continued on next page)*

## Confronting Consumer Issues (Continued)

the funds from their checking, savings, or investment accounts. These transactions can be completed in a matter of minutes—long before a victim could discover them in a monthly statement. Some thieves have successfully stolen hundreds of thousands of dollars in this way.

Suppose you think you have $1,000 in your checking account and write checks against that money to pay your bills. But a thief pretending to be you withdrew all your money first. Every one of your checks will bounce, resulting in hundreds of dollars in overdraft fees from your bank, and every one of your creditors will now think you are a "deadbeat." Ultimately, the bank will be responsible for the money that was stolen, but straightening things out may take months. In the meantime, you still owe your creditors, and you may have no way to pay them.

### Question for Thought & Discussion #8:

Do you know anyone who has been a victim of identity theft? What problems did this person experience? How successful was the person in repairing his or her credit history? Does the person know how the thief obtained his or her identity?

## HOW TO PROTECT YOURSELF FROM IDENTITY THEFT

A seemingly endless number of sources, both online and in print, offer information about protecting yourself from identity theft. The Social Security Administration (SSA) maintains a fraud hotline at (800) 269-0271 that provides information about ways you can protect yourself from identity theft and reclaim your identity if you become a victim. The SSA's Web site at http://www.ssa.gov provides helpful advice. The Federal Trade Commission also maintains an identity theft hot line at (877) 438-4338. Each of the credit bureaus identified in Table 4.1 maintains a Web site that offers suggestions for combating identity theft. Perhaps the most comprehensive Web site is maintained by the National Fraud Information Center. It lists the following steps that consumers may take to protect themselves from identity theft:

- *Account information.* Don't give out a credit- or debit-card number unless you're making a purchase with that account. If someone contacts you claiming to be from a business that already has your account number but is requesting it again, be suspicious. Contact the company directly to ask why the number is needed.
- *ATM, credit, and debit cards.* Don't leave these cards lying around your home or office. Carry only those that you

plan to use. If you have accounts that you don't use anymore, close them and cut the cards up.

- *Bills, bank statements, and other records.* Don't leave bills, statements, and other personal records around in plain sight. Keep important documents in a locked file cabinet. Shred information you don't need to keep.
- *Credit reports.* Check your credit report once a year to make sure it does not list accounts that you didn't open. Follow the instructions for disputing accounts you don't recognize or other problems you may spot.
- *Mail.* Remove incoming mail from your mailbox promptly. Send bill payments from the post office or a public mailbox, not from home. If you are going to be away, ask the post office to hold your mail.
- *Online payments.* Look for clues about security when providing account numbers online. At the point where you are asked to provide your financial account information, Social Security number, or other sensitive personal data on a Web site, the letters at the beginning of the address bar at the top of the screen should change from *http* to either *https* or *shttp*. Your browser may also show that your information is being *encrypted*, or scrambled, as it's sent so that it cannot be read by anyone who might intercept it. Look for a symbol at the bottom of your computer screen such as a broken key that becomes whole or a lock that closes. Don't provide sensitive information by e-mail, as it is generally not secure. Although your information may be safe in transmission, that doesn't guarantee that the company will store it securely. See what the Web site says about how your information is safeguarded. Card issuers may offer substitute numbers, a password to verify who is using the account, or other measures to increase security.
- *Passwords.* Don't give your passwords to anyone. Memorize them—don't write them down where others may find them. If someone claiming to be from a company with which you have a password asks to verify it, be suspicious. Contact the company directly to ask why the password is needed.
- *Personal information.* When you do business with companies, look for information about their privacy policies. Tell them when you don't want your personal information shared with other companies. Don't provide information on order forms, warranty forms, and registration forms that isn't necessary to complete the transaction.
- *Preapproved credit-card offers.* If you don't use credit offers that you receive, dispose of them by shredding.
- *Social Security number.* Don't give out your Social Security number unless it is needed for a legitimate purpose such as applying for credit, opening a bank account, or going

to work for someone. Never have the number printed on checks or write it there yourself.

- *Wallets and pocketbooks.* Don't tempt thieves by leaving your wallet or purse in plain sight at work. Don't leave your wallet loose in your back pocket or your purse hanging from a chair at a restaurant or another public place. Use purses that close securely.

- *Work.* Make sure your employer locks and limits access to personnel records. For your personal safety and the safety of your personal information, your employer should provide proper security and prevent strangers from wandering around the workplace.

## PROTECTING YOURSELF MAY NOT BE ENOUGH

Even when consumers take every recommended step to try to protect themselves from identity theft, they can't be sure it won't happen anyway. In 2001, Attorney General Eliot Spitzer of New York announced the arrest of a New York State Insurance Fund employee who was charged with systematically pilfering personal data from state files and selling them to crooks who used the information to steal individual identities. According to Spitzer, the case "shows how personal data can be misappropriated and used for criminal purposes. The result is ruined credit ratings for individuals, significant monetary losses for the defrauded companies and a shameful violation of trust by a public employee." The total amount lost in this scam exceeded $500,000. In this case, there was no possible way the victims could have protected themselves from identity theft fraud.

## WHAT TO DO IF YOUR IDENTITY IS STOLEN

Unfortunately, there is no way to be totally protected from identity theft. If you discover that someone is using your name and credit history to defraud others, here's what you should do:

- *Report the ID theft to the three major credit bureaus* listed in Table 4.1. They will put a "fraud alert" on your credit file so that if someone applies for credit using your personal information, the creditor will take extra care to ensure that it's really you. They will send you a free copy of your report so that you can check for any accounts that you didn't open.

- *Report the ID theft to law enforcement agencies.* The police may not be able to investigate every individual case of identity theft, but making an official report can help you as you fight to clear your name, and the information you provide may be used to stop the thief from victimizing others.

- *When a financial account is involved, contact the bank immediately.* If your credit card, debit card, ATM card,

or checks have been lost or stolen, or if you suspect that someone has obtained your account number for fraudulent purposes, inform the financial institution promptly and ask what you need to do to protect your money.

- *Contact the Federal Trade Commission's ID theft hot line.* This toll-free number, 877-438-4338, was established at the direction of Congress to provide a central source of advice for identity theft victims. Victims can also go online at **http://www.consumer.gov/idtheft** to report the problem and get helpful advice. The information that victims provide is useful to the FTC and other government agencies in investigating and tracking identity theft. The FTC will send you a comprehensive booklet with step-by-step instructions for actions you may need to take, including contacting the major credit bureaus, and forms that you can use to make the process easier. You can download an ID Theft Affidavit from a government Web site at **http://www.consumer.gov/idtheft/affidavit.htm** that you may use to officially register as a victim of identity theft with financial institutions. A copy of the first page of this document appears in Figure 4.3 on the next page.

**Should You Buy Identity Theft Insurance?**   In the past few years, many insurance companies have begun to offer identity theft insurance. These policies are designed to compensate victims of identity theft for the time and money needed to repair their credit histories and regain their lost funds. A few companies such as the Chubb Insurance Company automatically include this type of protection with their homeowners' or renters' policies. Others charge a premium for this coverage. Travelers Property Casualty Company charges $25 annually for identity theft insurance, but some insurers charge as much as $70 or more, which may be an excessive amount to pay for coverage that is available for less from other sources.

**In the End, It's Your Responsibility**   At the risk of repeating ourselves, consumers should remember that it is impossible for the government, or any private organization, to protect them in every situation. Your best protection from identity theft, or any other scam, lies in your ability to look out for you own interests. Even if you believe you shouldn't have to do this, the fact is, you will be better off if you do. Always open and inspect bills and bank statements when they arrive. Immediately contact the billing or banking organization if you find an entry you do not understand. Order copies of your credit report at least once a year and review them carefully. Guard your personal information. Assume that unsolicited requests for your Social Security, bank account, or credit-card numbers are

**Figure 4.3** ID Theft Affidavit

## Victim Information

(1)  My full legal name is _____
         (First)                    (Middle)                    (Last)              (Jr., Sr., III)

(2)  (If different from above) When the events described in this affidavit took place, I was known as
     _____
         (First)                    (Middle)                    (Last)              (Jr., Sr., III)

(3)  My date of birth is _____
                              (day/month/year)

(4)  My Social Security number is _____

(5)  My driver's license or identification card state and number are _____

(6)  My current address is _____

     City _____ State _____ Zip Code_____

(7)  I have lived at this address since _____
                                            (month/year)

(8)  (If different from above) When the events described in this affidavit took place, my address was
     _____

     City _____ State _____ Zip Code_____

(9)  I lived at the address in Item 8 from _____ until _____
                                             (month/year)          (month/year)

(10)  My daytime telephone number is (____)_____

      My evening telephone number is (____)_____

## DO NOT SEND AFFIDAVIT TO THE FTC OR ANY OTHER GOVERNMENT AGENCY

dishonest until someone proves otherwise to your satisfaction. You cannot make sure that you will never be scammed, but you can take steps to reduce the possibility that this will take place. Just be a rational consumer.

## Key Term:

**Identity Theft**   A crime in which a thief uses a person's Social Security number and other information to assume the person's identity as a consumer, borrower, saver, and investor.

## Chapter Summary

1. Consumer fraud involves making a false statement of fact with knowledge of its falsity, or with reckless indifference as to its truth, with the intent to cause someone to rely on such a statement and therefore give up property or a right that has value.

2. Proof of fraud is often difficult to establish. The best protection against fraud is to be aware of the various fraudulent schemes practiced by crooks in the marketplace and guard against them.

3. Many consumers are dissatisfied with their appearance, concerned about their health, or disappointed in their physical condition. Such people are often vulnerable to firms that offer products or services that claim they can make a person healthy, robust, or energetic. The FDA has published a list of "flags" consumers should recognize as being likely indicators of fraudulent offers.

4. Health-insurance fraud takes two basic forms. In one form, policies are marketed that do not actually exist. When claims are made, they are not paid. In the other type of fraud, dishonest medical-care providers file claims for services that were not needed or were never provided. These types of fraud cost many billions of dollars each year and can endanger the lives of consumers.

5. Many consumers become victims of fraud when they have high-tech products serviced or repaired. Most consumers are not qualified to know when or how these products need repair. They must rely on the honesty of technicians who are most often, but not always, honest. When having a high-tech product repaired, it is a good choice to use an established business that has a favorable reputation in your community.

6. Land sale frauds are easy to avoid if you use logic and reason. Never pay for property that you have not seen or adequately investigated. Another common type of fraud is the pyramid scheme that relies on the "greater fool" assumption. The only way a person can profit from a pyramid scheme is to recruit other people to join who will necessarily lose. In recent years, tax frauds have become more common as people try to find ways to avoid paying taxes. Not only do these schemes charge expensive fees, but they do not work and the taxpayers end up owing the government.

7. Consumers can help to protect themselves from fraud by being assertive, understanding the laws that protect them, and following the FTC's suggestions for online and mail-order shopping.

8. The development of the Internet has opened the door for many new types of scams. Among the most common Internet scams are fraudulent auctions, bogus work-at-home schemes, advance-fee loans, and dishonest credit-repair and debt-reduction services. Consumers can usually avoid being victimized by Internet scammers by using common sense and by refusing to act on any Internet offer without carefully evaluating it first. Any Internet offer that promises unusual value for a relatively low cost is likely to be dishonest.

9. Possibly the most damaging type of consumer fraud is identity theft. When this happens, a thief assumes an individual's identity by using his or her Social Security number and other personal information to open credit accounts and then charge purchases to them. In other cases, the victim's personal savings, checking, or investment accounts may be raided and depleted. Once a person's identity has been stolen, it can be very difficult to reestablish a good credit record.

10. The best way to guard against identity theft is to keep your personal information secret. If you become a victim of identity theft, you should immediately contact your bank and other financial institutions to tell them of the problem. The FTC will provide you with a free set of instructions for what you should do to regain your identity and protect your property.

## Key Terms

cooling-off period **80**

fraud **71**
identity theft **90**

pyramid scheme **78**

## Questions for Thought & Discussion

1. Have you, or a member of your family, ever purchased a dietary supplement in the hope that it would help with a health problem? What led to the choice to purchase the product? Did use of the product have the expected result? Would you ever buy such a product in the future?

2. Choose a female (possibly yourself) you know well. What types of cosmetics does this person buy? What is this person trying to accomplish with the cosmetics she uses?

Do you believe the cosmetics she buys are worth the amount she pays?

3. Have you ever considered using a weight loss product? What factors caused you to decide to purchase, or not purchase, such a product? Why do so many U.S. consumers spend their money to purchase weight loss products?

4. Why do most land sale scams target older consumers? Why wouldn't you be a good prospect for such a scam?

5. What types of people are most likely to fall victim to an income tax scam? Why is the government able to investigate almost all of these cases and warn consumers about them?

6. Do you shop online? If you do, how do you keep track of goods or services you have ordered? Have you ever been dissatisfied with a purchase? Do you believe you were a victim of fraud?

7. Why are people who are having financial difficulties often easy prey for scammers?

8. Do you know anyone who has been a victim of identity theft? What problems did this person experience? How successful was the person in repairing his or her credit history? Does the person know how the thief obtained his or her identity?

## Things to Do

1. Check through several tabloid-type magazines. Make a list of the types of products advertised in these publications. Do many of the ads promise amazing improvements in personal appearance, health, and the like? Do you believe any of these products can do what they advertise? Why or why not?

2. Assume you have been contacted by telephone or on the Internet by a person who offers to sell you equipment that will address and sort solicitations for charitable donations. You would have to pay $500 to purchase the needed equipment, but then, you are told, you could earn at least $600 per week by working from your home. The offer is quite persuasive. How would you react? How could you check out the offer?

3. Visit a local health-food store and select a product that is promoted as being beneficial to your health. Read the label carefully and then ask the clerk to explain the product's claims and how they have been substantiated. Do not be rude or abusive. Investigate the product by looking up information at several Web sites on the Internet. Use the information you find to evaluate the product. Are there differences between its claimed benefits and reality? Do you believe the sale of the product constitutes fraud? Explain your findings in an essay.

4. Ask your friends or relatives whether they have ever been victims of identity theft. If you find someone who has, ask this person to describe what happened and how the situation was resolved. If you do not find a victim, ask several people what they do to avoid becoming victims of identity theft. Based on what they tell you, evaluate the probability that they will be successful in protecting themselves from identity theft in the future. In either case, summarize your findings in an essay.

## Internet Resources

### Finding Consumer Information on the Internet

The following Web sites have been selected for their relevance to topics discussed in this chapter. Search these sites to locate information that can add to your knowledge of Internet fraud. Remember, Web addresses change frequently. If any of these addresses no longer function, find similar sites to investigate using any of the search engines available to you.

1. The Securities and Exchange Commission (SEC) offers advice on how to avoid Internet investment scams and also describes recent cases in which the SEC took action to fight Internet fraud. Visit its Web site at **http://www.sec.gov/investor/pubs/cyberfraud.htm**.

2. You can find information about the major types of Internet fraud at a Web site established by the U.S. Justice Department. It is located at **http://www.internetfraud. usdoj.gov**.

3. The Federal Bureau of Investigation (FBI) has set up an Internet Fraud Complaint Center. You can find it at **http://www.fbi.gov/hq/cid/fc/ifcc/ifcc.htm**.

## Shopping on the Internet

The following Web sites have been selected because they offer consumers services similar to those described in this chapter. These are commercial sites that are designed to market products. They do not represent a comprehensive or balanced description of all credit-repair services available online. How would you evaluate the services offered at these Web sites? Do you believe they are honest or likely to be examples of fraud? Remember, Web addresses change frequently. If any of these addresses no longer function, find similar sites to investigate using any of the search engines available to you.

1. The Web site for Credit Repair Services–Princeton Law says, in its own words, "Do you want good credit fast? See results in forty-five days. We do all the work and provide you with your very own legal assistant. Only $35 per month. Money-back guarantee." You can find this site at **http://www.myprinceton.com**.

2. ICR Services offers finance and telecommunications services and the opportunity to become an independent representative. Its Web site is at **http://www.icrservices.com**.

3. Universal Debt Reduction offers to help consumers get out of credit-card debt through debt negotiation, debt consolidation, consumer credit counseling, credit repair, and debt relief. Its Web site is at **http://www.4debtreduction.com**.

## InfoTrac Exercises

Purchasers of new copies of this text are provided with access to the InfoTrac Web site. This Web site links students to thousands of recent articles published in hundreds of periodicals. Use the key words **identity theft** or other terms from this chapter to conduct a key-word search. Choose one article that is of particular interest to you and write a brief essay describing what you have learned from the article. Be sure to cite the author and title of the article and the name and date of the publication in which it appeared.

---

## Selected Readings

"Avoid Fraud When Shopping Online." *USA Today Magazine,* December 2001, pp. 7–8.

Boss, Shira J. "Taken for a Ride." *Good Housekeeping,* October 2002, p. 53.

Culberg, Katya. "Regulating the Proliferation of Spam." *Journal of Internet Law,* September 2002, pp. 18–19.

Friedman, Monroe. "Coping with Consumer Fraud." *Journal of Consumer Affairs,* Summer 1998, pp. 1–11.

"FTC Cracks Down on Internet Scams." *Travel Weekly,* November 9, 2000, p. 11.

"FTC, Friends Target Advance Fee Loan Scams." *FTC Watch,* July 10, 2000, p. 18.

"FTC Says Fraud Threatens Growth of Internet." *Direct Marketing,* August 13, 2001, p. 108.

Goodwin, Bill. "MasterCard Blocks Net Fraudsters." *Computer Weekly,* March 6, 2003, p. 4.

Henkel, John. "FTC Program Aims to Cut Consumer Fraud." *FDA Consumer,* July 2001, p. 37.

"How to Spot Health Fraud." *FDA Consumer,* November 1999, p. 22.

Lee, W. A. "Amex to Help Web Stores Battle Card Fraud Online." *American Banker,* September 27, 2000, p. 9.

Lieberman, Janice. "So You Want to Work at Home?" *Redbook,* March 2002, pp. 86–87.

Rayport, Jeffery, and Bernard J. Jaworski. *Cases in E-Commerce.* New York: McGraw Hill, 2001.

"Seeing Double." *FTC Watch,* January 28, 2002, p. 10.

Taylor, Chris. "Giving Credit Where Credit Is Not Due." *Time,* December 9, 2002, p. 100.

# CHAPTER 5

# Protection for the Consumer

After reading this chapter, you should be able to answer the following questions:

- How has the government's role in consumer protection grown since 1900?

- What federal agencies monitor and enforce consumer protection legislation?

- What are the basic functions and powers of the Federal Trade Commission and the Consumer Product Safety Commission?

- What sources of consumer assistance and protection exist in the private sector?

- How can consumers use small claims court to protect their rights?

- What is the difference between express and implied warranties?

- Where can consumers turn for help in protecting their rights?

S tudents of consumer economics will benefit from a basic knowledge of the history of consumer protection. The earliest forms of consumer protection in the United States were really attempts to make sure that markets functioned efficiently. In competitive markets, there are many buyers and sellers, so no single individual is able to influence the price or quality of a particular good. Even when consumers are familiar with the products they are bargaining for, it can be difficult for them to be sure they receive the amount of a product for which they have paid. Therefore, from the earliest recorded times, governments have been involved in setting uniform standards for weights and measurements. The quality of products has also been regulated by many governments. The quality of flour exported from the American colonies, for example, was regulated by governments in North America before the United States became a nation. Even earlier, in ancient Greece, it was a crime for a merchant to add water to wine that was offered for sale.

After standards of weights and measures and quality are established, the next problem is enforcement. This second responsibility of consumer protection, policing standards, has caused the courts and other administrative bodies of our government to become involved in the consumer protection system. For consumer protection to work, standards must be set, they must be enforced, and those who violate them must be apprehended and punished.

> "For consumer protection to work, standards must be set, they must be enforced, and those who violate them must be apprehended and punished."

# Sources of Consumer Protection

There are several important and distinct sources of consumer protection. Table 5.1 lists the most significant laws passed specifically to protect consumers. In addition to federal legislation, each state has a consumer protection bureau, either as a separate state office or as a division of the state attorney general's office. County and municipal governments also often have consumer affairs offices, and numerous private groups—such as the Better Business Bureau—help consumers solve problems relating to product sales and services. Most major industries in the United States have also established consumer divisions to deal with complaints about their products. Indeed, today's consumer, with just a telephone or access to the Internet and a few basic consumer reference materials, can get nearly instant help with almost any consumer problem.

Effective consumer protection helps businesses as well as consumers. Businesses are protected from unfair competition by those who would cheat, or commit fraud against, consumers. After all, money that consumers pay to fraudulent businesses, and resources used to create inferior products, are money and resources that cannot be used by honest enterprises. Remember, businesses buy resources in markets, too. Standards established by the government protect firms from being taken advantage of in those markets. Businesses, then, have two reasons to support the enforcement of uniform standards:

1. To protect themselves when they go into the market to buy resources.
2. To protect themselves against competition from fraudulent businesses.

## Antitrust Laws

Further regulation of the marketplace was undertaken by the federal government in the late 1800s when **antitrust laws** were enacted to limit monopoly-like powers held by some businesses. The rights of buyers and other competitors had

**Antitrust Laws** Laws designed to prevent business monopolies. Antitrust laws are part of government antitrust policies that are aimed at establishing and maintaining competition in the business world to assure consumers of fair prices and goods of adequate quality.

| TABLE 5.1 | Major Federal Consumer Legislation |
|---|---|
| **STATUTE OR AGENCY RULE** | **PURPOSE** |
| **Credit and Finance** | |
| Consumer Credit Protection Act (Truth-in-Lending Act) (1968) | Offers comprehensive protection covering all phases of credit transactions |
| Fair Credit Reporting Act (1970) | Protects consumers' credit reputations |
| Equal Credit Opportunity Act (1974) | Prohibits discrimination in the extending of credit |
| Fair Credit Billing Act (1974) | Protects consumers in credit-card billing errors and other disputes |
| Fair Debt Collection Practices Act (1977) | Prohibits debt collectors' abuses |
| Counterfeit Access Device and Computer Fraud and Abuse Act (1984) | Prohibits the production, use, and sale of counterfeit credit cards or other access devices used to obtain money, goods, services, or other things of value |
| Home Equity Loan Consumer Protection Act (1988) | Prohibits lenders from changing the terms of a loan after the contract has been signed; requires fuller disclosure in home equity loans of interest rate formulas and repayment terms |
| Truth in Savings Act (1993) | Requires all depositor institutions to report interest earnings in the same way using the Annual Percentage Yield (APY) |
| Fair Credit Reporting Act amended (1994) | Increases consumer access to credit firms |
| **Health and Safety** | |
| Pure Food and Drug Act (1906) | Prohibits adulteration and mislabeling of food and drugs sold in interstate commerce |
| Meat Inspection Act (1906) | Provides for inspection of meat |
| Federal Food, Drug and Cosmetic Act (1938) | Protects consumers from unsafe food products and from unsafe and/or ineffective drugs (superseded Pure Food and Drug Act of 1906) |
| Flammable Fabrics Act (1953) | Prohibits the sale of highly flammable clothing |
| Poultry Products Inspection Act (1957) | Provides for inspection of poultry |
| Child Protection and Toy Safety Act (1966) | Requires child-proof devices and special labeling |
| National Traffic and Motor Vehicle Safety Act (1966) | Requires manufacturers to inform new car dealers of any safety defects found after manufacture and sale of autos |
| Wholesome Meat Act (1967) | Updated Meat Inspection Act of 1906 to provide for stricter standards for slaughtering plants dealing with red-meat animals |
| Public Health Cigarette Smoking Act (1970) | Prohibits radio and TV cigarette advertising |
| Consumer Product Safety Act (1972) | Established the Consumer Product Safety Commission to regulate all potentially hazardous consumer products |
| Department of Transportation Rule on Passive Restraints in Automobiles (1984) | Requires automatic restraint systems in all new cars sold after September 1, 1990 |
| Toy Safety Act (1984) | Allows the Consumer Product Safety Commission to quickly recall toys and other articles intended for use by children that present a substantial risk of injury |
| Smokeless Tobacco Act (1986) | Prohibits radio and TV advertising of smokeless tobacco products |
| Drug-Price Competition and Patent-Term Restoration Act (Generic Drug Act) (1984) | Speeds up and simplifies Food and Drug Administration approval of generic versions of drugs on which patents have expired |
| Clean Air Act (1990) | Revised act of 1970 to establish new pollution limitations including standards for automobile, power plant, and cancer-causing substance emissions |
| Dietary Supplement Health and Education Act (1994) | Requires manufacturers to submit information to the Food and Drug Administration to show their products are safe |

long been protected by **common law** based on traditions because courts refused to enforce monopolistic contracts. But before the Sherman Antitrust Act was passed in 1890, there had been no federal law that specifically made monopoly and price fixing illegal in the U.S. economy. Although the original intent of this

| TABLE 5.1 | Major Federal Consumer Legislation (Continued) |
|---|---|
| **STATUTE OR AGENCY RULE** | **PURPOSE** |
| **Labeling and Packaging** | |
| Wool Products Labeling Act (1939) | Requires accurate labeling of wool products |
| Fur Products Labeling Act (1951) | Prohibits misbranding of fur products |
| Textile Fiber Products Identification Act (1958) | Prohibits false labeling and advertising of all textile products not covered under Wool and Fur Products Labeling Acts |
| Hazardous Substances Labeling Act (1960) | Requires warning labels on all items containing dangerous chemicals |
| Cigarette Labeling and Advertising Act (1965) | Requires labels warning of possible health hazards of cigarettes |
| Child Protection and Toy Safety Act (1966) | Requires child-proof devices and special labeling |
| Fair Packaging and Labeling Act (1966) | Requires that accurate names, quantities, and weights be given on product labels |
| Smokeless Tobacco Act (1986) | Requires labels disclosing possible health hazards of smokeless tobacco |
| Food Labeling Act (1990) | Directed the Food and Drug Administration to establish new labeling standards for food products that would be more meaningful to consumers |
| **Sales and Warranties** | |
| Federal Trade Commission Act (1914/1938) | Prohibits deceptive and unfair trade practices |
| Interstate Land Sales Full Disclosure Act (1968) | Requires disclosure in interstate land sales |
| Odometer Act (1972) | Protects consumers against odometer fraud in used-car sales |
| FTC Door-to-Door Sales Rule (1973) | Federal Trade Commission rule regulating door-to-door sales contracts |
| FTC Rules of Negative Options (1973) | Federal Trade Commission rules regulating advertising of book and record clubs |
| Real Estate Settlement Procedures Act (1974) | Requires disclosure of home-buying costs |
| Magnuson-Moss Warranty Act (1975) | Provides rules governing content of warranties |
| FTC Vocational and Correspondence School Rule | Federal Trade Commission rule regulating contracts with these types of schools |
| FTC Used-Car Rule (1981) | Federal Trade Commission rule requiring dealers in used-car sales to disclose specified types of information in "Buyer's Guide" affixed to auto |
| FTC Funeral Home Rule (1984) | Federal Trade Commission rule requiring disclosure by funeral homes regarding prices and services |
| Children's Advertising Act (1990) | Directed Federal Communications Commission to establish rules to limit TV advertising directed toward children |

legislation was to protect the interests of competing producers in the market, it had consumer implications as well. By preventing firms from unfairly harming their competition, the Sherman Act and other antitrust laws protected consumers' access to a wider selection of products that were offered at lower prices.

At the turn of the twentieth century, consumer protection, as it is understood today, did not exist. Today's concept of consumer protection evolved during the years between 1900 and 1918. This was the time of the "muckrakers"—politicians, writers, journalists, and others who fought against corruption and the abuse of power in the economy and the government. Their efforts led to the passage of the first wage-and-hour laws, the first women's and minors' protective legislation, and the first federal laws designed specifically to protect consumers. Possibly the most important of these early laws was the Pure Food and Drug Act of 1906, which dealt with the production, transportation, and sale of food and drugs in this country.

**Common Law** The unwritten system of law governing people's rights and duties, based on custom and fixed principles of justice. Common law is the foundation of both the English and the U.S. legal systems (excluding Louisiana, where law is based on the Napoleonic Code).

**Question for Thought & Discussion #1:**
Do you believe you could protect your consumer rights without government assistance? Explain why or why not.

> "Many U.S. consumers came to feel that they did not understand the products they bought and that the businesses they dealt with no longer cared about them as individuals."

## Food and Drug Act

Although the federal government made fraud through the mails illegal in the 1870s, it emphasized consumer protection only in terms of transactions at the retail level of the marketplace. Upton Sinclair's book *The Jungle* alerted the general public to the fact that consumer protection meant more than making sure that accurate information was provided at the point of sale. In his book, Sinclair graphically described the squalor and unsanitary practices that existed in the meat-packing industry. In response, groups began seeking some form of "consumer protection" in the processing of products before they arrived at the marketplace. Eventually, the government reacted by passing the Pure Food and Drug Act of 1906, as well as the Meat Inspection Act of the same year.

More progress in consumer protection came less than ten years later in 1914 when the Federal Trade Commission Act was passed. This act provided administrative machinery to enforce the antitrust laws and spelled out unfair methods of competition, including deceptive advertising. It was not until 1938, however, that the Food, Drug and Cosmetic Act was passed to strengthen the protective features of the 1906 legislation. This act, for the first time, required manufacturers to prove that their drugs were safe before marketing them. Also, with the 1938 act, the Food and Drug Administration no longer had to prove that a firm manufacturing a harmful product had intentionally committed a violation before it could take action against that company.

The passage of the 1938 legislation was the last significant federal activity on the consumer protection front until the 1960s and 1970s, when a flood of legislative activity occurred at federal, state, and local levels. Between 1965 and 1975, more than twice as many consumer protection laws were passed than had been enacted in the previous ninety years.

## A Renewed Interest in the Consumer

What happened to rekindle interest in consumer protection in the 1960s and 1970s? Some people attribute this renewed interest to Ralph Nader, whose 1965 book *Unsafe at Any Speed* focused public attention on the issue of automobile safety. But other books, such as Upton Sinclair's *The Jungle* and Stuart Chase's *Your Money's Worth* of the 1930s, did not lead to a sustained consumer protection movement. So apparently something else was operating in the 1960s and 1970s that had not existed in U.S. society before. This something else may have been the growing complexity of modern economic life.

By the early 1960s, the U.S. public had felt the impact of the technology explosion that had transformed production, transportation, and information systems. The development of plastics, frozen foods, and freeze-dried foods had made preprocessing and prepackaging an everyday fact of U.S. life. The automobile had become a complex, accessory-loaded machine that many buyers could no longer understand. Consumers found themselves having to cope with ever-increasing amounts of information. Accompanying this expansion of the need for knowledge was a depersonalization of the marketplace. Many U.S. consumers came to feel that they did not understand the products they bought and that the businesses they dealt with no longer cared about them as individuals.

**THE NEED FOR GOVERNMENT HELP**   In making buying decisions today, consumers often spend much time seeking and evaluating information. In a simpler economic system, one might know enough about the products and their producers to feel comfortable about making a good decision. But with our complex technology, no one can possibly know enough about every product to feel confident of making the best or even satisfactory consumer decisions in every field. Furthermore, many of the businesses consumers buy from are so large that, in the consumer's mind, they have taken on the characteristics of a huge, uncaring machine. Consequently, consumers in the 1960s began to feel the need for more government protection. Government regulations multiplied for standards of packaging and for disclosure of information that would enable consumers to compare manufacturers' claims more easily. In addition, many consumers began to agitate for government-enforced standards of safety.

**INDIVIDUAL CONSUMERS HAD LITTLE POWER** Individual consumers have little control over producers because the few dollars they might withhold from a large corporation will have no meaningful impact on the firm's profits. Moreover, until the significant changes of the 1960s, the legal system was not suited to handle the problems of millions of individuals with small sums of money at stake; each sum was important to the individual, but no one amount was large enough to warrant the costs of taking a firm to court. The legal concept of *caveat emptor,* "let the buyer beware," therefore seemed to rule the economy.

A mounting sense of helplessness and frustration led consumers to look to the government for a new form of consumer protection: protection after the fact. The new emphasis in consumer protection became **consumer redress:** the right of every consumer to air legitimate grievances and to seek satisfaction for damages suffered from the use of a product or service. This was not the same as the earlier consumer protection against fraud. Consumers could now ask for redress, not because they had been deliberately defrauded, but because the complexity of the marketplace had made it impossible, in their eyes, for them to protect themselves adequately before the act of purchase.

**Consumer Redress** The right of consumers to seek and obtain satisfaction for damages suffered from the use of a product or a service; protection after the fact.

# The Presidents Speak Up

In the early 1960s, the government responded to consumer needs. In 1962, President John F. Kennedy sent Congress a consumer protection program calling for the recognition of four fundamental consumer rights:

1. *The right to safety*—protection against goods that are dangerous to life or health.
2. *The right to be informed*—disclosure laws to allow consumers not only to discover fraud but also to make rational choices.
3. *The right to choose*—a restatement of the need to have many firms in a competitive market and to have protection by government where such competition no longer exists.
4. *The right to be heard*—the right of consumers to have their interests heard when government policy decisions are made.

To these four rights, subsequent presidents have added others:

5. *The right to a decent environment*—assurance that consumers' health, property, and quality of life will not be harmed by actions that damage the environment.

6. *The right to consumer education*—through government programs specifically created for that purpose.

7. *The right to reasonable redress for physical damages suffered when using a product*—the right to receive just compensation for losses that result from the use of defective or dangerous products.

During the 1960s and 1970s, these rights were supported by a host of consumer protection laws passed by Congress and by the creation of federal agencies to administer and enforce them.

**A SHIFT TO LESS REGULATION**   The election of Ronald Reagan to the presidency in 1980 heralded a change in the executive attitude toward the consumer movement. According to the school of economic thought guiding the Reagan administration, if the government stepped back and gave businesses a freer hand to operate as they saw fit, competition and the marketplace would eventually eliminate many market imperfections. Although President Reagan stressed the importance of the consumer in the health of the economy, agencies involved in the administration and enforcement of consumer protection legislation faced significant reductions in their powers and budgets.

**STRIKING A BALANCE**   To some extent, the Reagan administration's approach to consumer protection was a response to the growing resentment felt by many Americans against increased and perhaps unwarranted intrusion by the government into their lives. It is also possible that the success of the consumer protection movement in the 1960s and 1970s reduced the need for government action during the Reagan era. In any case, the level of federal involvement in consumer protection began to grow again after 1990. Although some of this can be seen in a slow increase in the funding of consumer protection agencies, more is apparent in the expanded and changed regulations enforced by existing organizations. Amendments to the 1970 Clean Air Act, 1957 Poultry Products Inspection Act, and 1966 Fair Packaging and Labeling Act gave the Federal Trade Commission (FTC) and the U.S. Department of Agriculture (USDA) increased power to set and enforce standards for consumer products. The Consumer Product Safety Commission (CPSC) was given new power to keep track of businesses found guilty of producing or selling hazardous products. Government efforts to protect consumers continued to expand through the 1990s. Funding for the FTC, USDA, and CPSC almost doubled, and the number and types of products regulated grew significantly. By the 2000s, the federal government was accepting a greater responsibility in this area.

# Consumer Responsibilities

**Question for Thought & Discussion #2:**

Do you believe you are a responsible consumer? Describe several events that demonstrate the type of consumer you are.

No president has yet produced a list of consumer responsibilities to accompany the consumer rights listed earlier, but some obvious ones exist:

1. *The responsibility to give correct information*—when, for example, filling out an application for a loan or trading in a used car. To put it bluntly, consumers, like salespeople, shouldn't lie.

2. *The responsibility to report defective goods*—both to the seller and to the manufacturer. This way the seller and manufacturer can warn other con-

sumers of defective and dangerous products, stop selling them, and perhaps recall them. This responsibility is particularly important when automobiles and electrical equipment are involved. The consumer has a responsibility to society in this area.

3. *The responsibility to report wrongs incurred in consumer dealings*—either to the appropriate government agencies or to private organizations responsible for monitoring various aspects of the marketplace. Again, this is for the protection of other consumers.

4. *The responsibility to keep within the law when protesting*—a fundamental responsibility of consumers to protect the rights of others as they exercise their own First Amendment right to free speech.

5. *The responsibility to accept the consequences of one's consumer decisions*—for example, if a person chooses to use a cigarette lighter to look into a gasoline tank, it is her or his responsibility if it blows up.

# Enforcement of Federal Consumer Protection Laws

The passage of consumer legislation is only the first step in consumer protection. Legislation must be administered effectively. In 1964, President Lyndon Johnson made a gesture in this direction when he appointed the first special presidential assistant for consumer affairs. Although this person, a member of the staff of the Office of the President, had no direct authority, the fact that such a position existed ensured that consumer interests would have some representation at the federal policy level. The office was continued by President Nixon until 1973, when it was transferred to the Department of Health, Education, and Welfare, now the Department of Health and Human Services.

## Coordination of Consumer Programs

A continuing challenge to the government is the coordination of the various departments and agencies that are charged with different aspects of consumer protection. Different administrations have approached this problem in several ways. In 1979, President Jimmy Carter created the Consumer Affairs Council (CAC), which was charged with coordinating all federal efforts to protect consumers. Each federal department and agency was required to designate at least one person who would be responsible for addressing consumer issues and reporting to the CAC. This structure lasted through the presidency of Ronald Reagan and into the first Bush administration. The Clinton administration did not revive the CAC but did put more emphasis on the coordination and enforcement of consumer protection legislation than had been done under either Reagan or George H. W. Bush (at least in terms of dollars spent). Again, under the administration of George W. Bush, there is no single structure responsible for coordinating consumer protection efforts. There is, however, a central Internet site, at **http://www.consumer.gov**, that directs consumer inquiries to the appropriate government department or agency.

Two of the most important agencies responsible for enforcing consumer protection legislation are the Federal Trade Commission and the Consumer Product Safety Commission.

> "The passage of consumer legislation is only the first step in consumer protection. Legislation must be administered effectively."

## THE ETHICAL CONSUMER

## How Far Should the Government Go to Protect Children?

Today, it is possible to find almost any type of information or image on the Internet. Most of what is available is useful and socially beneficial, but some is not. With only a little effort, you can learn how to buy drugs, build a bomb, or view pornography on your home or school computer. As an adult, you probably feel that you are able to handle anything you might find on the Internet. But do you trust others to exercise the same responsibility, particularly if they are children? In recent years the federal government has passed a series of laws that were intended to prevent children from viewing certain types of materials on the Internet.

In 1998, Congress passed the Child Protection Act, which required Internet service providers, such as AOL and MSN, to in effect censor the Web sites they hosted. Given the hundreds of thousands of sites, this task was clearly impossible, at least in any comprehensive sense. The Third Circuit Court of Appeals recognized this when it suspended enforcement of the law in 1999.

In December 2000, Congress tried again by passing the Children's Internet Protection Act, which required public libraries and schools to install filters that would automatically screen materials on Web sites and deny access to sites that contained materials deemed inappropriate for children. At first glance this law appeared to be an improvement over the 1998 law because it relied on software programs, instead of people, to evaluate materials. In practice, however, this methodology resulted in many problems.

The first problem was funding. The cost of purchasing filtering programs for all of the nation's libraries and schools was estimated to be at least $1.3 billion per year, but the legislation provided no funding, so the expense would be borne by local governments and school districts.

A greater problem was that early versions of the filters simply did not work satisfactorily. To filter material on a Web site, the program reads its text and looks for words or phrases that indicate that access should be denied. The word *breast,* for example, is programmed as an indicator of pornography. As a result, when students try to look up information about breast cancer, they will be denied access to medical Web sites. Or suppose that a student is doing a report about HIV/AIDS for her health class. Aware that the nation's homosexual population has suffered disproportionately from this disease, the student types the words *AIDS* and *homosexual* into her search engine and is denied access. Is this what we want? As a final example, if a high school student doing research on the Vietnam War uses the words *bomb* and *torture* in his search, he will be denied access. The question is, should the government, or librarians, or teachers be limiting what anyone in the United States, young or old, can find on the Internet? In 2003 the Supreme Court ruled that libraries do have this right if they choose to use it. Do you agree with the Supreme Court's decision?

**If you had the power to decide, what would you do? Would you limit what can be seen on the Internet? If you would, how would you enforce your rules?**

## Federal Trade Commission

The Federal Trade Commission (**http://www.ftc.gov**) is foremost in federal consumer protection activities. Created in 1914 as a result of the Federal Trade Commission Act of that year, the FTC has five commissioners appointed by the president for terms of seven years. The commissioners have extensive enforcement responsibilities for a number of federal statutes.

The FTC's activities are divided among three bureaus—the Bureau of Consumer Protection, the Bureau of Competition, and the Bureau of Economics. Through its Bureau of Consumer Protection, the FTC can stop any "unfair or deceptive acts or practices" that are used to influence, inhibit, or restrict consumers unfairly in their purchasing decisions. Violations are punishable by law. For example, the FTC might issue a **cease-and-desist order** that prohibits an automobile manufacturer from advertising fuel-economy levels achieved by pro-

**Cease-and-Desist Order** An administrative or judicial order commanding a business firm to cease conducting the activities that the agency or a court has deemed to be "unfair or deceptive acts or practices."

fessional drivers without disclosing the drivers' professional status. If the firm fails to obey the order, the company may be fined $10,000 for each subsequent illegal advertisement. The FTC can also impose legal penalties against other automakers if they advertise similar mileage tests without disclosing that professional drivers have been used.

The Bureau of Competition is the arm of the FTC that, in conjunction with the Justice Department, enforces antitrust laws, such as price discrimination legislation that makes it illegal for one seller to sell the same product or service to two different buyers at two different prices (unless this difference can be justified in terms of varying costs). The FTC's Bureau of Economics provides economic advice to the other two bureaus and conducts economic studies concerning government regulations and their effect on consumers and the U.S. economy generally.

The FTC has established a schedule for reviewing its rules that extends ten years into the future. Table 5.2 lists some of the rules that have already been evaluated or soon will be.

Regional offices of the FTC are located in major cities across the country. These offices act as "mini–FTCs" and assist consumers whose complaints or problems fall under the FTC's jurisdiction.

## Consumer Product Safety Commission

As a result of 1970 recommendations of the National Commission on Product Safety, the Consumer Product Safety Act was passed in 1972, creating the Consumer Product Safety Commission (**http://www.cpsc.gov**) to regulate all potentially hazardous consumer products. The Consumer Product Safety Act was the outcome of product-safety legislation that had begun in 1953 with the enactment of the Flammable Fabrics Act. Over the next twenty years, Congress continued to enact legislation regulating specific classes of products or product design or composition. The Consumer Product Safety Act, however, is concerned with the overall safety of all consumer products.

The 1972 act states that "any article, or component part thereof produced or distributed for sale to a consumer for use in or around a permanent or temporary household or residence, a school, in recreation or otherwise, or for the personal use, consumption, or enjoyment of a consumer" shall be subject to regulation by the CPSC. The CPSC can set safety standards for consumer products as well as ban the manufacture and sale of any product deemed

| TABLE 5.2 | Selections from the FTC's Revolving Regulatory Review Schedule |
|---|---|
| **Topic** | **Year of Review** |
| Appliance Labeling Rule | 2004 |
| Credit Practices Rule | 2005 |
| Guides for Jewelry Industry | 2007 |
| Guides for Environmental Marketing Claims | 2008 |
| Guides for the Advertising of Warranties and Guarantees | 2009 |
| Care Labeling Rule | 2010 |

SOURCE: Taken from the FTC's Web site, **http://www.ftc.gov/05/2002/03/16cfrch1.htm**, on January 30, 2003.

Digital Vision/Getty Images

Playgrounds are designed for children's fun and enjoyment, but they figure into many of the accidents requiring a visit to the emergency room. How can parents and the builders of playgrounds work together to ensure that playgrounds are safe?

## Question for Thought & Discussion #3:

Have you ever purchased a product that you thought was dangerous? What did you do about it?

hazardous to consumers. The commission has the authority to remove from the market products that are deemed imminently hazardous and can require manufacturers to report information about any products already sold or intended for sale that have proved to be hazardous.

By using data obtained from hospital emergency rooms, the CPSC annually determines which products are the most hazardous—that is, which products are related to the most injuries (those treated in emergency rooms). Table 5.3 shows the results of the most recent study.

Many other government agencies contribute to the effort to protect consumers. These include the Food and Drug Administration, which administers federal laws that regulate the quality of foods and drugs offered for sale, and the Securities and Exchange Commission, which is charged with protecting consumer rights in many financial markets.

## State and Local Governments and Consumer Protection

While federal action clearly illustrates the national importance of an issue, the adoption of a federal policy is often the result of prolonged activity at the state and local government levels or in the private sector of the economy. This has been especially true of consumer protection policy. In fact, some states, localities, and private groups have gone far beyond the limits now set by federal policy.

State and local governments have always been involved in setting standards of weights and measures and marketing standards, as well as standards that define the term *fraud*. Even today, enforcement of consumer fraud statutes is left largely to state and local governments. Many areas of fraud are commonly dealt with under state and local criminal fraud statutes arising out of the criminal fraud case decisions of earlier years. Furthermore, in the areas of credit, insurance, health and sanitation, and all issues concerning contract rights, it has been mainly the state governments that have enacted legislation dealing with consumer problems. In fact, state responses have sometimes come much earlier than the federal response. For example, as early as 1959, both New York

| TABLE 5.3 | Selected Consumer Products Often Associated with Accidents |
|---|---|
| **Product** | **Total Accidents, 1998** |
| 1. Stairs, ramps, landings, and floors | 1,976,070 |
| 2. Bicycles | 577,621 |
| 3. Knives | 454,246 |
| 4. Beds | 437,980 |
| 5. Doors | 342,302 |
| 6. Cans and other containers | 289,453 |
| 7. Chairs | 286,020 |
| 8. Bathtubs and shower stalls | 181,837 |
| 9. Shop tools | 141,133 |
| 10. Swimming pools | 81,079 |

SOURCE: *Statistical Abstract of the United States, 2001*, p. 118.

and California had legislation on the books to protect the rights of consumers in credit transactions. And not until Massachusetts passed the first truth-in-lending law was federal action on this important issue able to succeed. When the federal Consumer Credit Protection Act was passed in 1968, Massachusetts became, in effect, a pilot case for the national legislation.

# Using Small Claims Courts

All states provide another path by which consumers can protect their rights and receive satisfaction for defective products or unfair treatment: the small claims court. Suppose you believe that your former landlord cheated you by keeping your security deposit when you moved out of an apartment. Or a dry cleaner ruined your best suit. Perhaps an insurance company refused to pay what you feel was a legitimate claim, or maybe a firm failed to respect the warranty on its product by refusing to replace an "unbreakable" plastic chair that broke as soon as you got it home.

If you've ever felt helpless in these or similar situations, you needn't have. To right such wrongs, you can take your case to a small claims court,[1] where you may receive quick justice without the cost of hiring a lawyer. Before pursuing such a course, however, you should exhaust other possibilities because using a small claims court requires time, patience, and effort. And, you must remember, it is possible that the judge will not rule in your favor. You could lose and also be forced to pay the defendant's court costs. Alternatives to small claims court include consumer hot lines provided by the government or other organizations, and consumer advocates who can often be found through local media (television or radio stations and newspapers). If, after considering your other choices, you still feel the need for judicial help, you may want to take your case to small claims court.

## What Are Your Rights and Responsibilities?

Although the basic structure of small claims court is much the same from state to state, there are important differences in the specifics, which consumers need to investigate. For example, recently the maximum amount for which a **plaintiff** could sue ranged from as little as $1,000 in Virginia to as much as $10,000 in Tennessee. There would be little point in suing for a loss of $10,000 if the most you could be awarded was $1,000. In some states, defendants are permitted to bring a lawyer, whose knowledge of the law might put the plaintiff at a disadvantage. Both **litigants**—defendant and plaintiff—have a right to appeal in most states. In those states, winning in small claims court at relatively low cost does not guarantee that you won't be taken to a higher court where you will need to hire a lawyer. Another factor that must be considered is the filing fee and the cost of having the court serve papers on the defendant. Depending on the state, these fees can be as high as $50 and $25, respectively. It would make little sense to sue a person for $100 if the cost of bringing the case to court was $75. The best way to obtain this type of information is to call the clerk of your local small claims court. Laws governing small claims court change frequently. Therefore, it is important to make this call even if you believe you know the rules for your local court.

## Cyber consumer
### Online Advice about Recalled Products

Every year hundreds of products are recalled by their manufacturers either voluntarily or under order from the Consumer Product Safety Commission. But how can you find out whether any products you own have been recalled? One way is by periodically reviewing lists of recalls that you can find on the CPSC's Web site at http://www.cpsc.gov. This site is updated daily and allows consumers to use a site search feature to look for specific products that concern them.

### Question for Thought & Discussion #4:
How would you feel about filing a suit in a small claims court? Do you believe you have the skills to be successful?

**Plaintiff**   One who initiates a lawsuit.

**Litigants**   Those people involved in a lawsuit; that is, in the process of litigation.

---

1. Called a *magistrate's court* in some areas and a *court not of record* or *conciliation court* in others.

Information about small claims courts is provided by all states and can be accessed or ordered on line.

## You Have to Watch Out

Complications can arise in small claims court proceedings. In many states, the defendant can automatically and routinely have a case transferred to a regular civil court. In most civil courts, your efforts are useless unless you have an attorney; if a case in which you are the plaintiff is transferred to the civil court, you must incur the expense of an attorney or drop the suit.

Further, a small claims judgment in your favor does not mean you will get full satisfaction for your loss. The judge may order the defendant to pay you $200 on a $350 claim (which, of course, is still $200 more than you started with). *But no matter what the defendant is ordered to pay you, the small claims court does not act as a collection agency.* The judgment merely gives you the legal right to your claim. You may be able to obtain what is called a *writ of execution* from the small claims court if you can show that the defendant is not paying you, but this writ against the defendant's property, bank account, or wages is often ineffective.[2]

Additionally, you must realize that you probably will have to make several trips to the courthouse, and if the court has no evening session in your area, you may miss time from work. Plaintiffs—those bringing the lawsuits—spend between ten and thirty hours on court-related activities, such as filing papers, preparing the case, and so on. Going to court—even small claims court—takes time and energy.

## Does Your Case Make Sense?

Before you go to the cost and bother of filing an application for a grievance to be heard in a small claims court, you should ask yourself if pursuing your case makes sense. From a strictly logical point of view, having suffered a loss is not enough cause, by itself, to file. You need documentary evidence to support your claim. To be successful, you must be able to prove that a legal contract has been broken, that your property or person has been harmed by the other party's negligence, that a legal right you are entitled to has been violated, or that you have purchased a defective good or service. Even when one of these situations exists, there is little reason to sue someone who clearly has no assets and no income with which to pay you. If you have purchased a defective product for cash from someone who sells goods from the trunk of his or her car, there is little reason to sue. You almost certainly will not be paid, even if the court rules in your favor. Judgments made by small claims

---

2. Even if a debt is not collectible now, however, it stays on the records. Thus, if the person who lost the judgment in small claims court and owes you money comes into some assets in the future, you can activate the judgment at that later time.

courts are most likely to be paid only by individuals or firms with a reputation worth defending in the community.

## How These Courts Work

The first thing you do is ask the clerk of the small claims court in your area whether the court can handle your kind of case. For example, some large cities have special courts to handle problems between renters and landlords. While you're at the courthouse, it might be helpful to sit in on a few cases; that will give you an idea of what to expect when your day in court arrives. Then make sure that the court has jurisdiction over the person or business you wish to sue. Usually, the defendant must live, work, or do business in the court's territory. If you're trying to sue an out-of-town firm, you may run into problems. You probably should go to the state government, usually the secretary of state, to find out where the summons should be sent. Remember that the small claims court does not act as a collection agency; if you're filing suit against a firm that no longer is in business, you'll have a very difficult time collecting.

Make absolutely certain that you have the correct business name and address of the company being sued. Frequently, courts require strict accuracy; if you don't abide by that requirement, the suit is thrown out.

Once you file suit, a summons goes out to the defending party, either by registered mail or in the hands of a sheriff, bailiff, marshal, constable, or, sometimes, a private citizen. When a company receives the summons, it may decide to resolve the issue out of court; about one-quarter of all cases for which summonses are issued are settled this way. Many times, however, the defendant company may not even show up for the trial, in which case you stand a good chance of winning by default (but it is usually hard to collect from no-shows).

**Question for Thought & Discussion #5:**
Do you know anyone who has used a small claims court to resolve a consumer complaint? How did it work out?

## Preparing for Trial

How should you prepare for trial? Obviously, if you know a lawyer, seek advice from her or him. In any event, you should have on hand all necessary and pertinent receipts, canceled checks, written estimates, contracts, and any other form of documentary evidence that you can show the judge. Set the entire affair down in chronological order, with supporting evidence, so you can show the judge exactly what happened. Make sure that your dates are accurate; inaccuracies could prejudice the case against you. Make sure you have a copy of the "demand letter"—similar to the one shown in Figure 5.3 later in this chapter—that you sent to the offending party. This document should be no more than two double-spaced, typewritten pages and should clearly summarize the facts as well as your demands on the other party. It is important that you be able to hand this letter to the judge on trial day. It will not only present your version of the story but will also demonstrate your reasonable approach to the situation.

If you are disputing something such as a repair job, you may have to get a third party—generally someone in the same trade—to testify as an "expert." It is often difficult to get persons to testify against others in their own profession, but they may be willing to give written (notarized) statements, which sometimes are considered acceptable evidence. If possible, when you are suing over disputed performance or repairs, bring the physical evidence of your claim into court. If, for example, your neighborhood dry cleaner shrank a wool sweater of yours to a size 3, be sure to show the sweater to the judge.

## What Happens in Court

The judge generally will let you present your case in simple language without the help of a lawyer. In fact, in some states neither the plaintiff nor the defendant may have a lawyer present. You may receive the judge's decision immediately or by notice within a few weeks. In some states, you can appeal the decision, but often the small claims court plaintiff does not have that right. Remember, whatever action you decide to take after the judgment should be weighed against the costs of that action. Your time is not free, and the worry that may be involved in pursuing a lost case further might detract from the potential reward of eventually winning.

If your opponent tries to settle the case out of court, make sure everything is written down so that the settlement can be upheld if the other party later fails to comply. You should sign all written documents and file them with the court so that the agreement can be enforced by the court. It is best to have your opponent appear with you before the judge to outline the settlement terms. Generally, if you win or if you settle out of court, you should be able to get your opponent to pay for the court costs.

# Sources of Private-Sector Consumer Protection

How does the private sector of the economy fit in with the public activities for consumer protection? As you might expect, activity in the private sector has been highly variable and, in many cases, short lived. But in some specific areas, private activities have been significant.

## Organizations That Report on Consumer Products

One of the most successful private forms of consumer protection has been in the area of product testing and reporting by not-for-profit organizations. Although the federal government has only recently begun to test products and to reveal test results in a form that aids consumers in their purchasing decisions, private product-testing and reporting groups have been active for a long time. Consumers Union and Consumers' Research, Inc., are two examples.

## Consumers Union

Consumers Union (**http://www.ConsumerReports.org**), the publisher of *Consumer Reports,* is a nonprofit organization chartered in 1936 under the laws of the state of New York. The objective of Consumers Union has been to bring more useful information into the seller/buyer relationship so that consumers can buy rationally. The first issue of *Consumer Reports,* in May 1936, went to three thousand charter subscribers, who were told about the relative costs and nutritional values of breakfast cereals, the fanciful claims made for Alka Seltzer, the hazards of lead toys, and the best buys in women's stockings, toilet soaps, and toothbrushes. Consumers Union's policy has always been to buy goods in the open market and bring them to its lab for objective testing.

Approximately 3.5 million subscribers and newsstand buyers now read *Consumer Reports* every month to receive advice on purchasing credit, insurance, drugs, and more. A priority of Consumers Union's testing is automobiles: which are the best buys, which are safe, which have good brakes, which have safety defects, and so on. Consumers Union has also published articles on such ecological topics as pesticides, phosphates in detergents, and lead in gasolines. (It also strongly criticizes government agencies when they act against consumer interests.) To avoid possible financial pressure from advertisers, Consumers Union accepts no advertising in its magazine.

> **"We mustn't place too much stock, however, in the 'seals of approval' that appear on numerous products."**

## Consumers' Research, Inc.

Consumers' Research, Inc. (**http://CRmag@aol.com**), which was founded in 1929, publishes *Consumers' Research Magazine,* with a monthly readership of several hundred thousand. It contains product evaluations, ratings of motion pictures and compact discs, and short editorials. In its reports on consumer goods, *Consumers' Research Magazine* strives to use only results from products that were selected at random rather than ones that were specifically prepared for test purposes by manufacturers. Consumers' Research often restricts its reporting to brands or goods that are nationally distributed, whereas Consumers Union sometimes tests brands that are distributed only in specific high-population density localities. It is the policy of Consumers' Research to service its national and international audience rather than give any special attention to products or brands sold in specific geographic areas. Further, *Consumers' Research Magazine* does not give brand names as "best buys" as *Consumer Reports* does. Both publications pride themselves on stressing safety and efficiency in products, and both have identified potentially unsafe items that escaped the government's attention.

**Question for Thought & Discussion #6:**
What do you think might happen if *Consumer Reports* and *Consumers' Research Magazine* accepted advertising?

## Other Product-Testing Groups

Other product-testing groups also provide information that consumers can use even though they were formed for other purposes. The American Standards Association (ASA), for example, is a private agency that was organized in 1918 to develop standards and testing methods for manufacturers. By setting a uniform level of performance, these standards and testing methods can protect manufacturers against unfair competition. At the same time, by ensuring the safety of products, the standards also provide protection to the consumers who buy products. Using the standards developed by the ASA, other private laboratories or testing groups may certify the efficiency and/or safety of such items as electrical and gas appliances, textiles, and many other products. In addition to product testing at the manufacturing level, a wide range of product testing is conducted by retailers who are eager to perform a consumer service and to provide themselves with a competitive advantage. We mustn't place too much stock, however, in the "seals of approval" that appear on numerous products. Their value depends on the organization that issues them. An alternative is to review information available at the Web site (**http://www.nist.gov**) of the National Institute of Standards and Technology (NIST), which provides consumers with similar but unbiased reports.

## Better Business Bureaus

Local private agencies are also active participants in the public area of consumer protection. Probably the best known of these private agencies is the Better Business Bureau (**http://www.bbb.org**), which is business supported. The National Better Business Bureau has been in existence since 1916 and has local affiliates in all major cities and most counties. Its purposes are:

1. To provide information to consumers on the products and selling practices of businesses.
2. To provide businesspeople with a source of localized standards for acceptable business practices.
3. To provide a technique for mediating grievances between consumers and sellers.

Because the Better Business Bureau has no enforcement powers, all actions must be voluntary. And because the Better Business Bureau is dependent on the business community for its membership, it cannot afford to antagonize that community. The weaknesses in the voluntary system were felt most strongly when the consumer movement began to press for protection not only against fly-by-night, illegal, fraudulent firms, but also against marketing practices that were generally accepted by the business community. Once consumers began to seek redress for damages suffered from exaggerated advertising, ineffective warranties and guarantees, safety hazards, and poor consumer choices due to market structure, the private business organization was unable to police its members effectively. Nevertheless, the Better Business Bureau continues to thrive as it seeks to improve communication with consumers. For example, the Better Business Bureau's arbitration program has been expanding in an attempt to deal more formally with the issue of consumer redress for grievances against sellers and producers of goods and services.

Don't get the impression, though, that the Better Business Bureau in your community is a truly effective consumer agency. In many communities, the Better Business Bureau simply keeps files of consumer complaints about businesses.

## Role of the Media

Although the Better Business Bureau is the oldest of the private agencies that seek to mediate grievances, it is by no means the only one. Newspapers, radio stations, and TV stations have all been in the forefront of the movement to help consumers who have legitimate complaints by providing column space or air time for consumer action, and they have been highly successful in obtaining results for consumers able to make use of them. Affiliates of commercial networks have run regular consumer reports and consumer action series, as have public television stations. These programs typically use publicity as a powerful weapon to resolve consumer grievances.

# Private Industry's Self-Regulation

Many industry associations offer help that can protect consumers in a variety of ways. They often set standards for an industry's products or services, publish these standards, and work to enforce them on their members. Another growing trend is for industry associations to establish formal methods that consumers and businesses may use to resolve disagreements. Depending on the industry, responsibility for dispute resolution may rest with a trade association, a service council, or a consumer action program. The National Automobile Dealers Association (NADA), for example, has established the Automotive Consumer Action Panels (AUTOCAP) system to resolve disputes between consumers and automobile dealers. This organization may be reached on the Internet at **http://www.nada.com**. In a similar way, the Major Appliance Consumer Action Panel (MACAP) was established by the Association of Home Appliance Manufacturers (AHAM) to resolve consumer disputes. This organization does not maintain a Web site but may be reached at (312) 984-5858 or at 20 N. Wacker Drive, Suite 1231, Chicago, IL 60606. Other consumer action panels can be located by conducting an Internet search using the name of the product and *consumer protection* as key words.

Most major corporations also have a consumer relations department and often maintain both a Web site and a consumer hot line that may be used by consumers who have a complaint or need information about one of the firm's products. The increasing willingness of industries to regulate themselves may be a result of the government's increased efforts to protect consumers. Apparently, businesses prefer self-regulation to government regulation.

> "The increasing willingness of industries to regulate themselves may be a result of the government's expanded efforts to protect consumers."

# Warranties and Consumer Protection

Many times consumers buy equipment that turns out to be defective and needs repair or replacement. Of course, we are all used to buying products bearing labels that promise a "money-back guarantee" or "full satisfaction guaranteed." But such guarantees may not be worth the paper on which they are printed. In the early 1970s, a survey by the Major Appliance Consumer Action Panel revealed that many warranties did not state the name and address of the warrantor, did not mention the product or part covered, did not indicate the length of the warranty, and did not indicate what the warrantor would actually do and who would pay for it. Instead, they presented the coverage in "legalese" that was difficult for the average consumer to understand.

## Magnuson-Moss Warranty Act

The Magnuson-Moss Warranty–Federal Trade Commission Improvement Act of 1975 closed many of the loopholes that manufacturers had included in their warranties. The act does not require that a manufacturer provide a written warranty or a guarantee; but if a warranty is offered, it has to comply with the following legal provisions:

1. Any warranty on a product that costs $15 or more must include a simple, complete, and conspicuous statement of the name and address of the warrantor, a description of what is covered and for how much, a step-by-step procedure for filing warranty claims, an explanation of how disputes between the parties will be settled, and the warranty's duration. This must be available to the consumer as prepurchase information.

2. Manufacturers cannot require as a condition of the warranty that the buyer of the product use it only in connection with other products or services that are identified by brand or corporate name. In other words, the maker of a flashlight cannot require that the purchaser use only Duracell batteries in that flashlight for the warranty to be effective.

**Question for Thought & Discussion #7:**
When you purchase consumer goods, how carefully do you read their warranties? Do warranties influence your spending decisions?

**Figure 5.1** Example of a Limited Warranty

### Apple Limited Warranty - iPod

**WARRANTY COVERAGE**
Apple's warranty obligations for the iPod are limited to the terms set forth below:

Apple Computer, Inc. ("Apple") warrants the iPod product against defects in materials and workmanship for a period of one (1) year from the date of original purchase ("Warranty Period").

If a defect arises and a valid claim is received by Apple within the Warranty Period, at its option, Apple will (1) repair the product at no charge, using new or refurbished replacement parts, (2) exchange the product with a product that is new or which has been manufactured from new or serviceable used parts and is at least functionally equivalent to the original product, or (3) refund the purchase price of the product.

If a defect arises and a valid claim is received by Apple after the first one hundred and eighty (180) days of the Warranty Period, a shipping and handling charge will apply to any repair or exchange of the product undertaken by Apple.

Apple warrants replacement products or parts provided under this warranty against defects in materials and workmanship from the date of the replacement or repair for ninety (90) days or for the remaining portion of the original product's warranty, whichever provides longer coverage for you. When a product or part is exchanged, any replacement item becomes your property and the replaced item becomes Apple's property. When a refund is given, your product becomes Apple's property.

**EXCLUSIONS AND LIMITATIONS**

This Limited Warranty applies only to the iPod product manufactured by or for Apple that can be identified by the "iPod" trademark, trade name, or logo affixed to it. This Limited Warranty does not apply to any non-Apple hardware product or any software, even if packaged or sold with the iPod product. Non-Apple manufacturers, suppliers, or publishers may provide a separate warranty for their own products packaged with the iPod product.

Software distributed by Apple under the Apple brand name is not covered under this Limited Warranty. Refer to Apple's Software License Agreement for more information.

Apple is not liable for any damage to or loss of any programs, data, or other information stored on any media contained within the iPod product, or any non-Apple product or part not covered by this warranty. Recovery or reinstallation of programs, data or other information is not covered under this Limited Warranty.

This warranty does not apply: (a) to damage caused by accident, abuse, misuse, misapplication, or non-Apple products; (b) to damage caused by service performed by anyone other than Apple; (c) to a product or a part that has been modified without the written permission of Apple; or (d) if any Apple serial number has been removed or defaced.

TO THE MAXIMUM EXTENT PERMITTED BY LAW, THIS WARRANTY AND THE REMEDIES SET FORTH ABOVE ARE EXCLUSIVE AND IN LIEU OF ALL OTHER WARRANTIES, REMEDIES AND CONDITIONS, WHETHER ORAL OR WRITTEN, EXPRESS OR IMPLIED. APPLE SPECIFICALLY DISCLAIMS ANY AND ALL IMPLIED WARRANTIES, INCLUDING, WITHOUT LIMITATION, WARRANTIES OF MERCHANTABILITY AND FITNESS FOR A PARTICULAR PURPOSE. IF APPLE CANNOT LAWFULLY DISCLAIM OR EXCLUDE IMPLIED WARRANTIES UNDER APPLICABLE LAW, THEN TO THE EXTENT POSSIBLE ANY CLAIMS UNDER SUCH IMPLIED WARRANTIES SHALL EXPIRE ON EXPIRATION OF THE WARRANTY PERIOD. No Apple reseller, agent, or employee is authorized to make any modification, extension, or addition to this warranty.

TO THE MAXIMUM EXTENT PERMITTED BY LAW, APPLE IS NOT RESPONSIBLE FOR DIRECT, SPECIAL, INCIDENTAL OR CONSEQUENTIAL DAMAGES RESULTING FROM ANY BREACH OF WARRANTY OR CONDITION, OR UNDER ANY OTHER LEGAL THEORY, INCLUDING ANY COSTS OF RECOVERING OR REPRODUCING ANY PROGRAM OR DATA STORED IN OR USED WITH THE APPLE PRODUCT, AND ANY FAILURE TO MAINTAIN THE CONFIDENTIALITY OF DATA STORED ON THE PRODUCT. APPLE SPECIFICALLY DOES NOT REPRESENT THAT IT WILL BE ABLE TO REPAIR ANY PRODUCT UNDER THIS WARRANTY OR MAKE A PRODUCT EXCHANGE WITHOUT RISK TO OR LOSS OF PROGRAMS OR DATA.

FOR CONSUMERS WHO HAVE THE BENEFIT OF CONSUMER PROTECTION LAWS OR REGULATIONS IN THEIR COUNTRY OF PURCHASE OR, IF DIFFERENT, THEIR COUNTRY OF RESIDENCE, THE BENEFITS CONFERRED BY THIS WARRANTY ARE IN ADDITION TO ALL RIGHTS AND REMEDIES CONVEYED BY SUCH CONSUMER PROTECTION LAWS AND REGULATIONS. TO THE EXTENT THAT LIABILITY UNDER SUCH CONSUMER PROTECTION LAWS AND REGULATIONS MAY BE LIMITED, APPLE'S LIABILITY IS LIMITED, AT ITS SOLE OPTION TO REPLACEMENT OR REPAIR OF THE PRODUCT OR SUPPLY OF THE REPAIR SERVICE AGAIN.

**OBTAINING WARRANTY SERVICE**
Please review the online help resources referred to in the accompanying documentation before seeking warranty service. If the product is still not functioning properly after making use of these resources, access the online website: www.apple.com/support for instructions on how to obtain warranty service.

**Note: Before you deliver your product for warranty service it is your responsibility to backup all data, including all software programs. You will be responsible for reinstalling all data. Data recovery is not included in the warranty service and Apple is not responsible for data that may be lost or damaged during transit or a repair.**

# Full versus Limited Warranties

If a warranty meets minimum federal standards, it can be designated a *full warranty*. If it doesn't, it must be explicitly designated a *limited warranty* and state how it is limited. An example of a limited warranty is given in Figure 5.1. Under a full warranty, the consumer merely informs the warrantor that the product is defective, does not work properly, or doesn't conform to the written warranty. The warrantor must then fix the product within a reasonable time and without charge. In fact, to obtain the designation "full warranty," the warrantor must pay the consumer for all incidental expenses if there are unreasonable delays or other problems in getting the warranty honored.

Further, the Federal Trade Commission now has the power to set a limit on the number of unsuccessful repair attempts possible under a full warranty. If, after a reasonable number of repairs, the product is still defective, the customer can choose between a refund or a replacement. The replacement must be made free of charge. If the refund option is chosen, the warrantor can deduct an amount for "reasonable depreciation based on actual use."

Full warranties apply to both initial purchasers and to those who buy the product secondhand during the warranty period.

# Settling Disputes

Under the current law, consumers who are dissatisfied with what the warrantor has done for them must try an informal settlement procedure first. Then, if still dissatisfied, the consumer may sue and is entitled to the recovery of purchase costs, damages, and attorneys' fees if the suit is won.

If many consumers feel they have been victimized by a fraudulent warranty, they may engage in a federal class-action suit. At least one hundred consumers with a minimum claim of $25 each must be involved, and the total amount in controversy must be at least $50,000.

# Implied Warranties

Up to this point, we have been discussing what are called *express warranties* because they are *expressly* pointed out in some written document, such as on a label or a card enclosed with an instruction booklet for a consumer durable. In addition, the law derives several types of *implied warranties* by implication or inference from the nature of the transaction. The Magnuson-Moss Warranty Act does not cover implied warranties. They are created according to the Uniform Commercial

Code, which is the code of law governing sales of products in the United States. There are basically two types of implied warranties: one of merchantability, the other of fitness.

## IMPLIED WARRANTY OF MERCHANTABILITY

An **implied warranty of merchantability** arises in every sale of goods made by a merchant who deals in goods of the kind sold. Thus, a retailer of ski equipment makes an implied warranty of merchantability every time she or he sells a pair of skis, but a neighbor selling skis at a garage sale does not.

Goods that are merchantable are "reasonably fit for the ordinary purposes for which such goods are used." They must be of at least average, fair, or medium-grade quality—not the finest quality and not the worst. The quality must be comparable to quality that will pass without objection in the trade or market for goods of the same description. Some examples of nonmerchantable goods include light bulbs that explode when switched on, pajamas that burst into flames on slight contact with a stove burner, high-heeled shoes that break under normal use, and shotgun shells that explode prematurely.

The implied warranty of merchantability imposes *absolute* liability for the safe performance of their products on merchants when dealing in their line of goods. It makes no difference that the merchant might not have known of a defect or could not have discovered it.

## IMPLIED WARRANTY OF FITNESS

An **implied warranty of fitness** for a particular purpose is made whenever any seller (merchant or nonmerchant) knows the particular purpose for which a buyer will use the goods and knows that the buyer has relied on the seller's skill and judgment to select suitable goods.

A "particular purpose of the buyer" differs from the concept of merchantability. Goods can be *merchantable* but still not fit the buyer's particular purpose. For example, house paints suitable for ordinary walls are not suitable for painting stucco walls. A contract can include both a warranty of merchantability and a warranty of fitness for a particular purpose that relates to the specific use or special situation in which a buyer intends to use the goods. For example, a seller recommends a particular pair of shoes, *knowing* that a customer is looking for mountain-climbing shoes. The buyer purchases the shoes *relying* on the seller's judgment. If the shoes are found to be suitable only for walking and not for mountain climbing, the seller has breached the warranty of fitness for a particular purpose.

A seller does not need "actual knowledge" of the buyer's particular purpose. It is sufficient if a seller "has reason to know" the purpose. The buyer, however, must have relied on the seller's skill or judgment in selecting or furnishing suitable goods for an implied warranty to be created. For example, Judy Josephs buys a short-wave radio from Sam's Electronics, telling the salesperson that she wants a set strong enough to pick up Radio Luxembourg. Sam's Electronics sells Judy Josephs a model XYZ set. The set works, but it will not pick up Radio Luxembourg. Judy wants her money back. Here, since Sam's Electronics is guilty of a breach of implied warranty of fitness for the buyer's particular purpose, Judy will be able to get her money back. The salesperson knew specifically that she wanted a set that would pick up Radio Luxembourg. Furthermore, Judy relied on the salesperson to furnish a radio that would fulfill this purpose. Sam's Electronics did not do so. Therefore, the warranty was breached.

**Implied Warranty of Merchantability** An implicit promise by a seller that an item is reasonably fit for the general purpose for which it is sold.

**Implied Warranty of Fitness** An implicit warranty of fitness for a particular purpose, meaning that a seller guarantees the product for the specific purpose for which a buyer will use the goods, when the seller is offering his or her skill and judgment as to suitable selection of the right products.

# Confronting Consumer Issues: Finding Help for Consumer Problems

**M**any organizations are designed to help U.S. consumers protect their rights. These organizations will do you little good, however, if you do not know how to use them or that they even exist. Knowing what kinds of services are available is the first step toward taking advantage of consumer service agencies. Generally, government agencies and private voluntary and business groups provide the following four types of consumer services:

1. *Information to consumers before a purchase is made.* This service includes standard setting, inspection, investigation of marketing techniques, product testing, labeling and other disclosure legislation, publication of results, and formal teaching.

2. *Aid to consumers after a purchase is made, generally through the enforcement of public policy to prevent repeated unsatisfactory or fraudulent practices.* This service includes accepting complaints, investigating the complaints, possibly instituting legal proceedings followed by a judgment, and imposing either an **injunction** against the action or a penalty for breaking the law. This kind of action does not help individual consumers make up their own losses.

3. *Redress to individual consumers for their individual losses as a result of purchases.* This involves a complaint, an investigation, possibly publicity or mediation, sometimes a settlement, or a legal action followed by a judgment and enforcement of it.

4. *Representation of consumers in issues of consumer interest before legislative bodies, government administrative agencies, and private business organizations.* Here again, this generally concerns a complaint, investigation and research regarding the complaint and the problem it reflects, and the subsequent development and publication of plans for its remedy. Often this results in changes in legislation.

## BE PREPARED

Obviously, the best way to deal with consumer problems is not to have them in the first place. Although this may be an impossible goal, you can take a large step toward it by having the following consumer resources

> **Question for Thought & Discussion #8:**
> When you make important consumer decisions, what steps do you take to protect your consumer rights?

immediately available in your home. Much of this help is available online.

1. *Consumer Reports.* The dollars you pay for a subscription to this publication, which was discussed earlier in this chapter, will be well spent.

2. *Consumer Action Handbook.* This booklet is prepared and updated periodically by the U.S. General Services Administration (GSA). A hard-copy version can be ordered from the Federal Citizen Information Center, Department 532-G, Pueblo, CO 81009, and it can be viewed online at **http://www.pueblo.gas.gov**. This handbook lists addresses, phone numbers, Web sites, and names of staff members of hundreds of federal and state organizations, corporate consumer relations agencies, private consumer action organizations, and arbitration and dispute-settling groups. It is an invaluable guide to the resources available to anyone who needs information about consumer assistance of any kind. For example, if you wish to call a local Better Business Bureau to check on the reputation of a firm you may want

**Reading publications such as these can help you keep up with consumer issues and protect your rights.**

to do business with, this *Handbook* has a complete listing of the addresses, phone numbers, and Web sites of the more than 170 bureaus throughout the United States. And if you need help in solving a dispute with a seller, you can locate the appropriate agency or organization to call within seconds.

3. ***Consumer Information Catalog.*** This free publication, which is also available online at the Federal Citizen's Information Center Web site, lists hundreds of free or next-to-free federal publications of interest to consumers. Food, nutrition, health, quackery, mortgages, credit, automobiles, travel, education—these are some of the areas covered by these federal publications.

By having this small, but effective, library at your fingertips—in addition to a telephone book—you will find it much easier to be an "informed consumer" and to avoid many potential pitfalls when purchasing consumer goods or services.

## WHEN YOU DO HAVE A PROBLEM

Whenever you have a complaint about a product or service you have purchased, the logical person to contact first is the seller from whom you bought it. And for this you must be prepared.

**A Strategy**   First, you should always try to figure out a strategy for settling your grievances without going to an outside party. After all, it takes additional time and effort to get a third party involved in your disputes with a seller. Thus, whenever you buy anything, *keep a receipt* if you are worried that there may be problems later. But even before you make the purchase, be certain to *have everything put in writing* about any take-back provisions, warranties, or guarantees.

Figure 5.2 shows the chain of complaint. Imagine, for example, that you buy something and it falls apart a week later. A friendly, polite call to the store and a talk with the salesperson will tell you immediately whether you will have problems. Many times, reputable stores will either give you an identical article that is in good working condition, repair the one you have, or refund your money. If the salesperson does not agree, then seek out the manager or the owner. If you still do not get satisfaction and if you are dealing with a nationally advertised product or with a large chain store, you may want to write the president or the chairperson of the board directly to complain.[3]

Figure 5.3 shows an example of an appropriate letter of complaint. Personal letters to the presidents of large companies get quick responses surprisingly often. But sometimes they do not. If your letter-writing effort fails, what is your next recourse?

**Figure 5.2** Chain of Complaint

When you purchase a product, save the sales receipt and write down the salesperson's name. Then, if the product proves to be defective, you should:

1. Visit or call the store and speak to the salesperson to try to resolve the problem.
2. If you are unsuccessful in obtaining a refund, repair, or replacement from the salesperson, speak to the manager to try to reach a satisfactory resolution.
3. If unsuccessful with the manager, take the following three steps simultaneously:
   a. When the product in question is nationally advertised or was purchased from a large retail chain, visit the firm's Internet Web site to file a complaint and find the name of the person in charge of consumer relations. Contact this person in a letter or by e-mail to explain the problem and request a solution.
   b. Contact local and state consumer groups about your problem, enclosing a copy of the letter you sent to the firm's director of consumer relations.
   c. Contact national consumer assistance agencies and consumer action panels such as AUTOCAP.
4. If you do not receive satisfaction, lodge a complaint with your state attorney general's office.

**Seeking Help**   When you need consumer assistance, it is best to work first at the local level. Is there a consumer affairs agency or office in your local government? The telephone book is probably your nearest source of this information; in most major cities today, a Yellow Pages entry under "Consumers" lists the major public agencies that provide consumer services. A call to the administrative officer of your county or city should also give you information on the availability of public consumer services.

If you do not find a local consumer agency, look under the state listings; if no listing there looks promising, call the state attorney general's office. That office works closely with consumer agencies because much of the consumer fraud uncovered by consumer agencies is prosecuted through the attorney general's office. Table 5.4 lists the addresses of state consumer protection agencies. The home page of the National Association of Attorneys General (NAAG, **http://www. naag.org**) provides Internet links to the office of the attorney general in each state.

---

3. In your local library, you can consult the *Consumer's Register of American Business* and the *Directory of Foreign Manufacturers in the United States.*

*(Continued on next page)*

## Confronting Consumer Issues (Continued)

### Figure 5.3 How to Lodge a Complaint

> Your Address
> Date
>
> Addressee
> Company Name
> Street Address
> City, State, Zip Code
>
> Dear Sir or Madam:
>
> I am writing this letter to inform you of my dissatisfaction with [name of product with serial number or the service performed], which I purchased [the date and location of purchase].
>
> My complaint concerns [the reason(s) for your complaint]. I believe that in all fairness you should [the specific action you desire for satisfaction] in order to resolve this problem.
>
> I sincerely look forward to your reply and a speedy resolution to my complaint. I will allow two weeks before referring this complaint to the appropriate consumer agency.
>
> Yours truly,
>
> Your Name
> Enclosures (include copies, not originals, of all related records)

**Private Organizations**  You might also want to contact a private consumer action group at the local level. Many local newspapers and radio and television stations in the United States today have an "action line" or "hot line" service to help consumers. These groups often can be extremely effective in bringing about a rapid and satisfactory resolution to consumer problems. To locate such services, check with your local newspapers, radio and TV stations, or library. There may be other local organizations available also. You doubtless will have heard of any successful ones, which probably will be listed in the telephone book. You might even consider joining one of these groups if you want to prevent a recurrence of your problem.

**Legal Assistance**  Getting your money back may depend on private legal action. If you have to go beyond the small claims court—which we discussed earlier in this chapter—to a higher court, you must be prepared to pay legal fees. But even if you can't pay them, don't give up; in many cities, the traditional legal aid society has been augmented by special legal services for low-income families, and these organizations often place strong emphasis on consumer problems. In some states, too, group legal practices have been approved, and you might obtain help through your union or some other organization that has contracted with such a service. Some colleges offer legal services that enable students to pursue solutions to their consumer problems.

A number of legal self-help and consumer advocacy groups exist that may be able to help you. For information relating to self-help, shopping for attorneys, and court procedures and requirements, you can write to HALT, 1612 K Street NW, Suite 510, Washington, DC 20006, or go to **http://www.halt.org**.

## SPECIFIC INDUSTRY AGENCIES: THIRD PARTY INTERVENTION

An increasingly popular way to resolve consumer problems is to allow third parties to settle a dispute between you and the seller of a product or service. Local agencies, trade associations, and individual companies are setting up mediation procedures, most of which are free to consumers. There are basically three types of third party interveners:

1. *A conciliator,* who simply brings together the parties in a dispute to get them to resolve their differences.

2. *A mediator,* who is in a stronger position and can make nonbinding recommendations and proposals.

> ### Question for Thought & Discussion #9:
> Have you ever filed a complaint with a manufacturer about a defective product? How did you do this, and what were the results?

3. *An arbitrator,* who makes a decision that can be binding on some or all of the parties. A binding decision can be enforced in a court of law.

Obviously, from the consumer's point of view, the best of all possible worlds is a dispute-settling process that is binding on the business but not on the consumer. This may sound unrealistic, but the arbitration procedures, for example, of Ford Motor Company's Consumer Appeals Board (HOW) operate that way.

**Where to Go**  Where you go depends on the kind of problem you have. If you are having a dispute with a local company, going to a local agency makes sense. Approximately

| TABLE 5.4 | State Consumer Protection Agencies |
|---|---|

**Alabama**
Assistant Attorney General
Office of the Attorney General
Consumer Affairs Section
11 South Union St.
Montgomery, AL 36103
(334) 242-7335
(800) 392-5658 (in state)
http://www.ago.state.al.us

**Alaska**
Consumer Protection Unit
Office of the Attorney General
1031 West 4th Ave., Suite 200
Anchorage, AK 99501-5903
(907) 269-5100
http://www.law.state.ak.us

**Arizona**
Chief Counsel Consumer
Protection and Advocacy Section
Office of the Attorney General
1275 West Washington St.
Phoenix, AZ 85007
(602) 542-3702
(800) 352-8431 (in state)
http://www.ag.state.az.us

**Arkansas**
Deputy Attorney
General Consumer Protection Division
Office of the Attorney General
323 Center St., Suite 200
Little Rock, AR 72201
(501) 682-2007
http://www.ag.state.ar.us

**California**
Director
California Department of
Consumer Affairs
400 R St., Suite 3000
Sacramento, CA 95814
(916) 445-4465
(800) 852-5210 (in state)
http://www.dca.ca.gov

**Colorado**
Consumer Protection Division
Colorado Attorney General's Office
1525 Sherman St., 5th Floor
Denver, CO 80203-1760
(303) 866-5079
(800) 222-4444

**Connecticut**
Commissioner
Department of Consumer Protection
165 Capitol Ave.
Hartford, CT 06106

(860) 713-6300
(800) 842-2649 (in state)
http://www.state.ct.us/dcp

**Delaware**
Director
Consumer Protection Unit
Office of Attorney General
820 North French St., 5th Floor
Wilmington, DE 19801
(302) 577-8600
(800) 220-5424 (in state)
http://www.state.de.us/attgen

**District of Columbia**
Senior Counsel
Office of the Corporation Counsel
441 4th St. NW, Suite 450-N
Washington, DC 20001
(202) 442-9828

**Florida**
Director of Division Consumer Services
Terry L. Rhoads Building
2005 Apalachee
Tallahassee, FL 32399
(850) 922-2966
(800) 435-7352 (in state)
http://www.800helpfla.com

**Georgia**
Governor's Office of Consumer Affairs
2 Martin Luther King, Jr. Dr., Suite 356
Atlanta, GA 30334
(404) 656-2790
(800) 869-123
http://www2.state.go.us/gaoca

**Hawaii**
Executive Director
Office of Consumer Protection
Department of Commerce and
Consumer Affairs
235 South Beretania St., Room 801
Honolulu, HI 96813
(808) 586-2636

**Idaho**
Deputy Attorney
General Consumer Protection Unit
Idaho Attorney General's Office
650 West State St.
Boise, ID 83729-0010
(208) 334-2424
(800) 432-3545 (in state)

**Illinois**
Chief Consumer Protection Division of
the Attorney General's Office
100 West Randolph, 12th Floor
Chicago, IL 60601

(312) 793-2852
http://www.ag.state.il.us

**Indiana**
Chief Counsel and Director
Consumer Protection Division
Office of the Attorney General
Indiana Government Center South
401 W. Washington St., 5th Floor
Indianapolis, IN 46204
(317) 232-6201
(800) 382-5516 (in state)
http://www.in.gov/attorneygeneral

**Iowa**
Assistant Attorney General
Consumer Protection Division
Office of the Attorney General
1300 East Walnut St., 2nd Floor
Des Moines, IA 50319
(515) 281-5926
http://www.iowaAttorneyGeneral.org

**Kansas**
Deputy Attorney General
Consumer Protection Division
Office of the Attorney General
120 SW 10th, 4th Floor
Topeka, KS 66612-1597
(785) 296-3751
(800) 432-2310 (in state)
http://www.ink.org/public/ksag

**Kentucky**
Director
Consumer Protection Division
Office of the Attorney General
1024 Capital Center Dr.
Frankfort, KY 40601
(502) 696-5389
(888) 432-9527 (in state)
http://www.kyattorneygeneral.com/cp

**Louisiana**
Chief Consumer Protection Section
Office of the Attorney General
301 Main St., Suite 1250
Baton Rouge, LA 70801
(800) 351-4889
http://www.ag.state.la.us

**Maine**
Division Chief
Public Protection Division
Office of the Attorney General
6 State House Station
Augusta, ME 04333-0006
(207) 626-8800
http://www.state.me.us/ag

(Continued on next page)

## TABLE 5.4 State Consumer Protection Agencies (Continued)

**Maryland**
Chief Consumer Protection Division
Office of the Attorney General
200 Saint Paul Place, 16th Floor
Baltimore, MD 21202-2021
(410) 576-6550
http://www.oag.state.me.us/consumer

**Massachusetts**
Director
Executive Office of Consumer Affairs and
Business Regulation
10 Park Plaza, Room 5170
Boston, MA 02116
(617) 973-8700
(888) 283-3757 (in state)
http://www.state.ma.us/consumer

**Michigan**
Consumer Protection Division
Office of Attorney General
P. O. Box 30213
Lansing, MI 48909
(517) 373-1140

**Minnesota**
Manager
Consumer Services Division
Minnesota Attorney General's Office
1400 NCL Tower
445 Minnesota St.
St. Paul, MN 55101
(612) 296-3353
(800) 657-3787
http://www.ag.state.mn.us/consumer

**Mississippi**
Director
Consumer Protection Division
Mississippi Attorney General's Office
P. O. Box 22947
(601) 359-4230
(800) 281-4418 (in state)

**Missouri**
Deputy Chief Counsel
Consumer Protection and Trade
Offense Division
P. O. Box 899
1530 Rax Court
Jefferson City, MO 65102
(573) 751-6887
(800) 392-8222 (in state)
http://www.ago.state.mo.us

**Montana**
Chief Legal Counsel
Consumer Affairs Unit
Department of Administration
1424 Ninth Ave.
Box 200501
Helena, MT 59620-0501
(406) 444-4312

**Nebraska**
Assistant Attorney General
Department of Justice
2115 State Capitol
P. O. Box 98920
Lincoln, NE 68509
(402) 471-2682
(800) 727-6432 (in state)
http://www.nol.org/home/ago

**Nevada**
Commissioner
Nevada Consumer Affairs Division
1850 East Sahara, Suite 101
Las Vegas, NV 89104
(701) 486-7355
(800) 326-5202
http://www.fyiconsumer.org

**New Hampshire**
Consumer Protection and
Antitrust Bureau
New Hampshire Attorney General's Office
33 Capitol St.
Concord, NH 03301
(603) 271-3641
http://www.state.nh.us/nhdoj/
Consuerm/cpb.html

**New Jersey**
New Jersey Department of Law and
Public Safety
Division of Consumer Affairs
P. O. Box 45025
Newark, NJ 07101
(973) 504-6200
(800)242-5846
http://www.state.ny.us/lps/ca/home/htm

**New Mexico**
Director
Consumer Protection Division
Office of the Attorney General
P. O. Drawer 1508
407 Galistero
Santa Fe, NM 87504-1508
(505) 827-6060
(800) 678-1508 (in state)
http://www.ago.state.nm.us

**New York**
Bureau Chief
Bureau of Consumer Frauds
and Protection
Office of the Attorney General
State Capital
Albany, NY 12224
(518) 474-5481
(800) 771-7755 (in state)
http://www.oag.state.ny.us

**North Carolina**
Senior Deputy Attorney General

Consumer Protection Division
Office of the Attorney General
P. O. Box 629
Raleigh, NC 27602
(919) 716-6000
http://www.jus.state.nc.us/cpframe.htm

**North Dakota**
Director
Consumer Protection and Antitrust
Division
Office of the Attorney General
600 East Boulevard Ave., Department 125
Bismarck, ND 58505-0040
(701) 328-3404
(800) 472-2600 (in state)
http://www.ag.state.ne.us/
ndag/cpat/cpat.html

**Ohio**
Ohio Consumer's Counsel
77 South High St., 15th Floor
Columbus, OH 43266-0550
(614) 466-8574
(877) 742-5622 (in state)
http://www.state.oh.us/cons

**Oklahoma**
Oklahoma Attorney General
Consumer Protection Unit
4545 N. Lincoln Ave., Suite 260
Oklahoma City, OK 73105
(405) 521-2029
(800) 448-4904
http://www.oag.state.ok.us

**Oregon**
Financial Fraud/Consumer
Protection Section
Department of Justice
1162 Court St. NE
Salem, OR 97310
(503) 378-4732
(877) 877-9392 (in state)
http://www.doj.state.or.us

**Pennsylvania**
Director
Bureau of Consumer Protection
Office of Attorney General
14th Floor, Strawberry Square
Harrisburg, PA 17120
(717) 787-9707
(800) 441-2555 (in state)
http://www.attorneygeneral.gov

**Puerto Rico**
Secretary
Department of Justice
P. O. Box 902192
San Juan, PR 0902
(787) 721-2900

| TABLE 5.4 | State Consumer Protection Agencies (Continued) |
| --- | --- |

**Rhode Island**
Director
Consumer Protection Unit
Department of Attorney General
150 South Main St.
Providence, RI 02903
(401) 274-4400

**South Carolina**
Senior Assistant Attorney General
Office of the Attorney General
P. O. Box 11549
Columbia, SC 29211
(803) 734-3970
http://www.scattorneygeneral.org

**South Dakota**
Directory of Consumer Affairs
Office of the Attorney General
500 East Capitol
State Capitol Building
Pierre, SD 57501-5070
(605) 773-4400
(800) 300-1986 (in state)

**Tennessee**
Director
Division of Consumer Affairs
5th Floor
500 James Robertson Parkway
Nashville, TN 37243-0600
(615) 741-4737
(800) 342-8385 (in state)

**Texas**
Assistant Attorney General and Chief
Consumer Protection Division
Office of Attorney General

P. O. Box 12548
Austin, TX 78711-2548
(512) 463-2070

**Utah**
Director
Division of Consumer Protection
Department of Commerce
160 East 300 South
Box 146704
Salt Lake City, UT 84114-6704
(801) 530-6601
http://www.commerce.state.ut.us

**Vermont**
Chief Public Protection Division
Office of the Attorney General
109 State St.
Montpelier, VT 05609-1001
(802) 828-5507
http://www.state.vt.us/atg

**Virginia**
Senior Assistant Attorney General
Office of the Attorney General
Antitrust and Consumer Litigation Section
900 East Main St.
Richmond, VA 23219
(804) 786-2116
(800) 451-1525
http://www.oag.state.va.us

**Washington**
Consumer Resource Center
Office of the Attorney General
900 Fourth Ave., Suite 2000
Seattle, WA 98164-1012
(206) 454-6684

(800) 551-4636 (in state)
http://www.wa.gov/ago

**West Virginia**
Deputy Attorney General
Consumer Protection Division
Office of the Attorney General
812 Quarrier St., 6th Floor
P. O. Box 1789
Charleston, WV 25326-1789
(304) 558-8986
http://www.state.wv.us/wvag

**Wisconsin**
Administrator
Division of Trade and
Consumer Protection
Department of Agriculture
2811 Agriculture Dr.
P. O. Box 8911
Madison, WI 53708
(608) 224-4953
http://www.datcp.state.wi.us

**Wyoming**
Office of the Attorney General
Consumer Protection Unit
123 State Capitol Building
Cheyenne, WY 82002
(307) 777-7874
(800) 438-5799 (in state)
http://www.attorneygeneral.state.wy.us

170 local Better Business Bureaus across the country provide arbitration. Your county office of consumer affairs will also be able to help. If your dispute is with the manufacturer of a product, then you should find out if there is an arbitration panel for that industry. The names and addresses of such services are given in the chapters of this text as we discuss specific products and services. There are arbitration panels for new automobiles, automobile repair services, new-home construction, major appliances, and so on.

**American Arbitration Association** The American Arbitration Association has been remarkably successful in settling disputes before they reach the courts. If there is a branch of this organization in your area, you may find it useful. For information or assistance, write to Public Relations Director, American Arbitration Association, 335 Madison Avenue, Floor 10, New York, NY 10027, or call (212) 716-5800. Or go to **http://www.adr.org**.

## CONSUMER HOT LINES

For a specific complaint, you may wish to call the consumer hot line that deals with consumer complaints concerning a particular industry or problem. Following is a list of hot line numbers that have been established by industry associations or the government to assist consumers.

● *Advertising:* Federal Trade Commission, Marketing Practices Division, (202) 326-2222 (**http://www.ftc.gov**); Council of Better Business Bureaus, National Advertising Division, (800) 955-5108 (**http://www.bbb.org**).

● *Air safety:* Federal Aviation Administration, (800) FAA-SURE (**http://www.faa.gov**).

● *Airline passenger complaints:* Department of Transportation, (202) 366-4000 (**http://www.dot.gov**).

*(Continued on next page)*

## Confronting Consumer Issues (Continued)

- *Appliances:* Major Appliance Consumer Action Panel (MACAP), (800) 621-0477.
- *Auto problems:* Automotive Consumer Action Panels (AUTOCAP), (703) 821-7144 (**http://www.nada.com**).
- *Auto safety:* National Highway Traffic Safety Administration, (800) 424-9393 (**http://www.nhtsa.dot.gov**).
- *Credit:* Office of Consumer Affairs, (202) 634-4140 (**http://CAffairs@doc.gov**).
- *Fraud:* Federal Trade Commission, Marketing Practices Division, (202) 326-3128 (**http://www.ftc.gov**).
- *Insurance:* Health Insurance Association of America (life and health insurance), (800) 635-1271; Insurance Information Institute (property and liability insurance), (800) 221-4954 (**http://www.hiaa.org**); National Association of Insurance Commissioners, (816) 842-3600 (**http://www.niac.org**).
- *Mail fraud:* Postal Service Inspector, U.S. Postal Service, (800) 275-8777 (**http://www.usps.com**).
- *Mail-order problems:* Federal Trade Commission, Marketing Practices Division, (202) 326-3128 (**http://www.ftc.gov**).
- *Product safety:* Consumer Product Safety Commission, (301) 504-0990 (**http://www.cpsc.gov**).
- *Safety at work:* Occupational Safety and Health Administration, (800) 321-6742 (**http://www.osha.gov**).
- *Stocks and bonds:* Office of Consumer Affairs, Securities and Exchange Commission, (800) 732-0330 (**http://www.sec.gov**).
- *Surgery:* Department of Health and Human Services, (800) 696-6775 (**http://www.dhhs.gov**).
- *Travel:* American Society of Travel Agents, (703) 739-2782 (**http://www.astanet.com**).
- *Unwanted mail:* Direct Marketing Association, (212) 768-7277 (**http://www.thedma.org**).
- *Warranties:* Federal Trade Commission, Marketing Practices Division, (202) 326-2222 (**http://www.ftc.gov**).

## FEDERAL GOVERNMENT

If your problem concerns a product sold nationally or if it affects a large number of people nationwide, you should appeal to a federal agency.

### Food and Drug Administration (500 Fishers Lane, Rockville, MD 20857)

The Food and Drug Administration (**http://www.fda.gov**) has regional offices in many cities, and each office includes a person specifically charged with consumer services. Many of the FDA's employees are technical experts working in specific fields under FDA jurisdiction. Any complaint about a food, drug, or cosmetic that you purchase should be made either to your regional office or to the above address. The agency will ask you for complete information, and its staff is particularly interested in examining the questionable food or drug about which you are complaining or the container in which it was sold. If the agency believes your complaint is justified, a member of the staff will visit the firm in question to observe its production and packaging procedures. If you do not have the product, you still may complain. The FDA is always interested in receiving consumer reports, even though the complaints may have no legal consequence. Through such reports, the FDA often discovers new problems or new incidences of old problems. In those areas in which the FDA sets and/or enforces standards, consumers can play a very important role. But unless the agency hears from consumers, it may be making avoidable mistakes.

### Federal Trade Commission (Washington, DC 20580)

The FDA largely enforces standards of products and performance, but the standards enforced by the Federal Trade commission (**http://www.ftc.gov**) are essentially those that are established for normal competitive businesses. In addition, the FTC handles complaints that are subject to federal credit and federal warranty legislation.

The FTC has regional offices and consumer service representatives in major U.S. cities. It has established a special office to serve consumers, and it provides a wide range of informative pamphlets for them. If you have a complaint for the FTC, you may address it to a regional office or to the Washington, D.C., headquarters. If the FTC believes you have a valid complaint, it will send an investigator out to check with both you and the firm. Typically, the FTC works in two ways. First, it investigates whether a particular seller or advertiser has violated a specific law that the agency enforces; if its findings are positive, it takes actions which may include seeking an injunction against the firm, to stop the practice by the single company. Second, the agency looks for new patterns of practice that may mislead consumers; if it finds such patterns, the FTC may attempt to act against an entire industry, rather than a single firm, to stop the practice altogether.

### U.S. Department of Agriculture (Washington, DC 20250)

Although the U.S. Department of Agriculture (USDA) primarily provides services to farmers, it also protects consumers in very important ways, notably by inspecting and grading meat, poultry, and fish. In recent years, the agency has also become a primary source of information for consumers on the best ways to spend their food dollars. The USDA does this through its Cooperative Extension Service, operated in conjunction with land-grant universities throughout the United States. Any complaint you have about the grades of meat you buy or the quality of poultry that is shipped interstate should be reported to your local health department or your local department of agriculture.

The USDA provides other services, including the Agricultural Research Service, the Animal and Plant Health Inspection Service, the Economic Research Service, the Food and Nutrition Service, the Forest Service, and the Rural Development Service. If you are interested in any of these services, you can get the appropriate telephone numbers and addresses from the Office of Information, USDA, Washington, DC 20250. Or go to **http://www.usda.gov**.

### U.S. Postal Service (Washington, DC 20260)

The U.S. Postal Service (**http://www.usps. gov**) is responsible for investigating mail fraud, unordered merchandise, obscenity in the mails, and other mail-related problems.

### U.S. Department of Housing and Urban Development (Washington, DC 20410)

The Department of Housing and Urban Development (**http://www.hud.gov**) is responsible for numerous federally subsidized housing programs. If you are experiencing problems relating to fair housing, you can call (800) 424-8590 toll free for assistance. This department also regulates interstate land sales. You can contact the department's consumer affairs coordinator if you have any relevant problem.

### Consumer Product Safety Commission (Washington, DC 20207)

If you think there is an unsafe product on the market, or if you have any questions about product hazards and safety, you may want to get in touch directly with the CPSC (**http://www.cpsc.gov**). The number, toll free from anywhere in the United States, is (800) 638-2772. You can also write to the Consumer Product Safety Commission, Washington, DC 20207, and explain your concern about a particular product or products.

### Obtaining More Information on Federal Consumer Services

To obtain more information on the availability of federal consumer services, send for the *Guide to Federal Consumer Services,* publication number (OS) 76-512, available from the Superintendent of Documents, Government Printing Office, Washington, DC 20402. You may wish to contact the Federal Information Center and request information about the vast number of federal agencies and programs. Probably the easiest way to find assistance from the federal government is to visit its unified Web site at **http://www.consumer.gov**.

### Key Term:

**Injunction**  A legal order requiring that an activity be stopped, corrected, or undertaken.

# Chapter Summary

1. Consumer protection is not new. Standards of weights and measures and standards for quality have long existed to benefit both consumers and businesses in the market-place. By 1900 antitrust laws had been enacted that were aimed at preventing firms from using monopoly-like powers to harm consumers or other businesses by fixing prices or controlling the allocation of resources and production and services.

2. The first federal law passed specifically to aid consumers was the Food and Drug Act of 1906. Although other consumer protection laws followed, there was little federal activity in this area after the passage of the Food, Drug and Cosmetic Act of 1938 until the 1960s, when public and government interest was renewed.

3. Strong government support for consumer protection activities in the 1960s and 1970s declined under the administrations of Ronald Reagan and George Bush in the 1980s. President Clinton was generally more supportive, increasing funding for consumer protection agencies. In recent years the Internet has provided easy access to federal help.

4. Two of the most important federal agencies that monitor and enforce consumer protection legislation are the Federal Trade Commission and the Consumer Product Safety Commission.

5. State and local agencies have introduced various measures to help consumers. Local government mediators act in disputes between consumers and sellers and often refer parties to appropriate agencies if laws have been violated. State governments also provide small claims courts through which consumers may pursue their legal rights at relatively low cost.

6. The best-known private agencies are Consumers Union and Consumers' Research, Inc.; both were established primarily to provide information to consumers. In addition, branches of the Better Business Bureau attempt to help consumers as well as businesspeople, but they are not effective consumer protection agencies because they are supported by the businesses they represent.

7. Many industry associations help resolve problems between their member companies and consumers. The media—newspapers, television, and radio—often assist consumers who have legitimate complaints.

8. Legislation passed in 1975 tightened the definition of *warranty* that manufacturers can use. The Uniform Commercial Code stipulates that every good sold by a merchant has an implied warranty of merchantability; some goods may also have an implied warranty of fitness for a particular purpose.

9. Consumer activists seek to participate in the functioning of the marketplace in the same way that large corporations and major trade unions do. This kind of consumer activism is significant because it gives consumers a voice in government.

10. When consumers seek assistance to protect their rights, they generally can rely on government agencies and other consumer-oriented groups to provide four types of services: (1) information made available to consumers before a purchase is made; (2) aid given to consumers after a purchase is made; (3) redress provided to consumers for losses that result from their purchases; and (4) representation of consumers in issues of consumer interest before a legislative body, governmental agency, or private business group.

11. By keeping a few basic consumer references in their homes or accessing them on the Internet, consumers may be better able to avoid problems and to determine what they should do if a difficulty occurs. Consumers should develop a logical strategy to follow when they have a complaint.

12. Third party intervention agencies can be very helpful in settling disputes between consumers and manufacturers.

13. A variety of federal governmental departments and agencies can be contacted for help with consumer complaints. Consumers are free to call the various agencies or use the Internet to reach the federal government's unified Web site.

# Key Terms

antitrust laws **95**
cease-and-desist order **102**
common law **97**

consumer redress **99**
implied warranty of fitness **114**
implied warranty of
   merchantability **114**

injunction **121**
litigant **105**
plaintiff **105**

# Questions For Thought & Discussion

1. Do you believe you could protect your consumer rights without government assistance? Explain why or why not.

2. Do you believe you are a responsible consumer? Describe several events that demonstrate the type of consumer you are.

3. Have you ever purchased a product that you thought was dangerous? What did you do about it?

4. How would you feel about filing a suit in a small claims court? Do you believe you have the skills to be successful?

5. Do you know anyone who has used a small claims court to resolve a consumer complaint? How did it work out?

6. What do you think might happen if *Consumer Reports* and *Consumers' Research Magazine* accepted advertising?

7. When you purchase consumer goods, how carefully do you read their warranties? Do warranties influence your spending decisions?

8. When you make important consumer decisions, what steps do you take to protect your consumer rights?

9. Have you ever filed a complaint with a manufacturer about a defective product? How did you do this, and what were the results?

# Things to Do

1. Call your local Better Business Bureau and ask for its booklet describing the bureau and all of the areas in which it is active. If there is no Better Business Bureau in your area, contact the local chamber of commerce.

2. Call your local television and radio stations and newspaper office or offices to find out if any of them devote time or space to consumer action services.

3. Look at the warranties of any consumer products you have recently purchased or are about to purchase. Do any of them give full warranties? If they have limited warranties, under what conditions can you have the product repaired or replaced?

4. Call one or more of the toll-free hot lines identified in this chapter. Ask for information about the organization's services. Evaluate how helpful the information you receive would be to you or other consumers.

5. Draw up a list of consumer agencies in your area by checking under "consumers" in the Yellow Pages of your local phone book or on the Internet, or by contacting the nearest office of your state attorney general.

6. Identify a particular good or service that you believe should be more closely regulated by the government to protect consumers. Investigate this product, and use the information you find to write an essay that explains why consumers need protection in their use of the product. Suggest the type of protection or regulation that you believe would be appropriate. In your essay, explain the benefits consumers would receive and any costs they would pay as the result of your proposal.

# Internet Resources

## Finding Consumer Information on the Internet

The following Web sites have been selected for their relevance to topics discussed in this chapter. Search these sites to locate information that can add to your knowledge of consumer protection services provided by state governments.

Then complete an Internet search to find the office that provides consumer protection services in your state. Compare what these sites offer. Remember, Web addresses change frequently. If any of these addresses no longer function, find similar sites to investigate using any of the search engines available to you.

1. The New York State Consumer Protection Board, a consumer watchdog agency, is online at http://www.consumer.state.ny.us.

2. The California Department of Consumer Affairs provides information for California consumers at http://www.dca.ca.gov.

3. The Florida Division of Consumer Services offers help to consumers at http://www.800helpfla.com.

## Shopping on the Internet

The following Web sites have been selected because they offer consumers services similar to those described in this chapter. These are commercial sites that are designed to market products. They do not represent a comprehensive or balanced description of all lawyers who offer consumer protection assistance online. Investigate these Web sites to find what services are offered and how they are promoted. Remember, Web addresses change frequently. If any of these addresses no longer function, find similar sites to investigate using any of the search engines available to you.

1. You can find the Law Offices of Jerril J. Krowen, who concentrates in consumer protection law and the hotel and tourist trade, at http://www.krowenlaw.com.

2. Waddell Raponi Lawyers offers services in all types of civil litigation, including consumer protection. Its Web site is at http://www.waddellraponi.com/practice.html.

3. Barrett Law Firm PLLC in Charleston, West Virginia, also offers consumer protection services. You can find it online at http://www.barrettlawfirm.net/biograph.html.

## InfoTrac Exercises

Purchasers of new copies of this text are provided with access to the InfoTrac Web site. This Web site links students to thousands of recent articles published in hundreds of periodicals. Use the key words **small claims court** or other terms from this chapter to conduct a key-word search. Choose one article that is of particular interest to you and write a brief essay describing what you have learned from the article. Be sure to cite the author and title of the article and the name and date of the publication in which it appeared.

# Selected Readings

*Antitrust Enforcement and the Consumer.* Available from the Department of Justice, 666 11th St., N.W., Room 910, Washington, DC 20530.

Bowden, Bill. "60,000 Wal-Mart Grills Recalled." *Arkansas Business,* September 9, 2002, p. 12.

Cohen, Lizabeth. *A Consumer's Republic.* New York: Knopf, 2003.

Consumers Union. *Consumer Reports Annual Buying Guide.* Published annually in December.

"FBI Pursues Online Bad Guys." *TechWeb,* May 25, 2001, p. 3.

*How to Complain and Get Results.* Available from the Federal Trade Commission, 6th St. and Pennsylvania Ave. N.W., Washington, DC 20580. http://www.ftc.gov.

Kandra, Anne. "Solve Big Hassles in SEC." *PC World,* August 2002, pp. 41–44.

Mader, Robert. "CPSC Recalls 35 Million O-Ring Fire Sprinklers." *Contractor,* August 2001, p. 1.

Nallin, Judith. "Consumer Protection—Magnuson-Moss Warranty Act." *New Jersey Law Journal,* April 1, 2002, pp. 71–74.

"Recall Resources." *Newsweek,* May 17, 1999, p. 90.

"Staking Small Claims." *Business Week,* January 31, 2000, p. F28.

*Warranties.* Available from the Federal Trade Commission, 6th St. and Pennsylvania Ave. N.W., Washington, DC 20580. http://www.ftc.gov.

Photodisc/Getty Images

## CHAPTER 6

# The Consumer as a Wage Earner

After reading this chapter, you should be able to answer the following questions:

- How can you invest in yourself?
- What factors determine the value of your labor and how much you earn?
- What rewards might you receive for going to school?
- Why do some workers earn more than others?
- How does inflation affect the purchasing power of our earnings?
- How do we measure the rate of inflation?
- What steps can consumers take to identify and pursue their career objectives?

**Human Capital**  The skills and abilities humans have that allow them to produce goods and services from other productive resources.

**Investment**  The act of giving up something of value at present to be able to receive something else of greater value in the future.

In the U.S. economic system, the amount people earn depends largely on the value of their contribution to the production of goods and services. The value of this contribution depends in part on an individual's basic intelligence, competence, and aptitude, and in part on the demand for the type of labor that the individual can provide. To achieve the highest possible productivity and income, people need to develop their personal aptitudes into highly demanded skills through education, training, and job experience. What we are talking about here is **human capital,** or the ability of a person to produce goods and services from other productive resources. Developing human capital is the key to increasing income.

## Investment in Human Capital

Few would consider it unfair to pay a person with a college degree more than one with only a grade school education. The better-educated person's labor is expected to be more valuable to an employer than that of the person with less education. This basic fact of life is probably one reason why you are making an investment in yourself by pursuing your education.

One way to define **investment** is to say it is the act of giving up the use of something of value now so that you may receive something else of greater value in the future. When you invest in a business, for example, you allocate, say, $10,000 to buy corporate stock in the hope that your stock will be worth more in the future. Thus, going to school is a type of investment in yourself—in your own human capital. By paying tuition and giving up income you could earn now, you hope to increase your future income. Usually, the longer you go to school, the more knowledge you will obtain, and the better your thinking ability will become. You may find it easier to solve problems, to direct and motivate other people, to organize, and to plan. Employers will be willing to pay you more because they believe your labor will have greater value to them than if you had not received those additional years of education. This is, at least, the theory of the labor market. Exceptions always exist.

Of course, education does not automatically guarantee you a higher income. To be marketable, your skills must be in areas of high demand. No amount of obscure services you could supply would induce others to hire you at high wages; what determines the individual's wages or income is the supply and demand for different types of labor.

Ultimately, then, your investment in your human capital requires careful planning. It should increase your productive capacities in areas that the economy demands. For example, it appears that very few people will own and use manual typewriters in the future. Most people who have reason to type letters or reports will use computers. It makes more sense, then, to study computer maintenance and repair than the repair of manual typewriters.

So choosing the right occupation may require that you become informed not only about current marketable job skills but also about future demands for different types of jobs. Just as you want to be careful about investing cash that you have saved, you also should be careful about investing time and effort in the development of your own human capital. You might want to seek the aid of a career counseling service, either on or off campus, in order to make the best possible choices.

# Rate of Return for Education

Although the evidence that an education is valuable is overwhelming, the old saying "Get all the education you can" does not apply equally to everybody. Perennial students, after all, are not big earners. We can offer a general rule, though: acquire more education as long as the expected benefits at least cover the costs.

Of course, our general rule has its limits because it is always more expensive to acquire greater education. The main cost of going to college is not tuition and books, but forgone income—that is, the *opportunity cost* of not working. In other words, if you had decided not to go to college, you could be working full-time at some average salary during those years. But even with the costs of forgone earnings, tuition, and books, the rate of return for investing in education (if you are successful at college) is at least as good as the rate of return for investing in the stock market and certainly higher than putting your savings into a savings account.

Table 6.1 shows the annual mean earnings for workers with varying amounts of formal education. As you can see, there is a direct relationship between the amount of education you acquire and your future income prospects. The difference in earnings between a person without a high school diploma and a person who has completed a bachelor's degree is $47,902 for men and $24,698 for women. Table 6.2 on the next page illustrates the relationship between education and income over time in an **age/earnings profile.**

Experts in the field of education now put much more emphasis on the nonmonetary benefits of going to college. There is, of course, no way to put a monetary value on "the educated person" or "the whole person" or on the fact that college introduces you to new people, new ideas, and new ways of thinking. But college does undoubtedly give you intellectual flexibility and fosters self-discovery. College grads do a lot of things differently from those who have not graduated from college. For example, they read more books, vote more often, participate more in civic organizations, and are more likely to report being satisfied with their jobs.

# The Changing Job Scene

In the 1970s, a college degree could almost guarantee its recipient a managerial job or a professional position. Unfortunately for those seeking a college education, that is no longer the case. The reason for this change is, of course,

**Age/Earnings Profile** The profile of how earnings change with your age. When you're young and just starting out, your earnings are low; as you get older, your earnings increase because you become more productive and work longer hours; finally, your earnings start to decrease.

| TABLE 6.1 | Education and Annual Mean Earnings for Full-Time Male and Female Workers Eighteen Years Old and Over, 1999 | |
| --- | --- | --- |
| **Years of Schooling** | **Mean Earnings** | |
| | **Male** | **Female** |
| Fewer than 12 years | $18,908 | $12,057 |
| Completed high school | 30,414 | 18,092 |
| Some college–no degree | 33,614 | 20,291 |
| Associate's degree | 40,047 | 25,079 |
| Bachelor's degree | 66,810 | 36,755 |

SOURCE: *Statistical Abstract of the United States, 2001,* p. 140.

**"Although a college degree no longer guarantees a good job, the lack of one may bar you from being considered for higher-paying, higher-status jobs."**

| TABLE 6.2 | Age/Earnings (Mean Income) Profile for Selected Levels of Education, 1999 | | | |
|---|---|---|---|---|
| Age Group (Years) | Fewer than 12 Years | Completed High School | Associate's Degree | Bachelor's Degree |
| 25–34 | $16,916 | $24,040 | $28,088 | $39,768 |
| 35–44 | 18,984 | 27,444 | 35,370 | 50,153 |
| 45–54 | 19,707 | 28,883 | 37,508 | 54,922 |
| 55–64 | 22,212 | 27,558 | 35,703 | 50,141 |
| 65+ | 12,121 | 18,704 | 17,609 | 30,624 |

SOURCE: *Statistical Abstract of the United States, 2001*, p. 140.

**Question for Thought & Discussion #1:**
In your opinion, what are the most important nonmonetary benefits of investing in an education? Explain your choices.

the increase in the *supply* of college graduates between 1972 and 2000. As Figure 6.1 illustrates, in just the last quarter-century, the proportion of workers with a college background in our economy has almost doubled.

Does this mean that you shouldn't bother about getting a college degree? No. Although a college degree no longer guarantees a good job, the lack of one may bar you from being considered for higher-paying, higher-status jobs. According to the Bureau of Labor Statistics, the fastest-growing jobs are likely to be in executive, managerial, professional, and technical fields that require the highest levels of education and skill. To compete in the marketplace for these kinds of jobs requires, at a minimum, a college degree. Those with only a high school education, or less, will face increasingly limited opportunities in the future.

Another advantage of having a college degree is that it keeps your lifetime options open; it allows for more flexibility in the job market in response to

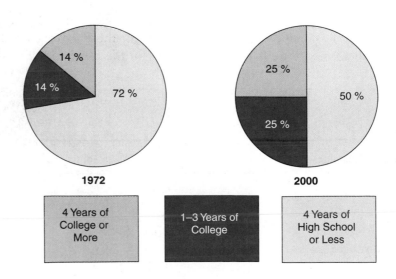

**Figure 6.1 Proportion of Workers with a College Background, 1972 and 2000**

1972

2000

4 Years of College or More

1–3 Years of College

4 Years of High School or Less

SOURCE: *Statistical Abstract of the United States, 2001*, p. 140.

changing economic conditions. This is an important consideration, especially in view of the fact that U.S. workers change jobs, on average, eight times during their lives.

# Occupational Wage Differentials

At the top of the income ladder shown in Table 6.3 are the so-called professions—medicine, dentistry, law—and managerial careers. Does that necessarily mean you should decide to study medicine, dentistry, law, or management? Obviously not. You could be wasting your time. For example, unless you're able to get into an accredited medical school (and, of course, graduate from it), you cannot legally practice medicine in the United States. The ratio of applicants to acceptances in most medical schools is astounding. Therefore, unless your father or mother is a doctor, or you are an extremely good student in an extremely good school, the odds are against your admission to medical training.

The same is not true of law, however. There are numerous law schools you can attend; you can even learn law at home by mail or on the Internet. Of course, you should not look only at the high salaries in law. To obtain a law degree, you must take at least three additional years of training after college, and, if you attend during the day, that means three more years of not earning any income. This additional cost means that the rate of return on becoming a lawyer may be no higher than if you chose another career.

Moreover, you may receive a relatively low salary for a number of years before you become a partner in a law firm. Even doctors earn relatively low incomes when they start their practices. So even though the average salary for a particular occupation is high, don't anticipate that your impressive amount of schooling will make you a nice sum of money right away. To see why this isn't necessarily unfair, we must look at the reasons behind the shape of the typical age/earnings profile, as represented in Figure 6.2 on the following page.

**Question for Thought & Discussion #2:** Why are computer programmers among the most highly paid workers?

## Wages and Ages

When you first start a job, or return to the labor force after a long absence, you may lack the skills you need to carry out your job responsibilities. If you need a lot of on-the-job training, your employer won't be inclined to pay you as much as a more experienced worker. Gradually, as you become better

| TABLE 6.3 Median Full-Time Income, by Occupation and Gender, 1999 | | |
|---|---|---|
| Occupation | Male | Female |
| Managerial and professional | $55,261 | $36,141 |
| Technical and administrative support | 41,700 | 30,001 |
| Precision production and repair | 34,429 | 24,946 |
| Operators, assemblers, and inspectors | 29,156 | 18,928 |
| Service occupation | 24,289 | 16,306 |
| Farm, forestry, and fishery | 18,949 | 13,230 |

SOURCE: *Statistical Abstract of the United States, 2001*, p. 404.

**Money Income**  The total amount of actual dollars you receive per week, per month, or per year.

**Psychic Income**  The satisfaction derived from a work situation or occupation; nonmonetary rewards from doing a particular job.

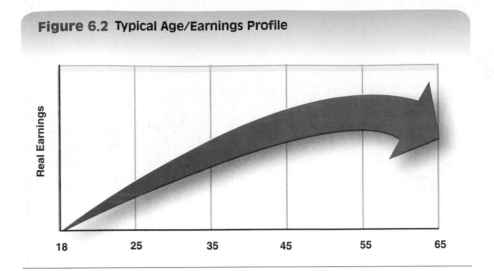

**Figure 6.2  Typical Age/Earnings Profile**

---

**TABLE 6.4**
**Average Starting Salary,**
**by Degree, 2000**

| Bachelor's Degree | |
| --- | --- |
| Computer engineering | $50,182 |
| Physics | 42,455 |
| Mathematics | 41,761 |
| Engineering (civil) | 37,932 |
| Accounting | 36,710 |
| Business (general) | 36,357 |
| Chemistry | 35,942 |
| Marketing | 33,373 |
| Humanities | 33,117 |
| Social sciences | 30,933 |
| **Master's Degree** | |
| Computer science | $61,377 |
| Chemistry | 46,389 |
| Marketing | 45,593 |
| Engineering (civil) | 44,587 |
| Accounting | 39,839 |
| **Doctorate** | |
| Computer science | $71,846 |
| Chemistry | 62,901 |
| Mathematics | 60,237 |
| Engineering (civil) | 54,588 |
| Physics | 46,500 |

SOURCE: *Statistical Abstract of the United States, 2001*, p. 174.

---

trained and more productive, your wage should increase (even corrected for inflation). Your employer gets more and more information on your productivity and your reliability from your continuing work record.

Your earnings may peak at age forty-five to fifty-five and then slowly decline until retirement, when you cease work altogether. The age/earnings profile eventually shows a downturn because older people generally work fewer hours per week and usually are less productive than middle-aged people.

## Occupational Choice and Income

Not only are there vast differences in the wages for different occupations, as shown in Tables 6.3 and 6.4, there are also vast differences in the qualifications, amount of training, and type of work required in each occupation. In an occupation with highly variable periods of employment, the average wage rate is relatively higher than in occupations that offer steadier employment; the higher wage rate compensates for the periods of unemployment.

**Money income** alone is not going to determine whether you make the right career choice. If you value independence, you certainly won't be satisfied doing paperwork in a large insurance office, and if you have a sense of adventure, you'll be restless as a salesclerk. Therefore, you may finally choose an occupation that promises you a lower wage rate than some others but a more acceptable work situation. After all, most of us work the better part of our lives; if we hate our jobs, we won't be very happy, even if we earn a good income. In other words, the total income you make from an occupation includes more than just money income. It also includes **psychic income,** or the satisfaction derived from your work situation or occupation. And psychic rewards from a job can be more important for some people than monetary payment.

You also have to decide whether you want to live in one area for a long period of time. If you become a junior executive in a company that has a history of switching its executives around the country every eighteen months, you'll be very unhappy if you dislike moving. If you want to see the country while you're young, however, you might be very satisfied with this transient lifestyle.

In many ways, your choice of occupation depends on your values and your desired lifestyle. The occupation you choose may even determine the nature of your consumption—that is, the house you live in or the clothes you wear. Your choice of occupation will also determine how much leisure you will have. As you've already seen, very few things come free of charge. If you want a job with more leisure, you generally will have less income to spend than if you take a job that offers less leisure. If you want a job that is highly stable and risk free, you will pay for it in the form of a lower income.

## Nonmoney Income

In figuring out what your standard of living would be in different types of occupations, you must also look at the nonmoney income that might be available. Nonmoney income, as opposed to the nonmonetary psychic rewards just discussed, refers to goods and services that individuals can obtain without paying money for them. The following are several sources of nonmoney income that can make our lives more satisfactory and/or comfortable:

1. *Material goods produced at home.* Such goods include those that come from growing our own produce, sewing our own clothes, and cutting firewood from the family lot, for example.

2. *Income from services in the form of food, clothing, or housing.* For example, farm laborers may receive housing accommodations in addition to money income. Ministers are often given food and housing. In fact, nonmoney income includes the whole category of fringe benefits for wage earners.

3. *Services provided by family members.* Full-time homemakers provide services to other members of the family for which the latter do not pay directly. Certain members of the family may do auto repairs, chores around the house, and lawn mowing without pay, thus providing nonmoney income to other family members.

4. *The implicit income or pleasure received from owned items, such as a house, a car, or furniture.* Specifically, if you own a house, you receive considerable pleasure from living in it. You could approximate the value of that pleasure by seeing what it would cost to rent the house.

5. *Barter income.* If you are able to exchange goods for services, goods for goods, or services for goods without resorting to the use of money or the marketplace, this can constitute part of your nonmoney income. Farmers, for example, can grow crops on their land and then trade them with other people for, say, furniture or clothes.

6. *Social income.* Such income is available largely at public expense and includes public health clinics, libraries, parks, public education, roads, and fire and police protection.

To estimate your total income, it is necessary to add up monetary income, the value of psychic income, and the value of all the nonmoney income you receive.

## Gender and Jobs

In the 1950s, a majority of Americans felt that a woman's place was in the home. And that's where most women were. Since 1960, however, the female work force has more than doubled. Today, almost three out of four adult women hold jobs, and more than 70 percent of women with children are in

## Cyber consumer
### Career Guidance Online

The Kuder organization was formed in 1938 to help individuals identify their career interests. In 2003, consumers could sign up online at http://www.kuder.com to take a $14.95 test. In just twenty minutes, they could complete the test and receive a list of careers that were well suited (according to the test) to their aptitudes and interests. According to the Web site, "The Kuder® inventories are widely recognized for their high degree of reliability . . . and [are] the latest in a series of career assessment tools." Visit this Web site and investigate what it has to offer. Would you consider spending $14.95 to take a career preference test? Why do many thousands of young and old career seekers sign up to take this test each year?

### Question for Thought & Discussion #3:
What nonmonetary benefits do you hope to receive from your career?

> **"**Although the gender gap in wages is closing, it is doing so at a very slow pace. If this rate continues, we will not achieve pay equity until nearly 2050.**"**

| TABLE 6.5 | Male and Female Shares of the Labor Force, 1960, 1990, and 2000 | | |
|---|---|---|---|
| | **1960** | **1990** | **2000** |
| Men | 66.6% | 54.9% | 52.5% |
| Women | 33.4% | 45.1% | 47.5% |

SOURCE: *Statistical Abstract of the United States, 2001*, pp. 411, 413.

the labor force. The Bureau of Labor Statistics reports that in recent years almost 60 percent of new entrants into the labor force have been women. This means that females make up a larger share of workers today than they did in the past. By roughly 2015, there are expected to be as many working women as men. Table 6.5 demonstrates this growth in female employment.

Changing views on the role of women in society, as well as civil rights legislation prohibiting discrimination against women in the workplace, have opened to women an increasing number of occupations once largely restricted to men. There is a growing number of female physicians, executives, airline pilots, construction workers, police officers, mechanical engineers, and firefighters—positions traditionally held by men. A great many more women are also establishing and managing their own business firms.

Lest we paint too rosy a picture, however, note that many women enter the labor force not out of choice but out of necessity—it is now an economic fact of life that, to meet living expenses, more households must have two incomes. Between 1980 and 2000, for example, the cost of owning a home increased significantly. Most couples cannot afford to buy a house without income from both husband and wife. Table 6.6 shows how the cost of home ownership has increased.

In spite of their aggressive pursuit of equality in wages, women continue to earn less than men holding similar positions, as Tables 6.1 and 6.3 earlier in this chapter illustrate. In 1999, women earned approximately 75 cents for every dollar earned by men holding similar jobs. Although the gender gap in wages is closing, it is doing so at a very slow pace. If this rate continues, we will not achieve pay equity until nearly 2050.

## Problems in the Workplace

Numerous federal and state laws exist to protect employees against job discrimination, hazardous or unsafe workplaces, sexual harassment, and other unfair employment practices. Some of the problems that consumers, as employees, might encounter in the course of employment are listed in Table 6.7 on page 134, along with the relevant federal legislation and the appropriate agency or agencies to contact. Since state laws relating to employment vary from state to state, only federal legislation is included.

Women continue to make progress in the nation's work force, as evidenced by the increasing number of female CEOs, lawyers, doctors, and entrepreneurs.

Photodisc/Getty Images

## Jobs through the Year 2008

Forget for the moment the current economic situation, for it may have changed by the time you use this text. The current economic situation is not really that important in making your career choice. Rather, you must look to the future to determine where the greatest demand will be for different occupations. The U.S. Department of Labor has devised a way to help you. The results of its study of employment through the year 2008 are presented in Figure 6.3 on page 137. Your choice of occupation may be influenced if you know where the jobs will be, but you still want to maximize the happiness factor in your work, so you should choose an occupation that you think you're going to enjoy from among those that are expected to be in demand.

| TABLE 6.6 Average Sale Price of New, One-Family Homes, 1980–2000 | |
|---|---|
| Year | Average Sale Price |
| 1980 | $ 64,600 |
| 1985 | 84,300 |
| 1990 | 122,900 |
| 1995 | 133,900 |
| 2000 | 169,000 |

SOURCE: *Statistical Abstract of the United States, 2001*, p. 598.

# The Problem of Inflation

**Inflation** is defined as a sustained rise in the weighted average of all prices. It must be considered whenever economic data are evaluated. For example, the amount of money a person earns must be evaluated in terms of what that income is able to buy. If a person in 2002 said, "When I started to work in 1959 I earned only 75 cents an hour," his statement would have little meaning for those who had no idea what 75 cents could buy in 1959. Between 1959 and 2002, the average price level in the United States increased by a little more than 420 percent. The 75 cents earned in 1959 had about the same purchasing power as $3.90 in 2002. This information regarding inflation is necessary to understand the true value of the person's 75-cent wage in 1959.

## How Inflation Is Measured

Although it is easy to recognize when prices are going up, it is much more difficult to measure inflation. The most common measurement of inflation is the **consumer price index (CPI)** that is prepared by the U.S. Bureau of Labor Statistics every month. This is a weighted average that measures price changes in what has been determined to be a "typical market basket" of goods and services U.S. consumers buy. The government has identified roughly four hundred products (both goods and services) that consumers typically buy. These include foods, clothing, automobiles, homes, rents, household supplies, medical care, sports equipment, legal services, transportation, utilities, and many other consumer products. The next step in completing the CPI is to determine what proportion of the typical consumer's income is spent on each type of product. If, for example, you buy a toothbrush every three months and twelve quarts of milk each week, a 10-cent increase in the price of the toothbrush would be much less important to you than a 5-cent increase in the price of milk. Therefore, a change in the price of milk is counted as a larger part, or is weighted more, than a change in the price of a toothbrush. These weights allow the government to count price changes in the CPI according to the share of consumer spending each one represents.

Each month statisticians from the Bureau of Labor Statistics measure the prices of the products in the typical market basket by collecting data from more than 18,000 households and 24,000 retail stores in 85 urban areas across the nation. The average price of each individual product is then multiplied

**Question for Thought & Discussion #4:**
What costs may society pay for discriminating against any group of workers in the labor market?

**Inflation** A sustained rise in the weighted average of all prices.

**Consumer Price Index (CPI)** A price index based on a fixed representative market basket of about 400 goods and services purchased in 85 urban areas.

## TABLE 6.7 Federal Legislation Protecting Employees

| Problem | Legislation | Agency to Contact[a] |
|---|---|---|
| **Discrimination** | | |
| Hiring, promoting, assigning jobs, enforcing dress codes, disciplining employees, firing | Civil Rights Acts of 1964 and 1991<br>Age Discrimination in Employment Act of 1967<br>Rehabilitation Act of 1973<br>Pregnancy Discrimination Act of 1978 | Equal Employment Opportunity Commission or U.S. Department of Labor, Office of Federal Contract Compliance Programs |
| Awarding pensions, retirement, and so on | Retirement Equity Act of 1984 | U.S. Department of Labor, Pension and Welfare Benefits Administration |
| **Health and Safety** | | |
| Workplace conditions | Occupational Safety and Health Act of 1970 | Occupational Safety and Health Administration |
| Refusing dangerous work | National Labor Relations Act of 1935 | National Labor Relations Board |
| Health care and insurance | Health Maintenance Organization Act of 1973 | U.S. Department of Health and Human Services, Social Security Administration |
| | Employee Retirement Income Security Act of 1974 | U.S. Department of Labor, Pension and Welfare Benefits Administration |
| | Pregnancy Discrimination Act of 1978 | Equal Employment Opportunity Commission |
| | Tax Equity and Fiscal Responsibility Act of 1982 | Equal Employment Opportunity Commission |
| Disability and death | Social Security Act of 1935 | U.S. Department of Health and Human Services, Social Security Administration |
| Ability to take leave from work to care for family members | Family and Medical Leave Act of 1993 | U.S. Department of Labor, Equal Employment Opportunity Commission |
| **Hours and Wages** | | |
| Hours, minimum wages, and overtime | Fair Labor Standards Act of 1938 | U.S. Department of Labor, Wage and Hour Division |
| Payment of wages (including deductions, assignments, garnishment, and collection) | Fair Labor Standards Act of 1938 | U.S. Department of Labor, Wage and Hour Division |
| | Equal Pay Act of 1963<br>Civil Rights Act of 1964<br>Age Discrimination in Employment Act of 1967 | Equal Employment Opportunity Commission |
| | Consumer Credit Protection Act of 1968 | U.S. Department of Labor, Wage and Hour Division |
| Retirement benefits | Social Security Act of 1935 | U.S. Department of Health and Human Services, Social Security Administration |
| | Employee Retirement Income Security Act of 1974 | U.S. Department of Labor, Pension and Welfare Benefits Administration |
| | Health Insurance Portability and Accountability Act | U.S. Department of Labor |
| | Consolidated Omnibus Reconciliation Act of 1985 | U.S. Department of Labor (COBRA) |

a. The addresses and telephone numbers of the federal agencies nearest you are listed in the white pages under "United States Government." If you are uncertain which agency to contact, go to the government's unified Web site at http://www.consumer.gov.

| TABLE 6.7 | Federal Legislation Protecting Employees (Continued) | |
| --- | --- | --- |
| **Problem** | **Legislation** | **Agency to Contact[a]** |
| **Hours and Wages** | | |
| Unemployment compensation | Social Security Act of 1935<br>Federal Unemployment Tax Act of 1954 | State Unemployment Compensation Agency |
| **Termination** | | |
| Bankruptcy of employee | Bankruptcy Act of 1978 | U.S. Department of Labor, Wage and Hour Division |
| Garnishment of wages | Consumer Credit Protection Act of 1968 | U.S. Department of Labor, Wage and Hour Division |
| Military service | Veterans' Reemployment Rights Act of 1974 | U.S. Department of Veterans Affairs, Veterans Employment Rights Office |
| Union activity | Labor-Management Relations Act of 1947 | National Labor Relations Board |
| Whistleblowing | Fair Labor Standards Act of 1938<br>Federal Water Pollution Control Act of 1948<br>Clean Air Act of 1963 | U.S. Department of Labor, Wage and Hour Division |
| | Civil Rights Act of 1964 | Equal Employment Opportunity Commission |
| | Occupational Safety and Health Act of 1970<br>Energy Reorganization Act of 1974 | U.S. Department of Labor, Wage and Hour Division |
| | Employee Retirement Income Security Act of 1974 | U.S. Department of Labor, Pension and Welfare Benefits Administration |
| | Federal Railroad Safety Authorization Act of 1980 | National Railroad Adjustment Board |
| Plant closing | Worker Adjustment and Retraining Notification Act of 1988 | U.S. Department of Labor, Wage and Hour Division |
| **Unions** | | |
| Organizing, collective bargaining, and strikes | Railway Labor Act of 1926 | National Mediation Board |
| | Norris-LaGuardia Act of 1932<br>National Labor Relations Act of 1935 | National Labor Relations Board |
| | Labor-Management Relations Act of 1947 | Federal Mediation and Conciliation Services |
| | Labor-Management Reporting and Disclosure Act of 1959 | National Labor Relations Board |
| **Other** | | |
| Credit reports | Fair Credit Reporting Act of 1970<br>Fair Labor Standards Act of 1938 | U.S. Department of Labor, Wage and Hour Division |
| Employing minors | Walsh-Healey Act of 1936<br>Fair Labor Standards Act of 1938 | U.S. Department of Labor, Wage and Hour Division |
| Polygraph testing | Employee Polygraph Protection Act of 1988 | U.S. Department of Labor, Wage and Hour Division |
| Veterans' rights | Veterans' Reemployment Rights Act of 1994 | U.S. Department of Veterans Affairs, Veterans Employment Rights Office |

## CONSUMER CLOSE-UP
# Equal Athletic Opportunities

It is no secret that the high costs of obtaining a college education force many students to seek some form of financial assistance. For a number of students, athletic scholarships open the door to education and a better life. Historically, the vast majority of athletic scholarships have been awarded to men. Most colleges had relatively few women's athletic teams and rarely offered athletic scholarships to women.

This inequity was addressed by Title IX, a 1972 amendment to the Equal Opportunity Act, that stated, "No person in the United States shall, on the basis of sex, be excluded from participation in, be denied the benefits of, or be subjected to discrimination under any educational program or activity receiving federal financial assistance." Although the meaning of this statement may appear clear, in practice, it left many questions unanswered. What were colleges expected to do? Were they required to spend the same amount on women's athletics as on men's? Did they have to have the same number of teams? Did they have to award the same number of dollars in scholarships to women and men? The new law did not explain how its objectives were to be achieved.

The United States Supreme Court answered some of these questions in 1992 when it ruled that punitive damages could be awarded when an institution intentionally failed to comply with the law's provisions. Then, in 1997, the Women's Law Center of Washington, D.C., filed complaints with the Office of Civil Rights against twenty-five National Collegiate Athletic Association (NCAA) Division I universities, charging that they were violating the law. To resolve these complaints, colleges and universities were required to increase the funds they allocated to women's sports and scholarships. They were also required to provide athletic "slots" to men and women in proportion to the distribution of the genders within their student bodies. In other words, if 60 percent of a college's students are female, then 60 percent of the openings on athletic teams must be available to female students.

Efforts by colleges and universities to comply with these Title IX rulings have increased the opportunities for women to participate on athletic teams and benefit from scholarships. Nevertheless, inequity persists. In 2002, 55.5 percent of college students were female, but only 43 percent of student athletes were women. The athletic scholarships awarded to women are, on average, much smaller than those given to men, and fewer of them are available. Typically, this difference amounts to several thousand dollars per scholarship, per year.

Colleges' efforts to follow the Title IX rules have had some unanticipated results. Many colleges moved to achieve equity goals by reducing spending for men's sports rather than increasing it for women's athletics. Hundreds of colleges have eliminated men's baseball, wrestling, and, in some cases, football or basketball. Where this has happened, the number of athletic scholarships available to men has declined without a corresponding increase in the number awarded to women.

Another concern is the impact of Title IX on college revenues. Most institutions of higher learning earn significant amounts of income from athletic contests. It is not uncommon for NCAA schools to fund all or a large part of their athletic programs with income earned from football and basketball games. Although women's sporting events are becoming more popular, they do not yet generate as much revenue as men's sports. To the extent that colleges reduce support for men's sports, they may also reduce the income they have available to fund all sports.

In 2001, President Bush established a commission to study Title IX. It finished its work in January 2003 and recommended that no basic changes be made in the law. One recommended change that could be important, however, would be to allow colleges to survey their student bodies to find out how many men and women would be interested in participating on athletic teams. Then, if only 20 percent as many women wished to be on a team as men, the college could meet the Title IX requirements by providing only 20 percent of its athletic slots to women. The recommendations of the commission are, of course, not binding. The law itself may be changed only by an act of Congress or new interpretations by the courts.

**What is your position on this issue? Do you believe institutions of higher learning should be required to offer equal athletic opportunities to women and men? How might such equity change educational and career opportunities in our nation?**

times its weight, and the answers are added together to reach a total. This total is compared with the total at some earlier time that is called the *base year*. In 2003, the base year was the average for prices in 1982 through 1983. The base year is assigned the value of 100. Increases in this average appear as percentage changes in the index. For example, the CPI was 181.3 in November 2002, meaning that the average weighted price of the typical market basket of goods and services had increased 81.3 percent between 1982–1983 and 2002.

**Figure 6.3** Employment Projections by Occupation, 1998–2008

| Occupation | Growth 1998-2008 |
|---|---|
| Computer engineers | 108% |
| Systems analysts | 94% |
| Paralegals and legal assistants | 62% |
| Medical assistants | 58% |
| Data processing equipment repairers | 47% |
| Medical records / health info. technicians | 44% |
| Dental hygienists | 41% |
| Correctional officers | 39% |
| Social workers | 36% |
| Occupational therapists | 34% |

SOURCE: *Statistical Abstract of the United States, 2001*, p. 383.

It is very easy to misunderstand the meaning of a change in the CPI. For example, if the CPI then increased to 186.3 by November 2003, you might think the rate of inflation between 2002 and 2003 was 5.0 percent (186.3 − 181.3 − 5.0). *But you would be wrong!* The 5.0 percent is based on prices as they were in 1982–1983, not as they were in 2002. To determine the percent of increase from one year to the next, you must divide the index from the first year into the index of the current year and then subtract 1, which represents prices as they were in the beginning year. Therefore, you find that the rate of inflation between 2002 and 2003 as measured by the CPI was 186.3/181.3 − 1 = .03 or 3.0 percent.

> **Question for Thought & Discussion #5:**
> Why are young people (as a group) harmed more by inflation than middle-aged people are?

## Problems with the CPI

Although the CPI is the best measure of inflation we have, it is not perfect, particularly when it is used to compare prices over a period of many years. The nature of many products we buy has changed dramatically over time. A 2003 automobile, for example, is substantially different from a 1973 model. It has safety features, pollution-control devices, and onboard computers that make our driving easier, safer, and less damaging to the environment. When the CPI is used to compare the prices of automobiles over time, it does not take these changes into account. We end up comparing apples to oranges, so to speak. This problem is made greater by the purchase of products that did not even exist at one time. For example, four out of every five U.S. families own computers today, a product that was not available for home use thirty years ago.

Another problem in the CPI is caused by our varying lifestyles and different spending patterns. People who are old or young, married or single, who live in the North or the South, or in urban or rural areas, are likely to spend their money in different ways. They will, therefore, be affected differently by changes in prices. For example, a person who lives in Idaho suffers from any increase in the cost of heating more than someone who lives in Florida. Housing prices have increased rapidly in recent years, but people who purchased their homes in 1973 have not been affected as much by these increased housing costs as people who are looking for a home to buy now. Differences like these cannot be completely accounted for in the CPI.

## Varying Effects of Inflation

Not everyone is equally affected by inflation. An increase in prices of 10 percent will have a limited impact on your life if your wages also grow by 10 percent. Inflation has its greatest effect on those whose incomes remain the same from year to year. Many retired people receive fixed pensions from their former employers that have not changed since they retired. In 1975, a monthly pension payment of $1,000 could provide a comfortable life in most parts of the United States. By 2002, the same annual $12,000 would place the recipient in poverty according to government measures.

Inflation also harms people who have saved money that earns a return smaller than the rate of inflation. If inflation was 10 percent (as it was in 1979) and you had $1,000 deposited in a bank account that paid you 6 percent interest, your savings would lose 4 percent of their purchasing power each year—before taxes are considered. That 6 percent interest would be taxed down to an amount between 3 percent and 5 percent for most people. High rates of inflation obviously tend to discourage people from saving.

But sometimes inflation can be beneficial. If you borrowed $1,000 and were required to pay interest at a rate lower than the rate of increase in your income, the debt would be easier for you to pay off. Suppose you agreed to pay the lender $1,080 at the end of one year (you would be paying an 8 percent rate of interest). If at the same time your income and prices increase by 10 percent, you would find it easier to pay your debt, and the money you paid back would have less purchasing power than the money you borrowed. In effect, you would gain while the lender would lose.

## Inflation and Consumer Decisions

Inflation may affect consumer decisions in a number of ways. The rate of inflation, for example, is not likely to be the same for all individual products. Therefore, relative prices of different goods and services will certainly change. If the price of beef increases by 20 percent when chicken's price goes up only 5 percent, the relative price of chicken will go down. People then will buy more chicken and less beef.

Beliefs about inflation can either encourage people to spend or discourage them. If you think the price of the type of car you want will increase $1,000 in the next few months, then you have an incentive to buy the car now before the price goes up. Even if you need to borrow more money now, the extra interest you would pay would almost certainly be less than the amount you would save by avoiding the price increase. In contrast, if you believe that price increases will make it difficult for you to buy more than the basic necessities of life in the future, you will hesitate to take on debt to buy a car or other expensive consumer goods. Most economists believe the effect of inflation is more often to reduce the total amount of goods and services that are sold and therefore to reduce production and employment in the economy.

Increased inflation often contributes to increased rates of interest. There are at least two reasons for this relationship. Lenders do not want to be paid back with dollars that have less purchasing power than those that they lent. Therefore, they are likely to demand higher rates of interest to compensate them for expected inflation. In addition, when there is inflation, an agency of the federal government, the *Federal Reserve System*, is likely to take steps that will force interest rates up to discourage borrowing and spending. The purpose of such a policy is to reduce spending and therefore the amount of inflation that actually does take place. In either case, consumers should not be surprised if inflation leads to higher interest rates that may affect the decisions they make.

## Real and Money Values

In Chapter 1, you learned about money prices versus relative prices. Related to these terms are two others, money values and real values. *Money values* are expressed in terms of the number of dollars that are involved in a transaction at a particular time. Your wage may have a money value of $9 an hour. Or the money value of your tuition may be $200 per credit hour. Money values provide information about current costs, but they have little meaning when used to describe changes over time—remember the example of the 75-cent wage rate paid in 1959.

**Real values** have been adjusted for inflation. When the 75-cent wage rate was adjusted for inflation, it became the equivalent of $3.90 in 2002. Real values provide more meaningful information than money values in most situations. If your uncle tells you he bought his first house for $38,000 in 1971, it tells you almost nothing. But if you know that $38,000 in 1971 dollars has a real value equal to $170,000 in 2003, you have a much clearer idea of the financial commitment he made. Consumers often need to consider the effect of inflation, and hence real values, when they make decisions. Workers should be more concerned with the real value of their wages than the money value.

**Real Values** Dollar values that have been adjusted for inflation.

**Question for Thought & Discussion #6:**
Suppose that in 2020 someone says, "I earn only $30 per hour." What information would you need to understand the meaning of this statement?

# Confronting Consumer Issues: Choosing the Right Career

The choice of a career will determine, to a large extent, your future income, but when you choose among career opportunities, you should not consider money alone.

## APTITUDE MAY DETERMINE YOUR CAREER

Many individuals have special aptitudes and abilities that lend themselves to specific careers. It would be futile to choose a career as a concert violinist if you had no aptitude for music. Virtually all specialty occupations that might be labeled "glamorous" or "artistic" require particular talents. This is also true for professional sports. Many individuals want careers in these areas but cannot and, indeed, should not seek them because they lack the appropriate abilities.

On the other hand, you can, with relatively little risk, try out a few of these areas. In effect, it is possible to test your aptitude when you are young. At this time, you can decide whether you should take the considerable risk of choosing a "glamorous" career.

You can also consider the possibility of choosing a less glamorous career in a glamorous field. If you would love to be in the theater but realized during your second year in college that you just don't have any natural acting talent, you can still enter that profession. You might train as a technician, an assistant producer, or a cameraperson.

In such careers as law, medicine, engineering, accounting, and others, aptitude is still crucial. The competition for good jobs (and even entrance to professional schools) is keen. If you are considering these careers, it would be appropriate to take aptitude tests well in advance. Most colleges and universities have services that can provide you with such tests, either free or for a small fee.

### Question for Thought & Discussion #7:

What aptitudes do you have that you could develop into skills that would help you achieve your career goals?

## FINDING INFORMATION ABOUT OCCUPATIONS

Many publications and Internet sources offer information about career opportunities. Among them are the following.

### The *Occupational Outlook Handbook*

The *Occupational Outlook Handbook* (OOH) is prepared annually by the U.S. Department of Labor. It surveys the

Having a successful interview with a prospective employer requires preparation. If you know something about the company, you will be able to relate your educational background and experience to their needs.

outlook for 250 occupational classifications and reports employment trends in these fields. The OOH is most useful for helping people identify occupations that they may wish to investigate further. You may order a copy of the OOH or review it online at **http://www.bls.gov/oco/home.htm**.

### O*NET

O*NET is an online service of the Bureau of Labor Statistics that provides a comprehensive listing of job characteristics and the skills and knowledge required for various types of employment. O*NET resources may be accessed at **http://online.onetcenter.org**.

### The *Occupational Outlook Quarterly*

The *Occupational Outlook Quarterly* (OOQ) is a federal government publication that provides information about specific careers. It contains in-depth articles about job opportunities and working conditions. Although the OOQ provides greater detail about specific careers, it does not cover as many occupations as do other government sources. The OOQ can be found at most college libraries and career centers or may be accessed online for free at **http://www.bls.gov/opub/ooq/ooqhome.htm**.

### The *Monthly Labor Review*

The *Monthly Labor Review* (MLR), another publication of the federal government, publishes articles about the labor market as a whole. Its articles are generally more useful for learning about trends in employment than for researching specific

careers. The MLR can be purchased or viewed online at
**http://www.bls.gov/opub/mlr/mlrhome.htm**.

### The *U.S. News* Work and Career Web Site

*U.S. News and World Report* maintains a commercial Web site at **http://www.usnews.com/usnews/work/wohome.htm**. This Web site provides access to *U.S. News* articles dealing with careers and employment. It also has links to a wide variety of other sites that you may find useful.

### Mapping Your Future

Mapping Your Future is a private organization that maintains a Web site at **http://www.mapping-your-future.org/undergrad**. This site furnishes extensive advice for job seekers. It also provides many links to other related sites.

Remember, Web addresses change frequently. If any of these addresses no longer function, find similar sites to investigate using any of the search engines available to you.

## OTHER SOURCES OF CAREER INFORMATION

In addition to the sources just described, there are many other resources that you may use to investigate your career interests. Be sure to evaluate these sources as well as the information they provide. Some commercial career advisement services may be more interested in earning a profit than in helping you.

### College or University Placement Centers

Although the exact title of the office may differ from college to college, every institution of higher learning has a service to help its students find employment either before or after graduation. Most placement centers have career consultants and vocational guidance counselors, as well as facilities for setting up interviews between students and recruiters from private employers and government agencies.

### State Employment Agencies

All states and the District of Columbia have employment offices that are operated in conjunction with the U.S. Department of Labor. These employment services charge no fee and make placements for many types of jobs; some even offer free career guidance and aptitude tests. You should recognize, however, that these employment offices list jobs only when employers contact them about openings. Many employers, particularly those who offer better-paying jobs, do not list their openings with state employment services.

### Help Wanted Advertisements

"Help wanted" ads in newspapers and professional or trade journals list vacancies for many jobs. Government agencies are often required by law to advertise their job openings. Private businesses, however, have no such requirement. Many do not advertise their "better" jobs because they already have many applications for employment on file from people who are qualified and looking for employment. Still, for entry-level jobs, newspaper advertisements can be a good place to begin.

### Private Employment Agencies

You can register with an agency and wait to be called, or you can apply directly for a job that is advertised in a periodical. Agencies generally require you to sign a contract that obligates you to pay a fee if the agency places you. Read these contracts carefully; the small print may reveal that you owe the agency the fee even if you're fired after one week. Agency fees may run from 5 to 15 percent of your annual starting salary. In the upper-income job brackets, agency fees can sometimes be as much as 30 percent of your first year's salary. Those agencies that receive their commissions from employers are usually free to applicants.

> ### Question for Thought & Discussion #8:
> Would you consider paying a private employment agency to help you find a job? What assurances would you require from the business? How would you investigate its claims?

### Office of Personnel Management

The federal government's Office of Personnel Management publishes information on job opportunities for government civilian jobs, both within the United States and overseas. For information, check with your local post office, write to the Office of Personnel Management, Washington, DC 20415, or contact the office online at **http://www.opm.gov**.

### Periodicals

Some periodicals, such as *Kiplinger's Personal Finance* and *Business Week,* offer job-outlook or jobs-in-demand sections each year in their January or February issues. These can be valuable sources for college graduates who want to know what to expect when job hunting, the kinds of jobs available, and what employers are looking for.

### Online Employment Services

A seemingly unlimited number of private and government-supported Web sites offer career guidance and employment services. Those operated by government agencies are generally free, but private services almost always charge a variety of fees. These sites can be found by searching the Internet with a Web browser using the key words *career guidance* or *employment opportunities.* Some Web sites are operated by employment agencies that actually offer specific job opportunities. Two of the largest online job-search services in 2003 were the Monster Board (**http://www.monster.com**) and Employment Guide (**http://www.employmentguide.com**).

*(Continued on next page)*

**Confronting Consumer Issues (Continued)**

# PREPARING A WINNING RÉSUMÉ

For almost all job applications, you must submit a **résumé.** Because personnel officers in corporations read thousands of résumés every year, yours should create the best possible impression in order to give you a competitive edge over other job seekers. Remember, your résumé is an advertisement for yourself.

**Keep It Brief**　　Since your résumé is, in large part, bait for the interview, it need not be an entire dossier, starting out with letters of commendation from your junior high school principal. Nor should it list your every accomplishment, information about your outside interests, or the backgrounds of your parents.

**Presentation of Your Résumé**　　Your résumé should be typed on one or more sheets of high-quality rag bond. A good résumé is usually professionally printed, not photocopied. Remember, the appearance of a résumé is like the appearance you will make for an interview: first impressions count in both cases.

**Format of Your Résumé**　　You needn't write a résumé as if it were an application for college. In other words, don't put the word NAME before your name. The fewer headlines, the better, but you can divide your résumé into sections such as the following for easy readability:

1. *Personal data.* Begin with your name, address, telephone number, and e-mail address (if you have one) at the top of your résumé.
2. *Employment objective.* Indicate the kind of job (or jobs, as long as they are within the same general area or industry) that you are seeking.
3. *Education.* List all schools you have attended and all other relevant academic information, as follows:

> ## Question for Thought & Discussion #9:
> What could you do now that would improve your résumé (and your chances of being hired for a job you want) in the future?

    a. Name and city/state of high school and the date you graduated (omit high school if you have a higher degree).

    b. Names and cities/states of colleges or universities you have attended, degrees received, and dates.

    c. Major and minor subjects and other courses related to your job goal.

    d. Scholarships, honors, and any extracurricular activities that may indicate social or leadership abilities.

4. *Work experience.* Any work experience you have could be a valuable asset when applying for a job. Depending on the job you are applying for, you may wish to put the work experience summary before the summary of your educational background.
5. *Miscellaneous.* Any abilities you have that may be appropriate to the job you are seeking should be listed. Depending on your prospective employer's needs, foreign language skills, for example, or the ability to operate special equipment may be a strong selling point.
6. *References.* Give the names, positions, and addresses of three persons who have direct knowledge of your work competence. If you are a recent graduate, you can list teachers who are familiar with your capabilities. It is courteous to obtain permission from those whom you wish to use as references.

**The Do's and Don't's of Résumé Preparation**　　Experts in the field of résumé preparation offer the following guidelines:

1. Make sure you proofread your résumé so there are no typographical or spelling mistakes.
2. Make sure there are no errors in grammar. When in doubt, ask someone who knows.
3. First impressions are important. Therefore, have your résumé professionally printed on good-quality paper. Don't use colored or perfumed paper. Don't include a picture of yourself on the front.
4. Describe yourself honestly. Don't exaggerate. If you're young, be candid about your experience—or lack of it.
5. Don't cram your résumé with useless information.
6. State very succinctly and clearly a job or career objective.

Figure 6.4 shows a sample résumé based on the preceding suggestions. There are other possibilities, of course, but the main things to keep in mind are clarity, brevity, and the relevance of your résumé's content to the kind of job you seek.

## LETTER OF APPLICATION

When you send your résumé to a prospective employer, you will also want to send a cover letter or a letter of application. This is the customary way to ask for a personal interview for a job. Figure 6.5 on page 144 shows a typical letter of application—in this case, for a sales job. Spend some time preparing the letter, making sure it is brief and to the point, yet personal. Whenever possible, address your letter to a specific person.

**Figure 6.4** A Sample Résumé

Jane D. Jones                                                       April 5, 2004
593 Ninth Avenue
Anytown, Ala 35204
(555) 422-2824
jdjones@aol.com

**Employment Objective:** Reporter, copy editor

**Education**

Standard State University, University City, Ala. B.S., cum laude, 2003
*Major:* Journalism. *Minor:* Psychology. *Other courses:* Beginning and advanced photography
*Honors:* Phi Kappi Phi
*Extracurricular activities:* Editor of college newspaper. Served earlier as copy editor and reporter.

**Experience**

September 2000–June 2003. Correspondent in University City for *Anytown Gazette*, Anytown, Ala.

June–August 1999. *Anytown Gazette.* Although working as a copy runner, I received a number of editorial assignments. Besides covering meetings and writing obituaries, I did a feature series with photographs on the county arts group. (Attached is a photocopy of stories I wrote.)

Summers, 1997 and 1998. Wilder Dress Shop, 215 Main Street, Anytown, Ala. Salesclerk.

**References**

Prof. J. W. Wynn, School of Journalism, Standard State University, University City, Ala. 34205

Mr. William T. Ryan, editor, *Anytown Gazette*, Anytown, Ala. 35204

Prof. Dora Cohen, School of Journalism, Standard State University, University City, Ala. 34205

SOURCE: Adapted from *Merchandising Your Job Talents*, U.S. Department of Labor.

Once your application letter and résumé have earned you an interview, consider these other pointers that can improve your chances of landing the job.

## HOW TO BE INTERVIEWED

Remember that the personnel officer of the company interviews many prospective employees. You must somehow convince the interviewer that you are as good as or better than anyone else who is being considered for the job. Basically, your interview should be constructed to convince the prospective employer that you will fulfill his or her needs. To do that, you must be prepared.

One skill that ranks high on the list of employers who are interviewing is the job candidate's ability to communicate well. If you know something about the company, it will help you to relate your background to the company's needs and to communicate effectively with your interviewer. Information about prospective companies can be found by looking at some of the following sources:

1. *Moody's* manuals
2. *Fitch Corporation* manuals
3. *Thomas's Register of American Manufacturers*
4. *MacRae's Blue Book*
5. Company annual reports

Here are some suggestions for a successful interview.

1. Be a few minutes early.
2. Come with a copy of your résumé.

*(Continued on next page)*

**Confronting Consumer Issues (Continued)**

**Figure 6.5** A Sample Letter of Application

John W. Doe
2422 Bay Street
San Francisco, CA 94102
(415) 778-0000
January 20, 2004

Mr. Wilbert R. Wilson
President, XYZ Company
3893 Factory Boulevard
Cleveland, Ohio 44114

Dear Mr. Wilson:

Recently, I learned through Dr. Robert R. Roberts of Atlantic and Pacific University of the expansion of your company's sales operations and your plans to create a new position of sales director. If this position is open, I would appreciate your considering me.

Starting with over-the-counter sales and order service, I have had progressively more responsible and diverse experience in merchandising products similar to yours. In recent years I have carried out a variety of sales promotion and top management assignments with excellent results.

For your review I am enclosing a résumé of my qualifications. I would appreciate a personal interview with you to discuss my application further.

Very truly yours,

John W. Doe

Enclosure

---

**3.** Always maintain eye contact and listen attentively.

**4.** Be honest and frank, but don't make derogatory comments about a previous employer.

**5.** Let your interviewer offer you information on benefits, salary, and agency fees (if any).

**6.** Dress appropriately; first impressions are important.

**7.** Remember, personality counts, too. Be well rested for the interview, alert and forthcoming in your responses, and courteous to, and thoughtful of, everyone included in the interviewing process.

**8.** Have answers ready. Try to imagine a variety of questions you may be asked during an interview and prepare a few answers in advance. Then rehearse your answers.

## SOME FINAL POINTERS ON JOB HUNTING

Think of job hunting as a full-time job (or part-time, depending on your work or school circumstances). Manage your time well and be methodical in your search. Consider your job search as work you are doing for yourself, and don't take "time-outs" from the job. After all, you're the employer!

The key to success in job hunting is motivation. If you are motivated, you will follow many of the suggestions in this section. If you feel that you need some more expert advice, consider seeking the services of a professional résumé writer, generally someone associated with a private employment counseling firm. If you need help with interviews, practice with a friend or with someone who works in the placement center at your college or university. Without a doubt, job hunting requires a great deal of effort.

### Key Term:

**Résumé** A brief summary of your education, training, and other achievements that you give to a prospective employer.

# Chapter Summary

1. Many income differences are the result of inherent differences in human beings, but they also are determined by the amount of training and education an individual has obtained, the amount of on-the-job training, and the riskiness of the occupation.

2. Going to school is an investment in human capital because it makes you, the human, more productive in the future. Generally, your investment in human capital will pay off in the form of a higher wage later on.

3. You should specialize in an activity that is in demand and for which there is not a surplus supply. Hence, choosing your occupation requires predicting both the demand and the supply for that particular occupation in the future.

4. Usually, individuals are paid according to their productivity. Therefore, anything that raises an individual's productivity may lead ultimately to a higher income.

5. The rate of return on education is as high as the rate of return on investing in other things. A college-degree holder may make two to three times as much income as a person who has only graduated from grade school.

6. The greatest cost of going to college is not the money spent for books and tuition, but rather the opportunity cost of not being able to work and to make an income during those years.

7. An individual's wages (corrected for inflation) are usually lowest when he or she first enters the labor force. That's because people are least productive then.

8. In determining your standard of living, it is important to include nonmoney income, which includes, but is not limited to, goods produced at home, services produced at home, and social income from government-provided goods and services.

9. Women are entering the labor force in increasing numbers. In spite of aggressive pursuit of pay equity with men, women still make approximately 75 cents for every dollar earned by men holding similar positions. If current trends continue, pay equity will not be achieved until nearly 2050.

10. Numerous federal and state laws protect employees against unfair or unhealthy employment situations.

11. Inflation is defined as a sustained rise in the weighted average of all prices. The most common measure of inflation in the United States is the consumer price index that is prepared by the Bureau of Labor Statistics every month. The CPI is determined by tabulating the weighted prices of a typical market basket of products and comparing the total to prices in a base year.

12. The CPI is the best measure of inflation we have, but it is not able to account for all changes in products or all differences in the way people spend their money.

13. Inflation affects people in different ways. It may encourage some spending decisions and discourage others. It hurts people on fixed incomes and those who have saved or loaned money. It can help those who are in debt.

14. The difference between money values and real values is that real values have been adjusted for inflation. In most cases, real values are more useful in making consumer decisions.

15. Many sources of career information can be located at most colleges, at public libraries, and on the Internet. College and university job-placement centers also have extensive selections of career information.

16. When consumers seek employment, they should keep their career objectives in mind. Preparing and planning to look for work is the best way to improve one's chances for success. Résumés should be brief and contain only information that is relevant to the job opening that is being sought. There are also many sources available that will help consumers complete job applications and carry out job interviews successfully.

# Key Terms

age/earnings profile **127**
consumer price index (CPI) **133**

human capital **126**
inflation **133**
investment **126**
money income **130**

psychic income **130**
real values **139**
résumé **144**

## Questions for Thought & Discussion

1. In your opinion, what are the most important nonmonetary benefits of investing in an education? Explain your choices.

2. Why are computer programmers among the most highly paid workers?

3. What nonmonetary benefits do you hope to receive from your career?

4. What costs may society pay for discriminating against any group of workers in the labor market?

5. Why are young people (as a group) harmed more by inflation than middle-aged people are?

6. Suppose that in 2020 someone says, "I earn only $30 per hour." What information would you need to understand the meaning of this statement?

7. What aptitudes do you have that you could develop into skills that would help you achieve your career goals?

8. Would you consider paying a private employment agency to help you find a job? What assurances would you require from the business? How would you investigate its claims?

9. What could you do now that would improve your résumé (and your chances of being hired for a job you want) in the future?

## Things to Do

1. Visit the career guidance center at your college or university. Ask the staff members to explain the services the center offers. Evaluate these services in terms of how helpful they would be to you.

2. List the costs and benefits of the education you are pursuing. What return do you hope to receive from your investment of time, money, and effort, and their opportunity costs?

3. Identify a career that interests you. Investigate this career using sources discussed in this chapter. Write an essay that summarizes the information you found and indicate where you located it.

4. Identify and describe at least three different ways you would be affected by a 10 percent increase in your cost of attending college.

5. Create a résumé for yourself that you might send to an employer who has job openings in a career that interests you. Do you think your résumé would impress the employer? What can you do now in terms of additional education, work experience, or a volunteer position that would improve the quality of the résumé you will send out in the future?

## Internet Resources

### Finding Consumer Information on the Internet

The following Web sites have been selected for their relevance to topics discussed in this chapter. Search these sites to locate information that can add to your knowledge of career information that is available on the Internet. Remember, Web addresses change frequently. If any of these addresses no longer function, find similar sites to investigate using any of the search engines available to you.

1. Mapping Your Future offers information on planning your career, selecting a school, and paying for your education. You can find it online at **http://www.mapping-your-future.org**.

2. Smartjobs.com offers career builder, career guidance, and job-search assistance. It can be found at **http://www.smartjobguides.com/benefits.cfm**.

3. Life Track offers career guidance and a plan of action for attaining your career goals. You may review this firm's Web site at **http://www.life-track.net**.

### Shopping on the Internet

The following Web sites have been selected because they offer consumers services similar to those described in this chapter. These are commercial sites that are designed to market products. They do not represent a comprehensive or balanced description of all résumé services available online. How would you choose among these résumé preparation services? Would you rather prepare your own résumé? Remember, Web addresses change frequently. If any of these addresses no longer function, find similar sites to investigate using any of the search engines available to you.

1. ResumeEdge.com offers résumé services from certified experts in forty industries. Its Web site is at **http://www. resumeedge.com**.

2. For résumé preparation assistance you may agree to pay $150 for personalized advice from e-resume.net. This organization's Web site is found at **http://www. e-resume.net**.

3. Résumé and cover letter services are provided by Employment 911 for $99. You may visit its Web site at **http://www.employment911.com**.

## InfoTrac Exercises

 Purchasers of new copies of this text are provided with access to the InfoTrac Web site. This Web site links students to thousands of recent articles published in hundreds of periodicals. Use the key words **career counseling** or other terms from this chapter to conduct a key-word search. Choose one article that is of particular interest to you and write a brief essay describing what you have learned from the article. Be sure to cite the author and title of the article and the name and date of the publication in which it appeared.

# Selected Readings

Anker, Richard. *Gender and Jobs.* Washington, DC: Bureau of Labor Statistics, 1998.

Bernstein, Alan. "Choosing a Career: What Color Is Your Career Path?" *Careers and Colleges,* January-February 2003, pp. 6–9.

Buckley, John E. "Rankings of Full-Time Occupations by Earnings, 2000." *Monthly Labor Review,* March 2000, pp. 46–57.

Cleaver, Joanne, Carrie Patton, and Lisa Holton. "Hot Commodities." *Working Woman,* February 2000, pp. 40–47.

Crocodilos, Nick. "How to Pick a Good Employer." *Electronic Engineering Times,* December 9, 2002, p. 95.

Crosby, Olivia, "Résumés, Applications, and Cover Letters." *Occupational Outlook Quarterly,* Summer 1999, p. 2.

"Dull Interview Skills Won't Cut It in a High-Tech Market." *Knight Ridder/Tribune Business News Service,* July 1, 2001, p. ITEM01182006.

Fisher, Anne. "My Company Just Announced I May Be Laid Off: Now What?" *Fortune,* March 3, 2003, p. 184.

"In Brief: Slower Price Rise Predicted for 03." *American Banker,* August 8, 2002, p. 11.

"Ingenuity Is Key to Landing a Position." *USA Today Magazine,* December 2002, p. 58.

Randinelli, Tracey. "The Ultimate Job Search Guide." *Careers and Colleges,* January-February 2002, pp. 42–47.

Schneider, Jodi. "Finding Work." *U.S. News & World Report,* February 24, 2003, p. 80.

Stern, Linda. "Money: New Rules of the Hunt." *Newsweek,* February 17, 2003, p. 67.

"Your Career Can Still Thrive in an Age of Flux." *Black Enterprise,* October 2002, pp. 163–169.

Photodisc/Getty Images

## CHAPTER 7
# Creating a Living Budget

**After reading this chapter, you should be able to answer the following questions:**

- What is the value of a budget?
- How can a budget be a part of an individual's or family's plan to achieve life-span goals?
- What special budget problems may college students have?
- What steps can consumers take to construct a useful budget?
- Why is record keeping important?
- What kinds of budgeting assistance are available to consumers?
- How may consumers budget their time and money for leisure and recreation?

American consumers are rich. That is, we are rich compared with the Hatian consumer, the Indian consumer, the African consumer, the Albanian consumer, or the Venezuelan consumer. The average per capita income in the United States is considerably higher than the per capita income in most other countries in the world.

But per capita income does not tell the story we want to tell. Table 7.1 shows the different percentages of U.S. families that make particular amounts of income, ranging all the way from poverty to extreme opulence. Most of us find ourselves somewhere in the middle range. We have income, so we're not destitute, but we're not millionaires either.

All of us, whether rich or poor, have this in common: we can't buy everything we'd like to buy. In other words, we all face the universal problem of scarcity. All of us operate with a limited income and amount of wealth we can spend. Because everyone faces this universal problem of scarce resources, personal money management is important for all of us, no matter what our income level.

> **"We can't buy everything we'd like to buy."**

## Consumers Need a Spending Plan

We've established the fact, then, that even extremely wealthy people need a spending plan. Consider the case of Donald Trump, who at one time controlled assets that included an airline, hotels, a railroad yard, and several casinos in Atlantic City. He was also $3.8 billion in debt and unable to make his payments on time. His creditors finally extended $60 million in additional credit, but only after Trump agreed to reduce his personal monthly spending from more than $580,000 to just $450,000. Although most people would be happy to "scrape by" on $450,000 a month, Trump's situation demonstrates the universal problem of scarcity that forces us all to make choices no matter who we are or what we own. A spending plan—more commonly called a budget—helps us make these choices.

Planning a budget and attempting to follow it force the issues of scarcity and opportunity cost out into the open. Budgeting also necessitates decision making and the establishment of priorities. If you include a trip to Mexico in your budget, you will realize that somewhere along the line another item or items must be cut out. A budget, then, helps you manage your money in a

| TABLE 7.1 | Money Income of Families, by Income Level, 1999 | |
|---|---|---|
| **Annual Money Income** | **Number of Families (Thousands)** | **As Percentage of All Families** |
| Less than $10,000 | 4,144 | 5.8% |
| $10,000–$14,999 | 3,485 | 4.8% |
| $15,000–$24,999 | 8,678 | 12.0% |
| $25,000–$34,999 | 8,550 | 11.9% |
| $35,000–$49,999 | 11,861 | 16.5% |
| $50,000–$74,999 | 15,236 | 21.2% |
| $75,000 and over | 20,076 | 27.8% |

SOURCE: *Statistical Abstract of the United States, 2001*, p. 439.

more or less systematic and rational manner. It is also a control mechanism that keeps you aware of the decisions you are actually making—decisions that are being made even if you don't wish to acknowledge them. Some of you may be able to easily determine the trade-offs and opportunity costs involved when you make a purchase. But most of us would benefit from a budget. With it, we may be able to hold in check undirected spending that can lead to unhappiness and, occasionally, to financial disaster. Your objective should be to create a budget that makes the most satisfying trade-offs in your life—a living budget.

## Democratic Decision Making

If, in your situation, more than one person is affected by how each month's income is spent, then you have to determine how decisions will be made. Will decision making be unilateral or democratic? That is, will everybody involved in the household participate? If not, some of those whose lives are affected may, at one time or another, feel cheated, left out, or imposed on.

In most families today, the "breadwinner" no longer automatically has all the financial decision-making powers. And, of course, in most U.S. households more than one person is employed. Moreover, the members of a household who perform chores such as cooking and cleaning are making a contribution to the welfare of their family no less than those who are working for an income outside the home. Hiring a housekeeper or a nanny, for example, is an expensive proposition that can cost $25,000 a year or more. The point to remember in preparing a budget, therefore, is that people should have a voice in it whether they contribute to the household by earning income or by completing household chores.

It is important to work out money problems within a family unit in a cooperative way. Surveys have shown that problems related to money are the most frequent cause of family disagreements. Unfortunately, people who fight over money are likely to fight over other matters as well, at least in part because it becomes difficult for them to communicate when they are already angry about their financial problems.

It goes without saying that getting everybody within a household to agree on budget allocations is not always easy—or even possible. And even in families that do engage in democratic decision making, usually the votes of the older and more experienced individuals (read: parents) carry the most weight. Nonetheless, the participation of all members of the family or household in the budget-making process has a very positive function. It forces each member of the group to consider the wants and needs of the others, and it acquaints each individual with the fact that not all wants and needs can be met. This knowledge can help a great deal in making sacrifice and compromise more acceptable, especially if each person's wishes and views have been treated fairly and reasonably by the others.

## The Family Council

Money is a sensitive issue in any family, and in today's busy world, the right moment for discussing the budget sometimes comes very infrequently. One way to overcome this difficulty is by holding a family meeting, or *family council,* to decide on budgeting issues. You may wish to start with a casual meeting. If this doesn't work, try a more formal setting. It's helpful, too, if

**Question for Thought & Discussion #1:**
How are important financial decisions made in your family? Do these decisions ever lead to family disagreements? How could such disagreements be avoided?

everybody in your household knows of the meeting a day or so in advance. This will allow each individual time for reflection on what he or she needs or wants, or wishes to complain about.

In some situations, communication on paper may be easier than verbal communication. If there is a quarrel about family spending priorities, you might suggest that each family member write down a list of individual priorities in descending order and then compare the lists. If the going gets rough, you might even want to record—and then listen to—the family money fights. Frequently, those who argue don't really hear themselves. In the negotiating process, contracts—between parents and teen-agers, for example—might be created.

## College and Budgeting

The last decade saw a steady increase in the cost of obtaining a higher education at the same time that the level of financial aid and scholarships offered at many institutions declined. In addition, a growing number of college students were parents and had to cope with family responsibilities along with their schoolwork. These trends continue today. Consequently, making and following budgets is even more important for college students now than in the past.

When college students prepare budgets, they need to plan the use of their time as well as the use of their money. Students can often increase their income by finding a job or by working more hours at a job they already have. They must remember, however, that their basic purpose in attending college (and in paying the associated costs) is to obtain an education that should increase their future income. Students need to be sure that working more hours while in school does not prevent them from achieving their full academic potential. When students find that they can't be successful in their courses and meet their financial needs at the same time, they may consider taking a semester off to earn more money or attending school part-time while they work more hours.

**AVOID THE LITTLE EXTRAS**   Many college students borrow to pay for all or part of their college expenses. This choice is often unavoidable for students who want to complete their education quickly and are unable to obtain other types of financial assistance. When students borrow, they should keep in mind that they will be required to repay their loans. It is easy to borrow more money than is necessary so that you can buy a few "extras" while attending school. These extras, however, will increase the amount of payments due after graduation. Often a better choice is to live with less now to avoid having to make do with less in the future. This is particularly true for people who have life-span goals that include borrowing for other purposes later on. For example, a person who has a $50,000 student loan to pay off may find it more difficult to qualify for a mortgage even years after graduation.

**WHAT DO CHILDREN REALLY NEED?**   When students are parents, they may be tempted to try to give their children all the toys, trips, clothes, and other material possessions other children might have. Students with children, however, need to keep the purpose of their education in mind. Although paying for tuition, books, and other supplies may prevent parents from buying as much for their children now, their education should enable them to give

**Question for Thought & Discussion #2:**
What special budget problems have you encountered because you are a student? How have you dealt with them?

Thinkstock/Getty Images

Holding a family council to make or review a budget can help prevent disputes over money.

their children a more financially sound future. Bringing older children into the budgeting process is often a good idea so that they can understand why they may not have all the toys or clothing they want at the present.

# Budget Making, Goals, and Values Clarification

In Chapter 2, we discussed values clarification—how you decide what your goals are and what your values are, and what they mean with respect to how you should spend your time. Ultimately, this all relates to what kind of life you want to lead. Now you can put this abstract problem into perspective by applying it to an actual dollars-and-cents decision-making process—budget formulation. When you sit down alone or with the other members of your household, you have to consider the values that you place on the various things you want to do with the income available. To be able to clarify your values, you first have to formulate your goals and those of your family as a whole. Then you must set *priorities* among your goals. These priorities will be related to three general types of goals you probably will set for yourself or your household—short-term, middle-term, and life-span goals.

## Achieving Life-Span Goals

Consider an example. Suppose you have set a life-span goal of studying French history at the Sorbonne in Paris. To do this, you will need to save or borrow $20,000, achieve an outstanding academic record so that you will be accepted at the Sorbonne, and learn to speak French. Your life-span goal can be reached only if you first set and achieve a series of short- and middle-term goals. Among your short-term goals are learning as much as you can in French 101 and European History 223 this semester. You might also join a French club at school or ask a French exchange student to help you with your pronunciation. And, of course, there is the matter of money. If you want to live in France for a year, you need to start to save now. Even if your short-term goal is only to set aside $20 or $30 a week, your funds will add up over time. By planning carefully and working to reach your short- and middle-term goals, you might be able to fulfill your life-span dream in five years.

Consider another possible life-span goal—that of acquiring a genuine appreciation of the arts for yourself and your family. Short-term goals, such as going to a museum once a month or purchasing art appreciation books, and a middle-term goal of saving enough money for season tickets to the opera might be involved in meeting this goal.

More basically, you may have the goal of seeing that your family is well nourished and adequately housed or that it has sufficient medical protection or safe, comfortable transportation. Here, your goals may require choices such as stinting on housing to provide adequate food or on transportation to provide medical insurance.

## What Makes You Happy

Essentially, everybody's main goal is to be happy. The problem is clarifying your values enough so that you can establish goals that, taken together, will spell happiness for you and for those around you. When you formulate a budget, you can see exactly what these goals cost. You are forced to rethink and to reformulate your values when you realize that they are either unattainable or extremely costly, in the sense that you must forgo other desired or necessary things.

Goal definition and values clarification are integral parts of budget formulation and may be considered the only ways to design a satisfactory budget that works. Remember, though, a budget should not be a straitjacket but rather an indication of direction that will change, depending on changes in your individual and family situations.

> **"A budget should not be a straitjacket but rather an indication of direction that will change."**

## How Does the Typical Household Allocate Its Income?

Averages sometimes can be deceiving. But it may be instructive for you to see how typical households in the United States allocate their disposable (after-tax) incomes to many competing demands. Figure 7.1 shows how an average U.S. family spends its take-home income. A large chunk usually goes for housing and household operation (including utilities and maintenance). Other large segments go for food, clothing, transportation, and insurance. These basic expenses often account for more than 80 percent of the average U.S. family's after-tax expenditures in any one year. Personal care and medical care constitute another big part of each family's budget in the United States. Medical care alone represents an increasingly large percentage of total U.S. consumption spending; we discuss the reasons for this increase and the future of medical-care expenditures in Chapter 16.

As income goes up, the percentage of total income spent on food generally falls, and the percentage spent on housing rises. That means that, as our incomes increase, we buy proportionally more housing and proportionally less food. After a certain point, then, housing is often considered to be a

### Figure 7.1 Typical Distribution of Personal Consumption Expenditures, 1999

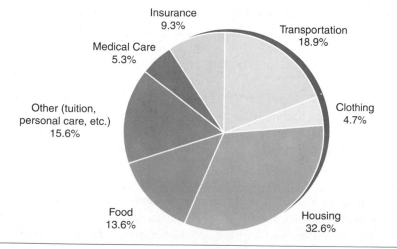

SOURCE: *Statistical Abstract of the United States, 2001*, p. 430.
*Payments made directly by individuals. Excludes payments made by insurance plans.

**Luxury Good** A good whose purchase increases more than in proportion to increases in income. Jewelry, gourmet foods, and sports cars usually fall into this category.

**Fixed Expenses** Expenses that occur at specific times and cannot be altered. Once a house is purchased or rented, a house payment is considered a fixed expense; so is a car payment.

**Flexible Expenses** Expenses that can be changed in the short run. The amount of money you spend on food can be considered a flexible expense because you can buy higher- or lower-quality food than you now are buying. These are also known as variable expenses.

luxury good, for people buy a disproportionate amount of it as they become wealthier. Food, in contrast, has the opposite characteristic and is, therefore, considered a necessity.

We have to be careful of such labels as *luxury* and *necessity*, however, because they have some subjective connotations. One person's luxury may be another person's necessity, and vice versa. Most of us have a hard time determining our own values and goals, let alone those of other people. But that's exactly what we attempt to do when we consider somebody else's spending to be "wasted" on so-called luxury items.

# Creating a Budget Plan

Very briefly, these are the basic steps in creating a spending plan, a budget. They will be discussed in greater detail in the following paragraphs.

1. Identify and prioritize the goals you would like to achieve through your budget.
2. Analyze your income and spending patterns by keeping records for a month or two.
3. Determine your **fixed expenses,** such as rent and any other contractual payments that must be made, even if they come due infrequently (such as insurance and taxes).
4. Determine your **flexible expenses,** such as those for food and clothing, which can be adjusted in the short run.
5. Balance your fixed and flexible expenditures with your available income. If a surplus exists, you can save it or apply it toward the cost of achieving more of your goals. If there is a deficit, you need to examine your flexible expenses with an eye to reducing or eliminating some of them. You may also reexamine your fixed expenses to determine if any of them may be reduced in the future.
6. Evaluate your current spending patterns. Ask yourself if they will help you achieve your life-span goals. Do they form the basis for a responsible long-term budget? What changes could increase the satisfaction you receive from the money you spend?

Note that so-called fixed expenses are necessarily fixed only in the short run. In the longer run, everything is essentially flexible or variable. You can adjust your fixed expenses by changing your standard of living, if necessary.

## The Importance of Keeping Records

**Question for Thought & Discussion #4:**
How do you organize and keep records of your financial transactions? Does your system work well? Why or why not?

A budget can be only as good as you make it—through good record keeping. Budget making, whether you are a college student, a single person living alone, or the head of a family, will be useless if you don't keep records. The best record-keeping system is one that is both convenient to maintain and efficient. Paying bills by check is one way to ensure both a record of your payment (in the check register) and a receipt (your canceled check, when the bank returns it to you). Those expenditures can then be listed easily on a ledger at the end of every month. A special note, or receipt, of all cash expenditures—placed routinely in a specific container in your home—can help you keep track of your cash outflow. Again, at the end of each month, these can be entered into your ledger to create, along with expenses paid by check, a complete listing of how you spend your money.

## Constructing a One-Month Budget Worksheet

In preparing a long-term budget, your first task is to construct a *one-month budget worksheet* to help you gain an understanding of your current spending patterns. First, you should list your sources and amounts of income. If you receive money only from your parents or relatives, or from a scholarship or fellowship, then you know what your income will be. If you earn all or part of your income, however, you must estimate what your net income for the month and year will be after withholding is taken from your gross pay for income taxes and Social Security. The easiest way to do this is to base your estimate on a recent paycheck stub.

Your next step is to identify and total the fixed expenses you must pay each month. Keep in mind that your budget worksheet should include a share of expenses that must be paid during the year, but not in each month. If your car insurance is $1,200 a year, for example, then you should be setting aside $100 each month to cover this expense even if you have already paid your bill for this year. The same sort of procedure may be used to set money aside for a future expense you know you will have to pay, such as the purchase of schoolbooks or equipment for a class you know you will take.

You also need to identify and estimate the amounts you believe you will spend on flexible expenses such as food, entertainment, and similar goods or services. For most students, these expenses make up roughly one-third of their spending, although special situations may change the share for you.

Finally, compare the amounts of your expected income and spending. If your spending is less than your income, you may be able to save money or allocate more of it to other types of flexible spending. If, instead, you realize that you have been planning to spend more than you will receive, you should review your expenses to see where you can make cuts.

## Developing an Annual Budget Worksheet

After completing your one-month budget worksheet, you need to see how it functions. Keep careful records of your income and spending and see if they are essentially the same as your budget. If they are, you may consider yourself an exception: most people encounter unexpected expenses that force them to make adjustments in their budgets. If the exhaust system on your car falls off, for example, you will need to have it repaired regardless of your budget. This may force you to cut down on entertainment expense. Or you may experience an unusually cold winter month, forcing your heating bills up higher than expected. These examples demonstrate why people should always try to have some unallocated funds in their budgets to cover emergencies.

Completing a series of three or four one-month budgets may be necessary for you to become confident that you understand your financial needs and resources. When you feel you have this comprehension, it is time to make a long-term, or annual, budget worksheet. This document should be directed toward achieving life-span goals in addition to covering day-to-day expenses.

In long-term budgets, many fixed expenses turn into flexible expenses. If you find that the rent on your apartment is so high that you are unable to save or buy other goods and services you need, you can look for a less expensive place to live or try to find another person to share your apartment and its costs. If your car loan payments and insurance bills are too large, you might consider trading down to a less expensive car or using public transportation instead of owning a car at all.

Long-term budgets may also help you evaluate your financial situation. For example, your long-term budget may cause you to see that you really need to earn more money if you are to accomplish your life-span goals. You might consequently choose to work more hours during your spring break or summer vacation than you would have if you had had no long-term plan. Long-term budgets tend to give people a sense of direction or purpose that helps them make more responsible consumer choices. You will be less likely to go out with your friends on weekends if you know you will be able to take a vacation or buy a new car by saving the money you would have spent on such outings.

# Reviewing Your Budget

Every successful budget requires a review of what has happened. You must be aware, however, that the money spent during the first several months in a particular category may be very different from the amount budgeted. That is to be expected; the budget will become more realistic as the process continues. Every few months, analyze your budget to see which categories are seriously out of line with reality.

It's also a good idea to rethink the budget process itself at least once a year. With time, you can predict relatively accurately what size each budget category should be. The next step is to determine for each category whether you are obtaining the maximum amount of satisfaction from the budgeted income. If you aren't, perhaps one category should be expanded and another one contracted. Thus, a budget will be continuously updated to reflect the individual's or household's understanding of the level of satisfaction received from each budget category. Additionally, family changes will necessitate modifications in budget categories. For instance, when children grow up, the family's expenditure patterns change, and when a homemaker takes a job, income patterns also change.

Making and keeping a long-term budget plan can help you achieve your life-span goals.

# Fitting It All into a Life-Span Plan

Today, there is much talk about retirement and the decisions that must be made if it is to be a happy period. There is also much talk about how increased leisure time will be spent and the need to purchase more leisure-related products. These matters should be considered as part of a life-span plan, one that is revised periodically to take into account changing values, income, and consumption patterns.

Life-span planning is actually based on the establishment and subsequent accomplishment of both mundane and lofty goals. As a consumer, you might begin your planning by drawing up monthly and yearly lists of goals, tasks, and ideas. The monthly list, for example, would tell you when to schedule maintenance on your car, when to have household appliances serviced, what days sales are coming up at various stores, and so on. The yearly list, of course, can do the same thing but probably will be more general and less specific.

## Life-Span Goals Are Not "Carved in Stone"

Your life-span goals are likely to change as time passes. Today, you may think studying history at the Sorbonne is the most important thing you could do in your life. Five years from now, this goal may no longer have much meaning for you. You could be married and starting a family. Then, saving for a home and your children's education could put studying in Paris on the back burner, so to speak. It is a good idea to set aside a period of time to just sit down and think about what you want from life. At a minimum, you should do this once a year. Those who are living with others should be sure to do this as a group. Remember, a household budget is more likely to succeed when everyone knows and agrees where the family is headed.

When you realize that your life-span goals have changed, you will also need to make adjustments in your short-term goals. If studying in France is no longer on your horizon, maybe you should spend fewer hours learning French and more hours working. Then you will be able to make a down payment on a home in a few years. And, if owning a home is your family's central goal, there should be agreement on what type of home will be best for all members of the household. While you might like a rustic log house in a rural area, your spouse might be more interested in a contemporary home near schools and shopping malls. Simply agreeing that "we want a home of our own" is rarely enough to assure harmony in carrying out a budget process. For most people, compromise is necessary.

## It's Better to Face Reality

Long-range planning is quite simple in concept but sometimes difficult to put into operation, mainly because people don't always like to face the reality of what is entailed in attaining certain goals. For example, the only way you can save is to consume fewer goods and services. Or, if you want to consume more goods and services, you will need to make more income. In most cases, the only way you can make more income is to become more productive in your job or to change jobs. That may involve going to night school, obtaining additional training, or working on weekends. If you're aware of such requirements, then you may be more willing to accept the cost of attaining a particular goal for yourself and/or your family.

# Budget-Assistance Programs

A number of resources are available to help you with your budgeting or to assist you if you are having difficulties meeting your financial goals.

## The Limits of Budget Software

Although many different software programs promise to ease you through the budget process, they have no value unless you use them responsibly. The user must gather and enter all of the data that a budget program needs to work. Omitting "just a few" payments can render your program useless and make the output it provides unreliable. Many people will tell you that budget software is little more than an automated ledger. If you choose to purchase such a program, recognize that you will end up doing most of the work that enables it to function properly.

## Consumer Credit Counselors

If you are having difficulties with budgeting or debt management, there is a nationwide organization you can turn to for advice and counseling. Consumer Credit Counselors, Inc., is a nonprofit organization that offers financial

**Question for Thought & Discussion #5:**
How have your life-span goals changed in recent years? How have these changes affected the way you budget your income?

# Cyber consumer
## Online Budgeting

There are literally hundreds of software programs that can help you construct and maintain a budget on your home computer. These programs may be purchased at many retail outlets including Barnes & Noble or online. One of the most popular programs in 2003 was Mvelopes Personal, which placed the following claim on its Web site (http://www.in2m.com):

Discover the real key to reaching your financial goals by using the most effective money management system ever. Mvelopes Personal is an online budgeting system that makes it easy to create an effective personal budget and track every aspect of your spending as it happens, providing you critical information to make better spending decisions wherever you go.

Investigate this and other similar Web sites. What claims are made about the offered products? How much would they cost to purchase?

counseling and budgeting services at no charge. If you need to do more than just trim your budget, this organization also has a debt-management program that can assist you. Although there is a fee for the latter service, it is often nominal and may be waived in some cases. If you wish to contact Consumer Credit Counselors, which has four hundred offices in the United States, look in the white pages of your local telephone directory, or write or contact

National Foundation for Consumer Credit, Inc.
8701 Georgia Ave., Suite 507
Silver Spring, MD 20910
http://www.NFCC.org

The National Foundation can direct you to the nearest local office.

# Confronting Consumer Issues: Budgeting for Recreation and Leisure

You may have heard the saying "All work and no play makes Jack a dull boy." The idea is that people won't have a rewarding life if they work so many hours that they have little time left to enjoy themselves. When consumers construct budgets, they often forget about planning for recreation and leisure time. Yet, like other consumer choices that are not carefully made, a spur-of-the-moment decision to buy a boat or take a vacation is unlikely to be the best use of your earnings or time. Budgets that include plans for recreation and leisure activities are more likely to result in personal satisfaction.

## Question for Thought & Discussion #6:

How do you decide what portion of your income to devote to recreation and leisure?

When constructing a budget, remember to plan for vacations and leisure activities.

## THE GROWTH OF RECREATION

When consumers use time for recreation, they make a trade-off with other uses they could have made of their time. In 1890, the average U.S. employee worked almost 55 hours a week. By 2000, average weekly hours at work had fallen to slightly more than 39 for full-time employees. This steady decline may be seen in Figure 7.2. Millions of other Americans work at part-time jobs that leave them even more time for other activities. Labor-saving devices, such as washing machines and microwave ovens, and other products, such as ready-to-eat and take-out foods and wash-and-wear clothing, have allowed the average U.S. consumer to enjoy a better standard of living and provided more leisure time for recreational activities.

U.S. consumers are also devoting a growing share of their spending to recreational products and activities. In 1933, only 4.7 percent of the average American's spending was used for recreation. As Figure 7.3 shows, by 2000 this share had reached 8.5 percent of consumer spending. In the sixty-seven years between 1933 and 2000, U.S. consumers

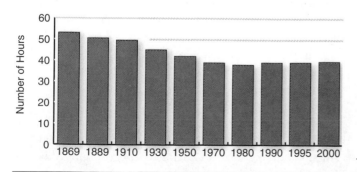

**Figure 7.2** Average Hours Worked per Week, 1869–2000

SOURCE: *Statistical Abstract of the United States, 2001*, p. 374.

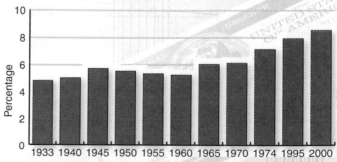

**Figure 7.3** Percentage of Personal Consumption Expenditures Devoted to Recreation

Back in 1933, only 4.7 percent of personal consumption expenditures were devoted to recreation. This percentage has been rising ever since, today reaching 8.5 percent or more. (The 2000 figure is an estimate.)

SOURCE: *Statistical Abstract of the United States, 2001*, p. 423.

*(Continued on next page)*

**Confronting Consumer Issues (Continued)**

increased their spending for recreation per person by more than 550 percent, even after adjusting for inflation.

Americans use their increased leisure time in a variety of ways. They travel more and participate in more organized sports. Many people have purchased equipment such as swimming pools and boats that allow them to enjoy their recreation without leaving their homes or communities. Numerous consumers use their time to visit national and state parks. The list goes on and on.

## VACATIONS

Between 1975 and 1999, spending on tourism in the United States grew from almost $95 billion to over $475 billion a year. In the same period, employment in the travel industry doubled. Although these figures show how tourism has grown, it is less clear that there has been a corresponding increase in people's enjoyment and satisfaction. Consumers benefit from careful shopping when they spend their recreation dollars. You can take many steps to control your costs and make your vacation a success.

**Investigate Your Alternatives**   If you plan to visit an unfamiliar place, before you leave be sure to investigate what there is to see and do at your destination.

**Travel Guides**   The American Automobile Association (AAA) publishes travel guides full of detailed information about points of interest, hotels and restaurants and directions for reaching these places. These guides are free to the AAA's members.

Another useful source that may be available free in most libraries is the various editions of the *Mobil Travel Guide,* which lists hotels and restaurants that have paid to have their names included. Although many fine establishments will not appear because they have not paid a fee, consumers can be reasonably sure that the ones included will meet minimum standards of quality and that their prices will be within the ranges indicated in these booklets. Many listed firms provide discount coupons in the *Mobil Travel Guide* as well as giving discounts directly to AAA members. Other travel guides with specific and useful information may be found in most libraries.

**Finding a Place to Stay**   Another way to find a place to stay is by using the toll-free numbers provided by many hotel and motel chains or by making reservations over the Internet. Virtually all hotels maintain their own Web sites or subscribe to an Internet listing service that will make reservations for them. All you need to do is to carry out an Internet search using the names of locations you wish to visit as your initial key words. These will almost inevitably lead you to listings of accommodations that often allow you to make reservations online. In most cases, the hotel or motel will provide a telephone number or e-mail address that you can use to obtain more information.

**Bed-and-Breakfasts**   One possibility you should not overlook is staying in a **bed-and-breakfast,** a private home where the owner rents rooms and also serves breakfast. The rooms often have a separate bath and entry. There are organizations of bed-and-breakfasts that provide brief descriptions of their members' establishments, both in print and on the Internet. Most libraries have books that tell about bed-and-breakfasts in different regions of the country and in foreign nations. Internet listings, however, are more up to date, and you can download photographs of the buildings, grounds, and rooms. An advantage of staying at a bed-and-breakfast is that it gives you an opportunity to gain personal knowledge of the area you are visiting and to plan the best use of your time.

**Rent a Condominium or Apartment**   Some consumers prefer to rent a condominium or an apartment instead of staying in a hotel, particularly if they expect to stay in one location for several days or more. Condominiums offer more space than most hotels, plus the use of a kitchen, which can save you the cost of some or all restaurant meals. A problem with many condominium rentals is that you may have to make your own beds and wash your own dishes. They do offer lower-cost vacations, however, particularly for large families. Travel agencies have access to listings of condominiums that may be rented, or you can write to local chambers of commerce.

**Time-Sharing Plans**   Another way of possibly reducing vacation costs is to purchase a resort **time-sharing** plan. Under these agreements, you buy the use of a resort facility for a specified period of time each year (most often two weeks). You are then allowed either to use your time in your own resort or trade your time with someone who owns a similar plan in another vacation spot.

In some time-sharing plans, you buy only the right to use a facility that belongs to someone else. In others, you actually become a part-owner of a condominium. Ownership plans can be quite expensive, typically requiring initial payments of $5,000 to $25,000. They also require the payment of a maintenance fee that may be as much as $500 a year. One advantage of these plans is the opportunity they provide to earn a capital gain if the property increases in value. Of course, they may also result in a loss if the property's value falls, and they may be difficult to sell quickly.

Generally, time-sharing plans are most appropriate for consumers who are financially able to commit a substantial amount of money over an extended period of time. Although many time-sharing plans are legitimate and offer consumers quality vacations at reasonable cost, some have involved deceit and fraud. Have a lawyer examine any time-sharing

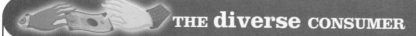

## THE diverse CONSUMER

### Going Home

Nearly half of the U.S. population is made up of people who are first-, second-, or third-generation immigrants. For many of these people, making a visit to the "old country" is one of their life-span goals.

Consider Chinese Americans, who make up roughly 1 percent of the U.S. population. Many Chinese Americans are descended from contract laborers who traveled to the United States at the turn of the twentieth century. Another large group came to the United States to escape communism when forces led by Mao Tse-tung defeated the Nationalist Chinese government in 1949. One such person is Lynn Pan, a naturalized U.S. citizen who has become a well-known author.

Pan came to this country as a young girl. Although she lived with her father for more than forty years, she never learned about her family in China. Her father refused to talk about his life there or what happened after he left. After his death, Pan decided to return to

China to learn about her family. While she was there, she visited cemeteries, former family homes and other buildings, and long-lost relatives. After her return from China, she wrote a book, *Tracing It Home: A Chinese Journey*, that described her travels and what she experienced.

Like Pan, millions of U.S. citizens are interested in learning about their families. Every year more than twenty million people from the United States take trips to other countries. Most of them just cross the border to Canada or Mexico, but at least three million visit countries on a different continent. Most of these people are probably on business or vacation, but some are looking for their roots.

**Where did your ancestors come from? Do you have a desire to visit the places where they lived? Would you consider such a trip a form of recreation or something else? How would you fit such a trip into your life-span plan?**

---

agreement before you sign it. For further information, write to the American Land Development Association, Resort Time-Sharing Council, 1000 16th Street, N.W., Washington, DC 20036, or visit **http://www.timesharecouncil.net**.

### Should You Use an Internet Travel Service?

There are many Internet travel services that offer quality transportation and accommodations at reduced prices. Although most Internet travel service offers are legitimate and provide good vacation deals to consumers, some do not. Before you use such a service, check it out with your state attorney general's office or the Better Business Bureau. Be sure to investigate the firm's service record and whether it has received many consumer complaints. The Internet offers an excellent way to accomplish this. Simply complete an Internet search using the name of the organization as key words or search an electronic periodical index such as InfoTrac. You will soon learn whether the firm has a good or bad reputation for dealing with its customers.

### Cutting Vacation Costs

Planning, and a willingness to cut some corners, may save you a significant

amount of your travel money. Here are some suggestions for reducing your vacation costs. Travel agents and budget travel books may provide others.

1. Take off-season vacations when possible. Hotel and resort rates may be reduced by as much as 50 percent. Trips to Europe in October, for example, are less costly than similar vacations in July. The historic buildings, museums, and theaters are the same regardless of when you go and are also less likely to be crowded with other tourists. Timing is a crucial factor in saving money on vacations.

2. Take advantage of airfare bargains. Always book your flights well in advance of the dates you intend to travel, and inquire about rates with several travel agents. If you can be flexible as to the time you are willing to fly, you often can find lower rates. Also, charter flights may be available through travel agencies or on the Internet at much lower costs than regularly scheduled flights. There are drawbacks to charter flights, however. They often depart at late hours and are well known for not leaving or arriving on time. Some charter organizations have gone out of business, leaving passengers with worthless tickets. Buying travel insurance may be worth the money if you use a charter airline.

3. If you are a student, obtain an international student identity card from the Council on International Educational Exchange, 777 United Nations Plaza, New York, NY 10017 (**http://www.ciee.org**). Many hotels, restaurants, theaters, and museums give discounts to people who have such cards.

### Question for Thought & Discussion #7:

How did you plan your last vacation? Were you satisfied with the results? What steps could you take to ensure that you have a better experience next time?

*(Continued on next page)*

**Confronting Consumer Issues (Continued)**

4. Investigate the benefits and costs of passes that provide unlimited travel on trains or buses. For travel in Europe, you can buy the Eurail pass (purchasable only here, before you travel). U.S. bus lines sometimes offer similar passes.

5. Consider staying in youth hostels, inexpensive dormitory-style accommodations that exist throughout the world. Fees in the United States range from $10 to $20 a day. In other countries, the fees may be somewhat higher or lower but will always be much less expensive than staying in a hotel. For information on youth hostels, write American Youth Hostels, AYH National Campus, Delaplane, VA 22025, or go to **http://www.hostelweb.com**.

6. Choose hotels or motels that do not charge extra for parking. Consider older, downtown hotels that most likely have lower rates than newer suburban motels.

7. Remember, it is not necessary to eat every meal in a restaurant. Picnics along the road or even in your car can be fun and are always less expensive than restaurant meals.

8. Shop around for the best travel deal. If you know where you want to go, invest the time and money to call several hotels or resorts to find out what each has to offer. It is not necessary to stay in a hotel that is on the beach to enjoy the beach, for example. Staying a few blocks away may save you money and allow you to get a better night's sleep because of less noise. And don't pay for more room than you need. Essentially, a hotel is mostly a place to sleep. An impressive lobby with extrathick carpets will not improve the quality of your rest.

## PARTICIPATING IN SPORTS

More and more Americans are using their leisure time to participate in various sports. The number of recreational golfers, for example, increased from about 17 million to 27 million between 1985 and 1999. Adult softball teams grew from 150,000 to 185,000 in the same time. The number of U.S. skiers almost doubled, and bicyclists increased by roughly 30 percent. Participation in many other sports has grown at similar rates.

### The Cost of Golf
Taking part in recreational sports requires a commitment of time and money. U.S. consumers spent almost $80 billion for sports equipment in 1999, up about 80 percent from the $48 billion spent in 1990. This growth in spending is not surprising. Consider the costs of being a golfer as an example. In 1999 a basic set of golf clubs cost at least a few hundred dollars, and a consumer could easily spend a thousand dollars or more on them. Greens fees and club memberships cost the average golfer another $750 a year. Almost 570 million eighteen-hole rounds were played in 1999, of which roughly half involved the rental of a golf cart. When the costs of food and beverages consumed before or after

playing golf are included, an estimated $50 billion was spent on this sport in 1999. Similar data can be found for other sports.

### Sources of Sports Equipment
Generally, sports equipment can be rented at a fraction of the cost of buying your own. For example, a set of downhill ski equipment that cost as much as $600 could be rented at most ski resorts for $40 to $70 a day in 1999. Many consumers have purchased expensive ski equipment before they were sure that they would continue with the sport or that the particular type of equipment they chose was the best for their needs. People who want to buy such sports equipment may benefit from investigating publications such as *Consumer Reports.*

Buying sports equipment is often the result of social pressure as much as need. Choosing the absolute best equipment may be rational for Olympic Games contenders, but it seems less reasonable for those who ski only a few times a year. In the 1990s, sales of skiing equipment grew by more than 50 percent, reaching an annual average of more than $700 million in 1999. Some of this money could certainly have been saved if consumers had rented equipment until they were sure they needed their own and learned what type was the best choice for their money.

## BUYING RECREATIONAL VEHICLES

Total sales for recreational vehicles—pleasure boats, motor homes, snowmobiles, and all-terrain vehicles and bicycles—increased by over 50 percent in the 1990s. In 1999, for example, 2.6 million new pleasure boats were sold for $13.6 billion. Although owning a boat or a motor home is often an enjoyable experience, it is also an expensive proposition. Consumers need to make the decision to buy such a product only after careful evaluation of the costs and benefits.

### Costs of Recreational Vehicles
Let's consider the choice of buying a motor home as an example. Many consumers picture themselves leisurely driving through the countryside, cooking their own meals, and staying in picturesque campgrounds with all the comforts of home. And they imagine having these benefits at a fraction of the cost of renting a room in a hotel and eating out in restaurants. They probably fail to realize all the costs involved in this dream. Most experts believe that owning and using a motor home for vacations increases the cost of recreation.

A typical new motor home in 1999 cost in the range of $40,000 to $60,000 depending on its size and features. Motor homes are not just expensive to buy, however; they are very costly to own and operate. Insurance for a motor home may cost $1,200 or more a year. These vehicles are expensive to register in most states because they are heavier than ordinary cars. They get poor gas mileage and tend to be expensive to repair. When you park them in a commercial campsite, or even in a state or national park, you can expect to pay a fee of $10 to $30 a night. They are not convenient to park or easy to drive, particularly in cities.

The greatest expense of buying a new motor home for most people is the cost of depreciation. Motor homes do not hold their value well. They commonly lose as much as 50 percent of their value in their first two or three years. They are also often hard to sell.

### Don't Expect to Save Money

The average cost of owning a motor home has been estimated to range from $8,000 to $20,000 a year. You can stay in a hotel and eat out for many days for that amount of money. If you ever decide to own a motor home, consider buying a used one that has already suffered much of the rapid depreciation that takes place in the years after it is first sold. In any case, don't expect to save money unless you have a large family and spend many days on vacation every year.

## VISITING NATIONAL AND STATE PARKS

For generations, our national and state parks have been a popular destination for recreational trips. In 1999, patronage of national parks exceeded 285 million visitor days; another 800 million visitor days were recorded at state parks. U.S. consumers spent more than $800 million for entry fees to make these visits. Our parks have become so popular that some have become overcrowded and have even begun to limit the number of people who may visit on a given day. Traffic jams on roads in national parks commonly reduce the speed of traffic to little more than 10 miles an hour, causing some national parks to exclude private vehicles.

To make the best use of public parks, consumers may choose to travel at times that are not the most popular. You can call the park's information center (the number can be found in most travel guides) about the best times to visit. At the same time, you may make reservations at the many superior hotels and resorts that exist around our parks. It is usually impossible to stay in these facilities without making reservations months in advance. Even making advance reservations for a campsite is usually necessary.

One reason our parks are so popular is their relatively low prices, made possible because the cost of maintaining the parks has traditionally been shared between those who use them and taxpayers in general. In 1999, only about 14 percent of the cost of running our national parks came from entry fees. The remainder came from government appropriations and fees paid by businesses for the right to operate in the parks. Although entry fees have increased to cover a larger share of the cost of operating our parks, many consumers have opposed these changes because higher fees could limit the use of national parks to people who have substantial incomes.

### Question for Thought & Discussion #8:

Many college students find it difficult to make time for recreation and leisure. Do you make an effort to set aside time each week to relax? How do the pressures on your time affect your outlook on life?

## MAKING TIME FOR LEISURE

Many people who do not have the time for or can't afford vacations spend their leisure hours at or near their homes. They take walks in parks or play games with their children. Although staying near your home is often a less expensive use of recreational time, it may not be free. In 1999, for example, more than a million Americans purchased home swimming pools. The average price for an above-ground pool at that time was $4,500, and for in-ground pools, the average was just over $15,000. The chemicals and electricity to run a pump for a pool add another $250 or more to this cost each year.

Some consumers relax by working in their gardens, but even this can be expensive. In 1999 Americans spent more than $7.6 billion on nursery products. Other types of spending in 1999 that could be considered recreational included $9 billion on renting videotapes or DVDs, $5 billion on photographic equipment, and more than $10 billion on pets and pet supplies.

Recreation and leisure are necessary to refresh your body and your mind. Most people are more productive if they devote a reasonable share of their time to having a good time. Yet some people never seem to find time for recreation. When they are not working at a job to earn an income, they are engaged in tasks around the house. Such people need to take control of their lives by scheduling time for rest and recreation. It is desirable to find the balance between work and play that best fits one's needs and personality.

A budget can help in this regard. The planning process can help you put into perspective the kind of life you wish to lead. Devoting time and money to recreation and leisure requires a trade-off. Most people can't earn money or advance their careers while they are having fun. But is working and earning money all you want out of life? How much more is another dollar worth if you don't have time to spend it? The procedure of making a budget, and setting various short-term and life-span goals in the process, can help you sort out these lifestyle issues.

### Key Terms:

**Bed-and-Breakfast**   A business run by an individual who rents rooms in his or her home and provides breakfasts to travelers.

**Time-Sharing**   An agreement through which consumers purchase the right to use a vacation facility for a specified period of time each year.

## Chapter Summary

1. Even the richest among us do not have an unlimited budget and must, therefore, make choices.

2. Budgeting, or making a spending plan, forces you to realize that you face the constraint of a limited income and that you must make trade-offs among those things you desire to purchase.

3. Democratic decision making means involving all members of the household in the budget-making process. Holding a formal or informal family council is one way to allow each member of the family to participate in budget formulation. Learning about trade-offs and budget constraints is a valuable lesson in economics for children.

4. A typical household in the United States spends about 80 percent of its after-tax income on food, housing, clothing, transportation, and insurance. As a family's income goes up, the percentage spent on food decreases and the percentage spent on housing increases.

5. Goals should be set according to your priorities. Short-term, middle-term, and life-span goals must be realistic, and striving to attain them may involve making trade-offs.

6. The first step toward a realistic budget plan is keeping accurate records and constructing a monthly budget worksheet. Once this is done, longer-term budgeting can be undertaken and an annual budget formed.

7. The overall purpose of budgeting is to ensure that you get the best returns—in terms of health and happiness—from your income. This means your life-span plan must be the ultimate framework for your daily, monthly, and annual budgeting strategies.

8. A number of good software programs are available to assist you in your budgeting effort. Another source of assistance is Consumer Credit Counselors, Inc., a non-profit nationwide program to assist consumers in budgeting and debt management.

9. U.S. consumers now have more leisure time than in the past and are devoting a growing portion of their earnings to recreation. Spending on tourism has increased by more than 350 percent since 1975. Consumers should budget their time and money to obtain the greatest value for their recreation/leisure dollar.

10. A large number of consumers spend their leisure time at or near their homes. This may involve paying substantial amounts of money for swimming pools, sporting goods, or video equipment.

11. Consumers can gain more satisfaction from their lives by choosing the balance between work and leisure that best suits their personalities and financial situations. To accomplish this, they may choose to create budgets via the decision-making process.

## Key Terms

bed-and-breakfast  **163**

fixed expenses  **154**
flexible expenses  **154**

luxury good  **154**
time-sharing  **163**

## Questions for Thought & Discussion

1. How are important financial decisions made in your family? Do these decisions ever lead to family disagreements? How could such disagreements be avoided?

2. What special budget problems have you encountered because you are a student? How have you dealt with them?

3. How does gaining a college education support your chances of reaching a life-span goal?

4. How do you organize and keep records of your financial transactions? Does your system work well? Why or why not?

5. How have your life-span goals changed in recent years? How have these changes affected the way you budget your income?

6. How do you decide what portion of your income to devote to recreation and leisure?

7. How did you plan your last vacation? Were you satisfied with the results? What steps could you take to ensure that you have a better experience next time?

8. Many college students find it difficult to make time for recreation and leisure. Do you make an effort to set aside time each week to relax? How do the pressures on your time affect your outlook on life?

# Things to Do

1. With the help of your reference librarian, go back to the earliest publication you can find from the Department of Labor, Bureau of Labor Statistics, and see what the average U.S. family budget looked like then. How has it changed over the years? Are we spending more on food or less? On housing? What about taxes? A useful publication is *Historical Statistics of the United States,* published by the U.S. Government Printing Office.

2. Identify your most important life-span goal. Explain how your values, personality, and family situation have led you to make this choice. What funds have you allocated in your budget to help you achieve this goal?

3. Investigate forms of recreation that you or your friends participate in that were not available when your parents were your age. How are these activities related to our greater incomes and technological advances? What leisure activities do you think your children might participate in when they are your age?

4. Visit a motor home or boat dealership. What sort of sales pitch does the salesperson give you? Are you told about the money you can save or the fun you will have? What do you believe you are not told that you should know about?

5. Prepare a one-month budget worksheet for yourself by completing each of the following steps. Use the left-hand column for estimated amounts. Record the actual amounts at the end of the month in the right-hand column.

   **a.** *Estimate your income for the next month.*

   |  | Estimated | Actual |
   |---|---|---|
   | Money provided by relatives, or from scholarships or fellowships | $_____ | $_____ |
   | Expected net wages (after withholding) | $_____ | $_____ |
   | Other sources of income | $_____ | $_____ |
   | Total expected income | $_____ | $_____ |

   **b.** *Estimate your fixed expenses for the next month.*

   |  | Estimated | Actual |
   |---|---|---|
   | Rent or mortgage payments | $_____ | $_____ |
   | Utilities (heat, electricity, and so on) | $_____ | $_____ |
   | Loan payments (car, credit card, and so on) | $_____ | $_____ |
   | Insurance (½ of annual costs for car, renters', homeowners', life, and so on) | $_____ | $_____ |
   | College costs (½ of annual costs for tuition, books, lab fees, and so on) | $_____ | $_____ |
   | Other fixed expenses | $_____ | $_____ |
   | Total fixed expenses | $_____ | $_____ |

   **c.** *Estimate your flexible expenses for the next month.*

   |  | Estimated | Actual |
   |---|---|---|
   | Food in your home | $_____ | $_____ |
   | Eating out | $_____ | $_____ |
   | Household items | $_____ | $_____ |
   | Transportation (gas, oil, bus fare, and so on) | $_____ | $_____ |
   | Clothing | $_____ | $_____ |
   | Personal items | $_____ | $_____ |
   | Gifts for others | $_____ | $_____ |
   | Entertainment (movies, sports, and so on) | $_____ | $_____ |
   | Other | $_____ | $_____ |
   | Total flexible expenses | $_____ | $_____ |
   | Total fixed expenses (from above) | $_____ | $_____ |
   | Total expenses | $_____ | $_____ |

   **d.** *Compare your total expected income with your total expected expenses to determine if you should anticipate a shortage or surplus of money at the end of the month.* If a shortage seems likely, you may want to adjust the amounts you have allocated for flexible expenses. Keep careful records of your income and spending over the month and record them in the right-hand column. Compare your estimates with what actually happened. You may choose to make adjustments in your budget and try it again for another month.

## Internet Resources

### Finding Consumer Information on the Internet

The following Web sites have been selected for their relevance to topics discussed in this chapter. Search these sites to locate information that can add to your knowledge of sources of low-cost airfares. These are commercial sites that are designed to market products. They do not represent a comprehensive or balanced description of all such Web sites. Remember, Web addresses change frequently. If any of these addresses no longer function, find similar sites to investigate using any of the search engines available to you.

1. Priceline.com claims that you can name your own price and fly anywhere in the world for up to 40 percent less than the available retail fares. Check out its Web site at http://www.priceline.com.

2. Cheap Tickets offers to help you obtain low prices on airfare, cars, cruises, condo rentals, and more. You can find its Web site at http://www.cheaptickets.com.

3. SkyAuction.com conducts auctions where you can bid on airline tickets, hotel rooms, all-inclusive island getaways, student travel, African safari adventures, and more. Its Web site is at http://www.skyauction.com.

### Shopping on the Internet

The following Web sites have been selected because they offer consumers services similar to those described in this chapter. These are commercial sites that are designed to market products. They do not represent a comprehensive or balanced description of all budget-assistance programs available online. Remember, Web addresses change frequently. If any of these addresses no longer function, find similar sites to investigate using any of the search engines available to you.

1. Household Express offers help for those who are starting to budget for the first time. Its Web site is at http://www.householdexpress.com/budget/body_budget.html.

2. Budget Resource provides self-help kits for personal and household budget management. Its Web site is at http://www.budgetresource.com.

3. MyFinancialSoftware.com offers various software packages to help you prepare a personal or family budget. Its products include budgeting worksheets, spreadsheets for expense tracking, and financial statements for businesses. You can find its Web site at http://www.myfinancialsoftware.com.

### InfoTrac Exercises

Purchasers of new copies of this text are provided with access to the InfoTrac Web site. This Web site links students to thousands of recent articles published in hundreds of periodicals. Use the key words **time shares** or other terms from this chapter to conduct a key-word search. Choose one article that is of particular interest to you and write a brief essay describing what you have learned from the article. Be sure to cite the author and title of the article and the name and date of the publication in which it appeared.

## Selected Readings

Aron, Cindy S. "National Parks, Crowded Now, Crowded Then." *The New York Times,* August 14, 1999, p. A27.

Cameron, W. Bruce. "Downsizing at the Dinner Table." *Time,* September 17, 2001, p. 105.

"Food Fight." *Marriage Partnership,* Spring 2003, pp. 22–23.

Halperin, Roy. "Do We Really Have Less Time Today?" *Good Housekeeping,* July 2002, p. 129.

Hardin, Drew. "Camping Gear Guide." *Hunting,* December 2002, pp. 15–17.

Kimmelman, John. "How to Build a Budget." *On,* January-February 2002, p. 58.

Lehrer, Ed. "Bargain Travel!" *Insight on the News,* January 11, 1999, p. 10.

"Making and Keeping a Budget." *Ebony,* September 1999, p. 32.

"National Park Service Camping Guide." *Motorhome,* August 2002, p. 106.

Neibert, Angel. "Responsible Spending." *Better Homes and Gardens,* April 2002, p. 15.

Papmehl, Ann. "In Praise of Budgeting." *CMA Management,* February 2002, pp. 48–49.

Quinn, Jane Bryant. "Yes It's Later Than You Think." *Newsweek,* December 4, 2000, p. 47.

Triverio, Jennifer. "Traveling.com." *PC Magazine,* April 6, 1999, p. 253.

"Who Controls the Purse Strings?" *Fast Company,* November 2001, p. 64.

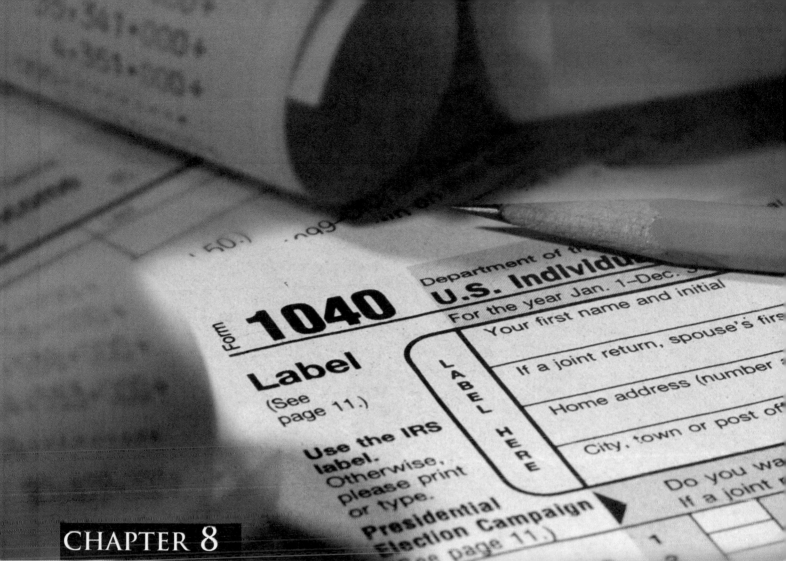

Photodisc/Getty Images

# CHAPTER 8

# Paying for the Government

**After reading this chapter, you should be able to answer the following questions:**

- Why are taxes necessary?
- On what principles are different taxes based?
- How has the federal income tax system evolved?
- How have the rules under which federal income tax liability is determined changed in recent years?
- How can consumers prepare their federal income tax returns so that their tax payments are no greater than is legally required?

**TABLE 8.1**
**Individual Income Taxes and Social Security Taxes as a Percentage of Federal Revenues, 1980–2003**

| Fiscal Year | Percentage of Federal Revenues |
|---|---|
| 1980 | 54.7% |
| 1985 | 63.6% |
| 1990 | 63.7% |
| 1995 | 61.6% |
| 2003 | 65.7% |

SOURCE: Based on data from *Economic Indicators*, December 2002, p. 33.

Our government provides numerous goods and services, such as a court system, police, firefighters, public schools, public libraries, and myriad other programs that most citizens could not afford to pay for themselves. These goods and services help the people of our nation live better and our economic system operate more efficiently. All government services, however, must be paid for in one way or another.

The government raises money to pay its expenses by shifting purchasing power from people and businesses to its own spending, either through taxes or borrowing. In Table 8.1, you can see that in most recent years an increasing portion of federal tax revenue has been paid by individuals through income taxes and Social Security taxes. In 1980, individuals paid 54.7 percent of total federal tax revenues. By 2003, this proportion was projected to be 65.7 percent. This means that, on average, Americans now work more days to earn money that they pay to the government in taxes than they worked in previous years.

In 2002, the Tax Foundation of Washington, D.C., estimated that to earn tax money, the average American worked from the first of the year until April 27, the date the Tax Foundation dubbed "Tax Freedom Day." Only after that date is the average American working for himself or herself. Figure 8.1 shows that in 1930 "Tax Freedom Day" fell on February 13. Of course, the government provided far fewer goods and services at that time. Figure 8.2 compares sources of federal revenue and types of spending in 1980 and 2001. What do these graphs show about changes in the role of the federal government in our lives?

# The Whys and Hows of Taxation

Governments—federal, state, and local—have various methods of taxation at their disposal. The best known, of course, is the federal personal income tax. At the state and local levels, property taxes make up the bulk of the taxes collected. In addition to these taxes, there are corporate income taxes, sales taxes, excise taxes, inheritance taxes, and gift taxes, not all of which can be investigated in detail here.

## First, a Little Theory

Naturally, everybody would prefer a tax that someone else pays. Because we all think that way, no tax could be devised that everyone would favor. Economists and philosophers have come up with alternative justifications for different ways of taxing. The two most prevalent principles of taxation are benefits and ability to pay.

**THE BENEFITS PRINCIPLE**   One widely accepted doctrine of taxation is the *benefits principle*. According to this principle, people should be taxed in proportion to the benefits they receive from government services. The more they benefit, the more they should pay; if they benefit little, they should pay little. This principle of taxation has problems in application, however. First of all, how do we determine the value people place on the goods and services the government provides? Can we ask them? If people think that others will pay their way, they will claim, on being asked, that they receive no value from government services.

**Cyber consumer**
**A Tax Watchdog**

In its own words, the Tax Foundation's mission is to "monitor fiscal [taxing and spending] policies at the federal, state, and local levels." It is a nonprofit, nonpartisan educational organization that gathers and analyzes information about government taxing and spending policies. Visit its Web site at http://www. taxfoundation.org to investigate its latest activities.

**Figure 8.1** Tax Freedom Day

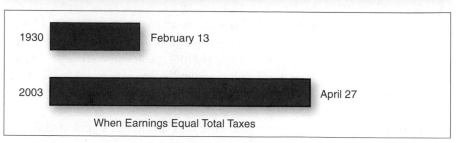

1930     February 13

2003     April 27

When Earnings Equal Total Taxes

**Figure 8.2** The Federal Government Dollar

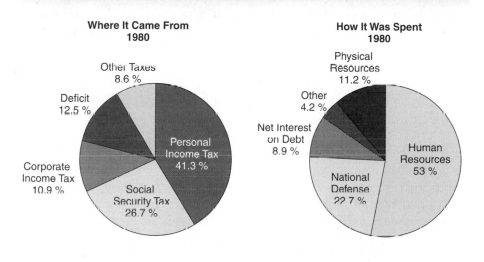

**Where It Came From**
**1980**

Other Taxes
8.6 %

Deficit
12.5 %

Corporate
Income Tax
10.9 %

Personal
Income Tax
41.3 %

Social
Security Tax
26.7 %

**How It Was Spent**
**1980**

Physical
Resources
11.2 %

Other
4.2 %

Net Interest
on Debt
8.9 %

Human
Resources
53 %

National
Defense
22.7 %

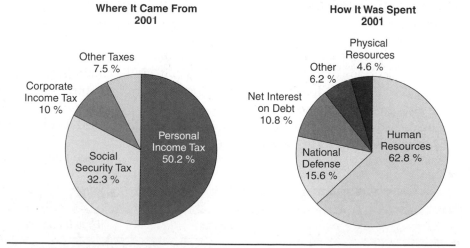

**Where It Came From**
**2001**

Other Taxes
7.5 %

Corporate
Income Tax
10 %

Personal
Income Tax
50.2 %

Social
Security Tax
32.3 %

**How It Was Spent**
**2001**

Physical
Resources
4.6 %

Other
6.2 %

Net Interest
on Debt
10.8 %

Human
Resources
62.8 %

National
Defense
15.6 %

*Note:* Part of Social Security taxes are paid by employers. This explains the difference between this figure and the percentage of tax revenue paid by individuals for income and Social Security taxes in Table 8.1.

SOURCE: *Statistical Abstract of the United States, 2001*, p. 305.

**Question for Thought & Discussion #1:**
To what extent is your ability to attend college dependent on government spending?

For example, they will say they are unwilling to pay for national defense because they do not want it and it is of no value to them. This is the **free-rider problem.** We are free riders if we think we can get away with it. If you think everybody else will pay for what you want, then you will gladly let them do so. The problem is schematized in Table 8.2. How much national defense will you benefit from if everyone pays except you? $300,000,000,000. How much will there be if you also pay? $300,000,000,300. Your $300 adds a negligible amount to the total. If you thought everyone else would pay, wouldn't you be tempted to get a free ride if you could?

One way out of this dilemma is to ensure that the higher a person's income, the more services he or she receives and, therefore, the more value he or she gets from goods and services provided by the government. If we assume that people receive increases in government services that are *proportional* to their incomes, then we can use this benefit to justify requiring them to pay more taxes.

**THE ABILITY-TO-PAY PRINCIPLE**   The second principle of taxation states that those who receive greater amounts of income should pay a larger proportion of their income in taxes than people who receive smaller amounts of income. In other words, those who are better able to pay more taxes should do so.

**PROPORTIONAL TAXES**   **Proportional taxes** are not examples of taxes based on ability to pay. They take the same share of everyone's income regardless of how much she or he makes. There are no true examples of proportional taxes in the United States. Even the Social Security and Medicare taxes, which take 7.65 of earned income, are not proportional. These taxes are not imposed on many types of income including dividends, interest, and profits earned from the sale of stock or other assets. Furthermore, the Social Security tax is applied only up to a certain level of earned income and not to amounts above that level. In general, people with smaller incomes pay a higher percentage of what they earn than people with larger incomes.

**PROGRESSIVE TAXES**   To follow the ability-to-pay principle, a tax must be progressive. **Progressive taxes** take a larger share of people's income as their incomes grow. Federal personal income taxes imposed on **taxable income** are progressive because as people earn more, the **marginal tax rate** they pay rises from 10 to 35 percent. Taxable income is the amount that taxes are applied to after various deductions and exemptions are subtracted from a person's gross or total income. The marginal tax rate is the share of the next dollar earned that is paid in taxes. A concise explanation of deductions and exemptions is provided toward the end of this chapter.

**Free-Rider Problem**   When individuals attempt to receive benefits from a good or service without paying their appropriate share.

**Proportional Tax**   A tax imposed in such a way that all people pay the same share of their income in tax.

**Progressive Tax**   A tax imposed in such a way that the greater your income is, the greater the share of that income you will pay in tax.

**Taxable Income**   A person's income that is subject to tax; computed by subtracting various deductions and exemptions from gross income.

**Marginal Tax Rate**   The share of the next dollar earned that must be paid in taxes.

| TABLE 8.2 | Scoreboard for National Defense | |
|---|---|---|
| | **If You Pay** | **If You Do Not Pay** |
| If Everyone Else Pays | $300,000,000,300 | $300,000,000,000 |
| If No One Else Pays | $300 | $0 |

**REGRESSIVE TAXES**   As you might expect, there is another type of tax that is the opposite of a progressive tax. A **regressive tax** takes a smaller share of people's income as their incomes grow.

Many people are surprised to learn that the majority of state and local taxes, as well as many federal taxes, are regressive. Consider the **excise tax** that is collected each time you purchase a gallon of gasoline, for example. Suppose this tax is $.45 per gallon in your state. Shelly is a corporate vice president who earns $100,000 each year. She buys 1,000 gallons of gasoline to operate her Cadillac and therefore pays $450 in gasoline tax each year. This tax is .45 percent of her income. Phil has a job cleaning Shelly's office at night. He earns only $20,000 each year. Phil drives a 1988 Ford Bronco that doesn't get good gas mileage. He also buys 1,000 gallons of gasoline each year and pays $450 in gasoline tax. But to Phil, this tax is 2.25 percent of his income. Phil, the person with the lower income, pays a higher percentage of his income in gasoline tax than does Shelly. Thus, gasoline taxes are regressive.

**Regressive Tax**   A tax imposed in such a way that the greater your income is, the smaller the percentage of that income you will pay in tax.

**Excise Tax**   A tax that is collected from the manufacturer of a product.

## How Our Progressive Income Tax Evolved

The Constitution gives Congress the authority "To Lay and collect Taxes, Duties, Imposts and Excises." At the time the Constitution was drafted, no reference was made to an income tax. But in 1894 the Wilson-Gorman Tariff Act provided for individual income taxes of 2 percent on incomes above $4,000. The country knew about income taxes from the period during the Civil War, when $4.4 million of such taxes were collected. Nonetheless, the concept of income taxation set forth by the Wilson-Gorman Tariff Act was vehemently challenged, and the controversy had to be settled by a Supreme Court decision in 1895. Finally, in 1913, the Sixteenth Amendment was passed:

> The Congress shall have power to lay and collect taxes on incomes, from whatever source derived, without apportionment among the several States, and without regard to any census or enumeration.

Section 2 of the Underwood-Simmons Tariff Act of 1913 provided for a 1 percent rate on taxable income, with an exemption of $3,000, plus $1,000 more to a married head of household. Notice the concept of exempting the first several thousand dollars of income from taxes. This has continued to the present time in the form of personal exemptions and standard deductions.

The Underwood-Simmons Tariff Act also provided for a surtax that was levied progressively on income over $20,000, with a maximum total tax rate of 7 percent on income over $500,000. These taxes may seem paltry in comparison with today's rates, but they were considered quite large in those times. The concept of progressiveness introduced in 1913 met with considerable debate, which continued for several years thereafter.

Undoubtedly, progressiveness is here to stay, at least in principle, but in fact, our personal income tax system is considerably less progressive than it once was. Before 1961 the maximum federal income tax rate was a whopping 91 percent; by 1980 it had been lowered to 70 percent; between 1981 and 1983, it was reduced to 50 percent; after 1986 it became effectively 33 percent; and in 1990 it was once again lowered, this time to 31 percent. These reductions were no accident, and they were not intended just as a tax break for the

**Question for Thought & Discussion #2:**
Why do you think relatively few taxes are progressive? Should this situation be changed? Why or why not?

**Question for Thought & Discussion #3:**
Do you believe that high maximum tax rates discourage wealthy people from working or investing in the economy? How important are tax rates to the success of our economy?

wealthy. The intent of lowering the maximum tax rate, and marginal tax rates in general, was to encourage people to work harder to earn and produce more, and thereby help our economy grow and become more productive.

In 1993, legislation was passed that increased the maximum tax rate to 36 percent for married Americans who earned over $140,000 in taxable income a year and charged taxpayers with taxable incomes over $250,000 an extra 10 percent, bringing their rate to 39.6 percent. This act reversed the trend of the preceding thirty years of lowering tax rates. Most recently, the Jobs Growth Tax Relief Reconciliation Act of 2003 lowered the maximum tax rate to 35 percent.

# Personal Federal Income Tax

As just explained, although the personal income tax structure in the United States is progressive, it is considerably less progressive than it used to be. Before 1986, there were fifteen different federal income tax brackets, ranging from a low of 11 percent on taxable income of $2,480 to a maximum of 50 percent on taxable income that exceeded $175,250. Starting in 1986, Congress passed a series of laws that reduced the number of tax brackets as well as the tax rates themselves. The Tax Reform Act of 2001 and the Jobs Growth and Tax Relief Reconciliation Act of 2003 continued this trend. They changed the tax structure to include six personal income tax brackets (10, 15, 25, 28, 33, and 35 percent) by 2003. President George W. Bush believed that if U.S. taxpayers were allowed to keep more of their income, they would spend more and stimulate the economy.

## A Complicated Tax System

When the federal personal income tax was first levied, it was a fairly straightforward, simple tax on income. Taxpayers had little difficulty calculating how much tax they had to pay the government. Since then, however, federal income tax laws have become an increasingly complicated morass of rules and regulations that confuse and frustrate most taxpayers. It is estimated that in recent years the cost of filling out federal income tax forms, measured in people's time, has been somewhere between $15 billion and $20 billion! And this doesn't include the fees many individuals pay to accountants and lawyers each year for tax assistance.

The creation of various **tax preferences,** or special rules that shelter certain income from taxation, is the major reason why our tax system has become, in the words of no less a personage than Albert Einstein, "the hardest thing in the world to understand." These preferences are shelters that have allowed individuals in high income brackets to take advantage of various business incentives, such as accelerated depreciation, the deduction of intangible oil-drilling expenses, reduced capital gains rates, and so on. Additionally, Congress has created numerous tax preferences to encourage specific types of activities. For example, consider all the energy tax credits that were available in the 1980s for individuals who incorporated energy-conserving technology when constructing or altering their homes and/or business properties. By offering the tax preferences, the government encouraged these individuals to conserve precious energy. Similar preferences are available today for people who buy electric cars.

**Tax Preferences**  A reduced tax rate applied to specific types of income; a legal method of reducing tax liabilities.

**CAPITAL GAINS**    One of the biggest preferences in personal income taxation concerns capital gains rates. A **capital gain** is the positive difference between the buying and selling price of a capital asset, such as a stock, a bond, or a house. If you buy 100 shares of Silver Syndicate Mining stock for $13 each and, being a financial wizard, are able to sell them at least one year later for $67, your capital gain is $5,400. Thanks to legislation passed in 2003, the maximum tax rate charged on this capital gain is 15 percent. Even if you pay income tax on ordinary income at a 35 percent rate, the most you will pay on your $5,400 capital gain is $810.

**MUNICIPAL BONDS**    Another tax preference is available for the **interest** earned on municipal bonds. Municipal bonds are sold by state and local governments when they need to borrow funds. These funds are then used to construct or maintain government equipment and facilities. Most schools, roads, and public buildings are financed through the sale of municipal bonds. The interest paid on municipal bonds is not taxed by the federal government or by the state in which they were issued. This is an important advantage for taxpayers who fall into higher marginal tax brackets. Suppose your marginal tax rate for federal income tax is 35 percent and you pay state taxes at a marginal rate of 8 percent. Together the rates total approximately 41 percent (the combined rate would be less than 43 percent because you could deduct the state taxes from your income before calculating your federal taxes). So, if you earn one more dollar in interest, you will keep only about $.59 of it. But, if you buy a municipal bond and earn a dollar in interest, you will keep it all. This tax preference benefits state and local governments by enabling them to borrow funds at lower interest rates. Lenders are willing to accept a lower rate because they don't have to pay taxes on the income that they earn.

## Taxpayers as Consumers of Government Services

In an ideal world, consumers would pay taxes to a government that provided needed goods or services in a timely and efficient manner. Although this often has happened (at least to some extent) in the United States, at various times U.S. consumers apparently have not received fair value for their tax dollars.

Consider the events that took place in our federal government in the winter of 1995–1996. Federal law requires the passage and implementation of a new federal budget by the first of October of each year. On a number of occasions, however, Congress and the president have failed to reach agreement on a spending plan by this date. In such situations, special legislation called a "continuing resolution" is typically passed to allow the government to continue to operate without a budget for a specific period of time. Without a continuing resolution, all "nonessential" federal services are stopped and the workers who provide these services furloughed (temporarily laid off). In the winter of 1995–1996, differences between President Bill Clinton's administration and the Republican-controlled Congress prevented passage of a budget for over six months and resulted in the federal government's twice closing part of its operations, for a total of more than three weeks. As a result, many U.S. taxpayers did not receive services to which they were entitled and for which they had paid.

**Question for Thought & Discussion #4:**
Do you believe tax preferences are fair? Do you think college expenses should be deductible on income tax returns?

**Capital Gain**    An increase in the value of something you own. Generally, you experience a capital gain when you sell something you own, such as a house or a stock. You compute your capital gain by subtracting the price you paid for whatever you are selling from the price you receive when you sell it.

**Interest**    The cost of using someone else's money.

If consumers have an ethical responsibility to report and pursue a just resolution to problems they have with commercial establishments, one could argue that they have no less of a responsibility to pressure government to provide, in a timely and efficient manner, the services for which it was created. Consumers of government services, for the most part, are also voters who may choose to exercise their constitutional right to petition government (tell their representatives what they believe) or to vote political leaders out of office. Ultimately, consumers can be sovereign in the political process just as they are in our economic system.

# Confronting Consumer Issues: Preparing Your Income Tax Return

I t is safe to say that most people don't enjoy paying income taxes. Whether we like it or not, though, income taxes are here to stay, and nearly 180 million Americans filed personal income tax returns in 2001. As taxpayers and citizens, we are ethically bound to pay our fair share of the cost of government. We have no moral or legal obligation, however, to pay more than the law requires. By keeping good financial records, choosing the appropriate tax forms, and completing the forms correctly, taxpayers can make sure that they pay no more than their legal share.

### Question for Thought & Discussion #5:

How do you maintain your financial records? Do you keep all the records you need to prepare your income tax return in one location where you can find them?

## KEEPING COMPLETE RECORDS

Although taxpayers are furnished with records of many types of income they receive, they are responsible for gathering other kinds of information they need to complete their tax returns. As a general rule, the more you earn and spend, the more important it is to have accurate and complete records.

**Information You Receive from Others**  At the end of the year, employers are required to provide each of their employees, and the federal and state governments, with copies of the **W-2 form,** showing employee gross income and amounts withheld for federal income tax, Social Security and Medicare, and state and local income tax. Employees must include one copy of this form with their tax returns so that the form can be compared with the copy sent to the government by the employer.

Banks report interest paid to taxpayers on Form 1099 INT, copies of which are sent to each depositor and to the Internal Revenue Service (IRS). It is important for taxpayers to report all of this interest income. The IRS enters amounts of interest reported by banks into computer files for each taxpayer. If you fail to report the correct amount of interest income you received, chances are very good that you will hear from the IRS.

Corporations report **dividends** paid to stockholders on Form 1099 DIV. Again, this information is provided to both stockholders and the IRS. Failure to report dividends accurately will almost certainly result in an audit of the tax return.

One way to ease the burden of paying taxes is to hire a tax accountant or professional tax preparer. Or, you can prepare your own return using one of the many computer programs that help organize and store financial records specifically for tax purposes.

**Information You Must Gather for Yourself**  Taxpayers themselves must keep records of many other types of income, such as self-employment earnings, rents, and interest paid by one individual to another, and must report it on their tax returns. Failure to do so is not as likely to result in an audit but still is against the law and can lead to the imposition of interest, penalties, and, in some cases, jail terms when tax evasion is proved.

It is also important for taxpayers to maintain records of their spending and their contributions to charities so that they will be able to determine if they should itemize their **deductions**—expenses that can be subtracted from income before figuring the amount of tax owed. To claim $150 spent on an eye examination and new glasses as a deduction on your income tax, for example, you must be able to prove you spent this money if you are audited. Keeping receipts will also remind you of your deductible spending when you

*(Continued on next page)*

**Confronting Consumer Issues (Continued)**

## CONSUMER CLOSE-UP
# Would You Give a Few Hours?

For many years, the Internal Revenue Service has maintained a Volunteer Income Tax Assistance (VITA) program that provides free basic income tax preparation assistance for individuals of low to moderate income, individuals with disabilities, non-English-speaking taxpayers, and the elderly. The VITA program is staffed almost entirely by volunteers, who are trained at no charge to complete basic income tax returns and receive support from income tax specialists employed by the IRS to handle more difficult situations. Volunteers are trained not only to help taxpayers complete the necessary forms but also to explain how the forms work so that the taxpayers can complete their own returns in future years.

The greatest difficulty the IRS has experienced in providing taxpayer assistance through VITA is a shortage of volunteers in many communities. In 2003, the VITA office in Pueblo, Colorado, had only three volunteers to serve an estimated 10,000 residents who qualified for assistance. As a result, the office offered no walk-in assistance. Taxpayers were forced to call weeks in advance to make appointments. The story was much the same in many other cities, including Phoenix, Tampa, and Philadelphia.

In an effort to address the shortage of volunteers, the IRS has been appealing to colleges to encourage their students to participate in the VITA program. The IRS emphasizes the value of community service and the interpersonal skills students may develop through this type of volunteer work. Figure 8.3 displays a recent promotional piece for VITA that appeared on the IRS Web site at http://www.irs.gov.

**Would you be willing to give a few hours of your time each week to help others complete their income tax returns? How might this experience benefit you?**

## Figure 8.3

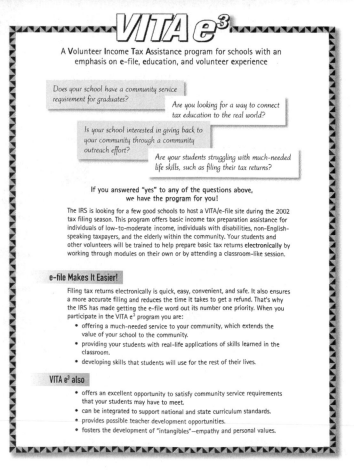

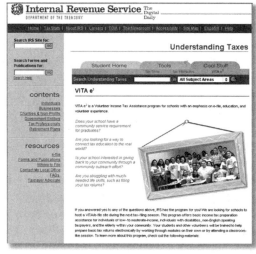

| **TABLE 8.3** | **Deductible Expenses—2002** |
| --- | --- |

- *Medical and dental expenses*—if amounts exceeded 7.5 percent of adjusted gross income.
- *Taxes paid*—state and local taxes paid, but not sales taxes.
- *Interest paid*—interest payments and certain other payments on home mortgages and some other types of interest, but not interest paid on most consumer loans.
- *Gifts to charity*—charitable contributions of either cash or property (additional forms required for large donations).
- *Casualty and theft losses*—uninsured losses resulting from accidents or theft if amounts exceeded 10 percent of adjusted gross income and had values over $100.
- *Moving expenses*—cost of moving that was necessary to accept new employment and was more than thirty-five miles from the former place of employment (additional forms required for this deduction).
- *Job expenses and most other miscellaneous deductions*—part of many job-related expenses and costs associated with completing tax returns if amounts exceeded 2 percent of adjusted gross income.
- *Other miscellaneous deductions*—for example, part of the loss from a failed pension plan if amount exceeded 2 percent of adjusted gross income.

*Note:* Adjusted gross income is income before most deductions and exemptions are subtracted.

prepare your tax return. The types of spending that are deductible are described in Table 8.3 on page 177.

## CHOOSING AND COMPLETING THE PROPER FORM

Taxpayers must choose among three basic forms when they file their tax returns. Each offers benefits and costs taxpayers must weigh to decide if the form is right for their personal situations.

**Forms 1040EZ and 1040A** The simplest way to file an income tax return is on Form 1040EZ. This form was given its name because it is relatively easy (EZ) to complete. To use this form in 2002, taxpayers could be either single or married but had to be under age sixty-five and could not be blind. They must have provided their own support so they could claim themselves as an **exemption**—taxpayers are allowed to reduce their taxable income for each exemption they claim; that is, for each person they support. Their taxable income had to be less than $50,000, and less than $400 of it could have come from interest income. They could have no adjustments to their income, they could owe no special taxes (such as for self-employment), and they could claim no tax credits (for low-income parents). Although this may seem like a long list of qualifications, about one-third of the returns filed in 2001 involved Form 1040EZ.

Many people who were over age sixty-five, had additional exemptions they could claim, or had income from sources that disqualified them from using Form 1040EZ could still file their returns on the relatively simple Form 1040A.

**Form 1040** The fact that a person is qualified to use Form 1040EZ or 1040A, however, does not necessarily mean

he or she should choose to do so. A form that allows itemized deductions (discussed shortly) may be more beneficial. If either Form 1040EZ or 1040A is used, the taxpayer is required to take the **standard deduction** for his or her **filing status**, or family situation (single, married, and so on). Table 8.4 shows the standard deductions that could be claimed on 2002 tax returns.

Form 1040 allows taxpayers to list, or "itemize," deductible expenses on Schedule A and to subtract them from income before figuring tax liability. If the amount of itemized deductions is greater than the standard deduction, this is the form to use. The list of expenses that could be claimed as deductions in 2002 appears in Table 8.3.

To itemize deductions, taxpayers must keep accurate and complete records of expenses. Records of most of these costs will not be provided by any organization, and the taxpayer's failure to keep these records may cause her or him to pay more tax than would otherwise have been legally required.

After taxable income is determined, you will find the tax owed in a table in the tax instruction booklet. Table 8.5 on the next page shows part of this table for 2002.

**If You Owe the IRS** If you end up owing the IRS for underpaid taxes, you can send a check, or you can make your payment by having funds automatically withdrawn from your checking or savings account and transferred electronically. You can also charge your tax bill to a credit card if you want. Be careful, though. Credit-card issuers generally charge a 2 percent fee for this service because the IRS refuses to pay a fee similar to the ones paid by businesses that accept credit-card payments.

| **TABLE 8.4** | **Standard Deduction for 2002 Federal Income Tax Returns** |
| --- | --- |

| Filing Status | Standard Deduction |
| --- | --- |
| Single (if you could be claimed as a dependent by someone else, this amount was $750 in 2002) | $5,850 |
| Married, filing joint return, or qualifying widow(er) with dependent child | $8,750 |
| Married, filing separate return | $4,825 |
| Head of household | $8,050 |

*(Continued on next page)*

**Confronting Consumer Issues (Continued)**

| TABLE 8.5 | Tax Rate Schedules for 2003 |
|---|---|

Here are the final rate schedules published by the IRS:

### Single persons:

| If your taxable income is: | Then your tax is: |
|---|---|
| $7,000 or less | 10% of your taxable income |
| Over $7,000 but not over $28,400 | $700.00 + 15% of the amount over $7,000 |
| Over $28,400 but not over $68,800 | $3,910.00 + 25% of the amount over $28,400 |
| Over $68,800 but not over $143,500 | $14,010.00 + 28% of the amount over $68,800 |
| Over $143,500 but not over $311,950 | $34,926.00 + 33% of the amount over $143,500 |
| Over $311,950 | $90,514.50 + 35% of the amount over $311,950 |

### Married persons filing a joint return, and qualifying widow(er):

| If your taxable income is: | Then your tax is: |
|---|---|
| $14,000 or less | 10% of your taxable income |
| Over $14,000 but not over $56,800 | $1,400.00 + 15% of the amount over $14,000 |
| Over $56,800 but not over $114,650 | $7,820.00 + 25% of the amount over $56,800 |
| Over $114,650 but not over $174,700 | $22,282.50 + 28% of the amount over $114,650 |
| Over $174,700 but not over $311,950 | $39,096.50 + 33% of the amount over $174,700 |
| Over $311,950 | $84,389.00 + 35% of the amount over $311,950 |

### Married persons filing separate returns:

| If your taxable income is: | Then your tax is: |
|---|---|
| $7,000 or less | 10% of your taxable income |
| Over $7,000 but not over $28,400 | $700.00 + 15% of the amount over $7,000 |
| Over $28,400 but not over $57,325 | $3,910.00 + 25% of the amount over $28,400 |
| Over $57,325 but not over $87,350 | $11,141.25 + 28% of the amount over $57,325 |
| Over $87,350 but not over $155,975 | $19,548.25 + 33% of the amount over $87,350 |
| Over $155,975 | $42,194.50 + 35% of the amount over $155,975 |

### Head of Household:

| If your taxable income is: | Then your tax is: |
|---|---|
| $10,000 or less | 10% of your taxable income |
| Over $10,000 but not over $38,050 | $1,000.00 + 15% of the amount over $10,000 |
| Over $38,050 but not over $98,250 | $5,207.50 + 25% of the amount over $38,050 |
| Over $98,250 but not over $159,100 | $20,257.50 + 28% of the amount over $98,250 |
| Over $159,100 but not over $311,950 | $37,295.50 + 33% of the amount over $159,100 |
| Over $311,950 | $87,736.00 + 35% of the amount over $311,950 |

SOURCE: Internal Revenue Service, Department of the Treasury.

## DO YOU NEED HELP?

Anybody who itemizes deductions faces at least a minimal ordeal in deciphering and applying current tax laws. One way to ease your tax burden is to hire a tax accountant or other tax preparer to help you in this task. But do you really require help? This is a question you need to ask yourself before tax season comes around. When you have only a few weeks left to file, it is too late to take many steps that could reduce your tax liability.

**Question for Thought & Discussion #6:**
Did you prepare your last income tax return yourself, or did you have assistance? Did you find the process challenging? Do you think you will need assistance in the future if your financial situation becomes more complex?

The complexity of your financial situation and the tax returns you will need to file are the most important factors in determining whether you should seek help in tax preparation. Simply having a large income does not necessarily mean you need assistance. A person might earn $250,000 but have a relatively simple tax return if the income is all from wages and there are no special deductions. (Such a person, however, would probably benefit from tax planning that would make his or her return more complex but would also reduce the taxes owed.) A self-employed person earning only $25,000 from a business operated out of her or his home might need to complete more complicated forms and require more help in tax planning than the individual with the larger income.

### Do You Need a Tax Preparer or a Tax Planner?

A tax preparer and a tax planner do not offer the same services. A *tax preparer* will complete necessary tax forms based on information supplied by his or her customers. A *tax planner,* however, helps customers arrange their financial situations to reduce the amount of their tax liability. If you hire someone to help you with your taxes, you need to decide which sort of service you need. A person with relatively little income and no special financial problems probably does not need the services of a tax planner.

Thousands of individuals and businesses offer both of these services. The best way to begin selecting one is to ask friends and relatives for their recommendations. Another source of information is the Yellow Pages of your phone directory. They list a selection of nationally franchised operations that are well qualified to complete relatively simple returns. H&R Block, which has roughly 8,000 offices throughout the nation, is an example. Also, many individuals offer tax preparation services; some are former IRS employees or at least have passed tests on tax preparation issued by the U.S. Treasury. These individuals will be advertised as *enrolled agents.*

### Help for More Complicated Returns

If you have more complicated needs, you may want to hire a certified public accountant (CPA) who has special training in tax matters. CPAs charge more for their services than firms such as H&R Block, but if they are able to save their customers taxes because of their specialized knowledge, they are worth the extra cost.

The most expensive tax help is offered by tax attorneys. Only people who have special needs or face legal problems related to their taxes should consider this type of help. Paying an attorney $500 to save $100 in taxes makes little sense. Indeed, any type of tax planning should be reserved for people who are wealthy enough, or are in situations complicated enough, to make it worthwhile.

## DO'S AND DON'TS OF HIRING TAX HELP

Following are some guidelines for selecting help in tax preparation or planning if you determine that you need it:

1. Be wary of tax preparers who promise to give you a check for your refund immediately. The preparer is probably offering you a loan on which you will pay interest.
2. Never sign a blank return.
3. Never sign a return prepared in pencil; it can be changed later.
4. Never allow your refund check to be mailed to the preparer.
5. Be wary of tax advisers who "guarantee" refunds, who want a percentage of the refund, or who supposedly "know all the angles."
6. Avoid a tax preparer who advises you to overstate deductions, omit income, or claim fictitious dependents.
7. Make sure the tax preparer signs the return he or she prepares and includes his or her address and tax identification number. (You, however, are legally responsible for virtually all errors on your return, no matter who fills it out, unless a blatant case of fraud is brought against the tax preparer.)
8. Be wary of preparers who claim they will make good any amounts due because of a mistake on your return. Usually, the preparer means that she or he will pay the penalty cost; the tax money due must come from you.
9. Find out the educational background of the preparer. Does the person have a degree in accounting?
10. Use only preparers who have permanent addresses so you will have no difficulty finding the preparer a few months later if problems develop.

## TAX GUIDES

If you hire somebody else to prepare your tax return, it is still a good idea to be somewhat versed in the tax laws affecting your financial situation. Even a minimal knowledge of these laws can help you get the most for your money from a tax preparer and help you take full advantage of the law if you prepare your tax return yourself.

Every year numerous tax guides are published and marketed. Some of the best known are Lasser's *Your Income Tax,* H&R Block's *Income Tax Workbook,* and Arthur Young's *Tax Guide.* The editors of *Consumer Reports* market an annual guide entitled *Guide to Income Tax Preparation.* You can also get *Your Federal Income Tax* free from the Internal Revenue Service in your area, and the IRS has numerous other free instruction booklets for every imaginable deduction for

*(Continued on next page)*

### Confronting Consumer Issues (Continued)

which you may be eligible. You may also find many tax guides online, but you should expect to pay for most of them.

## FILING ONLINE

At the beginning of 2003, the IRS reported that 47 million taxpayers had filed their 2001 personal income tax returns online. According to the IRS, these taxpayers benefited from greater accuracy, greater security, proof of acceptance, faster refunds, and an ability to use the e-file system to complete their state taxes as well. From the IRS's point of view, electronic filing has many advantages. It reduces the amount of paperwork for IRS employees—fewer envelopes to open and less data to enter. Further, e-filing requires taxpayers to use computer programs that will catch many of the mistakes they ordinarily make. The trend toward e-filing is saving the government millions of dollars each year.

> ### Question for Thought & Discussion #7:
> Although about 25 percent of taxpayers file online, 75 percent don't. Why do you believe they have chosen not to file online? Would you file your income tax return online?

**Ways to File Online**   A few years ago the only way to file an income tax return online was to pay an authorized IRS e-file provider a fee to file for you. This cost $30 or more in addition to the expense of having your return completed. As a result, relatively few taxpayers used the e-file option. This situation has now changed. Today, taxpayers can choose from many e-filing alternatives.

**Authorized IRS e-File Providers**   The original way to file online is still available. Thousands of authorized providers across the nation offer this service, but now they give you a choice. You may still choose to pay to have your tax return prepared and sent electronically to the IRS. Alternatively, you may complete your return yourself and then pay a fee to have your data entered into a computer to be sent to the IRS. This method costs less, but you have to do the tax preparation work yourself.

**Do It Yourself at Home**   Taxpayers may now prepare and e-file their individual income tax returns for free using commercial tax preparation software that may be acquired through a Web site at **http://www.firsstgov.gov**. For the majority of taxpayers, there is no cost for downloading this software to their computers. Although the software is free, you still must spend time completing the IRS forms correctly. Filing online does nothing to eliminate the complexity of many tax forms. This is probably why the

majority of e-filers are taxpayers that need to complete only the most basic tax forms.

**Other Online Advantages**   Probably the greatest advantage to filing online for most taxpayers is that they will obtain their refunds sooner. When paper returns are used, refund checks typically take six weeks or longer to arrive. Electronic filers can usually expect to receive their refunds in about two weeks (assuming no mistakes are made). You can shorten this time further by agreeing to have your refund deposited directly into your checking or savings account. Then you don't even have to wait for the mail to deliver your check. The amount simply appears in your account instantly after the refund is approved by the IRS.

## MINIMIZING YOUR TAXES— AND YOUR TAX TIME

Americans in general may be very honest. The IRS estimates that fully 92 percent pay their lawful due to the government. Being honest, though, does not mean you shouldn't take advantage of what is legally your right. That is, you can be very honest and report all income but, at the same time, make sure that you take all legitimate deductions. You owe that to yourself and to your family.

Consult current tax guides, or check with an accountant, to find out what deductions you can take to help reduce your federal income taxes. Whenever you are uncertain about the acceptability of the deduction, you might consider taking a chance. Many deductions are subject to interpretation by the IRS; that is, they are ambiguous, and, if you are audited, you stand as good a chance of winning your case as not winning it. At the most, because this action does not involve fraud or anything illegal, you pay only an interest penalty on the taxes due.

Remember, it's not worthwhile for you to spend weeks filling out tax forms to save a mere $25. You must figure out at what point you should *stop* trying to reduce your tax burden. This, of course, is a function of your tax rate. If you are in the 15 percent bracket, every extra dollar you can find as a legal deduction saves you on average only 15 cents. If it takes you an extra hour to find $10 more of deductions, the benefit to you of those $10 of deductions is on average only $1.50. Is your time worth more than $1.50 an hour?

## WHAT IF YOU CAN'T PAY?

One out of every four people who owe taxes on April 15 cannot pay them at that time. If this should happen to you, file your return anyway. When the Internal Revenue Service later sends you a bill for the taxes due, if you still don't have enough funds to cover the amount, pay what you can and make arrangements with IRS personnel to pay the remainder as soon as possible. You may want to take out a loan to pay the government its due. If you fail to pay, the IRS can attach your paycheck, your bank account, your car, and even your home, if necessary, to collect the taxes. Rarely, though, does the IRS resort to such drastic measures if an individual makes an earnest effort to pay the debt.

If you fail to file your return, you will face a fine of at least 5 percent (and possibly as much as 25 percent) of the amount you owe, in addition to about ½ percent charged each month for late payment.

Another alternative is to file Form 4868, which provides an automatic extension beyond April 15 for filing your return. This does not eliminate or postpone your tax liability, which must be estimated and paid. It does, however, provide extra time to obtain documents that will allow you to pay the smallest legal amount possible.

### How to Avoid a Tax Audit
Being audited by the IRS is time consuming and often traumatic, and it may cost you more tax dollars. Thus, astute taxpayers try to minimize the chances of being audited. Table 8.6 shows the percentage of individual returns that are audited. Note that the higher the income, the greater the chance of a tax audit. Overall, your chances of being audited are much less than 1 in 100. If your income is relatively low and you do not itemize deductions, your chances of being audited are almost zero.

### What Prompts an Audit?
The IRS computer system is programmed to select for an audit returns that don't conform to normal patterns of income and deduction levels. What those patterns are is a well-kept secret, however, and the IRS changes the patterns every few years to prevent taxpayers from "beating the computer." Frequent targets for audit by the IRS are tax returns claiming certain types of deductions, such as travel and entertainment expenses, self-employment deductions, charitable contributions, casualty losses (theft, for example), passive losses, and individual retirement accounts. Audits are also made when an individual who is not qualified to do so uses head-of-household tax tables. For example, two formerly married individuals with joint custody of their children can't both claim to be the head of the household.

A tax audit might also be conducted for the following reasons:

1. Savings and loan associations, employers, and Social Security wage reports revealed information that did not agree with the tax return.
2. More than one return was filed under the same Social Security number.
3. There was a discrepancy between the state income tax return reported by state tax agencies and the federal income tax return.
4. The wrong tax table was used.
5. There were computation errors.

### Backing Up Your Deductions
Using a computer, of course, does not mean that deductions on your claim will automatically be accepted. Human classifiers at the IRS are trained to spot deductions that are not normal. If, for example, you have extremely high medical expenses one year, make a separate schedule of them and attach copies of all the bills to your return. This will assist the classifier and perhaps save you a trip to the local field office. At the same time, though, it's inappropriate to overdo the extra schedules. To reduce your chances of being audited, verify, when you file, anything that might stand out and raise questions.

Human classifiers can quickly spot inconsistencies, so avoid such inconsistencies if possible. Your income after deductions must be enough to buy such essentials as food and clothing, so don't overdo the deductions. And business expenses must be appropriate for your occupation.

### Question for Thought & Discussion #8:
Has anyone in your family ever been audited by the IRS? How did this person prepare for the audit? Was this person satisfied with the result of the audit?

## HOW TO SURVIVE A TAX AUDIT
Should you receive in the mail a note from the IRS that says, "We are examining your federal income tax return for the above year(s) and we find we need additional information to verify your correct tax," don't despair. Many individuals are audited randomly, even though their returns seem to be in order. And approximately one-fifth of those who are audited leave the IRS office without owing more in taxes.

Prior to your scheduled audit, remember that personal attitude helps in a successful negotiation with the auditor. Therefore, you should:

1. *Be prepared.* You will be given at least six weeks to prepare for your interview with the IRS auditing agent. Use this time to study the facts in your case and, if possible, the law relating to the specific deductions being questioned.

| TABLE 8.6 | Percentage of Individual Returns That Were Audited, 1999 | |
| --- | --- | --- |
| **Nonbusiness Income** | | **Chances of an Audit** |
| Under $50,000 | | 0.36% |
| $50,000 to $99,999 (itemized) | | 0.37% |
| $100,000 and over | | 1.14% |

SOURCE: *Statistical Abstract of the United States, 2001*, p. 357.

*(Continued on next page)*

## Confronting Consumer Issues (Continued)

2. *Be businesslike.* Answer the letter from the IRS promptly and help the agent dispose of major issues quickly during the initial interview.

3. *Be cooperative.* Answer all questions, but do not volunteer unsolicited information, unless, of course, the agent has overlooked something that could alter things in your favor.

**Using Those Records** When you are audited, you realize how important record keeping is. An IRS auditor has the right to disallow completely unsubstantiated itemized deductions or to reduce them to what she or he might consider "reasonable." When you are asked to verify specific itemized deductions, provide only the information relating to those deductions, and provide it as completely as possible. You don't want to give the impression that you are hiding something.

For most individuals who are audited, negotiations with the IRS agent proceed smoothly. Once an agreement is reached, you will sign a form stating that you will pay the taxes you owe. See Table 8.7 for some do's and don'ts when you get called in for an audit.

**If You Disagree with the Agent** If there is still a disagreement after the audit is completed, you will get a copy of the agent's audit report, together with a letter telling you your various rights to appeal the findings.

At this point, you have thirty days to act. One way you might have your complaint resolved is by taking it to an IRS *problem resolution officer (PRO)*. The PRO must settle your claim within five working days or advise you as to its status and direct you to someone who can settle it for you. If the PRO or other IRS personnel cannot arrive at a satisfactory solution, you can request a conference with the IRS Appeals Division. The IRS appeals officer is allowed to weigh your appeal against the costs of possible court litigation. These conferences are usually informal, especially if the amount in question is $2,500 or less.

If the agent's position is upheld, you have two alternatives. You can pay the additional tax or wait for a ninety-day letter, which will mean that going to court is the only remaining method of continuing the case.

**Going to Court** If you decide to take your case to the Tax Court, you must file your petition within the ninety days. If the disputed sum isn't over $10,000, the Tax Court will handle your case informally as a small-case claim in its Small Claims Division. If the decision goes against you in this court, you will have to pay the additional taxes.

Think carefully before you go to the Tax Court, however. Statistics show that the chances of winning your case are not very great. Because of this, many taxpayers choose to have their case heard in a U.S. district court, where the chances of success are much higher. Before you use this court, however, you must pay the taxes in dispute.

Obviously, the best thing to do is to try to avoid a dispute in the first place by keeping the best records you can. And, if a dispute does arise, do all you can to settle it with the auditing agent, a PRO, or—if it gets to this point—the IRS appeals officer.

## Key Terms:

**W-2 Form** The form used by employers to report employee income and withholding to the employee and the government.

**Dividend** A share of profit paid by a corporation to its stockholders.

**Deductions** Different types of expenses taxpayers may subtract from their incomes before figuring their tax liability.

**Exemption** An amount of income that may be subtracted from income for each person a taxpayer supports.

**Standard Deduction** An amount all taxpayers are allowed to subtract from their incomes before figuring their tax liability. The amount varies, depending on the taxpayer's filing status.

**Filing Status** The family situation under which taxes are filed: single, married filing jointly, head of household, and so on.

| TABLE 8.7 | What to Do—or NOT to Do—When Audited |
|---|---|
| **Do** dress the way you normally do for business—whether in a finely tailored suit or in jeans. But don't flaunt wealth with expensive jewelry. | **Don't** antagonize the auditor by being late. |
| | **Don't** volunteer information, be chatty, or go to lunch with the auditor. |
| **Do** act natural—but if you can't help being visibly jittery, then tell the auditor you're nervous. | **Don't** walk in without any records—or try to overload the auditor with material. It can backfire on you. |
| **Do** take the audit seriously. Don't joke about it with the auditor or give flippant answers. | **Don't** rush the auditor or allow yourself to be rushed. |
| **Do** bring along a tax adviser, and let the adviser do the talking whenever you can. | **Don't** try for sympathy, plead that "everyone does it," or lash out at taxes in general. It's a waste of time. |
| | **Don't** underestimate the auditor. |

SOURCES: *U.S. News and World Report,* March 24, 1980, p. 81; and Paul N. Strassels (with Robert Wood), *All You Need to Know about the IRS* (New York: Random House, 1979).

# Chapter Summary

1. Taxes are paid to finance the expenses of government. The personal federal income tax, the most important source of government income in this country, accounts for about 50 percent of all federal revenues.

2. The two best-known principles of taxation are the benefits principle and the ability-to-pay principle.

3. Taxes may be classified into three types of structures: *proportional*, in which all taxpayers pay the same percentage of their incomes in tax; *progressive*, in which the percentage of income that is taken in taxes grows as income increases; and *regressive*, in which the percentage of income that is taken in taxes declines as income increases.

4. There are no examples of proportional taxes in the United States. The federal income tax is progressive for most people, and most state and local taxes tend to be regressive.

5. The personal federal income tax in the United States is a progressive tax. A federal income tax was authorized by the Sixteenth Amendment to the Constitution in 1913, and the progressive nature of our tax system was allowed for in the Underwood-Simmons Tariff Act of the same year.

6. Congress has created many tax preferences, either in response to the needs and lobbying efforts of special interest groups or to encourage specific types of activities. The most common preferences are the lower tax rate applied to most capital gains and the tax-free interest paid on municipal bonds.

7. Since 1986 a series of laws have reduced the maximum federal personal income tax rate from 50 percent to 35 percent. By 2003, there were six tax brackets of 10, 15, 25, 28, 33, and 35 percent, depending on each taxpayer's taxable income.

8. Taxpayers need to keep accurate and complete records of their income and expenses to help them complete their tax returns. Taxpayers must choose which of three basic forms to use when they file their returns. Although Forms 1040EZ and 1040A are less difficult to complete, taxpayers must take the standard deduction on these forms. People who have more deductible expenses than the standard deduction will benefit by using Form 1040 and Schedule A to file their tax returns.

9. Consumers should remember that tax preparers and tax planners do not provide the same services. Tax preparers may help people complete tax forms, but they will not be able to reduce a person's tax liability below what it otherwise would have been. In contrast, tax planners help people who have substantial incomes arrange their finances in a way that will reduce their tax liabilities.

10. If you decide you need assistance in preparing your tax return, there are numerous sellers of tax services available, including certified public accountants, storefront tax preparation operations such as H&R Block, and tax attorneys. Some knowledge of tax laws will be useful even if you employ tax help, because it will help you discuss your tax situation more intelligently. Annual tax guides published early in the year and free publications from the IRS can equip you to take maximum advantage of tax laws.

11. In preparing your tax return, remember that your time is valuable, too. Therefore, search for possible tax-saving deductions only as long as the dollars you save outweigh the cost of the time it takes to save them.

12. Nearly 50 million taxpayers file their federal income tax returns online. They benefit from greater accuracy and security and are able to receive refunds more quickly. Taxpayers may pay an authorized e-file provider to prepare and file their returns, or they may acquire software that will allow them to do it themselves. It is possible to pay taxes through automatic electronic withdrawals from bank accounts or to charge them to credit cards. Refunds may also be electronically credited to bank accounts.

13. If you are audited by the IRS, be prepared, businesslike, and cooperative. Bring records to the interview with the IRS agent so that you can substantiate all claims you made on your tax return. If you disagree with the IRS agent, you can take your complaint to an IRS problem resolution officer. If you are still unsatisfied, you can appeal your case to an IRS appeals officer. If IRS personnel can't satisfactorily resolve the problem, you can take your case to the Tax Court or to a federal district court.

# Key Terms

capital gain **173**
deductions **182**
dividend **182**
excise tax **171**

exemption **182**
filing status **182**
free-rider problem **170**
interest **173**
marginal tax rate **170**
progressive tax **170**

proportional tax **170**
regressive tax **171**
standard deduction **182**
tax preference **172**
taxable income **170**
W-2 form **182**

# Questions for Thought & Discussion

1. To what extent is your ability to attend college dependent on government spending?

2. Why do you think relatively few taxes are progressive? Should this situation be changed? Why or why not?

3. Do you believe that high maximum tax rates discourage wealthy people from working or investing in the economy? How important are tax rates to the success of our economy?

4. Do you believe tax preferences are fair? Do you think college expenses should be deductible on income tax returns?

5. How do you maintain your financial records? Do you keep all the records you need to prepare your income tax return in one location where you can find them?

6. Did you prepare your last income tax return yourself, or did you have assistance? Did you find the process challenging? Do you think you will need assistance in the future if your financial situation becomes more complex?

7. Although about 25 percent of taxpayers file online, 75 percent don't. Why do you believe they have chosen not to file online? Would you file your income tax return online?

8. Has anyone in your family ever been audited by the IRS? How did this person prepare for the audit? Was this person satisfied with the result of the audit?

# Things to Do

1. Review recent news articles to find one suggestion for changing federal taxes made by each of the two major political parties. Describe and evaluate each of these suggestions. How would each proposal affect consumers?

2. Find out whether your college provides a service to help students prepare their income tax returns. If it does, find out what assistance is available. If it doesn't, find out what would need to be done to provide such a service.

3. Visit the IRS Web site at **http://www.irs.gov**. Identify a particular tax instruction booklet that interests you. Download a copy or read the booklet online. Report on what you find.

4. Compare two tax preparation guides at your library or from a bookstore. Are they about the same, or are there important differences? If you were in the market to buy a tax preparation guide, which would you purchase? Explain your choice.

# Internet Resources

## Finding Consumer Information on the Internet

The following Web sites have been selected for their relevance to topics discussed in this chapter. Search these sites to locate information that can add to your knowledge of government sources of state income tax preparation advice. Compare the types of help provided by different states, including your own. How helpful is this assistance? Remember, Web addresses change frequently. If any of these addresses no longer function, find similar sites to investigate using any of the search engines available to you.

1. The New York State Department of Taxation and Finance offers tax forms, services, answers to frequently asked tax questions, and e-filing at its Web site. Visit it at **http://www.tax.state.ny.us**.

2. Through its Window on State Government program, the state of Texas offers tax forms and help with state and federal tax returns, including the federal earned income tax credit. Its Web site is at **http://www.window.state.tx.us/m23taxes.html**.

3. Oregon residents can download, view, and print Oregon tax forms and federal forms as well at **http://www.dor.state.or.us/forms.html**.

## Shopping on the Internet

The following Web sites have been selected because they offer consumers services similar to those described in this chapter. These are commercial sites that are designed to market products. They do not represent a comprehensive or balanced description of all tax preparation programs available online. Remember, Web addresses change frequently. If any of these addresses no longer function, find similar sites to investigate using any of the search engines available to you.

1. TurboTax for the Web says that its online tax software "asks easy-to-answer questions and double-checks for errors and missed deductions." You can find its Web site at **http://www.turbotax.com**.

2. TaxACT offers a standard version of its tax preparation software for free. The deluxe or state versions are available for a fee. Its Web site is at **http://www.taxact.com**.

3. TaxBrain.com says that its online tax help center will help you prepare and file your state and federal taxes online. Check out its Web site at **http://www.taxbrain.com**.

## InfoTrac Exercises

 Purchasers of new copies of this text are provided with access to the InfoTrac Web site. This Web site links students to thousands of recent articles published in hundreds of periodicals. Use the key words **Internal Revenue Service** and **audits** or other terms from this chapter to conduct a key-word search. Choose one article that is of particular interest to you and write a brief essay describing what you have learned from the article. Be sure to cite the author and title of the article and the name and date of the publication in which it appeared.

## Selected Readings

Arthur Young & Co. *The Arthur Young Tax Guide.* New York: Ballantine Books. Published annually.

"Deductive Logic." *Business Week,* February 3, 2003, p. 105.

Green, Pamela, and Robert McClelland. "Taxes and Charitable Giving." *National Tax Journal,* September 2001, pp. 433–454.

*How to Prepare Your Personal Income Tax Return.* Englewood Cliffs, NJ: Prentice Hall. Published annually.

"IRS Vigorously Pursing Preparers of Fraudulent Income Tax Returns." *Knight Ridder/Tribune Management News Service,* March 7, 2003, p. ITEM03066154.

Pack, Thomas. "Coping with Tax Law Changes." *Link-Up,* March 2002, p. 12.

Russell, Roger. "Outside Arbitrators Will Help IRS Resolve Disputes." *Accounting Today,* January 27, 2003, pp. 10–11.

*Your Federal Income Tax.* Available free from the Internal Revenue Service.

Digital Vision/Getty Images

# CHAPTER 9
# Choosing a Healthful Diet

**After reading this chapter, you should be able to answer the following questions:**

- How do Americans choose the foods they eat?
- How do government labeling and inspection requirements affect consumers?
- What types of additives are found in the foods we eat, and how might they affect consumers?
- What steps can consumers take to ensure that they are eating a balanced diet?
- What are the benefits and costs of convenience foods?
- Are brand-name foods better than store-brand or generic products?
- How might consumers safely reduce their weight?

Every year Americans spend a staggering amount of money on food products—almost $900 billion in 2003 alone. Americans consume approximately one-sixth of the world's total agricultural output. We are feeding a very large and hungry stomach. We buy our food products at more than 300,000 retail stores that carry an average of 25,000 different products on their shelves at any one time. The number of brands of different types of foods—canned peas, carrots, soups, cereals—is probably many thousands when you include all the regional specialties you can buy.

> "Americans consume approximately one-sixth of the world's total agricultural output."

## Choosing the Food We Eat

Even though we spend about $900 billion a year on food, that represents only 13.6 percent of disposable personal income in the United States. Figure 9.1 shows that the percentage of U.S. income spent on food has actually been falling. Is this surprising? Well, it shouldn't be. Ask yourself how much more food you would buy if you doubled your income. You could certainly buy better quality, and perhaps you could eat in restaurants more often. But there is a limit, at least for most of you, and that limit is a physical one. Your stomach can hold only so much at any one sitting, and your body will maintain its weight only if you do not take in more calories than you use. If people's expenditures for food had kept pace with their incomes over the past 150 years in the United States, we would be a nation of balloons, running into each other and having trouble sitting in chairs, driving cars, and getting on buses. (Americans do weigh more on average, though, than populations in other countries of the world.)

In 1856 a German statistician, Ernst Engel, made some budgetary studies of family expenditures and found that as family incomes increased, the *percentage* spent on food decreased—not the total amount of food, of course, but the percentage of income spent on it. A family making $80,000 a year certainly spends more on food than a family making $40,000 a year. But even though the wealthier family has an income twice as large as the other family's,

**Question for Thought & Discussion #1:**
When you choose your diet, what is the most important factor you consider—taste, nutrition, cost, or convenience?

**Figure 9.1** Percentage of Disposable Income in the United States Spent on Food

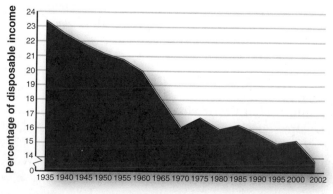

SOURCE: Statistical Abstract of the United States, 2001, p. 648

**Engel's Law** A proposition, first enunciated by Ernst Engel, that states that as a family's income rises, the proportion spent on food falls.

the richer family does not spend twice the amount that the other family spends on food.

**Engel's Law** has fairly universal applicability, not only through time but also across nations at any given moment. Richer nations spend a smaller fraction of their total national income on food than do poorer countries, and we can predict that, in the United States, if we become richer, our expenditures on food will become a smaller percentage of total expenditures. But for our purposes in this chapter, it is more important to discuss what we buy with our food dollars than how many of them we spend.

# Government Labeling and Inspection Requirements

Because food is an essential part of every consumer's budget and a determining factor in our health, the government has established a system of inspection and labeling designed to help us make wiser choices about the food products we buy. You may not be aware of it, but the government is constantly inspecting meat-packing houses and various food-processing establishments in an effort to ensure that our food is handled in a clean, bacteria-free environment so we won't suffer any harmful effects of improper food production. In addition, the government requires fair packaging and labeling of food products by means of the Fair Packaging and Labeling Act of 1966 and other federal laws.

## Fair Packaging

The Fair Packaging and Labeling Act of 1966 came about as a result of numerous complaints directed at packagers. The quantity of packaged contents was often inadequately or confusingly disclosed; there was no uniform designation of quantity by weight or fluid volume. For example, one producer would measure by ounces, while a competitor would measure by quarts or by quarts and ounces combined. In addition, packagers often used adjectives such as "giant" and "jumbo" or designated servings as small, medium, and large without indicating any standard of reference.

The Fair Packaging and Labeling Act applies to foods, drugs, devices, and cosmetics that are subject to the Federal Food, Drug and Cosmetic Act and to any other article customarily purchased through retailers for consumption by individuals or for "use by individuals for purposes of personal care or in the performance of services ordinarily rendered within the household." The act does not apply to tobacco, meat, poultry, or any other products already covered by federal laws.

The act authorized the secretary of commerce to limit "undue" proliferation of product package sizes; that is, a variety of sizes that "unreasonably" impair consumers' ability to make product comparisons. The secretary may ask manufacturers, packers, and distributors to participate in the development of a voluntary product standard for the packaging of a commodity. If no standard has been adopted within one year after such a request, or if the voluntary standard adopted is not observed, the secretary of commerce is required to report such determination to Congress.

# Food Labeling

Under the authority of the Fair Packaging and Labeling Act and other federal laws, the Food and Drug Administration (FDA) requires that every package of food, drugs, or cosmetics contain the following information on its label:

1. The name of the product.
2. The name and address of the manufacturer, packer, or distributor.
3. The net contents by weight, volume, or count.
4. Details of dietary characteristics, if applicable.
5. Whether the product contains artificial coloring, flavoring, or chemical preservatives.

**STANDARDS OF IDENTITY**  *Standards of identity* are provided by law for some food products, such as mayonnaise and bread, for which the ingredients have been traditionally well known. In foods with a standard of identity, certain ingredients must be present in a specific percentage—otherwise the food may not use the standard name. For foods without a standard of identity, ingredients must be listed on the label.

In 1991, the FDA announced new standards for labeling food products. The standards were finalized in 1992 and implemented in 1993. The goal was to make nutritional information provided on labels much more specific and usable than before (see Figure 9.2 on the next page). The FDA was authorized to establish these rules by the Nutrition Labeling and Education Act of 1990.

Prior to 1991, only about half of all packaged food products sold in the United States provided nutritional labeling. Under the rules established by the FDA, virtually every packaged food product carries nutritional information that lists calories per serving and amounts of carbohydrates, fat, fiber, protein, sugar, and salt, in addition to data on vitamins and minerals.

Standard serving sizes for many products are almost as important as nutritional information per serving. In the past, many manufacturers of food products were able to mislead consumers by labeling their products with serving sizes that were either unrealistically small (to make them seem low in calories or fat) or unrealistically large (to make them appear to be high in protein or vitamins and minerals). The 1991 FDA rules set standard serving sizes for 176 categories of food products; these standard sizes enable consumers to compare the nutritional values of similar items.

Another purpose of the rules was to encourage food producers to reformulate their products to make them more healthful. The reasoning was that businesses producing food products falsely marketed as being healthful and low in fat would probably choose to change their products to match the advertising rather than give up the advertising.

**COORDINATION OF GOVERNMENT EFFORTS**  The 1990 labeling law is also important because it was the first time the FDA and the U.S. Department of Agriculture (USDA) worked together to set standard labeling rules. For example, consumers can now compare the nutritional value of packaged pepperoni pizza (regulated by the USDA) against packaged cheese pizza (regulated by the FDA).

**Question for Thought & Discussion #2:**
Do you believe most consumers use information provided on labels when they buy food? Why or why not?

## Figure 9.2 A Translator's Guide to Food Labels

**Acidulants or acidifiers** Acids that have many food uses—as flavor-enhancing agents, as preservatives to inhibit growth of microorganisms, as antioxidants to prevent discoloration or rancidity, and as agents to adjust the acidity in some foods.

**Anticaking agents** Substances used to prevent powdered or granular foods from absorbing moisture and becoming lumpy. They help products like table salt and powdered sugar flow freely.

**Antioxidants** Preservatives that prevent or delay discoloration in foods, such as cut potatoes and sliced apples. They also help keep oils and fats from turning rancid. Examples: BHA, BHT, propyl gallate.

**Cholesterol** A fatlike substance found in foods of animal origin (meat, poultry, and dairy products) but not in foods from plants. Cholesterol is essential to body functions. But because the body can make what it needs, the amount in some people's diets is often excessive, increasing the risk of heart disease.

**Emulsifiers** Widely used in food processing, these agents stabilize fat and water mixtures so they will not separate. For example, in mayonnaise, egg yolks act as emulsifiers to keep the oil from separating from the acids (vinegar or lemon juice). Lecithin, derived from soybeans, acts as an emulsifier in such foods as chocolate and margarine.

**Fats** A major source of energy, they also play a key role as carriers of fat-soluble vitamins (A, D, E, and K). Fat is a constituent of most foods of plant and animal origin.

**Fatty acids** The major constituents of fat. Fats in foods are a mixture of saturated and unsaturated fatty acids. Fats with a high proportion of saturated fatty acids are solid or nearly solid at room temperature and are found in

larger amounts in foods of animal origin. Fats with mostly unsaturated fatty acids are liquid at room temperature and are found in largest amounts in plant oils, such as safflower, sunflower, corn, soybean, canola, and cottonseed oils.

**Fiber** Provides bulk or roughage in the diet. Fiber is found in such plant-derived foods as cereal grain products, vegetables, fruits, seeds, and nuts.

**Flavor enhancers** Help bring out the natural flavor of foods. Examples: monosodium glutamate (MSG), disodium guanylate, and disodium inosinate.

**Grains** Hard seeds of cereal plants, such as wheat, rice, corn, and rye. Whole grains contain the entire seed of the plant.

**Humectants** Chemicals such as glycerol, propylene glycol, and sorbitol that are added to foods to help retain moisture, fresh taste, and texture. Often used in candies, shredded coconut, and marshmallows.

**Hydrogenated** and **partially hydrogenated** Labeling terms that describe the process of adding hydrogen to an unsaturated fat to make it saturated; for example, oils may be hydrogenated to various degrees to make them suitable for use in products such as margarine. The more an oil is hydrogenated, the more saturated fatty acids it contains.

**Leavening agents** Substances such as yeasts and baking powders that are used to make foods light in texture by forming carbon dioxide gas in the dough.

**Niacin** A water-soluble B vitamin that is important for the health of all body cells. The body needs it to use oxygen to produce energy.

**Refined flour** A type of flour produced by milling grains to a fine white consistency. Refining removes bran, fiber, and some other nutrients. Enriched flour has iron and four B vitamins added to levels required by the FDA.

**Riboflavin** A water-soluble B vitamin that helps the body obtain energy from foods and aids in the proper functioning of the nervous system, in growth, and in digestion.

**Sequestrants** Chemicals used to bind trace amounts of metal impurities that can cause food to become discolored or rancid. EDTA is an example.

**Sodium** A chemical in some foods. Although it is essential for regulating body fluids and muscle function, excessive amounts have been linked with an increased risk of high blood pressure.

**Stabilizers** and **thickeners** Substances that give foods a smooth, uniform texture. They also protect foods from adverse conditions, such as wide temperature fluctuations and physical shock during distribution. The most common thickening agents are starches (cornstarch and wheat starch) and modified food starches. Other types include carrageenan, locust bean gum, agar, sodium alginate, gelatin, and pectin.

SOURCE: U.S. Department of Agriculture.

In addition to regulations of what food producers must include on their labels, there is an even longer list of terms they may not use except in special cases (see Figure 9.3). No product may be called "fresh" unless it has never been processed, frozen, or preserved. "More" can be used to describe a product's contents only when 10 percent or more of the desired ingredient has

> **Figure 9.3** Label Terms Established by the Agriculture Department and Food and Drug Administration
>
> **Free** Contains no more than an amount that is "nutritionally trivial" and unlikely to have a physiological consequence.
>
> **Fresh** Can refer only to raw food that hasn't been processed, frozen, or preserved.
>
> **High** A serving provides 20 percent or more of the recommended daily intake of the stated nutrient.
>
> **Less** May be used to describe nutrients if the reduction is at least 25 percent.
>
> **Light** May be used on foods that have one-third fewer calories than a comparable product. Any other use of "light" (or alternative spellings such as "lite") must specify whether it refers to the look, taste, or smell; for example, "light in color."
>
> **More** May be used to show that a food contains at least 10 percent more of a desirable nutrient, such as fiber or potassium, than a comparable food.
>
> **Source of** A serving has 10 to 19 percent of the recommended daily intake of the nutrient.
>
> **Calorie-free** Has fewer than 5 calories a serving.
>
> **Low calorie** Has fewer than 40 calories per serving or per 100 grams of food.
>
> **Reduced calories** Has one-third fewer calories than a comparison food.
>
> **Cholesterol-free** Has fewer than 2 milligrams of cholesterol per serving and has 2 grams or less of saturated fat per serving.
>
> **Low in cholesterol** Has 20 milligrams or fewer of cholesterol per serving or per 100 grams of food and has 2 grams or less of saturated fat per serving.
>
> **Fat-free** Has less than 0.5 gram of fat per serving and no added fat or oil.
>
> **Low-fat** Has 3 grams or less of fat per serving or per 100 grams of the food.
>
> **Low in saturated fat** Has 1 gram or less of saturated fat per serving, and not more than 15 percent of the food's calories come from saturated fat.
>
> **(Percent) fat-free** May be used only in describing foods that qualify as low fat.
>
> **Reduced fat** Has no more than half the fat of an identified comparison. *Example:* "Reduced fat, 50% less fat than our regular brownie. Fat content has been reduced from 8 grams to 4 grams." To avoid trivial claims, reduction must exceed 3 grams of fat per serving.
>
> **Sodium-free/salt-free** Has fewer than 5 milligrams of sodium a serving.
>
> **Low sodium** Has fewer than 140 milligrams of sodium per serving or per 100 grams of food.
>
> **Reduced sodium** Has no more than half the sodium of a comparison food.
>
> **Very low sodium** Has fewer than 35 milligrams per serving or per 100 grams of food.
>
> **Sugar-free** Has less than 0.5 gram of sugar per serving.
>
> SOURCE: U.S. Department of Agriculture.

been added. "Light" may be used to describe a product only when it has at least one-third fewer calories than comparable products. Producers cannot get around this rule by changing the spelling of the word—by calling a product "lite," for example. The list goes on and on.

Some producers have complained about the rules and restrictions. One consultant to the food industry reportedly said, "It's very hard to describe some products fairly and accurately and still make them salable." If you are producing "Cocoa Flavored Sugar Bomb Breakfast Cereal," for example, saying that it is bad for your health (which may be true) is unlikely to increase your sales.

Estimates are that the relabeling effort initially cost the food industry between $2 billion and $3 billion. Supporters of the rules, however, believe that over time consumers may save as much as $100 billion in reduced health-care costs as people choose to eat better diets. Since 1994, meat products must also be labeled with safe handling and cooking instructions similar to the example in Figure 9.4 on the following page.

> **Question for Thought & Discussion #3:**
> Have you ever become ill from food poisoning? How might you have prevented this from happening?

## USDA Grades

In addition to FDA labeling requirements for canned and processed foods and meat inspection, the USDA places grade marks on various meats and fresh produce. These grades are meant to inform the consumer about their level of quality.

**Figure 9.4** A Typical Safe Handling Instructions Label

# Safe Handling Instructions

THIS PRODUCT WAS PREPARED FROM INSPECTED AND PASSED MEAT AND/OR
POULTRY. SOME FOOD PRODUCTS MAY CONTAIN BACTERIA THAT CAN CAUSE
ILLNESS IF THE PRODUCT IS MISHANDLED OR COOKED IMPROPERLY.
FOR YOUR PROTECTION, FOLLOW THESE SAFE HANDLING INSTRUCTIONS.

 KEEP REFRIGERATED OR FROZEN.
THAW IN REFRIGERATOR OR MICROWAVE.

 KEEP RAW MEAT AND POULTRY SEPARATE FROM OTHER FOODS.
WASH WORKING SURFACES (INCLUDING CUTTING BOARDS),
UTENSILS, AND HANDS AFTER TOUCHING RAW MEAT OR POULTRY.

 COOK
THOROUGHLY.    KEEP HOT FOODS HOT. REFRIGERATE
LEFTOVERS IMMEDIATELY OR DISCARD.

**MEAT**  All fresh meats sold by retailers, with the exception of pork, are *voluntarily* labeled as to grade. Currently, the top grade available in supermarkets is "prime," followed by "choice" and then "select." Grading for quality should not be confused with inspection. All meat sold for human consumption is inspected for wholesomeness. Grading is voluntary and is usually a measure of the taste qualities of the meat, which have to do with the fat content or marbling of the beef. For example, prime cuts of beef have more fat than the choice and select grades, and that is one of the reasons prime beef is more palatable (and more expensive) than the leaner grades. Today, most (92 percent) of the beef available in supermarkets and restaurants is graded choice; only 3 percent is graded prime, and 5 percent (a small but rising percentage) is graded select.

**POULTRY**  Poultry, which was not under federal inspection standards until the Poultry Products Inspection Act of 1968, is also subject to USDA grading. Poultry is graded solely on physically observable characteristics, such as freezer burn, presence or absence of wing tips or other missing parts, and breaks in the skin. In other words, a grade A may appear on a turkey that is a tough old bird that just happens to have all its parts and suffers from no observable flaws. Similarly, a grade-B bird, though technically of a lesser grade, may be delicious but have a broken wing tip. Poultry grading does not guarantee quality, but you can be sure that a C rating means the product is likely to appear physically unmarketable.

This meat is graded by the USDA. Why do many experts believe inspection of meat products is more important to consumers than grading?

If you have any doubts about fish, meat, or poultry products you have purchased, you can visit the FDA's Center for Food Safety and Applied Nutrition Web site at **http://www.vm.cfsan.fda.gov/list.html**.

**OTHER FOOD PRODUCTS**  The USDA also grades other food products, such as eggs and milk and fresh fruits and vegetables, usually those that come prepackaged. In this area particularly, the grading is often misleading. For example, "U.S. No. 1" is the third grade for apples but the second grade for grapefruit and the first grade (top grade) for pears. Similarly, "U.S. No. 1 Bright" is the second grade of oranges, not the top grade, as might be assumed.

**Figure 9.5** The Radura

## Irradiated Meats

From time to time, U.S. consumers have been sickened by eating food products that were contaminated with harmful bacteria. The most common of these bacteria, *E. coli,* is often found in meat products. It has long been known that bacteria can be killed by bombarding food products with radiation. After a particularly widespread outbreak of *E. coli* food poisoning in 1999, the USDA approved the sale of irradiated foods. This ruling has the potential to virtually eliminate food poisoning from bacterial infections.

Some opponents of irradiation argue that possible side effects have not been properly investigated. Nonetheless, many supermarket chains now market both irradiated and nonirradiated meat. Irradiated products must carry a symbol called the "radura" (see Figure 9.5) so that consumers may identify them. As far as taste and nutritional value are concerned, there is no distinguishable difference between irradiated and nonirradiated meat. There is a good chance that you have consumed at least some irradiated meat products.

## Antibiotics in Our Food

For many years U.S. meat producers have routinely added antibiotics to the feed given to chickens, pigs, cattle, and other livestock regardless of whether the animals were sick or healthy. The purpose is to reduce sickness among the animals, limit animal losses, and increase food production. In 2002, more than 20 million pounds of antibiotics were administered to livestock in the United States.

At first glance this practice would appear to be a great success. The use of the drug Baytril, for example, has reduced the mortality rate among farmed chickens in the United States from nearly 30 percent in the 1960s to less than 5 percent today.

Unfortunately, there is also a downside to this policy. The pervasive use of antibiotics has resulted in the growth of new strains of drug-resistant bacteria that can no longer be controlled with conventional medications. These "superbacteria" can infect entire herds of livestock and may be passed on to consumers as well.

In 2002, the FDA imposed new rules on manufacturers of antibiotics that are administered to livestock. They are now required to determine whether the use of their drugs could result in the emergence of "disease-causing organisms that resist drugs used to treat humans." The FDA provided the manufacturers with a document spelling out the steps they must take to meet its new requirements. An alternative that has been discussed would be to allow medications

## THE diverse CONSUMER

### Genetically Engineered Bananas

It is difficult for U.S. consumers to avoid eating at least some genetically engineered (GE) foods. In fact, an estimated 70 percent of U.S. food products contain genetically engineered ingredients. Although GE foods have aroused relatively little concern among U.S. consumers, this is not the case in Europe or in many other nations of the world.

The European Union (EU), for example, has outlawed the growing or importation of food products that contain genetically altered ingredients. Europeans worry that GE foods could be harmful to people's health. In some cases, the importation of any U.S. food products, even those that do not contain genetically engineered ingredients, is outlawed. The EU assumes that if a food was produced in the United States, it must contain genetically altered ingredients regardless of what farmers or our government says. Some experts believe these laws are more concerned with protecting European farmers from competition than with protecting European consumers from real or imagined risks posed by GE foods.

Regardless of their intent, the EU's laws are having an important and often negative impact on people in developing nations. Consider the case of Uganda as an example. Bananas are a staple in the Ugandan diet. On average, Ugandans eat 500 pounds of bananas each year. They consume banana cake, banana bread, banana chips, banana soup, and banana beer. Unfortunately, in recent years

many of the banana trees in Uganda have been attacked by a variety of diseases. The most serious is a fungus disease called black sigatoka. Infected trees produce no bananas.

A few years ago a U.S. researcher, Romy Swennen, used genetic engineering to develop a strain of banana trees that could resist black sigatoka. When the Ugandan government heard of Swennen's work, it decided to plant his trees to assure an adequate supply of bananas for Ugandan citizens.

A problem soon arose, however. In addition to consuming vast quantities of bananas, Ugandans also export them to Europe. Hearing of the government's plan, the EU informed Uganda that if it introduced GE banana trees, Europeans would no longer buy any Ugandan bananas. The EU would even refuse to purchase bananas that were not grown on GE trees because there would be a possibility of cross-pollination between the plants. To avoid the loss of European sales, the Ugandan government has chosen not to plant GE banana trees. The result has been a shortage of bananas for Uganda's population.

A similar situation is happening in a number of other developing nations. At a time when many people in the world are going hungry, countries in Southeast Asia have chosen not to plant GE rice, and nations in South America are avoiding GE corn.

**Are you concerned about the health effects of GE foods? Do you believe their advantages outweigh their possible dangers?**

### Question for Thought & Discussion #4:

Do you consume genetically engineered foods? How have you formed your opinion of GE foods?

to be given only to animals that are sick. The costs of meeting these new rules will, of course, be passed on to consumers in higher prices. Do you believe the FDA's new policy is correct?

# Food Labels and the Consumer

All federal efforts to ensure that food products are truthfully labeled are aimed at providing consumers with information so that they may make rational choices. Nutrition labels are a special boon to consumers, who can see at a glance the nutritional content of the food and compare it with other products to make healthful choices. It takes little effort, for example, to detect from the nutrition labels which of two cereals contains more nutrients. Figure 9.6 explains the information you will find on a food label.

According to the FDA's Web site, consumers benefit from food labels by getting:

- Nutrition information about almost every food in the grocery store.
- Distinctive, easy-to-read formats that enable consumers to more quickly find the information they need to make healthful food choices.

## Figure 9.6 Food Label Requirements

**What's on a Label?**
The package must always state the product, the name and address of the manufacturer, and the weight of measure. The front label may also state information about sodium, calories, fat, or other constituents. Approved health claims may be made, but only in terms of total diet. Ingredients are listed in descending order of predominance.

**The Nutrition Facts Panel**

Thomson Learning

**Nutrition Facts**

Serving Size 1/2 cup (114g)
Servings Per Container 4

Amount Per Serving
**Calories** 90      Calories from Fat 30

| | % Daily Value* |
| --- | --- |
| **Total Fat** 3g | **5%** |
| Saturated Fat 0g | **0%** |
| **Cholesterol** 0mg | **0%** |
| **Sodium** 300 mg | **13%** |
| **Total Carbohydrate** 13g | **4%** |
| Dietary Fiber 3g | **12%** |
| Sugars 3g | |
| **Protein** 3g | |

| | | | |
| --- | --- | --- | --- |
| Vitamin A 80% | ● | Vitamin C 60% | |
| Calcium 4% | ● | Iron 4% | |

* Percent Daily Values are based on a 2,000 calorie diet. Your daily values may be higher or lower depending on your calorie needs:

| | Calories: | 2,000 | 2,500 |
| --- | --- | --- | --- |
| Total Fat | Less than | 65g | 80g |
| Sat Fat | Less than | 20g | 25g |
| Cholesterol | Less than | 300 mg | 300 mg |
| Sodium | Less than | 2,400mg | 2,400mg |
| Total Carbohydrate | | 300g | 375g |
| Dietary Fiber | | 25g | 30g |

Calories per gram:
Fat 9 ● Carbohydrate 4 ● Protein 4

**Serving size and calorie information**

**Percentage of daily value for nutrients**

**Reference Values**
This compares some values for the nutrients in a serving of the food to the needs of a person requiring 2,000 or 2,500 calories per day to show how the product fits into the daily diet.

**Calorie/gram reminder**

A container with fewer than 20 square inches of surface area can present fewer facts in this format.

**Nutrition Facts**
Serving Size 1/3 cup (85 g)**
Servings ?
Calories 111
  Fat Cal. 23
*Percent Daily Values (DV) are based on a 2,000 calorie diet.
**Drained solids only.

| Amount/serving | % DV* | Amount/serving | % DV* |
| --- | --- | --- | --- |
| **Total Fat** 3g | 5% | **Total Carb.** 0g | 0% |
| Sat Fat 1g | 5% | Dietary Fiber 0g | 0% |
| **Cholest.** 60mg | 20% | Sugars 0g | |
| **Sodium** 200mg | 8% | **Protein** 21g | |

Vitamin A 0% ● Vitamin C 0% ● Calcium 0% ● Iron 2%

- Information on the amount per serving of saturated fat, cholesterol, dietary fiber, and other nutrients of major health concern.
- Nutrient reference values, expressed as % Daily Values, that help consumers see how a food fits into an overall daily diet.
- Uniform definitions for terms that describe a food's nutrient content—such as "light," "low-fat," and "high-fiber"—to ensure that such terms mean the same for any product on which they appear.
- Claims about the relationship between a nutrient or food and a disease or health-related condition, such as calcium and osteoporosis or fat and cancer. These are helpful for people who are concerned about eating foods that may help keep them healthier longer.

- Standardized serving sizes that make nutritional comparisons of similar products easier.
- Declarations of the total percentage of juice in juice drinks. This enables consumers to know exactly how much juice is in a product.

It is important to remember that, although food packaging and labeling laws benefit consumers greatly, any increased information, regulations, and other changes in labeling policies mean added costs for food manufacturers. Much of this expense is passed on to the consumer in the form of higher prices for food items. In 2003, the FDA made plans to add trans fat to required food labels by 2006. Trans fats are compounds that are found in many prepared foods and have serious effects on the circulatory system.

# Nutrition

In the mid-1970s, studies by the FDA and various Senate committees created government concern about a "wave of malnutrition" in the United States. Much of this malnutrition was attributed to overconsumption of sugars and fats, especially fat from meat products. Since 1977, the Department of Agriculture and the FDA have recommended that Americans generally strive to reduce their "current intake of total fat, saturated fat, and cholesterol." These government efforts have been joined in the private sector by the American Heart Association, the National Cancer Institute, and numerous other health-oriented groups, as well as nutritionists.

Have these appeals been effective? Are Americans changing their dietary habits and consuming less fat? Yes—to some extent. Data indicate that between 1986 and 1997 Americans reduced their fat intake from 37 percent of the daily diet to about 34 percent. Still, they haven't reached the dietary goals set by the American Heart Association: a daily diet consisting of 30 percent fat, 15 to 20 percent protein, and 50 to 55 percent carbohydrates (some health experts recommend that fat consumption be reduced even further—to 25 or even 20 percent of the daily diet). That Americans continue to consume too much fat, especially saturated fat, was stressed by U.S. Surgeon General C. Everett Koop in his 1988 *Surgeon General's Report on Nutrition and Health*. His prescription was a familiar one: eat less fat-containing foods and more fruits and vegetables.

## Dietary Reference Intakes

To inform citizens of the nutrients required to maintain health, many countries have developed nutrient standards. In the United States, these standards are listed under the general title of **dietary reference intakes (DRIs).** The National Academy of Sciences has established DRIs based on extensive clinical studies carried out in the United States and many other countries. The United States and Canada have agreed to use the same DRI system, which is similar to the systems used in most other nations. DRIs include recommendations for consumption of energy (caloric intake), carbohydrates, dietary fiber, fat, proteins, and a wide variety of vitamins and minerals.

DRIs are expressed in two basic ways depending on the nutrient in question and whether people need to ingest the nutrient on a daily basis or over a longer period of time. The term *recommended dietary allowance (RDA)* is applied to nutrients that are needed by the body each day. *Estimated average requirements (EARs)* are used for nutrients that are needed over a period of days. EARs are more often applied to vitamins and minerals. These recom-

**Dietary Reference Intakes (DRIs)**
Nutrient standards established by the National Academy of Sciences and expressed as recommended dietary allowances (RDAs) or estimated average requirements (EARs).

mendations are provided in differing amounts depending on a person's weight, gender, and age. Lists of RDAs and EARs can be accessed on the USDA's Web site at **http://www.usda.gov**.

A condensed system of nutrient standards, called the **U.S. recommended daily allowances (U.S. RDAs),** was developed to indicate the *maximum amount* of each nutrient required for the following categories: infants, children, children over age four and adults, and pregnant and lactating females. Additionally, the U.S. RDAs recommend a larger quantity than the RDAs or EARs established by the National Academy of Sciences whenever evidence indicates that the population is lacking in a nutrient—for example, riboflavin. Food for infants or children uses only U.S. RDAs for that age group. Since 1992 FDA rules do not require extensive information about nutrient standards on food labels because many consumers found such lists confusing.

**U.S. Recommended Daily Allowances (U.S. RDAs)** A condensed system of nutrient standards that indicate the maximum amount of each nutrient needed for four broad categories of the population.

**Question for Thought & Discussion #5:**
Are you careful to limit the amounts of sugar and fat you consume? Why have you made this choice?

## Basic Food Groups

To calculate whether your daily diet contains the amount of nutrients prescribed by the U.S. RDAs is a difficult and time-consuming endeavor. To simplify your task, beginning in 1956, the Department of Agriculture developed dietary guidelines based on basic food groups. Originally, these groups were called the "basic four" and consisted of (1) vegetables and fruits; (2) breads and cereals; (3) milk and cheese; and (4) meat, poultry, fish, and beans. As you can see in Table 9.1, a fifth group (fats, sweets, and alcohol) has been added—to be used only as "extras" in the diet.

| TABLE 9.1 | Guide to a Balanced Diet—Basic Food Groups | | |
|---|---|---|---|
| **Food Group and Number of Servings** | **Count as a Serving** | **Nutritional Benefit** | **Use in Meals** |
| Fruit Vegetables 4 or more | • ½ fruit or typical portion as served<br>• 1 orange or banana<br>• ½ medium grapefruit or cantaloupe<br>• 1 wedge of lettuce<br>• 1 bowl of salad<br>• 1 medium potato | *Vitamin A*—Dark green and deep yellow vegetables.<br>*Vitamin C*—Citrus fruits (oranges, grapefruit, tangerines, lemons), melons, berries; tomatoes, dark green vegetables.<br>*Riboflavin, folacin, iron, magnesium*—Dark green vegetables.<br>*Fiber*—Unpeeled fruits and vegetables, especially those with edible seeds.<br>*Starch*—Potatoes (white and sweet), corn, green peas. | • Have at all meals and snacks.<br>• Serve raw or cooked.<br>• Use in salads or as side dishes.<br>• Use fruits as juice and occasionally in desserts (cobblers, pies, and shortcakes).<br>• Use vegetables in casseroles, stews, and soups. |
| Bread Cereal 4 or more | Products made with whole-grain or enriched flour or meal:<br>• 1 slice bread<br>• 1 biscuit or muffin<br>• ¼ to ½ cup cooked cereal, cornmeal, grits, macaroni, noodles, rice, or spaghetti; 1 ounce ready-to-eat cereal | *B vitamins and iron*—Most whole-grain and enriched breads and cereals.<br>*Zinc, magnesium, folacin, and fiber*—Whole-grain products.<br>*Protein*—A major source in vegetarian diets.<br>*Lower in calories*—If prepared and served with little or no fat and sweets. | • Have at all meals and snacks.<br>• Serve at breakfast as toast, muffins, pancakes, or grits, cooked or ready-to-eat cereals.<br>• Use at lunch or dinner as macaroni, spaghetti, noodles, or rice in a casserole or side dish, and breads in sandwiches—hot or cold.<br>• Use crackers or cereals as snacks.<br>• Have occasionally as a baked dessert, such as cake, pastry, or cookies made from whole grain or enriched flour. |

## TABLE 9.1    Guide to a Balanced Diet—Basic Food Groups (Continued)

| Food Group and Number of Servings | Count as a Serving | Nutritional Benefit | Use in Meals |
|---|---|---|---|
| **Milk**<br>**Cheese**<br>Teen-ager: 4 or more<br>Adult 2 or more[a] | • 8 ounces milk or yogurt<br>• 1 ounce Cheddar or Swiss cheese = ¼ cup milk<br>• 1 ounce processed American cheese = ⅔ cup milk<br>• ½ cup ice cream or ice milk = ¼ cup milk<br>• ½ cup cottage cheese = ¼ cup milk | *Calcium*—Major source in U.S. diets.<br>*Protein, riboflavin, and vitamins $B_6$, $B_{12}$, and A.*<br>*Vitamin D*—If product is fortified.<br>*Lower calories and fat*—Low-fat or skim milk products. Items fortified with vitamins A and D contain the same amount of nutrients as whole-milk products. | *Serve milk:*<br>• As a beverage at meals and snacks.<br>• On cereals.<br>• In soups, main dishes, custards, puddings, baked goods.<br>For variety, replace part of milk with:<br>• Yogurt.<br>• Cheese (plain, on crackers, or in sandwiches, salads, and casseroles). |
| **Meat, Poultry, Fish, and Beans**<br>2 or more | 2 to 3 ounces of lean cooked meat, poultry, or fish without bone. Equal to 1 ounce of meat:<br>• 1 egg<br>• ¼ to ½ cup cooked dry beans, peas, soybeans, or lentils<br>• 2 tablespoons peanut butter<br>• ¼ to ½ cup nuts or seeds<br>• 2 ounces of bologna | *Protein, vitamin $B_6$, and other minerals and vitamins.*<br>*Vitamin $B_{12}$*—Found only in foods of animal origin.<br>*Zinc*—Red meats and oysters are good sources.<br>*Iron*—Red meats are important sources.<br>*Magnesium*—Dry beans and nuts. | • Main dish.<br>• Ingredient in a main dish—soup, stew, salad, casserole, or sandwich. |
| **Fats**<br>**Sweets**<br>**Alcohol**<br>**Little or none required, unless for calories** | No serving size defined as no number of servings is suggested. Include:<br>• Butter, margarine, mayonnaise, salad dressings<br>• Sugar, candy, jams, jellies, syrups<br>• Soft drinks, other highly sugared beverages<br>• Wine, beer, liquor<br>• Unenriched, refined bakery products | *Calories.*<br>*Vitamin E*—Vegetable oils.<br>*Essential fatty acids*—Vegetable oils.<br>*Vitamin A*—Butter and margarine. | • Ingredients (sugar and fats) in recipes.<br>• Added to foods at table—sugar on cereals, dressings on salads, and spreads on bread.<br>• Expensive "extras"—candy, soft drinks, and alcoholic beverages. |

[a]Three servings for women who are pregnant and four servings for women who are nursing.

By selecting foods from these basic food categories in the recommended amounts per day, you can be fairly sure that you will obtain a nutritionally balanced diet. The nutrients will in most cases take care of themselves, and vitamin and mineral supplements should not be necessary—unless prescribed by a physician. To reduce fat intake, select products within each food group that are low in fat content—skim milk instead of whole milk, for example. To reduce sodium intake, select foods from each group that are low in sodium. The FDA recommends between 1,100 and 3,300 milligrams of sodium per day as a safe amount (for the record, a teaspoon of salt contains nearly 2,000 milligrams of sodium). Remember, too, that eating a variety of the foods within each group will help to ensure a healthy intake of nutrients.

In 1992, the USDA introduced an "Eating Right Food Guide Pyramid" to simplify the presentation of the four food groups to the U.S. public. As Figure 9.7 shows, this pyramid arranges the food groups in four layers. The bottom,

## Figure 9.7 The Food Guide Pyramid: A Guide to Daily Food Choices

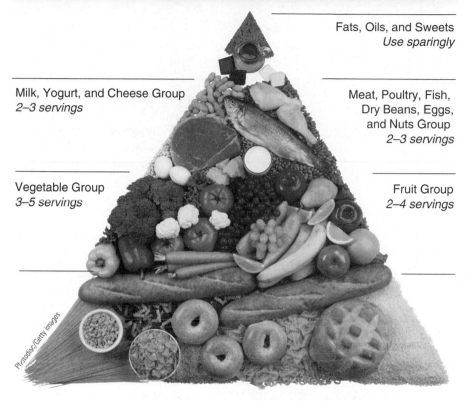

Fats, Oils, and Sweets
*Use sparingly*

Milk, Yogurt, and Cheese Group
*2–3 servings*

Meat, Poultry, Fish,
Dry Beans, Eggs,
and Nuts Group
*2–3 servings*

Vegetable Group
*3–5 servings*

Fruit Group
*2–4 servings*

Bread, Cereal, Rice, and Pasta Group
*6–11 servings*

SOURCE: USDA

and largest layer, is made up of complex carbohydrates such as bread, rice, and potatoes. The next layer contains fruits and vegetables. The third layer includes meat, fish, and dairy products. The peak is for sugar and fat. The idea is that the higher a food group is on the pyramid, the less of it you should eat each day.

**Question for Thought & Discussion #6:**
How closely does your diet follow the one recommended by the Food Guide Pyramid?

# Food Additives

More and more chemicals seem to be getting into the foods we eat, and many consumers, government officials, and nutritionists believe that such additives are causing health problems. Broadly defined, a food additive is any substance that is added to a food product either directly or indirectly. The FDA estimates that at least 2,800 substances are intentionally added to foods, and as many as 10,000 other compounds or combinations of compounds find their way into various foods during processing, packaging, and storing.

## Classification of Additives

Several government agencies have been granted authority to regulate the use of additives in products that cross interstate lines. Meat products are controlled by the USDA, and other food products are regulated by the FDA. Since

> **"Approximately 98 percent of all direct additives are in the form of sugar, salt, corn syrup, citric acid, baking soda, vegetable coloring agents, mustard, and pepper."**

1958 manufacturers of additives have been required to prove that the additives are safe for human consumption. All additives in common use prior to 1958, however, were exempted from this requirement. These additives are placed on one of two lists: Generally Recognized as Safe (GRAS) or Prior Sanction. Presumably, all the items on these lists are assumed to be safe because they have withstood the test of time. Saccharin was on the GRAS list, however, and sodium nitrate was on the Prior Sanction list. Studies show that both saccharin and sodium nitrate may cause cancer in laboratory animals when administered in doses larger than any consumer is likely to use.

Any additives tested after 1958 have been either rejected or placed on an Approved list. Whenever an additive on the market is found to be hazardous, it is placed on a Banned list. If serious doubts are raised concerning an item on the GRAS, Prior Sanction, or Approved lists, the item is placed on an Interim list.

## Direct Additives

Most of the substances added to food are for the purpose of improving its nutritional quality or to preserve its freshness for marketing purposes. Some direct additives are vitamins and minerals that might otherwise be lacking in a person's diet or that have been destroyed or lost in the processing of the food product. Other substances are added to make food taste better. Approximately 98 percent of all direct additives are in the form of sugar, salt, corn syrup, citric acid, baking soda, vegetable coloring agents, mustard, and pepper. According to the FDA, food flavors constitute the largest single category of food additives. Some consumers have questioned the necessity of using artificial flavors, but there has been less criticism of this form of food additive than of food-coloring agents, preservatives, and artificial sweeteners.

**FOOD COLORS**   The Color Additive Amendments, passed in 1960 (amending the Food, Drug and Cosmetic Act), require that all colors made from potentially harmful chemicals be certified by the FDA. Until the food color Red No. 2 was declared unsafe and banned in 1976, it was one of the most common food additives. Quite a few artificial colors are still used in foodstuffs for humans, even though questions about their safety exist. Two of the most widely used are Red No. 40 and Yellow No. 5, and both are under fire because of reports of possible health risks.

**PRESERVATIVES**   Many chemicals are added directly to food substances as preservatives. For example, sodium nitrate is a red coloring agent that inhibits the growth of botulism germs in hot dogs, bacon, and luncheon meats. Alone, sodium nitrate seems to be relatively safe in extremely small quantities. Nonetheless experiments have shown that when sodium nitrate joins with naturally produced proteins called amines, the result is nitrosamines. It is believed that nitrosamines may cause cancer in humans because they have been shown to cause cancer in animals. There is clearly a trade-off here: sodium nitrate preserves food longer, but it may cause cancer.

**ARTIFICIAL SWEETENERS**   Artificial sweeteners, such as saccharin (Sweet 'n Low), aspartame (NutraSweet, Equal), and acesulfame-K (Sweet One), are also used as direct additives in foods. The health risks posed by some of these artificial sweeteners have caused much controversy. Animal tests have produced evidence that both saccharin and cyclamate are **carcinogenic.** There

**Carcinogenic**   Cancer causing.

has been no evidence, however, that *persons* using saccharin or cyclamate have contracted cancer. The so-called Delaney Clause of the 1958 Food and Drug Amendment requires, nonetheless, that any direct additive found to be cancer causing (carcinogenic) in animals be banned from human food. In short, "No additive shall be deemed . . . safe if it is found to induce cancer when ingested by men or animals." Cyclamate was banned in 1970 on the basis of the Delaney Clause when that sweetener was found to cause bladder tumors in rats. In 1977, when Canadian tests showed that saccharin was carcinogenic in rats, a proposed ban was withheld because of public furor. Congress voted an eighteen-month delay on banning saccharin that has since been extended several times, and saccharin is still being used in food products.

When aspartame entered the market in the early 1980s, consumers were given an alternative. Aspartame is considered to be much safer for human consumption than saccharin, although products containing aspartame must bear a warning on their labels cautioning people with the rare genetic disease of phenylketonuria against its use. Acesulfame-k is the only artificial sweetener currently on the market that doesn't require a warning label—as yet.

## Indirect Additives

A potentially serious health problem arises with indirect additives, such as chemicals used on produce to control pests, affect growth, or preserve freshness and appearance. These indirect additives are not listed on any labels, and consumers usually have no way of knowing—short of extensive research—what produce has been subject to chemical treatment. Consumers must rely on the FDA's judgment as to what kind of chemicals, in what amounts, are safe to ingest. The FDA regulates indirect additives by setting up tolerance levels, expressed in parts per million or billion, of carcinogenic or other harmful chemical substances.

## How Safe Is "Safe"?

Some consumers believe that it is difficult—if not impossible—to determine "safe" amounts of any harmful substance and thus seriously question the value of the FDA list. One of the major problems with food additives is their unknown long-term effects. When taken in small quantities, they appear to be safe under most circumstances. When taken over a long period of time, they may be cancer causing. Cancer may take as long as forty years to develop, so we don't know very much about such long-term effects, particularly for additives that have only recently been used in or on foodstuffs.

## Consumer Confusion

Another problem is that consumers often receive mixed signals about the relative safety of certain additives. Growth hormones fed to beef cattle by farmers, for example, pose a significant health risk according to some health-advocacy groups. And the fact that the European Economic Community banned beef imports from the United States in 1988 because of these potentially dangerous chemicals created widespread concern. In contrast, the World Health Organization and the FDA both maintain that the tiny amount of growth hormone in beef poses no health threat. In the midst of this confusion was the concern expressed by the Centers for Disease Control (CDC) that the stir over beef obscured the more genuine threat caused by adding antibiotics to chicken feed

**Cyber consumer**

**Online Dietary Guidelines**

The USDA's Food and Nutrition Center maintains a Web site at **http://www.nal.usda.gov/fnic/dga** that provides vast amounts of information that consumers may access online. Among these materials are a variety of reports on nutrition guidelines for Americans. Visit this Web site and read several of the most recent reports on this topic. What developments have taken place regarding nutrition guidelines in recent years?

"Buying food, as with any other consumer choice, forces us to make trade-offs. The most tasty food products may be bad for our health."

to prevent disease. According to the CDC, the antibiotic regimen breeds antibiotic-resistant strains of the bacteria that cause salmonella and other serious illnesses. Certainly, reports of salmonella have been increasing, although thorough cooking will kill the bacteria. As mentioned earlier, the FDA has recently instituted new rules to try to curb the antibiotics problem.

U.S. consumers also received conflicting messages about the chemical daminozide (known by the trade name Alar), which in 1989 was being sprayed on an estimated 5 percent of the red apples marketed in the United States to produce uniform growth and ripening. Although the Environmental Protection Agency noted that the chemical seemed to pose a significant risk of cancer in humans, it nonetheless allowed the product to remain on the market for eighteen months until studies of the chemical's effects could be completed. In the meantime, an environmental group warned that preschoolers (who eat more apples than adults) could face as much as 910 times the acceptable cancer risk if they ate the daminozide-treated apples. Who or what is the consumer to believe?

## Dietary Supplements

Traditionally, dietary supplements were defined as products made of one or more essential nutrients, such as vitamins, minerals, and proteins, that were intended to enhance consumers' diets. In 1994 the Dietary Supplement Health and Education Act (DSHEA) expanded this definition to include any product intended for ingestion as a supplement to the diet, thereby adding herbs, botanicals, and other plant-derived substances. It is easy to identify these products because the DSHEA requires that they be labeled as *dietary supplements*. Unlike drugs, dietary supplements do not have to undergo extensive testing to prove their value. They must, however, be shown to be safe, and their labels must indicate that they have not been evaluated by the FDA. They may not claim to cure or prevent a specific disease. The fact that manufacturers are supposed to demonstrate that their products are safe does not necessarily mean that this will always happen. In February 2003, for example, Baltimore Orioles pitcher Steve Bechler died after taking the dietary supplement ephedra. Earlier, Jennifer Rosenthal, a twenty-eight-year-old mother from Long Beach, California, took usnic acid for seventeen days to control her weight. She then fell into a coma because this dietary supplement destroyed her liver. Rosenthal's life was saved by a last-minute liver transplant. Although she is alive, she will need to take antirejection drugs for the rest of her life.

In an apparent response to these tragedies the FDA proposed manufacturing guidelines for producers of dietary supplements in March 2003. Public hearings and time for public and corporate response were required before these guidelines could be finalized. Despite their widely publicized problems, the use of dietary supplements has become widespread. Sales of these products more than doubled during the 1990s. A sample label for a dietary supplement is shown in Figure 9.8.

## Trade-Offs We Make When We Buy Food

Buying food, as with any other consumer choice, forces us to make trade-offs. The most tasty food products may be bad for our health. Those that are easiest to prepare often cost more and may not be as nutritious as other foods that are less convenient. Finding the best food purchases often forces con-

**Question for Thought & Discussion #7:**
Do you or any member of your family take a dietary supplement? How was the decision to use the product made?

**Figure 9.8** Requirements for Dietary Supplement Labels

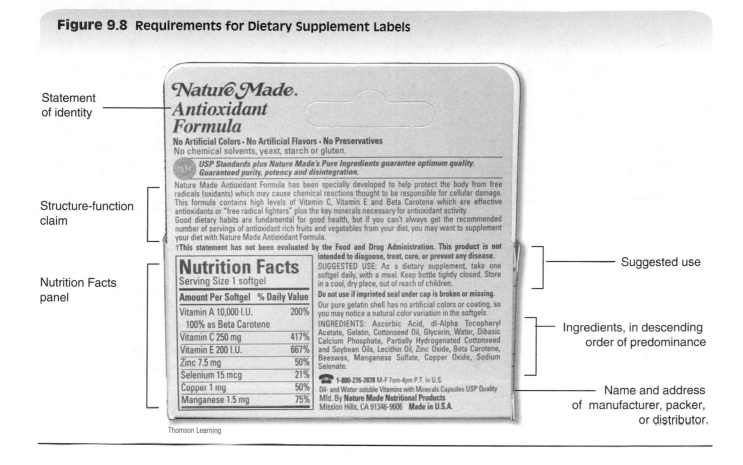

sumers to spend more time shopping. In the end, each consumer must decide what trade-off provides the most value for his or her food dollars and personal time and effort.

# Health, Natural, and Organic Foods

Many consumers worry about the possible harmful effects of eating foods that contain additives. As a result, they choose to trade off the cost of higher prices for the benefits of purchasing "health" or "natural" or "organically grown" foods, which are supposedly free of harmful substances. These three types of food are often lumped together, and thus definitions become hazy. The following broad definitions are fairly widely accepted, however:

● Health foods may include vegetarian and dietetic foods and other products not necessarily free of chemical additives.

● Natural foods do not contain artificial ingredients, preservatives, or emulsifiers.

● Organic foods are grown without the use of chemically formulated fertilizers or pesticides.

For many years, there were no federal rules that determined what foods could be labeled organic. Consumers did not know whether "organic" foods had been grown without chemical fertilizers and pesticides, or if they were

Are you willing to pay a premium price to purchase organic foods?

Thomson Learning

genetically engineered. In December 2000, this changed with the approval of the FDA's organic food labeling rules. These rules include:

- Foods labeled "100 percent organic" must contain only organically produced ingredients. These products cannot be produced using excluded methods.

- Foods labeled "organic" must consist of at least 95 percent organically produced ingredients. These products cannot be produced using excluded methods.

- Foods labeled as having been "made with organic ingredients" must contain at least 70 percent organic ingredients. These products cannot be produced using excluded methods.

- Foods that contain less than 70 percent organic ingredients cannot use the term "organic" anywhere on their principal display panels. However, they may identify the specific ingredients that are organically produced on their ingredients statements.

The fact that there are national organic labeling rules does not mean that all organic foods are good for you. Organic candy is still sweet and organic beef contains substantial amounts of fat. Nevertheless, it is more difficult to mislead consumers with fake organic labels today than it was in the past.

# The Growing Trend toward Convenience Food

The food industry is now providing us with buttered peas, frozen corn on the cob, stuffed baked potatoes, cheese in a spray can, complete frozen dinners, ready-to-eat tacos, and so on. You name it, and you can buy it already prepared. Just pop it into the oven (conventional or microwave) and wait.

## Why Convenience Foods?

Why are Americans buying so many convenience foods? The answer is quite simple: our high incomes prompt us to place a high value on our time. Americans are no lazier than other people. We simply are willing to pay more in both money and possibly lower food quality to save time. We prefer to use our time in ways other than food preparation.

Convenience foods require almost no preparation. You need none of the pots and pans and cutting utensils of old for something that is already chopped, frozen, buttered, and ready to eat after only heating. Certain convenience foods give you less nutrient value than if you spent the time making the dish yourself. Again, that is just part of the price you pay for convenience, and you should be aware of it. People more concerned about nutritional value and those who dislike consuming large quantities of food additives shy away from convenience food, but they pay a price for that. People who always cook meals from scratch spend more time in the kitchen than those who settle for TV dinners or canned foods. We cannot make an ultimate judgment about whether convenience foods are good or bad for the U.S. consumer. As long as you know what you are getting, then you can make the choice.

Perhaps it is incorrect to lump all convenience foods together. Although some, such as Tang, increase sugar in the diet, and others, such as frozen cream pies, are essentially nonfoods, others provide real convenience. Frozen string beans, for example, give you a reasonably fresh vegetable out of season with the time-consuming preparation job (that is, cutting and cleaning) already done.

The trend toward packaged foods is definitely on the upswing; restaurants, even some of the best ones, now have frozen convenience foods on their menus. It might surprise you to find that some parts of a $100 meal in an expensive French restaurant are frozen foods. Actually, you shouldn't be surprised; the cost of food-preparing labor in restaurants has risen so much that to stay in business, even the best restaurants have to cut corners. And one way to do so is to buy convenience foods.

The cost of convenience is often an important factor in consumer decision making. Although convenience foods may cost more than foods prepared at home, consumers using such foods pay a lower "time cost" in preparing them.

## Fast Foods Aren't Necessarily Junk Foods

In the first decade of the twentieth century, two items were marketed that were to change the course of U.S. life: the hamburger and the Model-T Ford. Since then, eating on the run has become increasingly common for Americans, and fast-food restaurants have met this demand with more and more offerings in more and more cities across the nation. Long criticized as sources of "junk food," these restaurants have changed their image. Starting with Wendy's in 1979, most fast-food chains introduced salad bars along with fish and chicken sandwiches as alternatives to the ubiquitous hamburger. By the late 1980s, in response to pressure from health-advocacy groups, Burger King, McDonald's, Hardee's, Wendy's, and Taco Bell had eliminated some, if not all, of the highly saturated fats they had been using in cooking.

Even nutritionists admit that it is possible to buy nutritious foods on the run, but they caution that consumers should be selective and watch the fat and sodium content of the fast foods they choose. This task has become much easier in recent years as many fast-food restaurants have begun to provide nutritional information about the products they sell. In all, the verdict remains generally the same as it has been for years: although an occasional meal at a fast-food restaurant won't harm anybody, a steady diet of such foods and too little in the way of nutrient-dense foods—those rich in minerals and low in calories—could eventually cause health problems.

## Getting the Most for Your Food-Shopping Dollar

We all have different buying habits, and to a great extent, the time we are willing to spend shopping for groceries determines our shopping tactics. As we noted in discussing convenience foods, our time has value for us, too. If we walk into a supermarket and buy the items we need without regard to price, we will obviously spend less time than the careful shopper who reads labels and compares food quality and costs. We will also very likely spend more money. But how much more?

To answer this question, the editors of *Consumer Reports* conducted an experiment. Two individuals were given an identical shopping list and asked to purchase all of the items on the list from a large, modern supermarket

> **"Even nutritionists admit that it is possible to buy nutritious foods on the run."**

**Question for Thought & Discussion #8:**
How often do you eat fast food? What are your costs and benefits when you consume these products?

Careful consumers can cut their food costs considerably if they are willing to use store brands, do some comparison shopping, and avoid impulse purchases.

chain store. One shopper was asked to buy impulsively, without regard to price; the other was instructed to choose carefully, comparing nutritional value and prices. The result? The impulsive shopper selected an array of mainly name-brand products and paid a total cost of $114.36. The careful shopper purchased mainly store brands and paid a total of $67.76. Clearly, there are savings to be had in shopping for groceries, if you want to take the time to comparison-shop.

In the same issue of *Consumer Reports,* a second study compared the cost of buying groceries in large amounts at a "warehouse" store instead of a typical grocery store. Again, careful shopping produced significant savings. The total cost of a "market basket" of goods at a warehouse club store was $76.71, while the exact same products in the same amounts cost $128.53 in a typical grocery store. Although costs of time and driving are associated with shopping at warehouse stores, the savings offered have caused more than 20 million Americans to join such organizations.

## Are Name-Brand Foods Better?

As a corollary to the preceding experiment, the editors of *Consumer Reports* also conducted a series of studies comparing the quality of name-brand food products with store and generic brands in the same food category. Foods were rated by taste experts as well as nutritionists, and nonfood products (such as aluminum foil and plastic wrap) were judged on the basis of durability, ease of handling, and other factors. The store brands selected for the study were those of Safeway and Pathmark. The results of the tests showed that the store brands were of equal or higher quality than name brands in seven of the ten food categories studied. The generic brands could compete in only one category—canned vegetables—in which they were rated equal in quality to the store brand.

Although this ratio may not hold with other store brands, which vary in quality from store to store, it does underscore what many consumers have suspected for some time—that the most significant difference between name-brand and store-brand foods may be the price, at least in the majority of cases. Why, then, do food buyers continue to purchase name brands? You probably already know the answer. For example, if you have eaten a certain Kellogg's cereal for years, that cereal is a known quantity to you. You can predict that it will contain certain nutritional values and have a specific taste and texture. This predictability may be important for you—it can save on shopping time and reduce your risk of ending up with a cereal not as good. Of course, exposure to name brands through advertising also induces many consumers to purchase such products, simply because the product names sound familiar. For the price-conscious consumer, however, being a little venturesome in the supermarket can cut food costs considerably.

## The Psychology of Selling Food

The wise consumer will also be on the alert for special marketing tactics employed by food sellers to lure customers to certain displays in their stores and tempt them to purchase their products. Say that you go to the supermarket to purchase a gallon of milk—at least, that's all you had in mind to purchase. But the milk is at the back of the store—and not by accident. To get to it, you must first pass through an aisle filled on both sides with food products, probably a frozen food section (since food marketers rate frozen foods as an "impulse buy"), and perhaps some end-aisle displays of special sale items. You might also need to resist purchasing the product promoted at a strategically placed "sampler station" or the bakery products that smell so good. In short, you will be exposed to numerous temptations to buy more than the milk—or whatever items you have on your grocery list.

Food marketers know that the rate of exposure is directly related to sales and that approximately 65 percent of all food sales result from impulse buying. So be prepared to have your senses of sight, taste, and smell all targeted for this exposure when you enter a supermarket.

> **"Food marketers know that approximately 65 percent of all food sales result from impulse buying."**

## Unit Pricing

An important tool consumers may use to compare prices and make better choices is *unit pricing.* Unit prices appear on tags below the product's location on the grocery store shelf. The tags identify the product, its total price, the size of the container, and its price per amount or unit. These units may be expressed in ounces, quarts, pounds, or many other types of measure depending on the product in question. The advantage of unit pricing to the consumer is that similar products can be compared easily regardless of the size of their containers. For example, one brand of detergent might come in a 48-ounce bottle at a price of $2.99, while another brand is offered in a 64-ounce bottle for $4.19. In this case, would you be able to do the math quickly and easily in your head to know which had the lower price? Most people couldn't. But if unit prices are given, you will know at a glance that the first product's price is 6.23 cents per ounce, while the second costs 6.55 cents per ounce.

## Electronic Price Scanning and the Consumer

Electronic price scanning, which has led to time-saving benefits for consumers and sellers alike, has also created a problem for consumers: the increasing lack of item pricing of food products. In some states, by law, each item on display must have a price tag on it. In states without such laws, however, many food sellers have stopped pricing each item. Of course, the price of each food product is shown somewhere nearby—on the shelf below it, usually—but often it is difficult to tell which product corresponds to which price marker.

There is also a question as to whether a scanned price is necessarily accurate. Although it may be true that computers don't make mistakes, the people who are supposed to enter prices in the computers that control scanning machines do. Recently, *Consumer Reports* wrote about a survey of various stores' scanning machines, carried out by the Food Marketing Institute, that was designed to see if these devices recorded correct prices. The study reported about a 3 percent rate of error, with some prices being too high and others too low. Although the low prices about balanced out the high prices, this would not make you happy if you paid $2.99 for a can of Parmesan cheese that actually was supposed to have a price of $2.29. So what should consumers do to protect themselves from scanned prices that are incorrect? About the only thing you can do is to record the prices of the products and keep track of them as items are run across the scanner. You should also note that some stores will give incorrectly scanned products to you free if you catch the mistake.

If these problems exist at the market or markets where you shop, complain to the store owners or managers. The more consumers complain, the more likely it is that the store or stores in question will change their policies—as they have in some areas in response to consumer concerns.

> **Question for Thought & Discussion #9:**
> When you shop, do you check to be sure that the prices rung up are the same as the posted prices?

# Confronting Consumer Issues: Good Health and Weight Control

Americans are trimming down (or trying to) with a vengeance—and for good reason. An increasing number of health problems are being linked to excess weight: heart disease, high blood pressure, kidney problems, diabetes, malnutrition, and complications of pregnancy—as well as psychological and social problems. More recently, cancer has been added to the list. Overall, it is now estimated that the mortality rate for people who are even 10 percent overweight is increased by 20 percent over a twenty-five-year period.

Whether for health or appearance reasons, we are in the midst of a diet craze. One recent study revealed that 90 percent of those Americans surveyed wanted to lose weight, and 35 percent wanted to lose at least fifteen pounds. In 2002, weight-conscious Americans spent over $35 billion for diet-related foods, drugs, and other products, up from $15 billion in 1993, only nine years earlier. Weight-reducing purchases ranged from the sublime (low-calorie but nutritious foods) to the near-ridiculous ("appetite-suppressant" sunglasses). According to the FDA, millions of consumer dollars were spent on gimmicks, potions, and other diet products that are largely worthless for effective and healthy long-term weight control. The FDA urges consumers to avoid products that promise the impossible and, instead, to follow a sensible diet plan. Most nutritionists agree that, although there may one day be a magic cure for the problem of being overweight, in the meantime the safest and surest way to lose weight is simply to count your calories.

With an astonishing array of low-calorie and reduced-calorie foods to choose from, counting calories is no longer the ordeal it once was. It is now possible to reduce your caloric intake substantially without changing your eating habits very dramatically.

## THE DIET-FOOD MARKETPLACE

Diet foods and beverages, though not new, have in the past several years increased in sales to the point where they now represent one of the fastest-growing segments of the U.S. food industry. A much greater choice of calorie-saving food is now available to consumers. You can buy low- or reduced-calorie versions of bread, cheese, mayonnaise, margarine, cake and pancake mixes, jams and jellies, syrups, salad dressings, ice cream, gelatin desserts, and frozen foods and dinners, as well as beer and wine; the list is seemingly endless.

These products and many more are available at supermarkets around the country. Some of them you will find in diet-

Photodisc/Getty Images

**How much would you like to weigh? Are** you willing **to change your eating habits to achieve** your ideal **weight?**

food sections in supermarkets or, in many cases, shelved with similar, nondiet foods, which makes comparison of price and nutritional and caloric content easy.

## DIET-FOOD LABELING

Prior to their regulation by the FDA, so-called diet (**dietetic foods**) often contained as many calories as their regular counterparts. But current FDA rules, which became effective in 1992, ended this confusion. Figure 9.9 shows a typical diet-food label today. Moreover, it is now possible for a consumer to know the nutritional and caloric contents of almost any food product, regardless of whether it is labeled "diet" or not. As explained earlier, the Nutrition Labeling and Education Act of 1990 led to the establishment of specific definitions for terms used to describe dietary foods. These were listed in Figure 9.3. The imposition of these rules on

### Question for Thought & Discussion #10:
Have you ever tried to lose weight? How did you go about doing this? To what extent were you successful?

**Figure 9.9** A Reduced-Calorie Food Product Label

This Product Is
Low Fat, Low Sodium

| | Heart-Safe Helpings Spring Vegetable Noodles | Daily Dietary Recommendations* |
|---|---|---|
| Sodium | 450mg | 2,400 |
| Cholesterol | 12mg | 300mg |
| Fat | 6g | 50g** |

*The National Research Council, *Diet and Health: Implications for Reducing Chronic Disease Risk* (Washington, D.C.: National Academy Press, 1989). Recommended maximum sodium intake of 2,400 mg per day; less than 300 mg cholesterol per day; daily fat intake not to exceed 30% calories from fat. Individual calorie needs vary; please note that it is important to consider your diet over time.

**50 g fat based on a 1,500 calorie per day diet.

| Nutrition Information | | Servings Per Container: 1 | |
|---|---|---|---|
| Calories | 240 | Fat | 6g |
| Protein | 16g | Polyunsaturated | 2g |
| Carbohydrates | 30g | Monounsaturated | 3g |
| | | Saturated | 1g |
| | | Cholesterol | 12mg |
| | | Sodium | 450mg |

Ingredients:
Tomato Puree, Cooked Enriched Macaroni Product, Part Skim Ricotta Cheese (Whey, Pasteurized Part Skim Milk, Vinegar, Carrageenan), Tomatoes, Spinach, Zucchini, Mushrooms, Carrots, Onions, Romano Cheese (Made from Cow's Milk), Sugar, Dehydrated Onion, Modified Food Starch, Spices, Granulated Garlic, Salt, Hydrolyzed Plant Protein, Olive Oil, Xanthan Gum, Autolyzed Yeast Extract, Calcium Chloride, Citric Acid.

food manufacturers has made the jobs of consumers less demanding but has not eliminated their need to exercise judgment when making food purchases.

## PRICES

Increased competition by manufacturers vying for a share of the growing diet-food market has lowered the prices of these foods considerably in the last several years. Many low-calorie food items are no more expensive than similar "regular" food products, as can be seen in Table 9.2 on the next page. Here we compare caloric content and prices for seven types of food products; in most cases the diet versions were the same price as the regular product.

By comparison shopping and planning your dietary needs carefully, you can now get a great deal for your diet-food dollar. Forming a rational weight control program can help

you maximize the benefits to be obtained from available diet products and prices.

## DON'T BE MISLED

When shopping for a healthful diet, watch for claims that might be misleading. "No cholesterol" on the label of a jar of peanut butter, for example, may tempt you to believe that that particular brand is healthier for you than the others on the shelf. What the label doesn't say is that no other brand of peanut butter has cholesterol, either. A box of cookies may be labeled "baked in pure vegetable oil"—which may lead you to assume that the product is low in saturated fat. In fact, it could easily contain one of the tropical oils, which are high in saturated fat and may be more damaging to cardio-vascular health than any other type of oil.

Careful reading of labels can help you avoid being misled to a certain extent. At one time, manufacturers were not required to disclose on food labels whether the fats contained in their products were saturated or unsaturated. Concerned about the link between nutrition and health, the National Research Council recommended fuller disclosure on food labels to encourage manufacturers to make more low-fat and low-sodium foods available. In the meantime, some consumer advocates who were not content to wait for congressional action embarrassed manufacturers—through full-page ads in leading newspapers and other tactics—into switching from the cheaper, highly saturated tropical oils to partially unsaturated vegetable oils (still not the best choice, since they are a source of unhealthy trans fats).

## A SENSIBLE DIET PLAN

The following suggestions may help you devise a healthful and effective diet plan:

1. *Determine your caloric needs.* To benefit from counting calories, you need first to know how many pounds you want to lose—or what weight you wish to maintain. Weight control is—theoretically, at least—a simple matter of addition and subtraction. A certain amount of caloric intake is necessary to maintain each pound of your body weight. How many calories are required depends on how active you are and how much energy (read: calories) you expend in daily exercise and activity. Table 9.3 on the following page lists the approximate number of calories required to maintain one pound of body weight for men and women and for three different metabolic rates—sedentary, moderate, and active. If, for example, you are a moderately active woman and your weight is 120 pounds, you will need to consume about 2,160 calories per day $(120 \times 18)$ to maintain that weight. If you want to lose 10

(Continued on next page)

**Confronting Consumer Issues (Continued)**

| TABLE 9.2 | Price and Calorie Comparisons of Regular and Reduced-Calorie Food Products[a] | |
|---|---|---|
| Smuckers Strawberry Preserves | 50 cal/tsp | 15.8 cents/oz |
| Smuckers Light Strawberry Preserves | 10 cal/tsp | 18.0 cents/oz |
| Zesty Kraft Italian Dressing | 100 cal/2 tbs | 19.9 cents/oz |
| Zesty Kraft Free Italian Dressing | 50 cal/2 tbs | 19.9 cents/oz |
| Hellman's Mayonnaise | 100 cal/tbs | 9.9 cents/oz |
| Hellman's Light Mayonnaise | 50 cal/tbs | 9.9 cents/oz |
| Log Cabin Syrup | 105 cal/oz | 15.3 cents/oz |
| Log Cabin Reduced Calorie Syrup | 50 cal/oz | 15.3 cents/oz |
| Breyers Yogurt | 220 cal/8 oz | 8.0 cents/oz |
| Dannon Light Yogurt | 120 cal/8 oz | 8.0 cents/oz |
| Land O Lakes Cheese Slices | 70 cal/.75 oz | 31.2 cents/oz |
| Healthy Choice Cheese Product Slices | 40 cal/.75 oz | 27.4 cents/oz |
| Lemon Jell-O | 80 cal/4 oz | 16.1 cents/oz |
| Sugar-Free Lemon Jell-O | 10 cal/4 oz | 16.1 cents/oz |

[a]Prices charged in Buffalo, New York, in January 2000.

pounds, you will need to eat fewer calories or exercise more—or both. Generally, whenever you consume 3,500 fewer calories than you expend through exercise, you will lose one pound. Thus, if you reduce your caloric intake by 1,000 calories per day (or by 500 calories per day and use up the equivalent of the other 500 calories in increased exercise), you will lose two pounds per week. In five weeks, you will have attained your desired weight. Since there are additional health benefits to be obtained by exercising, increased activity usually is recommended as a

| TABLE 9.3 | Daily Calorie Requirements for Levels of Activity | |
|---|---|---|
| **Metabolic Rate** | **Men** | **Women** |
| Sedentary[a] | 16 cal/lb of body weight | 14 cal/lb of body weight |
| Moderate[b] | 21 cal/lb of body weight | 18 cal/lb of body weight |
| Active[c] | 26 cal/lb of body weight | 22 cal/lb of body weight |

[a]Includes activities that involve sitting most of the day, such as secretarial work and studying.
[b]May include dancing, skating, and manual labor—for example, farm work or construction work.
[c]May include dancing, skating, and manual labor—for example, farm work or construction work.

SOURCE: Nutrition Search, Inc., *Nutrition Almanac*, 2d ed. (New York: McGraw-Hill, 1984), p. 283.

component of a healthful diet plan. Table 9.4, which lists the approximate calories "burned" per hour for a variety of activities, will help you relate exercise to energy expenditure expressed in calories.

2. *Plan for a moderate weight loss.* If you plan to lose weight, aim for a moderate weight loss of one to two pounds per week. This represents the maximum weight loss recommended by physicians for healthful and effective dieting. Beware of fad diets and diet products that promise a weight loss of "ten pounds in three days" or "five pounds overnight"; these usually result in only a temporary weight loss—if any—and often the weight loss is achieved through dehydration.

3. *Obtain a calorie counter.* To plan for your shopping and dietary goals, you will need to buy or borrow a nutrition manual or calorie guide that lists the caloric (and, optimally, the nutritional) contents for various foods. Nearly every major bookstore or online bookseller has a variety of such offerings to choose from, and many of these books now include a long list of fast foods, by brand name, as well.

4. *Know your eating habits.* The aim of a good diet plan is to change what you eat, in terms of calories, rather than when or how you eat. The key is to change your food-shopping habits so you have the right foods available when you need them. If you like to spend time cooking and preparing meals at home, shop accordingly. If you are a snacker, shop for low-calorie snack foods. If you eat on the run and prefer frozen dinners occasionally, substitute the low-calorie brands for the regular frozen dinners.

5. *Look for low-calorie alternatives when eating out.* Shop around for restaurants in your area that may offer low-calorie cuisine in addition to their traditional selections. Such restaurants are now increasing in number, and, if you live in a large metropolitan area, it won't be too hard to find one.

6. *Remember the basic food groups.* Malnutrition is, unfortunately, a not uncommon consequence of dieting for many consumers—particularly for those who opt to emphasize one food group over another. Most nutritionists agree that if you select your calories—sparingly—from the rec-

| TABLE 9.4 | Expenditure of Caloric Energy per Hour for Various Activities | | |
|---|---|---|---|
| **Activity** | **Calories Expended per Hour** | **Activity** | **Calories Expended per Hour** |
| Ballroom dancing | 330 | Preparing a meal | 198 |
| Bed making | 234 | Roller skating | 350 |
| Bicycling 5.5 mph | 210 | Running 10 mph | 900 |
| Bowling | 264 | Scrubbing floors | 216 |
| Bricklaying | 240 | Sitting and eating | 84 |
| Carpentry | 408 | Sitting and knitting | 90 |
| Desk work | 132 | Sitting in a chair reading | 72 |
| Driving a car | 168 | Skiing | 594 |
| Farm work in field | 438 | Sleeping (basal metabolism) | 60 |
| Gardening | 220 | Standing up | 138 |
| Golf | 300 | Sweeping | 102 |
| Handball and squash | 612 | Swimming (leisurely) | 300 |
| Horseback riding (trot) | 480 | Tennis | 420 |
| Ironing (standing up) | 252 | Volleyball | 350 |
| Lawn mowing (hand mower) | 462 | Walking (2.5 miles per hour) | 216 |
| Painting at an easel | 120 | Walking downstairs | 312 |
| Piano playing | 150 | | |

SOURCE: Nutrition Search, Inc., *Nutrition Almanac*, 2d ed. (New York: McGraw-Hill, 1984), p. 4.

ommended basic food groups discussed in the chapter, you can maintain a reasonably balanced nutritional input.

7. *Keep a record.* Because each individual's metabolic rate—the rate at which calories are expended through activity varies considerably, you can get a clear idea of your own particular caloric needs only through trial and error. A journal helps. After a few weeks of recording the calories you consume per day and the amount of exercise you undertake, you will be able to approximate fairly closely how many calories and how much exercise you need to attain, or maintain, your desired weight.

## WHEN YOU CAN'T DO IT YOURSELF

Many dieters find it difficult—if not impossible—to lose weight without some kind of external support. If you are among them, you may choose from among several diet programs that, for a price, offer support and in some cases special diet food and medical supervision.

The lowest-cost programs are TOPS and Overeaters Anonymous. These are informal, nonprofit groups that stress motivation and positive reinforcement. Overeaters Anonymous focuses less on weight than on learning how to stop eating compulsively. It is patterned after Alcoholics Anonymous and offers similar emotional and psychological support.

Weight Watchers involves a cost and is a more structured program. Nutrition specialists consider Weight Watchers to be among the best programs because it helps you develop eating habits that can keep your weight off once you have lost it.

The most structured and most expensive programs are Diet Center, Nutri/System, and the liquid-diet plans, such as Optifast. These programs are supervised by medical and psychological experts and require special food supplements or food products. The liquid-diet plans, which are the most expensive and fastest-growing diet programs, are supposedly safe—unlike those of the 1970s that nearly vanished from the marketplace after sixty people died of protein starvation. The strictest of the liquid-diet programs is Optifast, which requires a totally liquid diet for the first twelve weeks and then a gradual reintroduction of food by the dieter. Optifast is available only for those who are at least fifty pounds overweight.

### Key Term:

**Dietetic Foods**   Low-calorie or reduced-calorie food, or food intended for a special dietary purpose—for example, for low-sodium diets.

# Chapter Summary

1. The percentage of total U.S. income spent on food has been declining steadily since this nation was founded. This is characteristic of goods that are necessities as opposed to luxuries. As income rises, the percentage spent on necessities falls. A German statistician named Ernst Engel made this discovery in 1856, and it is now called Engel's Law.

2. Under the Fair Packaging and Labeling Act of 1966 and the Nutritional Labeling and Education Act of 1990, the labels of most food products must clearly indicate, among other things, the product's name, the name and address of its manufacturer, and the nutritional value of its contents per standard serving size. The manufacturer must also comply with limitations on the words that may be used to describe the product.

3. Nutrition labeling is required for products containing added nutrients or for which a nutritional claim is made. Nutrition labels must list the amount of nutrients contained in the product, as well as total calories, serving size, amount of calories and certain nutrients per serving, and so on. This kind of labeling has been especially helpful to consumers who want to compare nutritional values, as well as prices, when they shop for food.

4. Special labeling is required for diet or dietetic foods or any foods designated as low-calorie or reduced-calorie products.

5. The U.S. Department of Agriculture provides grades for meats, fresh produce, poultry, and other items. These grades reflect taste, not nutritional qualities. Grades for produce can be confusing owing to the use of similar names, such as "U.S. Fancy," to represent different grades for different fruit categories.

6. Labels and grades help consumers make more rational choices. Some consumers want even stricter labeling requirements and fuller disclosure by manufacturers as to product contents.

7. It is important for the consumer to be aware that increased labeling imposes a cost on food manufacturers—a cost that is usually passed on to the consumer in the form of higher food prices.

8. U.S. government agencies and private health groups have been increasingly concerned over the nutrition of Americans, particularly the overconsumption of fats and the linkage between fat intake and illness. Although Americans appear to be changing their dietary habits and consuming less fat, the unhealthy diet of many Americans continues to be a major issue.

9. Dietary reference intakes (DRIs) were developed by the National Academy of Sciences to measure the nutrient values of different foods for comparison purposes. A condensed system of nutrient standards, the U.S. recommended daily allowances, is used in product labeling.

10. A good way to ensure a nutritious diet is to select foods in the recommended amounts from the following basic food groups: (1) fruits and vegetables, (2) bread and cereals, (3) milk and cheese, and (4) meats, fish, poultry, and beans. Fats and sweets, a fifth group, affect only the taste of food and are to be eaten sparingly.

11. Food additives are substances added directly or indirectly to food products. Direct additives generally consist of flavoring, coloring, preserving, and sweetening agents. Indirect additives result from the use of chemicals or other substances in the growing or packaging process. Much controversy and uncertainty exist concerning the long-term effects of many of these additives.

12. Health, natural, and organic foods are easier for consumers to evaluate since FDA labeling rules were set in 2000.

13. There is a growing trend toward frozen convenience foods or fast foods because as Americans become richer, they are willing to pay more to save time. That is, their higher incomes lead them to place a higher value on their time, and they therefore rate the ease of using convenience or fast foods higher than the improved quality or lower cost of foods they prepare themselves.

14. By comparing the nutritional value and prices of food products, consumers can cut their food costs considerably and still achieve a healthy diet. Because comparison shopping requires additional time, some consumers prefer to purchase only familiar products (such as name brands), often paying a higher price for them.

15. An increasing number of health problems are being linked to overweight and obesity. It is estimated that the mortality rate for people even 10 percent overweight is increased by 20 percent over a twenty-five-year period. Reducing caloric intake is not difficult with the many low-calorie and reduced-calorie foods now available.

16. In 1990, Congress passed the Nutritional Labeling and Education Act that required the FDA to establish standards for nutritional information that must be included on food labels. These rules were announced by the FDA in 1991 and implemented on May 1, 1993. The result has been much more uniform and understandable labels that consumers can use easily.

17. Seven steps to successful, and healthful, dieting are suggested for weight-conscious consumers: (1) determine your caloric need by calculating what weight loss, if any, you wish to achieve; (2) plan for a moderate weight loss of one to two pounds a week; (3) obtain a calorie counter—a book or manual listing calories for different foods; (4) know your eating habits and shop accordingly; (5) look for low-calorie alternatives when eating

out; (6) remember to include foods from the recommended basic food groups to ensure adequate nutrition; and (7) keep a record of your caloric intake and expenditure to determine your own caloric needs.

18. Numerous weight reduction programs are available for those who find it difficult to control their weight without external emotional or psychological support or a structured plan.

## Key Terms

carcinogenic **200**

dietetic foods **211**

dietary reference intakes (DRIs) **196**

Engel's Law **188**

U.S. recommended daily allowances (U.S. RDAs) **197**

## Questions for Thought & Discussion

1. When you choose your diet, what is the most important factor you consider—taste, nutrition, cost, or convenience?

2. Do you believe most consumers use information provided on labels when they buy food? Why or why not?

3. Have you ever become ill from food poisoning? How might you have prevented this from happening?

4. Do you consume genetically engineered foods? How have you formed your opinion of GE foods?

5. Are you careful to limit the amounts of sugar and fat you consume? Why have you made this choice?

6. How closely does your diet follow the one recommended by the Food Guide Pyramid?

7. Do you or any member of your family take a dietary supplement? How was the decision to use the product made?

8. How often do you eat fast food? What are your costs and benefits when you consume these products?

9. When you shop, do you check to be sure that the prices rung up are the same as the posted prices?

10. Have you ever tried to lose weight? How did you go about doing this? To what extent were you successful?

## Things to Do

1. Write down what you eat for the next seven days and then answer these questions: How closely did you stick to the recommended minimum servings of the basic food groups? How many foods did you eat that could be considered convenience foods?

2. Select a certain food category, such as TV dinners or cereals, and compare the information on the labels of each brand available. Is there a wide variance? Do the prices of the brands have much to do with nutritional value?

3. Check out the fresh meat department at a major supermarket. How are the packages of meat labeled? Are there suggestions for preparation? Is the meat graded? Are irradiated meat products available?

4. Assume you have decided to lose ten pounds over the next five weeks through a combination of increased exercise and reduced caloric consumption. Make a list of activities you could undertake to burn more calories and types or amounts of food you would stop eating to achieve your goal. Remember, eating 1,000 calories less a day or burning 1,000 calories more a day equals a weight reduction of two pounds per week.

5. Visit the FDA's Web site at **http://www.fda.gov**. Read a recent publication concerning foods. Summarize what you learn and evaluate the importance of the information to the average consumer.

6. Visit a grocery store where you or another member of your family shops for food. Choose five food products that your family commonly uses and compare their unit prices with those of competing products. Does your family tend to consume food products that are more expensive or less expensive than most other similar items? What reasons can you give to explain your family's shopping habits?

# Internet Resources

## Finding Consumer Information on the Internet

The following Web sites have been selected for their relevance to topics discussed in this chapter. Search these sites to locate information that can add to your knowledge of food poisoning and ways to protect yourself from such illnesses. Remember, Web addresses change frequently. If any of these addresses no longer function, find similar sites to investigate using any of the search engines available to you.

1. The U.S. Centers for Disease Control (CDC) provides information about most common food-borne diseases, their diagnosis and treatment, and how food becomes contaminated at its Web site, which can be found at **http://www.cdc.gov/ncidod/dbmd/diseaseinfo/ foodborneinfections_g.htm**.

2. The California Poison Control System Web site can be used to review a publication, titled "Food Poisoning and Safety," that explains what *E. coli* food poisoning is and how it can be avoided. Visit this Web site at **http://www.calpoison.org/public/food.html**.

3. Public Education—Poison Control Articles is a Web site that provides current articles concerning food poisoning. It can be found at **http://www.poisoncontrol.org/ foodpoisoning.htm**.

## Shopping on the Internet

The following Web sites have been selected because they offer consumers services similar to those described in this chapter. These are commercial sites that are designed to market products. They do not represent a comprehensive or balanced description of all organic food products available online. How do the prices offered at these Web sites compare with prices on nonorganic products charged by your local stores? Remember, Web addresses change frequently. If any of these addresses no longer function, find similar sites to investigate using any of the search engines available to you.

1. California Flavors: Natural Gourmet Foods invites consumers to shop online from a great selection of natural and organic gourmet foods from California and the Pacific Northwest. Its Web site can be found at **http://www.california-flavors.com**.

2. Mail Order Catalog markets vegan and vegetarian foods and products, including meat substitutes, cookbooks, mixes, soy products, and pet food. Its Web site is located at **http://www.healthy-eating.com**.

3. Sunorganic Farm sells certified organic nuts, seeds, dried fruits, grains, and more. See what it has to offer at **http://www.sunorganic.com**.

## InfoTrac Exercises

Purchasers of new copies of this text are provided with access to the InfoTrac Web site. This Web site links students to thousands of recent articles published in hundreds of periodicals. Use the key words **weight loss programs** or other terms from this chapter to conduct a key-word search. Choose one article that is of particular interest to you and write a brief essay describing what you have learned from the article. Be sure to cite the author and title of the article and the name and date of the publication in which it appeared.

# Selected Readings

Abboud, Leila, and Patricia Callahan. "Food Industry Gags on Proposed Label Rules for Trans Fats." *The Wall Street Journal*, December 27, 2002, pp. B1, B4.

Allen, Andrea. "Organic Standards Go National at Last." *Food Processing*, May 1999, pp. 82–84.

Coorsh, Richard. "Zapped Meat." *Consumers' Research*, March 1999, p. 6.

Garrell, Diane. "Diet Riot Is Never Quiet." *Variety*, November 18, 2002, pp. 58–59.

Halwed, Brian. "Organic Gold Rush." *World Watch*, May-June 2001, pp. 22–23.

Kaiser, Emily. "Buying Good Fish." *Washingtonian*, September 2002, pp. 130–137.

Leak, Jessie A. "Herbal Remedies Pose Health Threats." *USA Today Magazine*, February 2002, pp. 1–2.

Mackenzie, Jane. "Unit Pricing, Consumers Do Want It." *Choice*, June 2000, p. 2.

Marcus, Mary Brophy. "Organic Foods Offer Peace of Mind, At a Price." *U.S. News & World Report*, January 15, 2001, pp. 48–50.

Mayur, Kumundini. "Obesity, A Growing Problem." *The Futurist*, October 1999, pp. 14–15.

"Miracle Weight Loss? Fat Chance." *Choice*, July 2002, pp. 6–7.

Sara, Chris. "Just the Facts." *Muscle and Fitness*, April 2003, p. 64.

Seligson, Susan V. "Wacky Weight Loss Products." *Good Housekeeping*, April 2001, p. 58.

Weinberg, Winkler, and Ted Geltner. "Beef Wars." *The Washington Monthly*, January-February 2001, pp. 45–49.

"Your Health." *USA Today Magazine*, February 2002, pp. 1–16.

Photodisc/Getty Images

# CHAPTER 10

# Purchasing Household Products

After reading this chapter, you should be able to answer the following questions:

- How can consumers use the rational decision-making process when choosing to buy nondurable goods or services?

- What trade-offs do consumers make regarding price, quality, and style when buying clothing?

- How can consumers judge the quality of different clothing items?

- How may consumers decide when to purchase consumer durable products?

- What similarities are there between the purchase of a consumer durable good and an investment?

- What criteria should consumers use when they choose among consumer durable products?

- How should consumers decide whether to buy energy-efficient products for their homes?

215

**"One of every three dollars you spend is used to buy food, clothing, or other products you will use within your home."**

**Question for Thought & Discussion #1:**
How important are the clothes you wear to your outlook on life and the way you feel about yourself? Explain your point of view.

You might be surprised to learn that according to government statistics one out of every three dollars you spend is used to buy food, clothing, or other products you will use within your home. Many of these products can be used for only a relatively short time. Clearly, the food you buy (the topic of Chapter 9) will not last long. Food, however, is not the only product whose value is used up quickly. The same can be said for most of the clothing you own, the newspapers or magazines to which you subscribe, the energy you use to heat or cool your home, and the medications you take to keep you healthy. Economists refer to products that have a useful life of less than three years as **nondurable goods.** The utility of these products is important, but it does not last.

Similarly, the value of most services consumers buy exists for only a little while. When you go to your doctor for a physical, you may learn that you are in the best of health. Although this knowledge is valuable to you at present, you cannot be sure that you will still be healthy six months from now. Or, if you are lucky enough to be able to pay someone to clean your home, the value of that service will not last past the next party you give. Paying an accountant to prepare your tax return this year will not eliminate your need to purchase this service again next year. With a few exceptions, such as preparing a will that may endure for many years, consumer services have value that lasts for only a short time.

This does not mean that consumers should avoid buying nondurable goods or services. It does mean they should be particularly careful to weigh the costs and benefits of such purchases. A new $500 suit may look and feel wonderful today, but in a year or two it may be out of style and have little value to you. As with all consumer purchases, it is worth taking the time to use the rational decision-making process when you consider buying these types of products.

# Clothing, Happiness, and Budget Considerations

After food, the most common type of nondurable good you are likely to buy is clothing. How you dress can enhance your satisfaction with life. For example, "dressing up" to visit friends, enjoy a party, or attend religious services may bring you enjoyment. If you are a parent, seeing your children in attractive clothing can bring you a sense of accomplishment and pride.

Although clothing can bring satisfaction, it may also be a source of conflict, particularly in family groups. When members of a family have different ideas about which fashions are appropriate or how much should be spent on clothing, there are likely to be disagreements. To avoid family arguments over clothing decisions, it is important to recognize and respect each person's values and point of view. As children mature, they should be given a greater voice in choosing what they will wear and how funds allocated to their clothing budgets are spent. Parents should not abdicate their responsibilities, but they should seek a reasonable balance between their values and financial limitations and the tastes and preferences of their children. This will require open lines of communication and a willingness to compromise. These and many other factors affect the trade-offs people make as they choose how best to allocate the dollars in their clothing budgets.

**Nondurable Goods** Products that have a useful life of less than three years.

The typical American has more clothes than are actually needed to provide physical protection from the elements. Today, most clothing is not so much a necessity as a consumer good that gives pleasure to the wearer because of its style or quality.

## Customs

The types of clothes we buy are often determined by customs, although this may be less true today than in the past. For example, it is custom that causes men most often to wear trousers, while women frequently wear skirts. In Scotland, though, some men regularly wear skirts called kilts. At formal occasions in our society, men often wear ties while most women do not. Customs tend to change over time. Today, it is socially acceptable for women to wear pants almost anywhere, whereas in the past this could have exposed them to ridicule or discrimination. In sixteenth-century France, men wore high-heeled shoes. Today, this style is worn almost exclusively by women.

How do your values help you choose the clothes you wear?

Customs are not created in a vacuum; most developed to satisfy a segment of the population. Although customs do change, the rate of change tends to be rather slow. Once established, customs help people complete the decision-making process by limiting the number of alternatives they feel compelled to consider and by providing a basis for evaluating each possible choice.

## Aesthetic Considerations

Clothing that pleases us as wearers, and that we think pleases those who see us, can positively affect our self-image and increase our confidence. One of the strongest motivations in selecting dress is to enhance the wearer's appearance to others. This is an aesthetic consideration; that is, it has to do with one's conception of beauty or good looks. Even among primitive tribes, self-adornment is often a stronger motivation for choosing clothing than protection from the elements, comfort, thrift, or durability. This reason for clothing selection need not have anything to do with snobbery. It may only indicate a positive self-concept that can be healthy and beneficial to individuals and society.

## Values

People often express their values, attitudes, and interests through their consumer choices. To a great extent, values determine our behavior and influence much of our decision making. In this sense, our clothing choices make a statement about who we are and what we believe. Some people, for example, choose to dress in ways that demonstrate their cultural or ethnic heritage.

Studies have found that clothing choices often reveal people's attitudes toward conformity, self-expression, aesthetic satisfaction, comfort, and economy. For example, if you value fitting in with others, you will choose clothing that reflects the fashion choices of your peers. If you are more concerned with aesthetic satisfaction, you might choose clothing for its style or fabric quality or for the pleasure you derive from being considered well dressed. If you are more concerned with economy, you will select clothing based on its practicality, durability, and price.

Even though our values often influence the clothing choices we make, many times our values conflict, forcing us to make compromises. A person may value conformity in dress, for example, but have a budget that places a constraint on that desire. The latest style that others are wearing may be too expensive. Clearly, in such cases a compromise must be made.

## Durability versus Price

**Question for Thought & Discussion #2:**
When you shop for clothing, what factors do you consider? Of these, which is the most important to you?

Most consumers realize that styles come and go. As a matter of fact, many people accuse the clothing industry of deliberately creating obsolescence. The ultimate in planned obsolescence may have been the dresses and other garments made of paper that were offered for sale in the 1970s. But, at least in that case, the consumer planned the obsolescence because anyone who bought such articles knew that they would soon be thrown away. Nobody was being fooled, for the manufacturers stated explicitly how many times the products could be worn.

Normally, durability in clothing is related directly to price. You usually have to pay more for garments that last longer, and you may find that their prices are higher than you are willing to pay. Depending on your tastes and your budget, you may be better off buying less durable clothes and replacing them more frequently—especially if the expense of cleaning is a major factor in their total cost. Whenever the maintenance costs of an item go up, you have a greater incentive to replace the item rather than to maintain or repair it.

With the knowledge that a product's cost is related to its durability, you can make more rational choices when buying clothing. You purchase a suit or a dress, for example, not only because you like it now but for the **service flow** it will yield over its expected life. Thus, if you pay $200 for a new coat that you will wear for three years, your expected cost per year is $67. This yearly cost is lower than if you paid $150 for a less durable coat that would last only two years ($75 per year).

In your own shopping forays, remember to consider the cost per year of owning a piece of clothing. It is also important to consider the maintenance costs of different fabrics. If you must dry-clean a gorgeous wool sweater, it will end up costing you much more than its original $100 purchase price. A cotton sweater labeled "machine wash cold, lay flat to dry" could be a much less expensive choice in the long run, even if it is priced at $120. This kind of information is important in determining the relative costs of different items of clothing (and is a good justification for labeling requirements imposed on manufacturers by the government).

## Indications of Quality

**Service Flow** The service that a product will yield over its expected life.

To get the most quality for every dollar you spend on clothing, it is important to recognize which garments will perform better and longer than others. How can you distinguish quality in clothing? Generally, there are two broad indicators—fabric content and the skill with which the garment was constructed.

# Fabric Content

You can tell the fabric content of any garment you might purchase by reading its label, which is required by law to be attached to every clothing item marketed in the United States. Higher-quality fabrics consist of either all-natural fibers (such as wool, cotton, or silk) or a blend of those with manufactured fibers (such as nylon, polyester, rayon, or acrylic). Manufactured fibers often lend durability and ease of care to fabrics. Cotton-polyester blends, for example, are popular today because they combine the comfort of cotton with the durability and ease of care ("wash and wear") of polyester.

Fiber content, however, won't tell you everything about fabric quality. The durability and overall performance of a particular fabric also depend on yarn size, tightness of weave, fabric finish, and other treatments that alter the fabric texture. In nylon products, for example, yarn size can create either one of the strongest fabrics available, such as that used in parachutes, or one of the most fragile, such as that used in hosiery. A loosely woven fabric is weaker and more likely to snag than one with a tighter weave. Finishes can also alter the performance of fabric. Cotton, for example, is one of the most absorbent fibers, but it can be made water resistant, as in the case of the polished cotton that is used in boat sails.

In sum, content labels, although helpful, do not indicate entirely how a given fabric will perform for you. Consequently, it is a good idea to save your clothing receipts until you've worn a garment a time or two and washed it (or had it dry-cleaned) at least once. It is also useful to know something about manufactured fibers, since they are being used increasingly in the garment industry. This type of information can be obtained from a variety of sources on the Internet.

> "What justifies the higher price charged for a shirt at an exclusive store? Generally, the higher price indicates a higher quality of craftsmanship in the shirt's construction."

# Tailoring

There is more to quality clothing than fabric. For example, if you compare shirts selling for $25 or $30 at a J. C. Penney's store with shirts priced at $100 or more at an exclusive men's store, you might find very similar, if not identical, fabric content. What, then, justifies the higher price charged for the shirt at the exclusive store? Generally, the higher price indicates a higher quality of craftsmanship in the shirt's construction. Labels won't tell you how well a garment is tailored—that is, the skill with which it is cut and sewn together. This is something you must judge for yourself, and it's not always easy to do unless you've taken a course in tailoring or done research on your own.

Some indications of quality craftsmanship, however, are easily visible and can be quickly checked before you purchase a garment.

1. Are the stitches neat and even along the seams, on the edges of the pockets, and behind the collar?
2. Are the inside seams well finished? A finished seam will be either stitched at the edge or cut with pinking shears to prevent unraveling of the fabric.
3. Is there a generous seam allowance? This is typical of well-made clothing, to allow for alterations that improve the fit of the garment.
4. In plaid, striped, or textured fabrics, do the patterns or textures match at the seams? Do they run in the appropriate direction?
5. In heavy-duty clothing, such as work clothes or jeans, are the seams flat-felled (doubled over and stitched twice on the outside)? This gives seams added durability under stress.

6. Is there double stitching at stress points in the garment, such as around the armholes and on the pockets?

7. Do the buttonholes fit the buttons tightly? Is the stitching around the buttonholes even and tight to prevent it from coming undone?

8. In jackets or blazers, does the collar have backing to help it retain its shape?

9. In shirts, are the front corners of the collar reinforced?

Perhaps the best way to familiarize yourself with quality in clothing is to visit an expensive store and try on some of the clothes. How do they feel? How do they fit? Look closely for the qualities just mentioned. If you have never done so, it might even be worth investing in one or two clothing items of superior quality to learn about their performance. Then, in the words of clothing designer Egon von Furstenburg, "having tried the best, you will be better able to achieve the best look at any price."

# The Powerful Fashion Industry

The immense size of the fashion industry and its apparent ability to influence our buying habits raise the eternal question of consumer versus producer sovereignty. Are fashions dictated by you, the consumer, or by the whims of designers employed in the fashion industry? Nobody will ever know who truly decides what will or won't be fashionable, but consumers obviously have some say in the matter. Any new style promoted by manufacturers will soon disappear if consumers refuse to buy.

## What's in a Name?

If you buy a pair of Tommy Hilfiger slacks, will their quality be better than another pair of slacks without a designer label? Possibly, but not necessarily. This is true because designers usually exercise only limited control over the actual production of garments that bear their names, regardless of the quality of the initial design. Most of the millions of designer products filling our shops, sold from mail-order catalogues, or marketed over the Internet are produced by manufacturers who contract to produce goods that bear a designer's name. As a result, quality control is often similar to that of many nondesigner garments consumers could buy (it has been estimated that about 10 percent of all new garments have manufacturing flaws).

When considering buying designer clothing, you should take time to make the same careful examination of the garment as you would for a nondesigner product. Through careful shopping, you may be able to purchase nondesigner garments of equal or superior quality for a lower price. You may even find garments (under different labels) that are identical to the more expensive designer products. Remember, a designer label by itself is no guarantee of quality.

When you buy designer clothing, part of the price you pay is for the label, in the sense that you absorb part of the royalty costs the manufacturer must pay for the use of the designer's name. In addition, the marketing of designer products requires expensive advertising. These costs are also covered by the price you pay.

Does this mean you should never buy designer clothing? Of course not. But you should be aware of what you are paying for. If quality alone is your

**Question for Thought & Discussion #3:**
How important are designer fashions to you? Do you feel a need to follow fashion trends?

concern, it can be found under other labels as well—labels that aren't as heavily advertised as name brands and designer labels. And, if you do wish to buy designer clothing, off-price stores offer a lower-cost alternative. Stores such as Loehmanns, T. J. Maxx, and Marshall's, among others, which purchase production overruns or end-of-season styles, can offer name-brand and designer labels at up to a 50 percent discount and still earn a profit. You may also purchase discounted designer clothing on the Internet.

## Government Regulation

The federal government imposes a number of standards on all firms that produce or market clothing in the United States. As mentioned earlier, all clothing manufacturers must affix labels to their products disclosing fabric content and required care. Labeling laws help manufacturers as well as consumers by protecting them from competition from dishonest businesses that could otherwise profit from misleading consumers about the quality of their products.

Federal labeling requirements have been set by three primary statutes: the Wool Products Labeling Act, the Fur Products Labeling Act, and the Textile Fiber Products Identification Act. All are enforced by the Federal Trade Commission (FTC). In addition, since 1972, the FTC has required all clothing to carry a label indicating the type of care required for proper maintenance.

Since 1953, the federal government has required garments sold in this country to meet flammability standards. In that year Congress passed the Flammable Fabrics Act in response to a public outcry over deaths and injuries caused by highly flammable fabrics used in some apparel. This act prohibited the "introduction or movement in interstate commerce of articles of wearing apparel and fabrics which are so highly flammable as to be dangerous when worn by individuals." A 1967 amendment to the act specifically prohibited the manufacture and sale of any product, fabric, or related material that fails to conform to an applicable standard or regulation.

Since the passage of the act, the government has created numerous regulations for its implementation. Flammability standards have been established not only for wearing apparel, but also for such items as mattresses, mattress pads, and carpets. To protect further against flammability in children's sleepwear, the government mandated in 1971 that fabrics used for such garments be treated with a flame retardant. Various chemicals have been used for this purpose. One of them, Tris, though effective, was banned in 1977 because it was thought to be carcinogenic.

## Choosing Durable Goods

American consumers spend almost as much for **durable goods** (goods that typically last three or more years) as they do for nondurable goods and services. Decisions involving durable goods are important because these products tend to be expensive, and once they are purchased, consumers generally live with them for a long time.

Consumer durables commonly found in our homes include washing machines, refrigerators, stoves, microwave ovens, furniture, carpeting, entertainment centers, and computer systems—to name only a few. (The next chapter examines another consumer durable owned by many Americans—

# Cyber consumer
## Shopping for Fashion on the Internet

Thousands of retailers offer to sell clothing online. You can find them by doing a search using the type of clothing you wish to purchase as a key word. Or you can use a Web site that links you to retailers such as http://shopping.yahoo.com. Visit this Web site and find out how easily you can purchase designer clothing online. Are the online prices lower, higher, or about the same as those you would pay at local retail outlets? Is the selection better? Is shopping online easier than visiting your local mall?

**Durable Goods** Goods that typically last three or more years.

**"Buying durable consumer goods may be regarded as a type of investment."**

the automobile.) To get an idea of just how durable some of these consumer products are, study Table 10.1, which shows the average life expectancy of a variety of consumer durable products. Before choosing to buy a durable good, consumers should search for and evaluate information that can help them identify and plan for their future consumer durable needs. If you own a washing machine that is more than ten years old, for example, you should expect to replace it in the near future. Consumer durable goods share many of the following characteristics:

1. They yield a service flow over many years.
2. They tend to be expensive.
3. Their purchase is often financed on credit.
4. They wear out or depreciate in value over time.
5. They require maintenance and repair to continue to be useful over time.
6. They reach a point when it is more economical to replace them than to repair them because the cost of maintenance exceeds their value.
7. They are typically insured through homeowners' or renters' policies.
8. They may represent a way of demonstrating one's wealth or status.
9. They are, in many cases, becoming more difficult to dispose of in environmentally responsible ways.

## Consumer Durables as an Investment

Unlike spending for nondurable goods or services, whose value is quickly used up, buying durable consumer goods may be regarded as a type of investment. You spend your money to buy a durable good because you expect to receive a flow of useful services over time that will have greater value than other uses for your money. Consumers should not think of such purchases as a cost that is paid in only one or even a few years.

Consider the decision to buy a washing machine as an example. Although the initial price of a washing machine may be $560, it probably will last eight years or more. The cost of doing your laundry each year, then, is $70 for the

**Question for Thought & Discussion #4:**
How did your family members go about choosing the television set they own? Are they satisfied with their choice?

| TABLE 10.1 | Average Useful Life of Various Appliances |
| --- | --- |
| **Type of Appliance** | **Expected Useful Life (in years)** |
| Gas range | 17 |
| Electric range | 16 |
| Refrigerator | 15 |
| Dishwasher | 12 |
| Room air conditioner | 11 |
| Microwave oven | 9 |
| Color TV | 9 |
| Washing machine | 8 |
| VCR | 7 |
| Cordless telephone | 7 |
| CD player | 5 |
| Personal computer | 3 |

SOURCE: *Consumer Reports 2000 Buying Guide,* p. 22.

The purchase price is only one factor consumers should consider when they evaluate consumer durable goods.

machine plus the cost of water and electricity to run it—say, another $60 a year, for a total annual cost of $130. That isn't the end of it, however; if doing your washing at a coin laundry for $5 each week is your alternative, buying the machine will save $130 each year ($5 × 52 weeks = $260 − $130 = $130). When you consider the cost of transportation to a coin laundry, and the value of the time you would spend there while your clothes are being washed, your savings are even greater. Similar savings may be realized when we purchase freezers, sewing machines, microwave ovens, or dishwashers. The next section of this chapter presents a method of evaluating the costs and value of appliances over their useful lives.

Even furniture, which generally does not provide a service of measurable value, may still be evaluated in terms of time. A new bed, for example, probably will not save you money or increase your income, but if it helps you sleep and feel better for the next 10 years, its $500 price may be worthwhile. After all, $500 divided by the next 3,652 days (10 years) is only about 13 cents a day. Would you be willing to pay 13 cents a day to feel better?

# Evaluating the Purchase of a Consumer Durable

In many respects, the purchase of a consumer durable good is just like a business's decision to buy a machine. Businesses determine the profitability of investing in different types of equipment through a cost-benefit analysis. We can use a similar analysis to determine whether a new consumer durable

**Energy Guide Label** A label that shows the expected energy costs of operating an appliance for a year; required by law for many consumer durables.

should be purchased. In this way, you can determine when and how much to invest in new furniture, a new washing machine, a new refrigerator, or any other consumer durable you might consider buying for your home.

You conduct this analysis by looking at the expected costs and benefits of a consumer durable item. Evaluate your possible purchase in terms of all costs and all benefits expressed in dollars per year. Consider a hypothetical example in which you are debating the advisability of purchasing a freezer that will allow you to store food in large quantities.

## The Cost Side of the Picture

The cost of any durable good purchase will, at minimum, include depreciation, operating costs, maintenance and repairs, and interest.

1. *Depreciation.* Take an example of a freezer that has a ten-year useful life expectancy, after which it is likely to have zero value and must be scrapped. Its full purchase and delivery price is $500, so the average annual depreciation will be $50 a year. Actually, this understates depreciation in the first year and overstates it in the later years of the freezer's useful life. It would be virtually impossible to sell a used $500 freezer at the end of one year for $450. Just as with automobiles (discussed in Chapter 11), durable goods used in the home depreciate more in the first year than in subsequent years of ownership. We'll ignore that fact for the moment.

2. *Operating costs.* Most durable goods, and particularly a freezer, require some form of energy to operate. In this case, it is electricity. The government requires the manufacturers of many consumer durables (including water heaters, refrigerators, and freezers) to attach **energy guide labels** to their products. These labels provide consumers with a range of the expected energy costs for operating the appliance for a year. An example of an energy guide label appears in Figure 10.1.

   Although individual consumers' costs will vary, all manufacturers must prepare these labels in the same way so that consumers can easily compare the energy efficiency of different products. For our example, we will assume that the freezer will use $65 worth of electricity each year.

**Question for Thought & Discussion #5:**
How important are energy guide labels to most consumers? Do you believe the average consumer knows what the labels mean? How could the government make them more useful?

3. *Delivery, installation, and take-away fees.* These fees are often charged when consumers purchase consumer durables. Many appliance or furniture stores bill consumers for delivery, particularly for heavy or bulky items. When electronic devices, dishwashers, or garbage disposals are purchased, there is almost always an installation fee, which can be substantial. Most firms also charge to take away old consumer durables that have been replaced. These fees reflect the cost of disposal, particularly for products such as refrigerators, freezers, and air conditioners that contain refrigerants that must be disposed of in ways that protect the environment. For the purchase of a freezer, additional fees may run as much as $100. This cost should be spread over the expected ten-year life of the product, or $10 per year.

4. *Repairs.* A freezer will probably have to be serviced occasionally over its ten-year expected useful life. For an initial payment of several hundred dollars, you might buy a service contract that will provide complete coverage for all replacement parts and labor for a number of years. Assume that you buy a service contract for $150 that provides this type of protection for ten years. Your annual cost is then $15.

5. *Interest.* If you borrow $750 to buy the freezer ($500 price + $100 in fees + $150 for a service contract), you will have to pay interest on the bor-

**Figure 10.1** Sample Energy Guide Label

General product category

Distinguishing size or feature

Relative energy efficiency, based on energy requirements, as compared with the best and worst in the same product category

Manufacturer or brand name

Specific model or style

Typical annual operating cost based on national averages

Refrigerator- Freezer
Capacity: 23 Cubic Feet

(Name of Corporation)
Model(s) AH503, AH504, AH507
Type of Defrost: Full Automatic

**ENERGYGUIDE**

Estimates on the scale based on a national average electric rate of 4.97¢ per kilowatt hour.

Only models with 22.5 to 22.4 cubic feet are compared in the scale

**$91**

Model with lowest energy cost
$68

THIS ▼ MODEL
*Estimated yearly energy cost*

Model with highest energy cost
$132

Your cost will vary depending on your local energy rate and how you use the product. *This energy cost is based on U.S. Government standard rates*

How much will this model cost you to run yearly?

| Cost per kilowatt hour | Yearly cost |
|---|---|
| | *Estimated yearly energy $ cost shown below* |
| 2¢ | $36 |
| 4¢ | $73 |
| 6¢ | $109 |
| 8¢ | $146 |
| 10¢ | $102 |
| 12¢ | $218 |

Ask your salesperson or local utility for the energy rate (cost per kilowatt hour) in your area.

**Important** Removal of this label before consumer purchase is a violation of federal law (42 U.S.C. 6302)

(Part No. 3710760)

---

rowed funds until the debt is paid off. This could add another $200 to the cost of the freezer. You may believe you could avoid this type of cost by paying cash. Not so! Paying cash prevents you from earning interest you could have received if you had deposited your money in a bank. Generally, the interest you could earn from a deposit is much lower than what you would pay for borrowed funds. Still, paying cash entails a cost that is measured in the value of forgone interest.

6. *Total costs.* We find, then, that the total expected annual cost, on average, of service from the freezer is:

| | |
|---|---|
| Depreciation | $50 |
| Operating costs | 65 |
| Delivery, installation, and take-away fees | 10 |
| Service contract | 15 |
| Interest | 20 |
| **Total** | **$160** |

## Benefits

The benefits of having a freezer include, but are not limited to, lowered food bills, convenience, reduced food spoilage, and less time and gas used for shopping trips.

1. *Lowered food bills.* Depending on the size of the freezer, the owner can take advantage of sales on meat, frozen fruit juices, vegetables, ice cream, fish, and other items that are either sold frozen or may be frozen and stored. Also, a freezer allows the owner who grows vegetables in a home garden to freeze and store them for the winter. Finally, some foods can be bought in bulk at reduced prices. By purchasing a side of beef, for example, as much as $.20 to $.30 per pound can be saved. Assume that from studying your food budget and spending patterns, you predict that you can save $150 each year if you buy and use a freezer.

2. *Saving from fewer shopping trips.* If owning a freezer enables you to spend less time going shopping, then you can place a value on your saved time and reduced cost of transportation. Suppose you typically go to the grocery store three times a week. You believe that owning a freezer will allow you to reduce your shopping to two trips per week, thereby saving you one hour of time and the cost of driving your car for the six-mile round trip. You estimate this saving has a value to you of $100 per year.

3. *Convenience.* Suppose, on the spur of the moment, you invite a friend home for dinner. If you do, it is certainly more convenient to remove an extra steak from your freezer than to make a special trip to the store. Or, maybe you feel a need for ice cream topped with strawberries late at night. Without a freezer, you may be out of luck. It is difficult to put a dollar figure on convenience, but it can be an important benefit of buying a freezer.

4. *Other benefits.* A freezer may allow you to reduce your cost of spoiled food by storing leftovers. You can also prepare larger quantities of food, eat part now, and freeze the rest for another meal later when you have less time to cook.

5. *Total benefits.* We can add up the total expected annual benefits as follows:

| | |
|---|---|
| Reduced food bill | $150 |
| Reduced transportation cost | 100 |
| Convenience | ? |
| Other benefits | ? |
| **Total** | **$250+?** |

If the items included as question marks in this calculation have an annual value of $60, you would expect to save $150 per year from buying the freezer. Even if they are disregarded, you would still save $90. This cost-benefit analysis shows that buying this durable good is a wise choice unless you have an even better use for your limited funds.

# Servicing Your Consumer Durables

Almost all consumer durable goods require some type of service and care throughout their useful lives. In fact, homeowners can attest that the more appliances there are in their households, the more time and money it takes to have them maintained or repaired.

## Pros and Cons of a Service Contract

When you purchase a new appliance—particularly washers, dryers, refrigerators, freezers, or air conditioners—you generally can also purchase a service contract that covers the cost of parts and labor for a specified time after the full or limited warranty runs out. For example, if you buy a refrigerator from Home Depot, Best Buy, or Sears, the store will likely offer to sell you a service contract for extended protection over an additional three to five years. In a sense, a service contract is really a form of insurance. You pay a predetermined amount to avoid the possibility of having to pay a larger amount later should your refrigerator require repairs. You are betting that you would have to pay more for repairs than the service contract costs. Or, you would rather pay a known cost now than face the possibility of incurring an unknown cost in the future. The seller of the service contract is betting that, on average, the repairs it makes will cost less than the total it receives from all the contracts it sells. Clearly, the sellers of service contracts must profit from these arrangements. Does that mean that you will always lose if you purchase such a contract? No, you may not lose. The product you buy may break and require expensive repairs during the time covered by the service contract. And, even if the product performs perfectly, during this time you benefit from the knowledge that you will not be required to pay for expensive repairs.

There is another positive aspect to purchasing a service contract. You may decide to have minor problems repaired immediately instead of waiting for them to turn into major problems. In so doing, you may extend the useful life of the appliance and delay the need to replace it. If, for example, your refrigerator develops an annoying vibration, you may have it serviced quickly because your contract will pay for the service call. The repair person may catch a problem with a compressor that needs to be lubricated before it has a chance to burn out. Had you been required to pay for this service yourself, you might have let it go, and your refrigerator would eventually have required a much more expensive repair.

## The "Ninety" Rule

Industry sources frequently cite what they call the "ninety" rule: if an appliance goes ninety days without breaking, it probably will not break for quite a while. Since most consumer durables are covered by warranties during this initial period, you might question the wisdom of paying more for a service contract. According to John Gooley, retired service manager for the National Association of Retail Dealers of America, the odds are that a product that performs well during its warranty period will continue to do so throughout the time covered by a service contract. *Consumer Reports* and many other consumer protection organizations generally recommend *against* the purchase of service contracts.

## Before You Call a Service Person

Service calls are costly, so before you call for help, go through the following checklist to see if you can solve a problem with a consumer durable yourself:

1. Make sure that you have read and followed all the manufacturer's instructions. Many instruction booklets for durable goods include a troubleshooter's checklist that may help solve your problem. Many dollars are wasted on unnecessary service calls. Paying a service person to clean the drain in the bottom of a refrigerator, something the owner could easily have done with the help of the instruction booklet, is a case in point.

2. Check to see that electric appliances are plugged in and that fuses have not been blown or that circuit breakers are closed.

**Question for Thought & Discussion #6:**
What type of appliance service would you try to complete yourself, and what would you hire a technician to do?

## CONSUMER CLOSE-UP
### Rent-to-Own Agreements

Consumers who are short of cash or who have been denied credit are sometimes tempted by rent-to-own offers promoted by some stores. Under these plans, a consumer "rents" an appliance or piece of furniture by agreeing to pay a weekly or monthly charge. At the end of a specified time, the consumer owns the product. Does this sound fair to you?

Research shows that consumers should avoid these agreements if they possibly can. Those who accept these agreements are likely to end up paying as much as 300 percent more than normal retail prices for the things they buy. And, if they miss a single payment, they run the risk of losing both the item they are "renting" and the money they have already paid.

Rent-to-own agreements were originally designed to allow businesses to avoid legal limits on interest rates and rules of disclosure for information about credit costs. Businesses argued that because they were "renting" instead of selling items they were not required to provide this information to their customers. After losing a number of court cases, however, rent-to-own stores are now required to provide interest rate information to customers. This has not stopped them from charging very high prices. Although interest cost information is now available, research shows that most people who make rent-to-own agreements are interested only in the amount of their weekly payment.

Consider one specific case. In 2002, a professor at Iowa State University visited a rent-to-own store in his community to find the cost of renting a personal computer. He was quoted a price of $32.99 per week that would be paid for two years. Doing a little math, he found that the total cost of the computer would be just over $3,430. He then visited a local retailer to price the same computer and found he could buy one for $800. Even if he had borrowed the $800 to buy the machine, his cost, including interest, would have totaled less than $900. No matter how he looked at it, renting the machine cost at least $2,500 more than buying one outright.

Roughly three million rent-to-own contracts were signed in 2001. The total value of these contracts was estimated at $5 billion. Rent-to-own customers have been surveyed to find out why they signed these contracts. The most often mentioned reason was a desire not to be tied down by an agreement. If they bought on credit, they would be obliged to continue to make their payments no matter what. But, with a rent-to-own agreement, they could end their financial commitment merely by calling the store and asking that the item be picked up.

If you really need a product you can't afford, it is generally better to purchase a used product than to sign a rent-to-own contract. Or, if you possibly can, you should simply do without the product until you can afford to purchase what you want.

**Do you have any friends or relatives who have decided to purchase household goods from a rent-to-own business? How satisfied were they with their choices?**

---

3. Make sure gas or water connections have been turned on correctly. If all else fails, call the service center, describe the problem, and give the model number of your appliance (taken from the nameplate). This way you will be less likely to pay for two service calls: one to see what is wrong and the second to bring a needed part that could have been in the service person's truck to begin with.

## The Special Case of Buying Furniture

**Question for Thought & Discussion #7:**
How did your family members choose the furniture they own? Are they satisfied with their choices?

In many ways the decision to buy furniture is more difficult than deciding to buy an appliance such as a freezer or a stove. The reason is that the value consumers receive from furniture is hard to quantify. You can determine what a new washing machine or freezer is likely to save you in terms of dollars and cents, but it is much more difficult to measure the monetary value of a new couch for your living room. You may like the way it looks and feels, but how much is it worth? Although we could get by with secondhand furniture that costs only a few hundred dollars, most of us would rather have better furniture. But why is this so? A chair is still a chair, whether it is made of molded plastic or hand-carved solid oak. What value do we receive from owning a piece of furniture, and how can we measure that value?

One of the best ways to measure the value of furniture is to think of its value in terms of other items you must give up in order to purchase it. If owning a new dining room table means you can't buy any new clothes this year or new tires for your car, you can view the table's value in terms of the value you would place on having those other products.

Like other durable goods, the value of furniture should be considered over time. A fine dining table may cost $3,000, but it could last the rest of your life. If this is true, it may be a better choice than a less expensive table that would last only a few years.

Attractive or stylish furniture is not necessarily well-constructed furniture. Be sure to evaluate the quality of furniture before you buy.

## Judging the Quality of Furniture

Most consumers cannot afford the most expensive, highest-quality furniture. Nevertheless, through careful shopping it is possible to purchase attractive, well-constructed furniture that will last many years. You can perform a number of simple tests that will improve your chance of getting a good deal for your furniture dollars:

1. Check the frame of any piece of upholstered furniture. It should not wobble, creak, or sag when you sit or push on it. The best frames are constructed from solid pieces of kiln-dried hardwoods. Corners should be reinforced with blocks that are glued and screwed in place.

2. Springs and padding should cover the frame so that you cannot feel the joints. There should be no lumps or wrinkles in the padding. Springs should be attached to each other so that they all give support and move at the same time. You should not be able to feel a spring "bottom out" when you sit on it.

3. Cushions that are not attached should be reversible so that they can be rotated to wear evenly. They should be made with high-density foam (at least 1.8 pounds per cubic foot). The foam should be wrapped in polyester fiberfill and covered with a fabric liner to improve wear. Cushions should fit snugly and not overlap the space they were designed to fill.

4. Fabric should be tightly woven and treated with chemicals to be stain-resistant. Patterns last longer when they are woven into the fabric rather than printed on. The texture of the fabric should be pleasing to the touch.

5. Wooden furniture should be made of solid pieces of wood with joints that fit closely. Drawers should slide smoothly on runners or rollers. Doors should swing easily and fit into their openings evenly. Hardware should be strong and well attached.

6. Painted or varnished finishes should feel smooth and contain no cracks or bubbles. Grain and coloration of the wood should be uniform. Joints should not have apparent filler or excess glue.

> **Before you buy any furniture, check out the store's service policies.**

Although this list of tests is not exhaustive, it will get you started in the right direction. Furthermore, any furniture that passes these tests is likely to be of high quality in other aspects as well.

## Choosing Where to Purchase Furniture

Whenever consumers buy a durable good, they are paying for the services they expect to receive from the dealer as well as the product itself. Your enthusiasm for a lovely, well-constructed, solid oak table will diminish if it is delivered with a large scratch on its surface that the store refuses to repair. Before you buy any furniture, check out the store's service policies. If a salesperson is unwilling to discuss service with you in specific terms with written guarantees, you should take your business elsewhere. Even when the product was defective when it was shipped from the manufacturer, the store that sold it is still responsible. The store should promise in writing to quickly repair, replace, or refund the purchase price of any defective product it sells regardless of who was originally responsible for the defect.

For example, a few years ago, Ethan Allen, a nationally known furniture manufacturer, produced a line of furniture called Cordova. Unfortunately, the upholstered backs of the dining room chairs kept falling off the frames until the design was changed several years later. In the meantime, irate customers demanded satisfaction from Ethan Allen dealers. Some customers went through as many as three sets of chairs before they received the quality for which they had paid. Would you be willing to pay a few dollars more for a dealership that is willing to make this kind of effort?

Consumers can be reasonably sure that if they don't pay much for furniture they buy, they will receive neither high-quality products nor good service. There is no assurance, however, that paying high prices will guarantee either of these. Consumers must rely on their own ability to gather, organize, and evaluate information to make the best decisions possible.

## Purchasing Furniture over the Internet

In recent years, many consumers have chosen to purchase furniture directly from manufacturers over the telephone or via the Internet. These methods of shopping offer lower prices and greater selection than can often be found in stores. The most obvious downside of catalogue or Internet shopping is the inability to see or try out purchases before they are delivered. If you consider ordering furniture directly from a manufacturer, you should be particularly careful to investigate the firm's guarantee for its products. Find out what must be done if a defective product is delivered or if it is damaged in shipping. Can you return furniture simply because you don't like how it looks in your home? How can you receive service for the product if it is needed? If you decide to return a piece of furniture, who will pay for the shipping? You should obtain the answers to these and other questions before you make any long-distance purchases of expensive products.

# What to Do When You Have a Complaint

If you have a complaint about a consumer durable, your first step should be to report your problem to the store or business where you made your purchase to give the firm an opportunity to resolve your problem. Businesses rely on their reputations to gain new and repeat customers. Most stores will work to satisfy their customers whether or not the product in question is covered by a specific warranty. A business can do this, of course, only if consumers let it know that they are dissatisfied. Blaming a store without informing it of the problem isn't fair.

If the seller of a product fails to satisfy your complaint, you may wish to contact the company that manufactured or distributed the good. A polite letter describing what has happened and what resolution you seek will usually bring some action. Address your letter to the customer relations department. Be sure you make a copy of your letter and enclose *copies* of the sales receipt, guarantees, signed agreements, canceled checks, and any other related documents that support your claim. You may wish to send your complaint by certified or registered mail, with a return receipt requested.

If you still do not receive satisfaction, you may contact independent arbitration organizations that are identified in the warranties offered with some products. You may also file a complaint with an office of the Better Business Bureau or at the office of your state's attorney general. Only as a last resort should you consider hiring a lawyer to sue the store or manufacturer, and then only if you have a reasonable chance of winning a settlement that would justify the cost of employing a lawyer and taking your case to court.

# Confronting Consumer Issues: Saving Money by Saving Energy

The energy we use to heat and light our homes is perhaps the best example of a nondurable good you could imagine. Once energy is used, it is gone—it cannot be retrieved, recycled, or reused for other purposes. In 1997 (the most recent year for which the federal government has gathered and published data), the average U.S. family spent $1,338 to purchase energy for its home. By now this amount is estimated to exceed $2,000 per year.

Most households could save a significant proportion of their home energy spending by taking a series of simple steps. The largest barrier to doing this for many people appears to be inertia—they just don't choose to make the effort. You, however, as an educated consumer, are in a position to make choices that can save you money as well as protect the environment and reduce our nation's dependence on foreign sources of energy.

Digital Vision/Photodisc

**How much could you save by using energy more efficiently?**

## HOW DO WE USE ENERGY?

Look around your home and consider the different ways your family depends on energy. The greatest use for most families is for heating or air conditioning. Nearly two-thirds of the typical household's energy spending falls into these two categories. Other important energy uses include water heaters, lighting, laundry, stoves, refrigerator/freezers, and home entertainment. The exact amount of energy you consume depends on your lifestyle and where you live. But no matter who you are, or how you live, you could probably save at least several hundred dollars a year in energy costs without making a large investment of either time or money.

## SIMPLE STEPS CAN SAVE YOU BIG-TIME

You might think that improving your home's energy efficiency would be a large and expensive undertaking. The truth is, it doesn't have to be. A few hundred dollars invested only once in simple energy-saving devices can save you hundreds of dollars in energy costs each year. Consider the following.

**Seal Your Home**    All houses leak—not water, but air. Air comes in, air goes out. When it's cold outside, a crack under your front door or around your windows can add 10 to 15 percent or more to your heating bill. The same is true of your air-conditioning spending if it's hot outside. Installing weather stripping and putting new caulking compound around your windows is likely to cost you between $50 and $100 depending on the size of your home. It will probably save you that much on your heating bill in

the first year. After that your energy savings can be viewed as "pure profit." Checking and repairing defective seals on your fireplace damper, or on your drier or kitchen fan, can also save you energy dollars.

**Wrap Your Tank**    Hot-water tanks lose heat as they sit in the basement even when no one is using hot water. The amount of heat that is lost can be reduced by purchasing a fiber-glass blanket for your water heater. You may also wrap hot-water pipes with fiber-glass ribbon where they can be reached. It will take several years to recover your costs for these steps, but if you live in the same home for five or more years, the money will be well spent.

**Turn It Down—Turn It Up**    Many people set their thermostats high in the winter and low in the summer. In doing so, they waste both energy and money. Experts recommend that water heaters be set to 120 degrees. At this temperature they will consume nearly 20 percent less energy on average than when they are set to 150 degrees. Dishwashers, which need higher water temperatures, have heating elements to raise the water temperature enough to kill bacteria.

Each degree you lower the temperature of your house will save you from 2 to 3 percent of your heating bill. (The same is true for higher temperatures for cooling bills in the summer.) Turn your thermostat down at night when you go to bed. Or better still, buy an automatic set-back thermostat for about $50; it can be programmed to turn your furnace down when you go to bed and back up before you get up in the morning. Then you can't forget to do it yourself.

## Question for Thought & Discussion #8:

Do you believe the members of your family lead energy-efficient lives? What small steps might they take to reduce their consumption of energy and save money?

### Change Your Lifestyle

How many people do you know who turn their television sets on the second they get home and don't turn them off again until they go to bed, even when they are not watching anything? Surveys show that a significant number of people do this. And what about turning the lights off when you leave a room? Depending on the energy costs in your community, a single 60-watt bulb costs between 5 and 8 cents to operate for eight hours. You might think that a few pennies couldn't make much difference, but ten bulbs left on an extra four hours per day costs $90 to $144 extra per year. Is it worth turning out your lights to save that kind of money? You might also consider replacing your bulbs with ones that have a lower wattage or with new fluorescent bulbs that last much longer than ordinary incandescent bulbs and use less energy.

You can make many other little adjustments that can save energy and money. Refrigerators do not need to be kept below 38 degrees, and freezers should be set between 0 and 5 degrees. You do not need to take thirty-minute showers. When you do the dishes, don't leave the hot water running, and be sure your dishwasher is full before you run it. There is no reason to preheat an oven for more than a few minutes before it is used. The list of energy- and money-saving measures you can easily take goes on and on. What other simple steps can you think of?

## TAKING BIGGER STEPS

If you are serious about saving energy and money, you may want to consider taking steps that involve greater effort and initial expenditures but also promise a bigger return in the long run. As you would expect, the greatest potential savings lie in the areas where people spend the most on energy.

### Consider a New Heating System

"Should I buy a new furnace?" If you rent, you probably have little choice in whether a new heating system is installed in your home, but if you are a homeowner, this is an important question to ask and answer. Clearly, if you need to replace an old heating unit that no longer functions, you should purchase the most fuel-efficient one you can afford. The few extra dollars a fuel-efficient furnace will cost will soon be repaid in lower heating expenses. But what if your old furnace still works? Then what should you do?

The energy efficiency of heating systems is rated using a scale known as the annual fuel utilization efficiency (AFUE). New fuel-efficient furnaces have AFUE ratings in the range of 90 to 95 percent. If the rating of your old furnace is significantly lower than this, you have the potential to save a large part of your heating bill. Some older furnaces have AFUEs as low as 50 percent. If you replace a 65 percent furnace with a 90 percent furnace, you should save 14 percent of your heating bill. For a family that pays $1,000 for heat, that would be $140 every year for the twenty-year expected life of the furnace, or $2,800. Not many other investments can earn that type of return. Heating contractors have tables that can be used to predict what your savings will be if you replace your furnace.

### Cost-Efficient Cooling

Energy efficiency is just as important for cooling systems as it is for heating. Central air conditioners and heat pumps are rated according to their seasonal energy efficiency rating (SEER). Many older central air-conditioning units have SEER ratings as low as 6 or 7. New energy-efficient units have ratings that fall between 11 and 12. A few are even higher. This means that new units are able to cool the same amount of air as an older unit using about 60 percent as much energy. If you spent $1,000 cooling your home last summer with a system rated 7, a new 12 system could accomplish the same cooling next summer at a cost of about $600. This, of course, assumes that the cost of electricity and the need for cooling remain the same. A $400 savings each year could quickly pay for the cost of a new air-conditioning unit.

### What about New Windows?

One of the most expensive energy upgrades homeowners can make is to purchase new windows. Depending on the number and size of the windows required, the cost of installing new high-efficiency windows can easily run to $10,000 or more. And for older homes, new windows are only the beginning of the cost. Many older homes had little insulation installed when they were built. A new, energy-efficient window will provide slight benefit if it is mounted in a wall that leaks heat like a sieve. To make a rational decision, you need to weigh all the costs and benefits of your choice.

Before you decide to replace your windows, you should consult with a reputable contractor who can provide a reasonable analysis of your potential costs and benefits. If new windows reduce your heating bill by $500 a year, it will take you twenty years to recoup your $10,000 investment, and that does not include the return you could have earned on your money if it had been put to some other use. Still, new windows will make your house a more comfortable place to live. Money is not the only factor to consider.

Replacing windows makes economic sense only if you expect to live in your home for many years. New windows

*(Continued on next page)*

**Confronting Consumer Issues (Continued)**

will increase the resale value of a house, but not nearly as much as the cost of the windows.

**Improved Insulation**   An alternative that is usually more cost-effective than new windows is having insulation added to your home. Many older houses have insulation that was neither well installed nor provided in sufficient quantities to meet today's building standards. The insulation in most homes can be significantly upgraded for only a few thousand dollars. The result is a sizable energy saving that typically amounts to 10 to 20 percent of the cost of installation per year. This is a good return on your investment. Again, homeowners who are considering improving the insulation of their homes should consult with one or more reputable contractors who can provide a reasonable analysis of the potential costs and benefits.

## CHOOSING ENERGY-EFFICIENT APPLIANCES

At one time it was difficult for consumers to know which appliances were the best buys. Many people simply selected the lowest-priced appliance that seemed to meet their needs. As a result, they often purchased products that were less expensive in the short run, but much more costly over time.

### Question for Thought & Discussion #9:
When members of your family shop for new appliances, how much do they care about the energy efficiency of the products they consider? Do they look for Energy Star products?

You have learned that three basic costs are associated with buying appliances: (1) the initial purchase price, (2) the cost of maintaining them over time, and (3) the cost for the energy needed to operate them. Until fairly recently, it was almost impossible for the average consumer to accurately determine what the cost of energy for an appliance would be. Energy guide labels introduced in the 1980s began to address this problem, but according to many experts, they still did not do enough to encourage the production and purchase of energy-efficient appliances. Beginning in 1992, a new program was undertaken to try to change this situation.

**Energy Star Products**   In 1992, the U.S. Environmental Protection Agency (EPA) introduced the **Energy Star program,** a voluntary labeling program designed to identify and promote energy-efficient products

to reduce greenhouse gas emissions (largely carbon dioxide, which is produced when fossil fuels are burned to generate electricity).

The first products to be given the Energy Star label were computers and their monitors. By 1995, residential heating and cooling equipment had been added, and at present the list includes most types of major home appliances, office equipment, lighting fixtures, and electronic equipment. Specific criteria have been established that each category of product must meet to qualify to use the Energy Star label shown in Figure 10.2. Home air-conditioning units, for example, must have a SEER rating of 12 or better to bear the Energy Star label. It is estimated that a new home built and equipped with only appliances that meet Energy Star standards would use one-third less energy than the same house with only average appliances.

**Buying an Energy Star Appliance**   The fact that an appliance meets Energy Star criteria for energy efficiency does not necessarily mean it is the right appliance for you. You should consider many other factors when you select an appliance.

Suppose you have decided to replace your washing machine and want to buy one that is energy efficient. As you begin to investigate your alternatives, you will quickly realize that there are many brands with various attributes that meet Energy Star criteria. You will discover that front-load, horizontal-axis clothes washers are more energy efficient than top-loaders because they use a relatively small amount of water for each cycle. A typical top-loading washing machine will use from ten to twenty gallons of water per cycle. Front-loaders use from three to five gallons per cycle.

**Figure 10.2** The Energy Star Logo

That means they need 75 percent less hot water. They are also easier to load (less lifting) and require less detergent.

Front-loaders do have two important disadvantages. They do not hold as much laundry as most top-loaders, and they cost about 75 percent more to buy. On average, a front-loader should be expected to recover its greater initial cost within about five years through energy savings.

### FTC Recommendations

The FTC has recommended that consumers follow these steps when they decide to purchase any appliance:

1. *Select the size or capacity that makes sense for you and your family.* A small washing machine, for example, may cost less to buy and operate for each cycle than a larger one. But if you must run it many times to do your family's laundry, you will use more time, energy, and water than if you had a larger model, and it will wear out sooner. Think of your future. Most washing machines last many years. You may not have any children now, but will you need a large machine in five years when you have several "little ones" to contend with?

2. *Know where to shop.* Don't buy the first appliance you see that seems to meet your needs. There are many appliance outlets, and don't forget the Internet. Even if you don't buy on the Internet, it is a good way to evaluate offers made by local dealers.

3. *Compare different models and brands.* All major appliance manufacturers maintain Web sites that provide extensive information about the different products they manufacture and their specifications. Your local appliance dealer may carry only three or four models even though the firm manufactures twelve others. You may do better ordering the one you want or looking for it at a different outlet.

4. *Evaluate the appliance's costs.* These include the initial purchase price, the projected cost of maintenance, and the cost of the energy the appliance will consume. The product with the lowest purchase price is often not the least expensive to own over many years. Such products tend to break more often and can be much more costly to operate.

5. *Ask about special energy offers.* Many states and some electric utilities offer special rebates to people who purchase Energy Star appliances. Special financing may be available as well.

### Question for Thought & Discussion #10:

Have your family members been offered an opportunity to change the firm that supplies their electricity or gas? If so, why did they choose to change or not change their supplier?

### BUYING CHEAPER ENERGY

Today, energy consumers in most states are able to choose where they buy their electricity and natural gas, although they must still use a particular firm to deliver energy to their homes. It would be terribly inefficient, for example, to have more than one set of electric lines and transformers in any one community. In Chapter 1 you learned that these businesses are called *natural monopolies.*

Although it may be necessary for energy-delivery firms to be monopolies, the suppliers of energy do not have to be. Once an electric distribution system is established, energy can be fed into the grid at any number of locations. You might live in Ohio but buy electricity that was generated by a firm in West Virginia. This situation has allowed consumers to look for low-cost sources of energy. It is important, however, to investigate any firm you consider using to supply your energy. In the late 1990s, some firms promised to supply low-cost electricity to consumers who made up-front payments. When these firms failed, the people who had paid for electricity in advance lost, too. Most of them got nothing for their money. With the present increased government oversight, however, such problems are less likely to occur in the future.

### ALTERNATIVE SOURCES OF POWER

The possibility of choosing your own source of power has opened up another opportunity for environmentally motivated consumers. In many locations, consumers can subscribe to power services that generate electricity from alternative sources such as solar, geothermal, and wind

**Figure 10.3 "Green Energy" from Sterling Planet**

*(Continued on next page)*

### Confronting Consumer Issues (Continued)

power. In 2003, Niagara Mohawk customers in upstate New York, for example, could sign up to buy electricity from Sterling Planet, a generator that sells only "green energy" (see Figure 10.3).

Power generated in alternative ways is more expensive than power generated by nuclear plants or by burning fossil fuel. Many consumers, however, are willing to pay a higher price to encourage development of alternative power sources. Alternative energy sources are nonpolluting and renewable, and they reduce our dependence on foreign sources of energy.

In the final analysis, the decision to conserve energy in your home is a personal choice, and there is a range of choices you can make. Taking even the most basic energy conservation steps can save you money. For renters, energy conservation in their homes is limited in many ways by the willingness of their landlords to cooperate with their wishes.

Another impediment for a number of people is that many forms of energy conservation involve a significant up-front cost and a payoff that only comes over time. People who have little money are rarely in a position to change their way of life to save energy, and they often live in the most energy-inefficient homes. Still, every cubic foot of natural gas, kilowatt-hour of electricity, or gallon of fuel oil that is not used saves someone money, reduces the impact we have on our environment, and makes this country less reliant on foreign suppliers of energy. Energy conservation is a goal worth pursuing.

## Key Term:

**Energy Star Program**   A voluntary labeling program initiated by the Environmental Protection Agency to identify and promote energy-efficient products and thereby reduce greenhouse gas emissions.

## Chapter Summary

1. When consumers buy nondurable goods or services, which have life spans of less than three years, they will benefit from using the rational decision-making process.

2. Perceived clothing needs can lead to conflicts in the family budgeting process. It is important to make compromises over personal tastes and the amount of spending allocated to each individual's wardrobe.

3. Clothing choices are determined by more than the need to protect ourselves from the environment. Customs, aesthetic considerations, and psychological factors also influence how we dress.

4. Generally, we pay more for more durable clothing. Clothing quality depends on fabric content and on how well garments are tailored. Many fabrics that contain a blend of natural and manufactured fibers are popular because of their durability and ease of care.

5. Designer labels do not necessarily indicate that a garment is of superior quality. Similar garments without designer labels may often be found at much lower prices because the prices do not need to cover the cost of royalties and extensive advertising.

6. There are many ways to buy quality clothing at relatively low prices. These include shopping for sale items and at factory outlet or off-price stores.

7. Consumer durables yield a flow of service over time. They are often purchased on credit, must be cared for, and eventually wear out and are replaced.

8. The purchase of a consumer durable good is similar to an investment because durable goods do not wear out immediately. To make a rational decision when buying a durable good, consumers should carry out a cost-benefit analysis.

9. Costs of durable goods typically include depreciation, delivery and installation, operation, repair, and the value of forgone interest. When making a purchasing decision, these costs should be weighed against the value of the potential benefits of owning the product.

10. When purchasing furniture, it is often difficult to measure the flow of value that is provided. One way to accomplish this is to consider the value of goods or services that must be given up to pay for the item of furniture.

11. Consumers can save by conserving energy in their homes. To accomplish this, they can take many small steps including installing weather seals, wrapping hot-water tanks, adjusting thermostats, and changing how they live.

12. People who are serious about energy conservation can take many steps that are more expensive but offer much larger savings. Among these are replacing heating and cooling systems with energy-efficient models, buying new windows, installing additional insulation, and choosing energy-efficient appliances.

13. In 1992, the EPA began the Energy Star program that allows manufacturers to use the Energy Star logo for products that meet the program's energy usage guidelines. Most consumers may purchase the energy they use in their homes from a variety of suppliers. This allows them to base their purchases on low price or on how the energy is generated. Alternative sources of energy including solar, wind, and geothermal generation are available to most consumers at somewhat higher costs.

## Key Terms

durable goods **221**

energy guide label **224**
Energy Star program **236**

nondurable goods **216**
service flow **218**

## Questions for Thought & Discussion

1. How important are the clothes you wear to your outlook on life and the way you feel about yourself? Explain your point of view.

2. When you shop for clothing, what factors do you consider? Of these, which is the most important to you?

3. How important are designer fashions to you? Do you feel a need to follow fashion trends?

4. How did your family members go about choosing the television set they own? Are they satisfied with their choice?

5. How important are energy guide labels to most consumers? Do you believe the average consumer knows what the labels mean? How could the government make them more useful?

6. What type of appliance service would you try to complete yourself, and what would you hire a technician to do?

7. How did your family members choose the furniture they own? Are they satisfied with their choices?

8. Do you believe the members of your family lead energy-efficient lives? What small steps might they take to reduce their consumption of energy and save money?

9. When members of your family shop for new appliances, how much do they care about the energy efficiency of the products they consider? Do they look for Energy Star products?

10. Have your family members been offered an opportunity to change the firm that supplies their electricity or gas? If so, why did they choose to change or not change their supplier?

# Things to Do

1. Visit an expensive clothing store and examine two or three examples of a type of garment you might consider buying. Evaluate the quality of their fabric and construction. Make a list of what you find. Then do the same for similar garments sold at an inexpensive clothing outlet. Compare the quality of the products. What do consumers receive when they pay higher prices? Do you believe these extra qualities are worth the extra cost? Explain your findings in an essay.

2. Identify a particular type and brand of consumer durable good that you might purchase for your home. Find three outlets that offer the product for sale. Be sure that at least one of the outlets is on the Internet and at least one is a local store. Compare the prices, service, and convenience of each outlet. Where would you choose to buy the product? Explain the reasons for your choice.

3. Examine a piece of furniture that has lasted for many years. You may find one in your home or in the home of a friend or relative. What characteristics of construction and quality can you find that would have indicated that this was a piece of quality furniture when it was new? Could you find similar qualities in new furniture today?

4. Identify a type of home appliance that you might like to own. Identify several models of this type of product, being sure that at least one of them bears the Energy Star logo. Compare the cost of buying and operating each of these models over the expected life of the product. Which will cost more to own over this time? Which would you choose to purchase? Explain your choice.

# Internet Resources

## Finding Consumer Information on the Internet

The following Web sites have been selected for their relevance to topics discussed in this chapter. Search these sites to locate information that can add to your knowledge of Energy Star housing. Investigate Energy Star housing efforts in your state. How much support does your state's government provide for this type of housing? Remember, Web addresses change frequently. If any of these addresses no longer function, find similar sites to investigate using any of the search engines available to you.

1. The Arizona Energy Star Labeled Homes Web site features central Arizona home builders that participate in the Energy Star program. Investigate this Web site at **http://www.arizonaenergystarhomes.com**.

2. The federal Energy Star program Web site describes the program and pollution-prevention partnerships that encourage the use of energy-efficient equipment to drastically reduce air pollution. This Web site is located at **http://www.energystar.gov**.

3. ENERGY STAR HOMES lists firms that construct Energy Star homes in the Northeast. Find out what this Web site has to offer at **http://www.energystarhomes.com**.

## Shopping on the Internet

The following Web sites have been selected because they offer consumers services similar to those described in this chapter. These are commercial sites that are designed to market products. They do not represent a comprehensive or balanced description of all furniture retailers that offer products online. How do the prices offered at these Web sites compare with the prices charged by your local stores for similar products? Remember, Web addresses change frequently. If any of these addresses no longer function, find similar sites to investigate using any of the search engines available to you.

1. GiGaFurniture offers a wide selection of furniture at prices that are said to be much lower than those charged

by local furniture retailers. Discover what this firm has to offer by visiting its Web site at **http://www.gigafurniture.com**.

**2.** World Design Center offers products fashioned after a variety of styles including Chippendale, Louis XVI, and American Victorian. Investigate the types of furniture marketed by this firm at **http://www.worlddesigncenter.com**.

**3.** Oak Plus Furniture specializes in oak, pine, and leather furniture. Many of its products have a rustic appearance. Find out about its products by visiting its Web site at **http://www.oakplus.com**.

## InfoTrac Exercises

 Purchasers of new copies of this text are provided with access to the InfoTrac Web site. This Web site links students to thousands of recent articles published in hundreds of periodicals. Use the key words **energy-efficient appliances** or other terms from this chapter to conduct a key-word search. Choose one article that is of particular interest to you and write a brief essay describing what you have learned from the article. Be sure to cite the author and title of the article and the name and date of the publication in which it appeared.

## Selected Readings

Anderson, Michael H., and Raymond Jackson. "A Reconsideration of Rent to Own." *Journal of Consumer Affairs*, Winter 2001, pp. 295–306.

"Appliance Makers Plan Smart Devices." *The New York Times*, January 17, 2000, p. C8.

Boss, Shira J. "Big Savings on Appliances." *Good Housekeeping*, April 2002, p. 175.

Consumers Union. *Consumer Reports Annual Buying Guide.* Published annually in December.

Evans, Sheldon. "Owners Can Be Stars." *Real Estate Weekly*, June 12, 2002, p. 53.

"Extended Service Contracts." *Consumer Reports Buying Guide 2000*, December 1999, pp. 21–22.

Gardner, Elizabeth. "Selling Designer Brands at a Discount." *Internet World*, May 24, 1999, p. 19.

Harder, Nick. "Consumers Should Ask Questions about Extended Warranties." *Knight Ridder/Tribune Business News Service*, June 6, 2002, p. ITEMPK1099.

_____. "Look Past Surface When Shopping for Furniture." *Knight Ridder/Tribune Business News Service*, May 9, 2002, p. ITEMPK5425.

Karon, Tony. "Anatomy of a Trend." *Time*, February 5, 2003, p. 10.

Lewis, Jake. "Renting to Own." *Multinational Monitor*, October 2001, pp. 16–17.

Mechanic, Michael. "High Fashion Gets Trashy." *Mother Jones*, July–August 2002, p. 20.

Samuels, Allison. "It's All in the Jeans: Designer Denim for the A Crowd." *Newsweek*, June 24, 2002, p. 90.

"Selecting and Buying Quality Furniture." *Consumers' Research*, December 1993, pp. 30–34.

# CHAPTER 11

# Satisfying Transportation Needs

After reading this chapter, you should be able to answer the following questions:

- What transportation alternatives are there to owning a car?

- What benefits do consumers receive and what costs do they incur as the result of owning a car?

- How might consumers decide whether to buy a new or a used car?

- How can you shop for a car over the Internet?

- How can you decide whether leasing a car is right for you?

- What choices are available to consumers who want to finance the purchase of a car?

- How can lemon laws help consumers?

- What steps can consumers take to reduce their chances of being cheated when their car or truck is repaired?

The twentieth century in the United States was called the Age of the Automobile. From a modest beginning at the turn of the century, when a few courageous souls drove around in Stutz Bearcats and Hupmobiles, until Henry Ford developed low-cost mass-production techniques to put out an $870 "Tin Lizzie," right up to the end of the 1990s, when fully 90 percent of all U.S. families owned cars, the automobile became a pervasive part of U.S. life. At the beginning of the twenty-first century, it appears that automobiles will continue to play a major role in the U.S. lifestyle.

**"One out of every six jobs in the United States is concerned with automobiles or trucks in some way."**

**Question for Thought & Discussion #1:**
Do you or any of your friends get along without a car? How does this lack affect one's life?

# We Depend on Cars in Many Ways

In the United States, over 200 million motor vehicles flood our roads and highways. In recent years, annual car and truck sales have amounted to about 17.6 million vehicles—for a value of roughly $370 billion in 2003. Although the U.S. automobile industry is shrinking, it is still huge. One out of every six jobs in the United States is concerned with automobiles or trucks in some way. Whether people are factory workers in Detroit or Tennessee, repair people in dealerships, advertising executives in New York City, or stevedores unloading ships from Japan, their jobs depend on our "love affair" with the automobile.

## Do You Need to Own a Car?

For many consumers, owning a car is part of their lifetime dreams. We may not know why, but countless Americans think of acquiring a car as a necessary step toward achieving their personal freedom—a sort of rite of passage. If this is the way you feel about owning a car, perhaps you should step back a moment and think again. Remember, at the most basic level, a car is just a means of transportation. It is a way to get to work or school and home again. You use it to go shopping, go to the doctor, or take trips. But, when you come down to it, how many of the trips you take must be made by driving? Most of the places people drive (at least in urban areas) could be reached in some other way—and usually at a much lower cost. Before you commit yourself to owning an automobile, take the time to evaluate your transportation alternatives and determine what owning a car would do to your financial situation.

## Transportation Alternatives

For people who live in rural areas, there may be few suitable alternatives to owning a car. For consumers who dwell in urban areas, however, cars are often unnecessary and are not the quickest way to get from one place to another. When you drive a car, you must deal with traffic, find a place to park, walk from where you park to where you really want to be, and contend with the possibility of being involved in an accident. Consider some of the transportation alternatives that could save you time and money.

**YOU COULD RIDE A BICYCLE OR A MOTORCYCLE** For several hundred dollars you could buy a bicycle to get from here to there. Think of the advantages—bicycles can take you around

Would riding a bicycle be a reasonable transportation alternative for you?

Photodisc/Getty Images

traffic jams, it is easier to find a place to leave them (most often for free), you can usually ride them right up to the place you are going, they cost nothing to insure, and using one may improve your health.

There are, of course, limitations on bicycle transportation as well. Bicycles cannot be used in bad weather; they cannot carry large packages; you may arrive at your intended location in need of a shower; and if you are involved in an accident, you are more likely to be seriously injured. Nevertheless, for nearly ten million adult Americans, a bicycle is their first choice for daily transportation.

Motorcycles share some of the advantages and disadvantages of bicycles. They are usually less expensive than a car to buy and operate, more maneuverable, and easier to park. But they cost almost as much as a car to insure, and people who ride motorcycles are much more likely to be injured in accidents than those who travel by car. Still, using a motorcycle can be an alternative worth your consideration.

**YOU COULD USE PUBLIC TRANSPORTATION**   All major cities, as well as many smaller ones, have some form of public transportation system. This may include buses, subways, trolleys, taxis, or some combination of these. The greatest advantage of using public transportation is probably the lower cost. Owning and operating a car in the United States costs around $8,500 per year on average. But in cities this cost is much higher. In urban areas the cost of buying insurance skyrockets. Then add the expense of parking and the costs of gas, oil, and maintenance, which are typically higher in cities. You will find that the average urban dweller must spend about $12,000 each year to own a car.

A round trip from your home to your place of work may cost you $3 or $4 per ride on public transportation. If you work 250 days a year, the most you would pay is $1,000. That's $11,000 less than owning a car. To be sure, you would have to go other places from time to time, but even if you took 1,000 bus trips per year, you would still spend only $4,000. That would leave lots of money to hire a taxi or rent a car when public transportation just won't do.

There are, however, problems with public transportation that you should consider. Buses and subways do not always run at convenient times and may shut down entirely at night or on weekends. They do not take you to the exact location you want to go, and to get across town you may have to change buses or subways. There is a limit to what you can carry on a bus or subway. If you do your grocery shopping once a week, you won't easily be able to carry six bags of food home with you on a bus. Still, public transportation is an alternative many millions of Americans have chosen.

Could you use public transportation to travel to work or school? If you did, what benefits and costs would you incur?

**YOU COULD RENT A CAR**   Competition has brought the cost of renting a car for a day or two down significantly in many urban areas. When you really need a car, you should be able to rent one for $100 a day or less, including insurance. That may seem like a lot of money, but renting twenty days a year would total only $2,000. That's much less than the cost of owning a car. It would leave you with $10,000 to take a bus most of the time. Furthermore, rental rates are often much lower than $100. Many major rental agencies, including Avis and National, offer special reduced rates on weekends. Enterprise advertises rates as low as $9.99 a day, plus insurance. If you can plan to use rented cars on weekends, you could rent a car ten to twenty times a year for only a few hundred dollars.

**YOU COULD JOIN A CAR-SHARE PLAN**   Consumers who live in large cities are often able to choose a different alternative. There is a good chance that one or more organizations may offer them a new way to use a car—they can sign a contract to share a car. Although car-share organizations have been successful in Europe for more than twenty years, they first appeared in the United States on the West Coast around 2000. Since then they have spread to many large cities.

If you need a car for only a few hours each week, or for weekend travel, and you can predict when you will need a car in advance, you may be able to save thousands of dollars every year by joining a car-share organization. In 2003, there were four large car-share organizations: Flexcar, Zipcar, I-Go Car, and E-Motion. Specific information about these businesses can be found on the Internet at **http://www.carsharing.net**.

Car sharing works by having each member reserve a car for certain days and hours during each month. The organization then fits individual members' reservations together into a master schedule. Members are required to pick up and leave off the cars at designated locations. They may make extra reservations when they need to have a car for a few days, and they may call to borrow a car on short notice in special situations. Additional fees are generally imposed in these cases. The organization maintains extra cars in case of breakdowns and cars that are returned late. Too many late arrivals can cause a member to be expelled from the organization.

The organization maintains the cars and pays for the gas and oil used. Members are required to pay an annual fee to cover the cost of the car and a mileage or per hour use fee, and they must arrange to purchase insurance. An annual cost of $4,000 to $5,000 typically covers fifty hours of use per month. This amount is not small, but it is less than the cost of owning a car of your own. If you lived in a large city, would you consider joining a car-share organization?

Let's assume that you have weighed all the benefits and costs of your transportation alternatives and have decided to join the ranks of car or truck owners who make up the majority of the U.S. population. A little background information will be helpful.

# The Trend toward Safer Automobiles

For decades, many have depicted the automobile as a four-wheeled rolling coffin because so many Americans—currently almost 45,000 a year—lose their lives in vehicle accidents. For younger Americans—those between five and thirty-four years of age—traffic accidents are the leading cause of death. In addition, nearly two million individuals are injured annually as a result of car accidents.

Many highway deaths could have been avoided had the drivers been more careful. But according to some observers, many others could have been avoided if additional safety features were required on automobiles. In 1958, the Ford Motor Company tried to sell additional safety features to the car-buying public, but sales declined and Ford lost money. Finally, after the exposé by Ralph Nader in his book *Unsafe at Any Speed,* Congress passed the National Traffic and Motor Vehicle Safety Act of 1966, the basis of most of the current safety requirements for automobiles.

## Safety Requirements

Some of the requirements imposed on car manufacturers by the National Highway Traffic Safety Administration include:

● Dual braking systems.

● Nonprotruding interior appliances.

● Over-the-shoulder safety belts in the front and rear seats.

● Head restraints on all front seats.

● Seat-belt warning systems and ignition interlocks.

● Collapsible, impact-absorbing steering columns and armrests.

● Impact-absorbing instrument panels.

In 1984, the National Highway Traffic Safety Administration took another major step toward increasing auto safety. Under its regulation of that year, all car manufacturers were required to include passive-restraint safety features—such as air bags or automatic safety belts—in their new cars, beginning with the 1987 models. This rule was implemented gradually until 1990 when all models were required to have passive restraints.

## Costs and Benefits of Safety Devices

Automobile safety devices raise the cost of automobiles, and you, the consumer, pay for the safety directly out of your pocket. Air bags are particularly costly, adding $600 or more to the price of your new car.

On the benefits side, however, safety features such as seat belts are responsible for reducing injuries and fatalities from auto accidents. A recent University of Chicago study found, for example, that the use of seat belts reduces the severity of injuries by roughly 60 percent—even more than previously thought. Because side-impact air bags are judged to be an important precaution against injury, many consumers are willing to pay a higher price for this safety. Although air bags protect most passengers in an accident, they can harm or kill children or smaller adults when they deploy. Manufacturers now offer consumers the option of having a cutoff switch installed to disable air bags when necessary.

## Safety and Sobriety

Drunk drivers are responsible for slightly less than half of all traffic fatalities each year, according to the National Highway Traffic Safety Administration. And for every death, twenty more people are injured in alcohol-related car accidents. Shocking as these figures are, they are lower than in the past. In 1985, for example, 26 percent of traffic accidents resulting in deaths involved drivers with a .1 percent alcohol level or higher; by 1999 this figure had dropped to 17 percent.

Several factors have been responsible for this decline in drunk-driving fatalities. Among them are the stiffer penalties now imposed on drunk drivers in nearly all states, as well as the fact that more of the population—either by

**Question for Thought & Discussion #2:**
Do you always use a seat belt when you travel in a car? Why or why not?

law or by choice—now wear seat belts or have some kind of passive-restraint system protecting them during accidents. There has also been a growing perception on the part of the U.S. public of the risks involved in driving while intoxicated—thanks largely to the efforts of a single organization, Mothers Against Drunk Driving (MADD), which has been campaigning for more stringent laws since 1980. Students Against Destructive Decisions (SADD) has joined this campaign in recent years.

Another reason for the decline in deaths related to intoxication is that all states have passed laws mandating twenty-one as the minimum age for purchasing alcoholic beverages. Studies have shown a direct relationship between the minimum drinking age and the number of traffic fatalities in the eighteen- to twenty-one-year-old age group. Nationwide, fatal traffic accidents have been reduced by an estimated 16 percent in this age group as a result of the higher drinking age required under state laws.

## The Social Costs of Driving

When you get into your car and fire it up, you incur, in addition to the private costs discussed next, the **social costs** of driving. You are all aware of them, particularly if you live in Los Angeles, New York, or Washington, D.C. One of the biggest social costs of driving has been air pollution. That engine does not just pull your car around. It also emits by-products that, when added together, may be dangerous to your lungs and mine. This is, of course, why the federal government as well as individual states regulate the pollution output of automobile engines.

Thus, you, the individual driver, are forced to take account of the social cost you impose on the rest of society in the form of pollution. When you have no alternative but to purchase automobile engines that have pollution-abatement equipment, you pay directly in the form of a higher purchase price.

Private automobile transportation has other social costs. One of them is congestion. Congestion on bridges, on highways, and in inner cities is a problem of social concern, even though private individuals, at least until now, have not been forced to pay the full price of driving their cars. That price includes making other people late for work or making them spend more time in their own cars. In other words, simply by driving onto a crowded bridge, you slow down everybody else somewhat. When you add up the value of everybody else's time, you see that you impose a pretty high cost. And the same is true of every other person on that bridge.

## The Private Costs of Driving

The cost of driving a car involves more than making a monthly payment, although that is part of it. The **private costs** of driving include such things as wear and tear, repairs, gas and oil, insurance, and taxes, in addition to the price of the car and interest payments, if the car is financed. Table 11.1 on the next page shows different types of vehicles and their estimated costs in 2001. It is difficult to say exactly what the per-mile cost of owning and operating a car would be for any individual because there is such a wide range of insurance costs and driving styles. Your driving costs probably fall somewhere between 50 cents and 80 cents a mile.

**Cyber consumer**

## Advocates for Highway and Auto Safety

Advocates for Highway and Auto Safety is an alliance of consumer, health, and safety groups and insurance companies and agents working together to make U.S. roads safer. Its Web site is located at http://www.saferoads.org. Visit this Web site and investigate the organization's efforts to encourage the adoption of federal and state laws, policies, and programs that save lives and reduce injuries.

**Social Costs**   The costs that society bears for an action. For example, the social costs of driving a car include any pollution or congestion caused by that automobile.

**Private Costs**   The costs that are incurred by an individual and no one else. The private costs of driving a car, for example, include depreciation, gas, and insurance.

| TABLE 11.1 | The Average Cost of Driving a New Vehicle 15,000 Miles per Year in 2001 | |
| --- | --- | --- |
| **Vehicle Type** | | **Annual Cost** |
| Subcompact automobile | | $6,648 |
| Midsize automobile | | 7,456 |
| Full-size automobile | | 8,873 |
| Full-size pickup truck | | 8,368 |
| Minivan | | 9,141 |

SOURCE: American Public Transportation Association, **http://www.apta.com**.

## The Repair Dilemma

Some consumer economists believe that preventive maintenance avoids large repair bills. This is true, but you must take account of the maintenance costs themselves. In the long run, it may be cheaper not to keep your car in perfect condition but rather to let some things (other than brakes, tires, and safety-related parts) wear out and replace them only when they do or to trade in your car every few years. Some state governments buy fleets of cars that they do not service at all for a year and then trade in. This seems to be less expensive than trying to maintain the cars; as the prices of repair services rise, this practice will become still cheaper by comparison. Thus, you have two choices: you can buy a car that you expect to keep for only a short period of time, or you can buy a car that has a reputation for very low service requirements. Each year, the April issue of *Consumer Reports* recounts its readers' experiences with the repair needs of different makes and years of cars. This is an important aid when you try to assess the annual cost of operating an automobile.

## The Cost of Depreciation

With the possible exception of paying for insurance, which is discussed in Chapter 17, depreciation is the greatest cost of owning an automobile. Cars lose value as they grow older even if they are driven only a few miles each year and have not worn out. They depreciate because dealers and people are willing to pay less money for an older car than for a newer model. Remember, the market value of any product is measured in terms of what someone else is willing to pay for it. Although different models of cars depreciate at different rates (see Table 11.2), a good rule of thumb is that a car will lose about one-fifth of its value each year you own it. Suppose you spend $20,000 on a new car. When it is one year old, it will probably be worth about $16,000, or one-fifth less than its purchase price. Its value will fall by roughly another fifth, or $3,200, to $12,800 in the second year you own the car. By the end of the third year, its value will be about $2,560 less, or $10,240. The longer you own a car, the less it will depreciate each year. However, there is another factor to consider: as automobiles become older, they also require more maintenance. A five-year-old car may not depreciate much in a year, but it will probably need more repairs than a new automobile.

**Question for Thought & Discussion #3:**
What would you have to give up to buy a new car? Would the costs be worth it? Explain your answer.

| TABLE 11.2 | Resale Value of Various Automobile Models | | |
|---|---|---|---|
| Model | 2000 Price | 2003 Resale Price | Percentage of Original Price Retained |
| Nisson Maxima | $24,570 | $17,200 | 70.0% |
| Pontiac Grand Am | 18,360 | 12,575 | 68.5 |
| Cadillac Seville | 41,750 | 28,475 | 68.2 |
| Mercury Mystique | 15,800 | 10,200 | 64.6 |
| Lincoln Town Car | 42,830 | 24,675 | 57.7 |

SOURCE: *NADA Used Car Guide,* February 2003.

## Getting Good Repairs

All car owners face the problem of having their autos repaired. Finding a good repair shop or an honest mechanic may be difficult. Car repairs can also be very expensive—as evidenced by the fact that Americans pay approximately $200 billion a year for these services. Methods consumers may use to assure the quality of car repairs they purchase will be presented later in this chapter.

# Deciding When to Buy

At what point does it make sense to stop spending money on repairs for your old car and purchase a new or different one? An older model, if it has been well serviced and maintained over the years, can sometimes go 100,000 miles before requiring a new transmission or other major repairs. If you have an older car and if your mechanic thinks it is in relatively good condition, don't panic at just one repair bill. Remember, it costs less to own and operate an older vehicle than it does a new one *if* that vehicle has been well maintained.

The decision to repair or replace ultimately depends on what your repairs are costing in terms of both money and inconvenience and on what your mechanic predicts about future repairs. Leil Lowndes, in *Shopping the Insider's Way,* has devised a formula you may wish to use in making repair-or-replace decisions. Lowndes suggests you divide the book value (BV) of the car into the repair estimates (RE) to obtain the replacement percentage (RP). For example, a $500 repair estimate on a $5,000 car works out to 10 percent. If the RP is 25 percent or less, Lowndes says go ahead with the repairs; if the RP is 40 percent or more, replace the automobile. For RPs between 25 and 40 percent, consumers should use their best judgment and their financial situation to decide whether to repair or replace the car.

# Deciding What You Want

Whether you choose to buy a new or a used car, or to lease your next vehicle, you first need to decide what kind of car or truck you want, the options you need, and the price you are willing to pay. Since buying or leasing a vehicle represents a substantial expenditure, you will want to gather sufficient information before you actually sign any agreement.

Though the majority of car buyers still like to visit a dealership to test-drive their potential purchases, the Internet allows them to do much of their research at home. Because of the large amount of information online, buyers are more informed than ever about options and costs.

## Gathering Information

One way to start your market research is by asking relatives and friends about their experiences with different makes and models of cars or trucks. Additionally, you can review reports in a wide variety of periodicals including *Motor Trend, Car and Driver,* and *Road and Track.* You would be well advised to check the most recent April issue of *Consumer Reports,* which contains a summary of recent test results for new models of cars and trucks.

You can get an idea of vehicles you can afford by finding out what dealers pay for different types of cars and trucks, as well as the cost of various options you would like to have. This type of information is available from print sources or (usually at a cost) over the Internet. One free source for this information is the Web site of the National Automobile Dealers Association at **http:// www.nada.org**. Another popular source is *Consumer Reports,* which can be reached at **http://www.ConsumerReports.org** or by calling 800-422-1079. Because *Consumer Reports* charges a fee for a report on a vehicle, consumers should narrow their search to a few models before seeking specific cost information. Knowing how much a dealer paid for a new vehicle will help you decide which vehicles you should consider, and it can help you negotiate a fair price when you decide to buy.

The December issue of *Kiplinger's Personal Finance Magazine* and *Edmund's New Car Prices* provide basic cost information for new cars and trucks. The advantage of these publications is that they can be examined for free at most public libraries or online. Once you have adequate information about new cars and trucks on the market, their options, and their costs, you are ready to find and evaluate your buying opportunities.

## What Type of Car to Buy?

Deciding which new car or truck to buy depends at least in part on how much money you want to spend. You should predict the yearly out-of-pocket costs you will incur for different types of vehicles and then decide what you are willing and able to pay. Remember, as you go up the ladder of car prices, you often are not buying any more safety, capacity, or power, but only styling, prestige, and the like. Be aware of the price you are paying for these qualities.

You should be sure to include operating and maintenance costs in your calculations. New cars do have warranties and are less likely to suffer breakdowns than older vehicles, but they are more costly to insure and require as much gas and oil to operate as an older car. Remember, car size and weight are related to operating costs, but cost is not the only important consideration. Compacts use less gasoline than full-size cars but they hold fewer people, can carry less baggage, and offer less protection in accidents.

**OPTIONS**   The options you choose depend on your tastes relative to your financial situation. Some options are wise purchases even if they don't add to your comfort or to the usefulness of the vehicle. Some cars, for example, offer side impact air bags as an option. Others allow you to turn off the front passenger air bag when a child or small person is occupying that seat. Buying air conditioning is advisable even if you personally like heat and never use it. Selling a used car without air conditioning is difficult.

Everyone would like to own a car that is easy to maneuver and control. Tire quality goes a long way toward determining a car's maneuverability. You shouldn't compromise on this important feature. Radial tires offer improved handling and longer life than most other tires. Today, most new cars are delivered with radials as standard equipment, but if the car of your choice comes without them, you should consider trading up to radials. The same could be said for limited-slip differentials, automatic braking systems, and all-wheel drive vehicles in snowy climates.

Figure 11.1 is a chart you can complete to compare the costs of different cars or trucks you might want to buy with different options. You can obtain the prices of specific options over the Internet or from dealerships. Recognize that dealers will probably quote prices that are significantly higher than their own costs.

> **Question for Thought & Discussion #4:**
> Is a car just a means of transportation to you, or is it something more? Explain your point of view.

## Figure 11.1
### Price Comparison Chart

| | Car 1 | Car 2 | Car 3 | Car 4 |
|---|---|---|---|---|
| **List Price** | | | | |
| **Options** | | | | |
| Power steering and brakes | | | | |
| Automatic transmission | | | | |
| Nonstandard engine | | | | |
| Air conditioning | | | | |
| Rear-window defogger | | | | |
| Special radio/tape deck | | | | |
| Limited-slip differential | | | | |
| Whitewall tires | | | | |
| Tinted glass | | | | |
| Vinyl roof | | | | |
| Tires—radial, oversized, or snow | | | | |
| Speed control | | | | |
| Power door locks | | | | |
| Power windows | | | | |
| Remote side-view mirrors | | | | |
| Compact disc player | | | | |
| **Total cost** | | | | |
| Subtract trade-in or down payment | | | | |
| **Total amount to be paid to dealer** | | | | |

It's a good idea to talk to others who have purchased a car from any dealer you are considering.

**DEALER ADD-ONS**   In addition to factory options, dealers may try to sell consumers special add-ons that most often increase the dealer's profit more than the value of the car. Almost all new cars sold in the United States have been built and treated to resist rust. Many new cars have "rust-through" warranties that last for as many as 100,000 miles or ten years. Nevertheless, some dealers try to convince consumers to spend an extra $300 to $500 to have their cars "rustproofed." Dealers are also likely to offer special waxes or leather treatments. They often encourage consumers to buy extended protection plans. In general, dealers benefit much more from these add-ons than consumers do.

# Getting the Best Deal on a New Car

Once you have obtained enough information to know what kind of car or truck you want to purchase and have identified your price range, your next step is to decide which method of buying you wish to use. Basically, you can shop for a new car in two ways—through a dealership or over the Internet. As with other consumer purchases, benefits and costs are associated with each alternative.

## Buying through a Dealership

If you decide to purchase your car or truck through a dealership, you should consider a number of criteria before choosing the dealer you will use. Among these are location, reputation, facilities and personnel, and the deal you are offered.

**LOCATION**   Where a dealer is located can have an important impact on the satisfaction you receive from the vehicle you buy. After all, you will probably take your car in for service, at least while it is under warranty. No matter how high the quality of a car's manufacture, it is likely to need a few adjustments even if only to change the height of a seat-belt attachment. Choosing a dealer near your home or place of work can make your life much easier. You should also investigate the firm's policy for providing loaners if you must leave your car at the dealership: Are they free, or will you be required to pay?

**REPUTATION**   Talk to customers who have bought from any dealer you are considering. Were they treated fairly? Did they have any problems with their purchases or financing provided through the dealership? What experiences have they had with the firm's service department? Were they satisfied with the quality of the work and the prices they were charged if their cars were no longer under warranty? You could also take your current car to the dealer to have it serviced to see what type of treatment you receive.

**SERVICE FACILITIES AND PERSONNEL**   To find out about a dealer's service facilities and personnel, ask specific questions such as the following: How long must a customer wait for service? How long is work guaranteed after it has been completed? What types of diagnostic equipment does the service department use? How are service technicians trained? What are the service department's hours? You should be sure to find answers to these questions before you decide to buy. Learning that a dealer's service department is not well run or is not open at convenient times will do you little good after you accept delivery of your car.

**Question for Thought & Discussion #5:**
How do you feel about dealing with salespeople? Do you believe you could negotiate successfully to buy a car you want at a fair price?

# THE diverse CONSUMER

## Women Buy Cars, Too

In 2001, 49 percent of new cars purchased in the United States were registered to women. There is, of course, no way to tell how many of these cars were actually purchased by men for women, but then, there is no way to know how many of the new cars registered to men were actually purchased by women either. Regardless, it is safe to say that millions of women negotiate the purchase of automobiles every year. But do automobile salespeople treat women differently than men, and, if they do, what can be done about it?

In 2001, a group of women were asked how they were treated by salespeople at automobile dealerships. Here are some of their responses:

● "The salesman ignored me until I went into his office and asked for help. Then he made it perfectly clear that he thought it was a waste of his time to talk to me. He tried to manipulate me into buying a car that I did not want."

● "I knew exactly what I wanted in a car, but the salesman paid no attention. When I protested, he actually called me 'little lady.'"

● "I should have taken my husband with me. I was forceful and demanding, but it made no difference. The salesman told me that the information I had downloaded from the Internet didn't mean a thing."

● "Two men walked into the dealership after me and were immediately approached by salesmen. I was totally ignored."

● "He tried to tell me I had to buy a car that day. When I said I was just looking and asked for information about the car, he asked me if I had 'some kind of a female problem.'"

As appalling as these experiences may seem, are they unique to women, or are men treated insensitively as well? If consumer surveys are to be believed, there is a double standard for men and women. Many salespeople who work in car dealerships apparently do not take women seriously.

If you are female, you can take several steps when you feel you are not being treated appropriately at an automobile dealership:

1. Come prepared. Know what questions you want to ask and demand answers to those questions. If you don't appear to be confused, you are less likely to be manipulated.

2. If you feel you are not being treated appropriately, jot down a list of your specific complaints that you can refer to later.

3. Ask the salesperson who the supervisor is, and then ask to speak to this person. Be prepared to explain your specific complaints and what you want done about them.

4. Contact the customer relations department of the manufacturer. These people are much less likely to be prejudiced against women and will not take your complaint lightly. Again, it is important to have a list of the specific problems you encountered.

5. In many places there is a local automobile dealers' association that you may contact. This group is not likely to act as a result of a single complaint, but if many women complain, the organization can put pressure on a dealership to change its ways.

6. Contact the consumer help reporter at your local newspaper or television station. Explain what happened. You may or may not receive assistance. If you do, the dealer will beat a path to your door. No business wants this type of publicity.

7. Exercise your right as a consumer to leave the dealership. Write a letter to the owner explaining your dissatisfaction and pointing out that you will not have pleasant things to say about the dealership when you talk to your friends and relatives.

None of these actions will change the situation overnight. But, in time, after enough women refuse to accept inappropriate treatment, things should improve.

**Do you believe gender makes an important difference in how people are treated? What would you do if you were treated in an inappropriate manner by a car salesperson?**

SOURCE: Carmen Tellez, "Winning Women Over in the Car-Buying Process: Part 1," *Edmund's*, April 23, 2001. http://www.edmunds.com/advice/buying/articles/45991/article.html.

**THE DEAL OFFERED**  Obviously, the deal you are offered is the central issue to consider when you choose a dealership. Although you may be willing to pay slightly more to buy from a dealer who has a superior reputation or is located near your job, there should be a limit to how much more you would pay. Another important factor to consider is the way you are treated by the

**High-Balling**  Offering an artificially high value for a product that is traded in and then inflating the price of the product that is sold.

**Low-Balling**  Offering to sell a product at a low price in a telephone conversation and then increasing the price when the consumer visits the firm to purchase the product.

**Bushing**  Adding unordered accessories to a product to increase its price.

dealer's sales representative. If you believe the salesperson is trying to "hustle" you, there is a good chance that other aspects of the firm's operation won't be what you would like them to be either. Even when you intend to buy directly from a dealer, it is a good idea to check out the offer by comparing it to offers made on the Internet.

## Sales Tactics

Automobile sales personnel have dubious reputations, which may or may not be justified in individual cases. Nonetheless, it pays to be aware of common techniques used to sell cars—and to extract more money from consumers than is justified.

**HIGH-BALLING**  With **high-balling,** the salesperson offers an unreasonably high trade-in value for an old car—far more than competitive dealers are willing to pay. The wily salesperson hopes the unwary customer will think he or she is getting a bargain, but actually the additional amount the consumer receives for the trade-in is included elsewhere in the price of the new car. To avoid being high-balled, assume that you can get only the trade-in value listed in the National Automobile Dealers Association's *Official Used Car Guide* (http://www.nada.org). This publication may also be examined at most banks or public libraries.

**LOW-BALLING**  With **low-balling,** a customer who calls the dealership is told over the phone that a particular car can be sold at a very low price—that is, a low-ball estimate. When the customer comes to the showroom, however, the salesperson explains that there was a mistake or that that particular car was just sold. The only cars left are more expensive, and there is no way to know when, if ever, another less expensive model will be available. But, the customer is told, no one really wants such a cheap car with no options anyway. Better to pay for a more luxurious car, whether the customer wants it or not. The purpose of low-ball estimates is to get consumers to the dealership, where pressure can be applied to increase the amount paid and profits for the dealer and salesperson.

**BUSHING**  **Bushing** occurs when a salesperson adds unordered accessories to a car without the customer's approval. When the customer arrives to pick up the car, the unwanted options or dealer add-ons have increased the price beyond the amount agreed on. The salesperson explains that she or he thought the customer had asked for the options and that any reasonable customer would want them. Regardless of what is said, they are now attached to the car and must be paid for unless the customer wants to wait weeks or months for a different car. The possibility of bushing is one reason to read a sales contract carefully before you sign to be sure it states exactly what will and will not be included on the car you agree to purchase. Demand a copy of the sales agreement to take with you when you make your deal. Doing so improves the chances that the car you are offered will be the same as the one you ordered.

In fairness, it should be emphasized that most automobile dealers are honest businesspersons who expect to earn a fair profit by providing quality products and good service. Any firm—car dealership or otherwise—that is consistently dishonest is unlikely to stay in business for many years. Word will spread and customers will take their business elsewhere.

# Car Shopping on the Internet

By 2003, it was estimated that fully one-third of new-car shoppers at least investigated online car offers before buying. Autobytel.com, one of the first Internet car marketers, has been joined by hundreds of other electronic auto-buying services. Most automobile dealerships also maintain Web sites where consumers may review the vehicles offered even if they are not able to complete purchase contracts online.

Most online car marketers do not actually sell cars themselves but simply represent dealers who make offers to sell online. Typically, it works like this: Consumers visit the organization's Web site and enter the brand and model of car that interests them and their desired options, color, and other features. They are given the telephone numbers of several dealers who are willing to sell the vehicle at a stated price. These dealers are not likely to be the ones closest to where the consumer lives. It is then the consumer's responsibility to contact the dealerships to conclude the deal. If a consumer has no car to trade, a cash deal can be completed quickly. A consumer who wants to trade in a car, however, will need to negotiate its trade-in price. This can be time consuming and frustrating.

**ARE ONLINE PRICES LOW?**   There is no guarantee that an online price will be lower than a price that could be negotiated in person. As a matter of fact, many experts have found that a lower price can often be obtained by going in person to a dealership. Still, there are advantages to shopping online. Reaching a sales agreement on the Internet can help people who don't like to deal with salespeople. It is also easy to compare the offers made by different dealerships without actually taking the time to visit them and deal with several salespeople. Finally, it is easy to end contact with a dealership that does not please you. All you need to do is to hang up your phone or go offline.

**OBTAINING SERVICE AFTER YOU BUY**
Although buying a car over the Internet may reduce the car's initial cost, Internet consumers may face other problems later. Dealers sell more than new cars—they also market service. Like a car bought through a dealer, a vehicle purchased over the Internet at a low price can develop problems and need to be taken to a dealer for warranty service. Although dealers are supposed to treat all customers equally, this does not always happen. A dealership may give preference to its own customers over an Internet buyer who purchased the car at a different dealership fifty miles away.

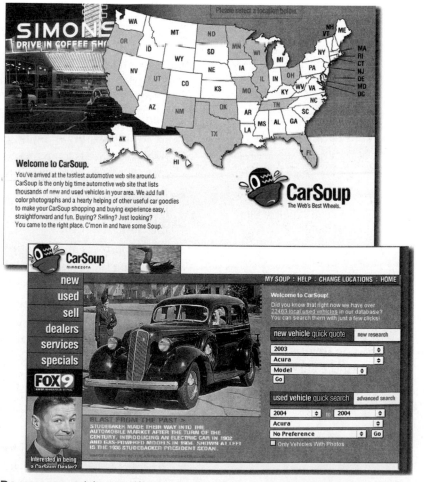

Do you expect to use the Internet when you next shop for a car? Would you only seek information, or would you be willing to complete your purchase online as well?

**Gross Capitalized Cost (GCC)**  The stated value of a new car that is leased.

**Residual Value**  The predicted value a car will have at the end of a lease agreement.

# Leasing versus Buying

In 2003, more than one out of every three new cars was leased rather than purchased. Many consumers choose to lease cars because leasing enables them to drive more expensive automobiles and make lower monthly payments than if they bought a car outright. Leasing can entail significant costs, however, that need to be considered. Many consumers, for example, who would automatically negotiate the purchase of a new car to obtain a good price make the mistake of accepting a lease offer without understanding its meaning or dickering to reduce its cost. Before you lease any vehicle, you need to do your homework. Otherwise, you will probably end up paying too much.

## How a Lease Works

Consumers who lease automobiles are paying for the use of the car. They own nothing. When a lease agreement is made, the manufacturer is paid for the car with money borrowed by the leasing business. Payments made by consumers and the money received from the sale of returned cars are used to repay this debt. Consumers often do not realize it, but leasing a car is much like taking out a loan. In either case they are required to make payments over time. The biggest difference is that at the end of the lease, they have nothing to show for their payments.

To understand lease agreements, consumers need to learn the meaning of special terms used in these contracts:

- *Gross capitalized cost* is the price you would pay for the car if you bought it outright. You should negotiate the **gross capitalized cost (GCC)** just as if you were buying the car.
- *Capitalized cost reduction* is the cash down payment and the value of any car you trade in.
- *Acquisition fee* is a payment to cover the cost of setting up a lease. It often includes a security deposit that should be refunded at the end of the lease.
- *Residual value* is the predicted value the car will have at the end of the lease. The **residual value** is also what you will have to pay if you purchase the car at that time.
- *Money factor* is an indication of the interest rate that you will pay for financing the lease of the car. Multiply it by the number of months in the contract to calculate the approximate annual interest rate.
- *Monthly payment* is the amount you will pay each month during the lease.
- *Mileage allowance* is the number of miles you can drive the car without paying an extra charge, which is typically $.12 to $.20 per mile. People who drive many miles each year should be particularly careful to evaluate the possible cost of excess miles.
- *Gap insurance* pays the difference between the actual value of a leased car and what is owed by the customer if the car is destroyed or severely damaged.
- *Excess wear fee* is charged for damage to the car that is beyond what could reasonably be expected. The amount of excess wear is determined by the dealer when the car is returned. Find out how any excess wear will be assessed before you sign a lease.

**Question for Thought & Discussion #7:**
Have you or anyone in your family leased a car? How satisfactory was the lease arrangement?

- *Disposition fee* is paid when the car is returned at the end of the lease. It is supposed to cover the cost of getting the car ready to sell to someone else.

## Steps to Leasing

Leasing involves six steps:

- *Step one—negotiate the price.* Consumers hardly ever pay the sticker price when they buy a new car—they typically negotiate a lower price. People who choose to lease a car should also negotiate the gross capitalized cost. Suppose you agree to lease a car with a GCC of $17,000.

- *Step two—agree on a residual value.* The residual value of the car may also be negotiable. You should make sure that it is reasonable, particularly if you plan to buy the car at the end of your lease. Otherwise, you might be expected to pay $9,000 for a used car that is worth only $8,000. The difference between the GCC and the residual value of the car is the amount you will pay through your monthly payments. If you agree on a residual value of $8,000, this amount will be $9,000 ($17,000 − $8,000 = $9,000).

- *Step three—make your down payment and security deposit.* Your acquisition fee includes a security deposit and a payment for the dealer's cost of setting up the lease agreement. The security deposit should be refunded to you at the end of the lease (although many consumers have reported difficulty in obtaining this refund). If your acquisition fee is $1,000 and includes a security deposit of $450, the final cost to you should be $550 ($1,000 − $450 = $550).

- *Step four—calculate your finance cost and monthly payment.* The amount you will finance is the difference between the GCC and the residual value plus the amount of your down payment (or capitalized cost reduction). In this case this amount is $10,000 ($9,000 + $1,000 = $10,000). Determine the approximate interest rate by multiplying the money factor (assume 0.0029) times 24. This tells you that you will be paying about 7 percent for the borrowed funds (0.0029 × 24 = 0.07 = 7%). The amount you should pay over a two-year lease should total roughly $10,700. Your monthly payment should be roughly $446 ($10,700/24 = $445.83). If you are asked to pay substantially more than this amount, you should ask for an explanation. Buying gap insurance, of course, would increase this payment.

- *Step five—check out your car before you return it.* Have a mechanic go over a leased car to find any significant problems before it is returned. If something is wrong, it could be less expensive to have the car repaired by someone other than the dealership. At a minimum, have the mechanic list any problems and their expected cost of repair.

- *Step six—return the car.* After the lease is over, the dealer will check the car for excess mileage and wear. You will be expected to pay whatever amount is determined and the disposition fee. Many dealers are willing to forgo all or part of these fees if you either purchase the leased car or agree to a new lease agreement.

Leasing is not a good idea for everyone. It often encourages consumers to choose more expensive cars than they can comfortably afford. People may consider only the lower monthly payment, forgetting that at the end of the lease they will own nothing and will have a significant end-of-lease payment to make. In general, leasing is not a good choice for those who can afford to pay cash, drive many miles each year, keep their cars for a long time, or are particularly hard on cars. For these people, buying a new or used car is probably a better choice.

# Should You Choose a Used Car?

If the cost of a new car is beyond your budget or if you want to use your limited funds in other ways, buying a used car may be the right choice for you. Used cars are offered by new-car dealers, by businesses that deal exclusively in used cars, by some car rental companies, over the Internet, and by private individuals. Buying a used car allows consumers to avoid the rapid depreciation new cars inevitably experience. The increasing number of returned leased cars has made it possible to buy cars that are only a few years old and in good condition at relatively low prices. But you should be careful no matter how good a deal seems to be at first glance. Any used car may have been mistreated or improperly maintained by its previous owner. It might have been in an accident, had a major mechanical failure, or been driven in harsh locations or weather conditions.

## Have a Mechanic Check It Out

One way to reduce the chance of experiencing a major problem with a used car is to have an independent mechanic check out the car before you commit to buying it. In most communities there are electronic diagnostic centers that specialize in identifying defects in automobiles. Some of these businesses do not even perform repairs. You will probably be charged $50 to $100 for this service, but the money will be well spent if it prevents you from purchasing a defective vehicle. Learning that a car's transmission is about to go or that there is a crack in the engine block could help you avoid hundreds or even thousands of dollars in unexpected repair costs.

In many cases untrained consumers can reject defective cars through a simple visual inspection without having to pay a mechanic to examine the vehicle. To tell whether a car has been in an accident, look for mismatched colors, ripples, or bumps in the paint. Body parts that do not line up are another good indication that a car has undergone major repairs. Finding oil or transmission fluid puddled under a car is a sign that something is seriously wrong. Cars with engines that are particularly noisy or transmissions that jerk when they are engaged should be avoided. Use your common sense to narrow your alternatives before you pay for a professional opinion.

## Investigate Your Used-Car Alternatives

Many sources of general information about used cars are available in print or on the Internet. The section on buying a used car in any annual *Consumer Reports Buying Guide* reports on the reliability of most recent makes and models of cars and trucks sold in the United States. It also suggests tests you can carry out on used vehicles you might buy. If you are considering a car that was traded in at a dealership, you may be able to get the previous owner's name and telephone number. Then call this person and ask about possible problems, defects, or advantages of the car.

Another way to protect yourself from major repair expenses is to buy from a dealer who offers a written warranty with used cars. Buying a used car with a warranty will almost certainly be more expensive than purchasing a similar vehicle that is not protected. What you are buying with your extra money is security—with the warranty you know that you will not have to pay for major breakdowns for a specific period of time or number of miles (six months or 6,000 miles is typical).

**BUYING USED CARS ONLINE** In recent years hundreds of entrepreneurs have tried marketing used cars online. Most of them have not succeeded. There is something about buying a used car that makes people want to do it in person. They want to examine and drive the car—"kick its tires," so to speak—and you can't do that with a car you have seen only online. How would you know whether it has been involved in an accident or if its engine leaks oil? When buying new cars online, consumers know they will get a brand-new car that is backed by a warranty. This is often not the case for used cars. Would most consumers be willing to drive fifty miles to see a car that "Aunt Camilla only drove to church on Sundays"? Probably not. But if you want to buy a specific make and model of a car with relatively few miles on it, online shopping may be your only choice.

Large online car marketers, such as Carsdirect.com and Autobytel.com (which only sold new cars at first), are able to offer virtually any type of used car you might want. You can specify the color, a leather interior, a sunroof, and a towing package for your boat. And even if you make a tentative deal online, you will not conclude it until you actually see the car or truck. Don't automatically reject the idea of online used-car shopping just because it's a new idea.

**ONLINE CAR AUCTIONS** In the last few years, used cars have become available through online auctions. In 2002, nearly 300,000 cars were auctioned through eBay.com. That is not a large part of the total used-car market, which involves nearly 100 million sales each year. Nevertheless, online auctions can offer advantages. The advantage for the buyer is the low price that can often be achieved. Sellers generally benefit from being able to get a higher price than they would if they traded in their used cars for newer models at a dealership. The seller is supposed to certify that the car is in good working order and has been inspected by a mechanic, but the buyer is still taking a risk. Would you be willing to buy a car you have never seen through an auction on eBay?

# Final Guidelines for Consumers

Once you reach an agreement to buy or lease a new or used vehicle, make sure that everything you and the salesperson have agreed to verbally is listed specifically on the contract you sign. Be sure that the agreement is countersigned by someone in a position of authority at the business where you are making your purchase. At a minimum, the following four assurances should be included in the agreement:

1. There will be no changes in price. The price indicated by the agreement is clearly stated to be the *total* cost that must be paid.

2. If you trade in a car, be sure that the price you have been offered is specifically listed in your agreement and that the car cannot be appraised after the deal has been made.

3. There can be no changes in the car that is ordered—its options cannot be added to or subtracted from.

4. The car will be delivered at a definite time, after which the agreement becomes void and any down payment will be totally refunded.

Finally, never fall for a "switch" that you might be offered a few days after you have signed a sales or lease agreement. You might be told that the car you wanted is not available, but by agreeing to pay a little more, you can get a better car right away. Never accept such a change without carefully evaluating the new offer. You have a legal right to receive the car you ordered. To ask you to accept a different vehicle, a dealer should make you a distinctly better offer.

# Financing That Purchase

A new or used car is usually such a major purchase that at least part of it has to be financed by credit. Do not automatically accept the credit that the dealer offers you when you decide to buy a car. Shop around for credit just as you shop around for anything else. Fortunately for you, the Truth-in-Lending Act of 1968 requires every lender to disclose the total finance charge and actual annual interest rate to be paid. Thus, you can easily compare the credit offered by the dealer with the credit offered by competing sources, such as banking institutions and finance companies. Remember, if you default on your car payment, your car may be repossessed. This is a real possibility: in some states, finance companies can take your car away from you without a judicial hearing. Do not buy a car that costs more than you know you can afford.

## Where to Borrow for a Car

There are numerous sources for automobile loans: life-insurance policies, savings and loan or commercial banks, credit unions, auto insurance companies, online lenders, and auto dealers themselves.

Automobile dealerships often offer credit to their customers—sometimes at very low rates. In 2003, for example, "0 percent financing" was a common marketing tool used by U.S. automobile manufacturers. Although 0 percent financing may sound like the best deal possible, it might not be. Dealers often offered large rebates as alternatives to low-interest loans. A $3,000 rebate could be a better deal if you can borrow the price of the car from your credit union at 6 percent interest.

Life-insurance loans are possible if you have a whole life policy (see Chapter 18) that has been in effect for at least a few years. It will have accumulated cash value, and you can borrow up to that amount from the company. The maximum annual percentage rate for a life-insurance loan is often limited to 8 percent by law. The only problem with taking out such a loan is that it reduces the amount of life insurance you have in effect.

A passbook loan is available if you have a savings account. The bank or savings and loan association may lend as much as 90 percent of the amount on deposit. The bank will freeze enough to cover the unpaid balance on a loan, but the entire account will continue to earn interest.

Generally, credit unions offer the most beneficial rates on automobile loans; if you are a member of one or can become a member without too much trouble, find out what you will be charged there.

Banks are the second most common source of automobile financing, after the auto dealers themselves. What your local banker will charge you depends on your credit rating, the amount of down payment or trade-in value on the car you are buying, and the general state of the economy.

Auto-insurance companies sometimes issue car loans through a bank or through their own subsidiaries. To find out if your auto-insurance company does this, call your agent.

Since the late 1990s, a growing number of Internet lenders have entered the credit market. Although many of those firms offer low-interest loans, they do not provide face-to-face contact with their loan officers. Be sure to carefully evaluate their contracts before you sign.

You will find that if you go to a small-loan company, you will pay the highest annual percentage rate for an auto loan.

**Question for Thought & Discussion #8:**
Why is borrowing to buy a car a rational decision for most consumers?

## What Length of Loan to Take Out

Most consumer experts recommend that automobile loans be taken out for the shortest period possible. They point out that you end up paying a relatively high total interest charge when you take out a three- or four-year car loan. Additionally, you end up having a hefty balance to pay if you are ready to trade in your car before the end of four years.

Does that mean that you should not take out a four-year auto loan? No, not necessarily. You really are asking a question about how much you should be in debt; the fact that it concerns an automobile is irrelevant. If you think you would be uncomfortable having a debt outstanding for four years, then that may be a reason to opt for a shorter time period. If you do so, however, you must use more of your discretionary funds to make the automobile loan's monthly payment, and you will, therefore, have less to spend on other items during that period. The fact that it costs you in additional charges to keep an auto loan outstanding longer should not be surprising. You are asking to use someone else's money for a longer period. As with all borrowing decisions, you must balance the benefits of having more cash available for other purchases against the increased cost for borrowing more or for borrowing for a longer period of time. Table 11.3 on the following page shows what it costs in finance charges per $1,000 of a car loan.

## Know Your Lemon Laws

All states have passed lemon laws that pertain to new-car sales and service. In effect, if your new car has a serious problem that can't be fixed in a reasonable time, either you get your money back or you get a new car. Most states say you've purchased a lemon if, within its first year or 12,000 miles, your car has a serious defect that an authorized facility hasn't been able to repair in four attempts or if the car has been out of service for thirty days. Generally, defects must be covered by the manufacturer's warranty and must substantially reduce the use, safety, or value of your vehicle.

If you are a new-car owner, you may want to visit the Center for Auto Safety's Web site at **http://www.autosafety.org** to download a copy of your state's lemon law. You can also call your state's attorney general's office for information on your state's lemon law and how to use it if you have purchased a lemon.

| TABLE 11.3 | What Your Car Loan Will Cost per $1,000 Borrowed | | | | | | | |
| --- | --- | --- | --- | --- | --- | --- | --- | --- |
| | One Year | | Two Years | | Three Years | | Four Years | |
| Annual Percentage Interest | Monthly Payment | Total Finance Charge | Monthly Payment | Total Finance Charge | Monthly Payment | Total Finance Charge | Monthly Payment | Total Finance Charge |
| 5% | $86 | $27 | $44 | $ 53 | $30 | $ 79 | $23 | $105 |
| 6 | 86 | 33 | 44 | 64 | 30 | 95 | 23 | 128 |
| 7 | 87 | 38 | 45 | 75 | 31 | 112 | 24 | 150 |
| 8 | 87 | 44 | 45 | 85 | 31 | 128 | 24 | 172 |
| 9 | 87 | 49 | 46 | 96 | 32 | 144 | 25 | 194 |
| 10 | 88 | 55 | 46 | 107 | 32 | 161 | 25 | 217 |
| 11 | 88 | 61 | 47 | 118 | 33 | 178 | 26 | 240 |
| 12 | 89 | 66 | 47 | 130 | 33 | 196 | 26 | 264 |
| 13 | 89 | 72 | 48 | 141 | 34 | 213 | 27 | 288 |
| 14 | 90 | 78 | 48 | 152 | 34 | 231 | 27 | 312 |

NOTE: Figures have been rounded to the nearest dollar.

## Arbitration Programs

Most states' lemons laws require an aggrieved new-car owner to notify the dealer or manufacturer of the problem and provide the opportunity to solve it. If the problem remains, the owner must then submit complaints to the program specified in the manufacturer's warranty before taking the case to court. Your owner's manual will tell you which arbitration program applies to your car.

Many major car companies use their own arbitration panels. Ford and DaimlerChrysler, for example, have the Ford Consumer Appeals Board and the DaimlerChrysler Customer Arbitration Board, respectively, to which lemon-law disputes are submitted. Some companies, however, such as General Motors, subscribe to independent arbitration services, such as those provided by the Better Business Bureau. These arbitration services are free to consumers, and lawyers are not involved. You simply explain your problem to the arbitration panel or board, in person or in writing, and submit any evidence you feel is necessary to back up your claim—such as an independent mechanic's evaluation of your car's problem. The board will rule within sixty days on whether your car is a lemon and whether it will be replaced, the price refunded, or other action taken. Often disputes are settled before they get to the arbitration stage. General Motors, for example, settles approximately 90 percent of disputes prior to arbitration.

Although decisions reached by these groups are usually binding on manufacturers, they usually are not on car owners, who are free to sue in court if they are not satisfied with the results of arbitration. Even if arbitration does not conclude with a consumer's receiving a new car or a refund of the purchase price, it often results in the complaint's being resolved. Automobile manufacturers tend to repair defects that have led to arbitration.

# Criticism of Industry Arbitration Panels

Although arbitration boards must meet state and/or federal standards of impartiality, some consumers have doubts that industry-sponsored arbitration boards are truly impartial in their decisions. In response to consumer pressure, most states now give consumers a choice of an industry arbitration board or one organized by the state. Once one of these boards is chosen, the consumer is then precluded from later appealing to the other. A comparison of the rulings made by government-sponsored and industry boards seems to justify consumer complaints of industry partiality: according to the Center for Auto Safety, government-sponsored boards rule in favor of consumers about six times as frequently as industry arbitration boards.

# Confronting Consumer Issues: Reputable Repairs for Your Car

Since the inner workings of an automobile engine are a mystery to many consumers, the likelihood of fraud is high in this industry. It is estimated that perhaps half of the nearly $200 billion spent each year on car repairs is wasted—through incompetence or fraud. Even if you know nothing about car engines, you can reduce your chance of falling prey to unscrupulous mechanics by taking certain precautions. The following tips may help you avoid getting "taken for a ride" the next time you have car trouble.[1]

## WHEN SOMETHING GOES WRONG

If your car begins to sound strange or performs significantly differently in any way, have it checked immediately. Don't ignore the problem in the hope it may disappear. Chances are, if you don't check it out while the car's still in running condition, you may face a breakdown at an inconvenient time or place.

Before taking your car to a repair shop, however, be sure to do the following:

1. *Check your owner's manual.* Sometimes you can diagnose the problem yourself by carefully reading your owner's manual. Such manuals often offer troubleshooting advice that may help you fix the problem yourself. If you don't have a manual for your model of car, try to obtain one from the nearest dealer.

2. *If the owner's manual can't help you, try to determine as closely as possible where the problem is located.* Don't take your car to a specialty shop if you think a local service station might be able to fix it.

3. *Before you take your car in for repairs, check your warranty, if you have one.* If some repairs and parts are covered by your warranty, you probably will be required to take the car to an authorized franchised dealership for your make of car.

## WHERE TO GO

Even if you don't have a warranty for repairs, you may consider taking your car to a dealership for service. Although a dealership is likely to charge more for services than some competing garages, it may be able to provide better-quality

Photodisc/Getty Images

**Although consumers can check the fluid levels in their cars, most need to employ a skilled mechanic when their automobiles need repairs.**

service. Most mechanics at dealerships have received training in repairing the particular types of cars their employer sells. The dealer also has its reputation to consider. If you receive inferior service, you are unlikely to return to buy a new car in the future. Records show that the number of consumer complaints about service received from dealerships is lower than for independent repair shops.

For reliable work at perhaps a lower price, however, you may want to shop around. The following suggestions may be helpful:

1. *Talk to your friends, family, and co-workers.* Find out where they take their cars for repairs. This is often the best way to obtain the name of a reputable mechanic. If a certain shop comes highly recommended, try it out—and let the owner know who recommended you. If the shop owner thinks you, or whoever recommended you, may return for future repairs, your chances of getting good service may increase.

2. *Call various local repair shops.* See if you can find a mechanic who is certified by the National Institute for Automotive Service Excellence (NIASE). This organization certifies auto mechanics on the basis of voluntarily

---

1. Many of the suggestions made here are based on an analysis of car-repair fraud complaints conducted by the Council of Better Business Bureaus, 1515 Wilson Blvd., Arlington, VA 22209, and published in its Consumer Information Series, Publication No. 311-03246.

taken competency tests in eight different repair specialties. Of course, this certification doesn't guarantee excellent repair service, but it does increase its likelihood.

3. *Call the American Automobile Association (AAA).* The AAA can refer you, whether or not you are a member, to mechanics in your area who are on its list of reputable car service mechanics. Again, the AAA doesn't guarantee the quality of these mechanics' work, but it does base its recommendations on objective information, including dealership testing and certification by the National Institute for Automotive Service Excellence.

4. *You may also want to drive to an independent diagnostic center if one is located in your area.* These centers use electronic equipment to test your car thoroughly and diagnose any problems. Many of these centers, however, are affiliated with repair shops that may have a vested interest in the diagnostic results.

### Question for Thought & Discussion #9:
How was the person who services your family's car chosen? Are your family members satisfied with the service they receive?

## AT THE REPAIR SHOP

Once you've located a repair service, you can take several steps to ensure against possible fraud:

1. *Have a list handy of the things you want done.* If you're not sure what's wrong with your car, describe the problem as clearly as you can. And if you don't know the difference between a differential and a distributor, don't advertise your ignorance.

2. *Get an estimate.* This is probably the single most important step you can take to prevent being overcharged for your repairs. Have the mechanic specify clearly what he or she thinks is causing the malfunction and what it will cost to repair it. If you have to leave your car for a time—to go to school or work—leave your phone number so the mechanic can call you with the estimate prior to doing any repairs. If the required repairs are major, get your estimate in writing.

3. *If you think the estimate is higher than it should be, or if you don't think the diagnosis is correct, get a second opinion.* Be especially wary if the mechanic claims you need to have any of the following parts replaced: ball joints, batteries, shock absorbers, or brakes. It's relatively easy for consumers to be conned by mechanics who can be persuasive about why these need to be replaced—when, in fact, they don't.

4. *When you decide to go ahead with repairs, check the service order carefully.* Make sure it includes either the specific repairs you have requested or those recommended by the mechanic who gave you the estimate.

5. *Stay with your car if possible.* If you can't, leave a number where you can be reached by phone in case the mechanic needs to talk to you.

6. *Check your odometer if you have to leave your car at a repair shop overnight—unless you know the mechanics well.* It is possible for mechanics to "road-test" a car for their own transportation purposes and at your expense.

7. *Last, but not least, clarify with the mechanic or repair shop manager the terms of payment.* Must it be in cash, or does the shop take credit? Does it take personal checks? Many consumers have found themselves in the uncomfortable position of being refused possession of their cars after repairs because they lacked the cash to pay the bill. In such cases, if you can't produce the payment within a reasonable time, the repair shop can legally place a **lien** on the car, sell it, and collect the amount due from the price—returning the balance to you.

## BEFORE PAYING FOR REPAIRS

Before you write out your check or pull out your credit card, be sure to examine your car thoroughly to make sure that everything was, in fact, repaired or that parts were really replaced. Don't be afraid to ask the mechanic to show you any replaced parts and explain specifically what he or she did. Ask questions until you understand what was done. After all, it's *your* car and *your* money. Check to make sure the actual charge is reasonably close to the estimate.

Also, request an itemized bill listing all repairs that were done and all parts that were replaced. If possible, get a written guarantee for the services performed. If certain kinds of repairs were made, such as installing brakes or a transmission, you may also want to road-test your car before paying. Be sure to keep all of your receipts in case problems develop in the future.

## STATE REGULATION

Some states have passed legislation specifically to regulate auto repair shops. If you live in California, for example, you have a right to a written cost estimate for repairs before any repair work is undertaken. The final cost of the work cannot exceed this estimate—unless you have authorized the increase. You also have a right to the return (or at least the inspection) of all parts that have been replaced and to a written invoice itemizing parts and labor.

If you are unsure whether your state has a comparable law, you can find out by calling the state consumer protection division (see Table 5.4 in Chapter 5).

*(Continued on next page)*

Confronting Consumer Issues (Continued)

## REPAIRS ON THE ROAD

Having your car break down in the middle of nowhere or on a freeway in the middle of a city can be a very costly nightmare. If you are not a member of an auto club, such as the AAA, your only recourse may be to call a towing service listed in the Yellow Pages. Be wary of mechanics or repair services recommended by tow-truck drivers, however. The driver may be more interested in kickbacks from repair shops than in your pocketbook or the quality of service you get. If you are unfamiliar with the area, your best bet is to arrange to have your car towed to a local dealership for diagnosis and minimum repairing. Save the major repairs, if possible, until you return home.

## WHEN THOSE NEW-CAR REPAIRS GET STICKY

When the going gets sticky on a matter of warranty or car performance, you can turn for help to an organization called the Automotive Consumer Action Panels (AUTOCAP). AUTOCAP was formed because of the numerous complaints consumers have made concerning car dealerships and repair services.

### Question for Thought & Discussion #10:

Have you or a member of your family ever been dissatisfied with a repair that was performed on a car? What was done about the situation?

## HOW DOES AUTOCAP WORK?

If your new car isn't performing satisfactorily and if the dealer isn't giving you the repair service you feel you should have, check your local white pages for the AUTOCAP nearest you. If you can't find one nearby, write to its headquarters at 8400 Westpark Drive, McLean, VA 22102, or call (703) 821-7000. You will be sent a complaint form, and when you return it, AUTOCAP will contact the dealer and urge the dealer to work out the problem with you. If this fails, the matter goes before the AUTOCAP arbitrating panel, consisting of four dealers and three public members.

It is a painless job to arrive at a "just" settlement when a dealer and a customer can agree. Obviously, the panel is not a court of last resort; it has no enforcement powers and relies on dealer cooperation to handle complaints satisfactorily. So far, about 40 to 50 percent of the complaints coming before AUTOCAP each year have been settled in favor of car owners. If you feel you are not getting fair treatment from AUTOCAP, you can go to the state motor vehicle agency or take private legal action.

### Key Term:

**Lien** A claim placed on the property of another as security for some debt or charge.

# Chapter Summary

1. The automobile industry in the United States is huge. Currently, Americans spend more than $370 billion for passenger vehicles and purchase more than 17 million cars and trucks per year.

2. There are many transportation alternatives to owning a car, particularly for consumers who live in urban areas. They may choose to use bicycles, motorcycles, public transportation, or rented or shared cars. Although these alternatives usually involve a cost in lost convenience, they all offer the benefit of lower expenditures for the consumer.

3. There are approximately 45,000 traffic fatalities a year in the United States, and traffic accidents are the leading cause of death for those between five and thirty-four years of age.

4. Beginning with the passage of the National Traffic and Motor Vehicle Safety Act of 1966, the federal government has been active in regulating the auto industry to promote safer vehicles by requiring that autos be equipped with certain safety devices. Safety features raise the prices of automobiles, and consumers thus pay for the additional protection they receive.

5. Stiffer penalties for drunk driving, laws mandating a higher drinking age in many states, and greater awareness by Americans of the risks of drunk driving have all contributed to a reduction in alcohol-related traffic fatalities and injuries.

6. Some of the social costs of driving are pollution and traffic congestion.

7. The private costs of driving a car will vary according to the age of the car, its size, insurance expenses, and driving styles. For most U.S. consumers, this cost will fall between 50 and 80 cents a mile.

8. When making a decision about replacing an existing car, you should consider the costs of owning and operating a new car versus your existing older car and the costs of repairs in terms of time and inconvenience compared with the book value of the older car.

9. Since a new car represents a substantial expense for most people, you should conduct some market research before you buy one. Knowing the list price that dealers pay for new cars and new-car accessories can help you shop comparatively among dealers.

10. In recent years many Internet businesses have come online offering to sell new or used vehicles. Although those firms may offer lower prices, they often do not accept trade-ins, and there can be a problem when their cars require warranty service.

11. Because new cars generally have problems that you will want the dealer to take care of, the proximity of the dealer and the dealer's willingness to handle such warranty difficulties are important considerations when deciding where to buy a new car.

12. Before you buy a new car, you should consider the benefits you might obtain by leasing one instead. Leasing a car offers several advantages and is often no more expensive in the long run than financing a new car.

13. The purchase of a used car requires as much shopping as for a new car, or more, for the mechanical condition of the car is now in question. If you wish to have a warranty, you can purchase a used car from a dealer offering a six-month or 6,000-mile warranty for a somewhat higher price.

14. Shop for automobile financing just as you shop for any other product. Shop on the basis of the down payment required, setup charges, actual finance charge, and actual annual interest, as well as the number of months required to pay. Remember, the sooner you pay off the loan, the less interest you pay. If your payments are high relative to your income, however, you will have less money for other purchases.

15. All fifty states plus the District of Columbia have some form of lemon law governing new-car performance. If you feel you have purchased a lemon, check with the Center for Auto Safety or with your state attorney general's office for information on lemon laws.

16. Before you agree to allow a mechanic to repair your car, you should check your owner's manual for troubleshooting advice and reread your warranty to see if the work in question is covered.

17. In searching for a reliable repair shop, first check with your friends and family. Other alternatives are to (1) find a shop with mechanics certified by the National Institute for Automotive Service Excellence, (2) go to a dealer or repair service recommended by the American Automobile Association, or (3) go to an independent diagnostic center for an electronic evaluation of your car's condition.

18. When you agree to have your car repaired, demand a written estimate and insist that you be consulted before any repairs beyond that estimate are made. Agree on how payment will be made, before work is done. Stay with your car if possible when it is being worked on. Insist on seeing any old parts that were taken out of your car. Drive your car before you make your payment.

## Key Terms

bushing **252**
gross capitalized cost (GCC) **254**

high-balling **252**
lien **264**
low-balling **252**

private costs **245**
residual value **254**
social costs **245**

## Questions for Thought & Discussion

1. Do you or any of your friends get along without a car? How does this lack affect one's life?

2. Do you always use a seat belt when you travel in a car? Why or why not?

3. What would you have to give up to buy a new car? Would the costs be worth it? Explain your answer.

4. Is a car just a means of transportation to you, or is it something more? Explain your point of view.

5. How do you feel about dealing with salespeople? Do you believe you could negotiate successfully to buy a car you want at a fair price?

6. Would you consider buying a car on the Internet? Explain your point of view.

7. Have you or anyone in your family leased a car? How satisfactory was the lease arrangement?

8. Why is borrowing to buy a car a rational decision for most consumers?

9. How was the person who services your family's car chosen? Are your family members satisfied with the service they receive?

10. Have you or a member of your family ever been dissatisfied with a repair that was performed on a car? What was done about the situation?

## Things to Do

1. Check with the Department of Motor Vehicles in your area to find out what penalties are imposed on drunk drivers and what the state law is regarding the use of seat belts.

2. Write, call, or visit the Web site of the consumer protection division or attorney general's office in your state and ask to have sent to you any information available on your state's lemon law and how it is enforced. Evaluate the information you receive. You may also obtain this information online.

3. Even if you are not in the market for a new car, try shopping for one over the Internet. Pick a particular make, body style, and set of accessories. Search at least three Web sites. See if you can get an actual quote. Are there any significant differences among the quotes?

4. Go to the library and get the latest December issue of the *Annual Buying Guide* of *Consumer Reports*. Look at the section on buying a used car. Could you perform the eight to ten on-the-lot tests given in that section? Have you ever tried the driving tests given in that section when you were looking for a used car?

5. Comparison-shop for a three-year, $20,000, new-car loan. How much difference is there in the interest rates you would have to pay to borrow from various lending institutions, in your monthly payments, and in your total cost of borrowing the money? What factors do you believe cause different lending institutions to charge different rates? Are there any factors you should consider in taking out a loan other than the interest rate? If so, what are they and why are they important?

# Internet Resources

## Finding Consumer Information on the Internet

The following Web sites have been selected for their relevance to topics discussed in this chapter. Search these sites to locate information that can add to your knowledge of automobile safety information that is available on the Internet. Remember, Web addresses change frequently. If any of these addresses no longer function, find similar sites to investigate using any of the search engines available to you.

1. The Department of Transportation maintains a hot line at 1-888-DASH-2-DOT (1-888-327-4236). It explains how this hot line works at its Web site, which can be accessed at **http://www.nhtsa.dot.gov/hotline**.

2. The Insurance Institute for Highway Safety gathers and provides data related to automobile and highway safety. Its Web site is located at **http://www.hwysafety.org**.

3. Crashtest.com is an organization that gathers insurance risk data and crash-test ratings on new and used cars, trucks, vans, and SUVs going back to the 1970s. You can investigate the information it provides at **http://www.crashtest.com**.

## Shopping on the Internet

The following Web sites have been selected because they offer consumers services similar to those described in this chapter. These are commercial sites that are designed to market products. They do not represent a comprehensive or balanced description of all businesses that offer automobile financing that is available online. How do the deals offered at these Web sites compare with financing deals offered by local financial institutions in your community? Remember, Web addresses change frequently. If any of these addresses no longer function, find similar sites to investigate using any of the search engines available to you.

1. Preapproved Online Auto Loans is a company that offers preapproved auto loans for all credit histories. You can apply online at **http://www.autonetfinancial.com**.

2. Driverslane.com offers car loans and bad-credit automobile financing to people who have poor credit histories. According to the firm's Web site, it provides "simple, nonthreatening financing for first-time buyers." Find out what this firm has to offer by visiting its Web site at **http://www.driverslane.com**.

3. E-LOAN offers auto loans for new and used vehicles. It says you can refinance your current loan and save money. Investigate this firm's offerings at **http://www.wheels.eloan.com**.

## InfoTrac Exercises

Purchasers of new copies of this text are provided with access to the InfoTrac Web site. This Web site links students to thousands of recent articles published in hundreds of periodicals. Use the key words **smart, air bags,** or other terms from this chapter to conduct a key-word search. Choose one article that is of particular interest to you and write a brief essay describing what you have learned from the article. Be sure to cite the author and title of the article and the name and date of the publication in which it appeared.

# Selected Readings

Amadio, Jill. "You're Covered." *Entrepreneur,* December 2002, p. 47.

Consumers Union. *Consumer Reports Annual Auto Issue.* Published annually in April.

Halliday, Jean. "No Haggle Has a Place." *Advertising Age,* April 19, 2001, p. S22.

Harris, Donna. "Service Contracts Thrive As Loans Replace Leases." *Automotive News,* November 11, 2002, p. 61.

Henry, Ed. "Find a Free Fix." *Kiplinger's Personal Finance Magazine,* September 2000, p. 105.

———. "Unload Your Lemon." *Kiplinger's Personal Finance Magazine,* September 2001, p. 126.

Henry, Jim. "Buy Here–Pay Here Dealerships Flourish." *Automotive News,* November 18, 2002, p. 40.

"Magnificent 7: Top Online Automobile Buying Services." *Automotive News,* October 2, 2001, p. 33.

Naughton, Kathi. "More Than Zero Financing Deals Confound Skeptics." *Newsweek,* September 16, 2002, p. 42.

"New-Car Leasing Continued Decent in 02." *Automotive News,* January 27, 2003, p. 28.

Nikkel, Cathy. "The Truth about Lemon Laws." *Motor Trend,* October 2000, p. 93.

"Old Car, New Warranty." *Business Week,* May 2002, pp. 16–18.

Stern, Linda. "Money: Car Cash." *Newsweek,* August 12, 2002, p. 65.

"Used-Car Buying without Fear." *Consumer Reports,* April 1999, pp. 67–71.

Wilson, Amy, Mark Rechtin, and Bill Ford. "Higher Incentives Are Here to Stay." *Automotive News,* February 10, 2003, p. 24.

# CHAPTER 12

# Choosing a Place to Live

After reading this chapter, you should be able to answer the following questions:

- What factors enter into our housing decisions?
- What are the benefits and costs of renting versus buying a home?
- How can you use the Internet to find housing information?
- What kinds of services do real estate agents provide?
- How do you negotiate a housing transaction?
- What are common sources and types of mortgage loans?
- What are closing costs?
- What factors should consumers consider when they choose to rent an apartment?

I f you happen to be an Inuit living in the Yukon Territory, putting a roof over your head is complicated but not impossible: you make an igloo. If you happen to live in the bush country of Tanzania, putting a roof over your head takes some time, but eventually your thatch hut will be just what you need. If you were a pioneer settling on some cleared land in the Old West, putting a roof over your head would have meant building a log cabin or a sod house.

Today, by way of contrast, if you are Mr. and Ms. Average American, to purchase a new three-bedroom, two-bathroom home and the land under it, you will have a pay about $180,000. In all, Americans today pay almost $850 billion a year to purchase new and previously owned homes.

> **"Most Americans live well because we really do like to cater to our fancies. We enjoy living the good life, and that includes a spacious home, often with unique features."**

## Why Are We Willing to Pay So Much?

Early Americans must have had just as many fanciful ideas about how they would like to live as we have. But today, some of us—in fact, most of us—live like kings compared to earlier Americans. Why is this so? Is it simply because we earn more money? No. The reason we live better than our ancestors is because our economic system has become more productive over the years. On average, in each of the last 150 years the quantity of goods produced in the United States per person has increased by more than 1.5 percent. A great deal of this extra production has been allocated to housing. Most Americans live well because we really do like to cater to our fancies. We enjoy living the good life, and that includes a spacious home, often with unique features.

It is also important to remember that when we buy or rent houses or apartments, we are also buying the services they yield—just as we do when we purchase clothing or cars or anything else that lasts. When we buy a house, we buy not simply the house but also the pleasure we derive from living in it. That pleasure is a function of the house's size, the convenience it offers, how scenic the view is, what the neighborhood is like, and everything else that can contribute to our happiness when we are home. Generally, the reason that there is "no place like home" is that all of us try to make our homes as special as possible so that we get maximum utility, or service, from them. It is important, then, to realize that when a consumer buys a $200,000 house, it is because this dwelling offers a perceived larger flow of services per month than could be obtained by purchasing a $100,000 house.

The actual price U.S. consumers pay for housing at any time and in any location depends on (1) the amount of money they have and are willing to spend on housing and (2) the quantity of housing units that exist. In Chapter 1 you learned that these are the forces of demand and supply. If, for example, you want to buy a house near the ocean in a nice neighborhood of San Diego, California, you should expect to pay more than for a similar house located in the frozen woodlands of northern Minnesota, far from any major city. The number of people who have money to spend and want to live in San Diego far exceeds the number of houses available there. Houses in cold climates, located away from job centers, are deemed less desirable and thus cost less.

> **Question for Thought & Discussion #1:**
> How do you decide how much of your income you will budget for housing?

## Renting versus Buying

At some point, you may be faced with the question of whether you should rent or buy a home. In this section, we look at some of the factors that you will want to consider in making your decision. We then address ways in which consumers may approach the purchase of a home. That does not mean that everybody should—or can—buy instead of rent. More space is given to the home-buying process simply because it is a more complicated procedure. Renting a home, however, also involves specific procedures and legalities with which you should be familiar. These are discussed in greater detail at the end of this chapter.

## Advantages of Renting

In many cases, your budget will determine whether you can choose to rent or to buy. But the budget is not the only factor in your decision. Consider, for example, the following advantages of renting, instead of buying, a place to live:

1. It is generally easier for people who rent to move quickly.
2. No down payment is involved, although a security deposit will be required to cover any breakage or cleaning costs.
3. Renters don't face the maintenance tasks and expenses that accompany home ownership.
4. The exact cost of housing services can be calculated easily for the period of the lease.
5. Future housing needs in terms of family size do not have to be estimated carefully.
6. Renters often have the use of common facilities—such as a swimming pool, a tennis court, or a sauna—that would be expensive to purchase when buying a home.
7. Renting offers new members of a community an opportunity to become familiar with an area before investing in a house.
8. There is no loss of interest on investment of savings.
9. There is no liability of ownership.
10. Renting may be the only affordable alternative for many consumers.

## Advantages of Buying

There are also, of course, important advantages to buying a housing unit:

1. You can remodel your home or make it into anything you want.
2. A home offers an investment option that historically has been a good hedge against inflation.
3. Home ownership causes you to save because part of your monthly payments creates equity interest in the housing unit.
4. Owning a home gives you the benefits of being able to deduct interest payments and property taxes from your income before paying taxes—although how much of a saving this will represent depends on your tax rate.
5. Owning a home provides people with an asset that may be borrowed against, making it easier for them to obtain credit for other purposes.

## Tax Benefits of Home Ownership

The tax benefits of home ownership are often a compelling factor in the decision of whether to rent or buy. If you buy a house and borrow the money to pay for it, all the interest you pay to the bank or mortgage company can be deducted from your income before you pay taxes. For most consumers, this represents a sizable savings.

For example, suppose you buy a $120,000 house and borrow $100,000 to help pay for it. Let's say that the interest you pay every year on that $100,000 comes to $7,000. You can deduct that $7,000 from your income before you pay taxes on it. If you are in the 28 percent tax bracket, you get a tax saving of $1,960; the interest on your loan is, in effect, costing you only $5,940. This is why, as people get into higher income tax brackets, it generally is more advantageous to buy a house instead of renting.

Taxes and housing are related in another way. Imagine that you bought or built a house for $100,000 that you and your family lived in for ten years. Then your employer transfers you to another city, so you are forced to sell your house. Because real estate values have increased, you earn a capital gain when you sell your home for $160,000. Unlike other types of capital gains, those earned on owner-occupied homes are not taxed until they exceed $500,000. Whether you save your $60,000 capital gain, use it to help buy a new house, or spend it for something else, all of this profit will be yours to use as you see fit.

An additional tax benefit may be obtained if you qualify for a *home-equity loan*, which is a second mortgage. In a home-equity loan, your equity interest in your home (what you have paid toward its purchase, aside from interest, plus any appreciation in its market value since you purchased it) is given as security for the loan. Under the Tax Reform Act of 1986, the interest you pay on a home-equity loan is fully tax deductible—in contrast to the interest paid on other types of consumer loans, which is not.

## How Much Can You Afford?

Before you go house hunting, you should determine how much you can pay. There is no point in looking at homes that cost more than you can afford. Remember, it's easy to get carried away when buying housing services. A nice house is something in which you and your family can take pride. Owning a home can also—although often unfairly—make you appear to others to be a more responsible and committed member of the community. The tax and other advantages of home ownership discussed earlier can also be persuasive. But, as mentioned earlier, although your budget shouldn't be the sole determining factor in deciding whether to purchase a home, it is a very important one.

**INITIAL COSTS**   Unless you are among the fortunate few, you will need to finance your home purchase by obtaining a **mortgage,** which is a loan for purchasing property in which the property is pledged as security. Only rarely, however, will the mortgage cover 100 percent of the purchase price. Most often 10 to 20 percent of the purchase price will be required as a down payment, and you will be expected to pay this at the time of purchase. In addition to your down payment, you will need to pay *closing costs*. These costs, which we discuss later in the chapter, usually range from 3 to 6 percent of the home's purchase price. In sum, if you want to buy a $100,000 house and finance it with a mortgage loan, you will need not only a good credit rating and a dependable source of income that is great enough to qualify for the loan, but also as much as $25,000 to cover the initial costs.

**Question for Thought & Discussion #2:**
Why are the tax advantages of home ownership less important to most young people than to older people?

**Mortgage**   A loan obtained for the purpose of purchasing land or buildings, in which the property is pledged as security.

### TABLE 12.1    Checklist for Calculating Total Housing Expenses

| Housing Expenses per Month | Current | Future |
|---|---|---|
| Mortgage payments | _____ | _____ |
| Property taxes | _____ | _____ |
| Insurance | _____ | _____ |
| Heating oil, gas, or wood | _____ | _____ |
| Gas, electricity, water, telephone, sewage service | _____ | _____ |
| Yard care, trash pickup, and so on | _____ | _____ |
| Savings fund for repairs, remodeling, and maintenance | _____ | _____ |
| Other | _____ | _____ |
| **Total** | _____ | _____ |

**MONTHLY COSTS**   The checklist in Table 12.1 can help you estimate the monthly expenses you will pay as a homeowner. You should expect your largest monthly expense to be your mortgage payment. The size of this payment will depend both on the amount of the loan and on the interest rate charged by the lender. Table 12.2 shows what your monthly payment would be for different amounts at different interest rates. For amounts above $100,000, add to the payment the amount of the payment corresponding to the additional amount. For example, for a $120,000 loan at 8 percent interest, your monthly payment would be $733.77 plus $146.75, or $880.52.

Some lenders pay taxes and insurance for the buyer. These costs are then included in the monthly payment. The payments given in Table 12.2 do not include taxes and insurance, so you will need to record them separately on your list of estimated monthly expenses. Taxes vary considerably, depending on the location of your home and other factors. The best way to anticipate what your taxes will be is to check either with the previous homeowner or with the local

### TABLE 12.2    Monthly Mortgage Payment

| Mortgage Loan (for 30 Years) | Interest Rate | | | | |
|---|---|---|---|---|---|
| | 6% | 7% | 8% | 9% | 10% |
| $  10,000 | $ 59.95 | $ 66.53 | $ 73.38 | $ 80.46 | $ 87.76 |
| 20,000 | 119.91 | 133.06 | 146.75 | 160.92 | 175.51 |
| 30,000 | 179.87 | 199.59 | 220.13 | 241.39 | 263.28 |
| 40,000 | 239.82 | 266.12 | 293.51 | 321.85 | 351.03 |
| 50,000 | 299.78 | 332.65 | 366.89 | 402.31 | 438.79 |
| 60,000 | 359.73 | 399.18 | 440.26 | 482.77 | 526.55 |
| 70,000 | 419.69 | 465.71 | 513.64 | 563.23 | 614.31 |
| 80,000 | 479.64 | 582.24 | 587.02 | 643.70 | 702.06 |
| 90,000 | 539.60 | 598.77 | 660.39 | 724.16 | 789.82 |
| 100,000 | 599.55 | 655.30 | 733.77 | 804.62 | 877.58 |

county assessor's office. To help determine what your insurance costs might be, refer to the discussion of homeowners' insurance in Chapter 17.

**OTHER COSTS TO CONSIDER** Several rules of thumb apply when buying a home. One is that you should spend no more than one-third of your household income each month on housing costs. Another is that the maximum amount you should pay for a house is two and one-half times your annual household income. Housing prices vary dramatically from region to region, though, so you will need to decide on the basis of your income and the housing market in your area what amount you can and should pay.

When estimating what it will cost to own a home, be sure to allow a wide margin (at least 2 percent of the purchase price per year) for repairs and maintenance. Houses have a tendency to fall apart just like anything else, and homeowners commonly face many unanticipated, and often expensive, repairs. The wage inflation of people who come to fix your sprinkler system, clogged drain, leaking roof, or broken furnace is among the highest in the nation. In addition to repairs, you will incur routine maintenance costs—for lawn and garden upkeep, for example, or for repainting the exterior of your house, its interior, or both.

# If You Decide to Buy

Part of the American dream is home ownership. Let us assume that you share that dream and are planning to buy a house. Let us further assume that you have determined how much you can afford to spend for a home, know the city or area in which you wish to locate, and have an idea of the approximate size and style of home you wish to buy. The next step, then, is to locate the specific house that suits you and your housing needs. How do you do this?

Looking through the classified ads in the newspaper is one way. These ads offer free information and can give you some general ideas of what you can obtain at what price. And, should you decide to buy a house offered for sale directly by the current owner, you could save money because the seller does not have to pay a 6 to 7 percent commission to a real estate broker for selling services—a cost that is passed on to you in the form of a higher purchase price. Frequently, however, the ads you see in the paper are placed by real estate agencies, so it may or may not pay you to spend hours and hours searching for information through the classified ads.

## Looking on the Internet

In the past few years, the Internet has become a basic tool for investigating housing choices. Thousands of real estate firms list homes for sale on their Web sites. These listing generally include photographs and may take you on a virtual guided tour of the dwelling. The advantages of this method of searching are clear. Internet housing searches allow you to complete a basic investigation of hundreds of houses without leaving your home. You don't even need to be in the same city. In most cases, listings are organized in a variety of ways including location, size, price, and special features such as swimming pools, air conditioning, or school system. To find useful Web sites, complete an Internet search using the name of the community you are interested in and the key words *real estate* and *listings*.

> **"When estimating what it will cost to own a home, be sure to allow a wide margin (at least 2 percent of the purchase price per year) for repairs and maintenance."**

**Dual Agency** When an agent represents both the buyer and the seller in a transaction.

## What Does a Real Estate Agent Do?

Essentially, real estate agents are information brokers. They provide buyers and sellers of houses with information. Information, remember, is a costly resource. This is particularly true with such a nonstandard product as a house. Every house is different from every other one, and it is difficult to get buyers and sellers together for such nonstandard products. Generally, for standard products, or even for nonstandard products that do not cost very much, there are no agents. Because a house is often the largest purchase a family will make, consumers who are considering buying a house are usually willing to spend more time and money gathering information. A real estate agent can save information costs by helping to match the wants of buyers with the products offered by sellers.

A real estate agent can help in other ways. The agent generally knows the area well and can furnish information on schools, sources of financing, repair services, and closing costs that you wouldn't be able to find easily by yourself. Since some agents are better informed than others, it is usually worthwhile to shop around for the right one. After an agent has shown you a few houses, you will sense how serious the broker is about helping you and whether she or he understands your tastes and preferences in housing.

## For Whom Does a Real Estate Agent Work?

There is a tradition in the U.S. economy that an agent should not represent both parties in a transaction or dispute. For example, if you were being sued by another person, you would never consider hiring the plaintiff's lawyer to represent you, too. You would recognize that this would create a conflict of interest, so you would seek assistance from a different attorney. This tradition, however, is often ignored when it comes to real estate sales, where one agent frequently represents both parties in a transaction. This practice is so common that many states have laws that recognize and regulate **dual agency.** New York, for example, passed a law in 1995 that requires real estate agents to inform buyers and seller when they represent both parties in a transaction. This situation leaves the consumer asking this reasonable question: How do I know if my agent is really working to obtain the best possible agreement for me?

**A REAL ESTATE AGENT'S DUTIES** Agents in all financial situations are supposed to provide six duties to their customers: (1) confidentiality, (2) obedience, (3) undivided loyalty, (4) full disclosure, (5) reasonable care, and (6) accounting for property received. When one person represents both sides in a transaction, it is doubtful that these duties can be fulfilled equally for the buyer and the seller.

On the surface it appears that the way real estate agents are compensated encourages them

Why are more than 90 percent of all homes sold through real estate agencies? What do agents do to earn their commissions?

to work more for the seller than the buyer. The commission, after all, is paid directly by the seller. At the same time, it is part of the purchase price that is paid by the buyer. It is possible, therefore, that the agent will be *most* interested in seeing a transaction take place to earn a commission regardless of whether the final deal is in the best interest of either the buyer or the seller.

**SHOULD BUYERS HIRE AGENTS, TOO?**  For these reasons, many consumer organizations recommend that buyers hire agents who represent only their side in a real estate purchase. This will increase the total amount of fees associated with a transaction because the buyer and seller will both pay agents. Nevertheless, having your own agent should increase the probability that you will receive good advice and end up buying a home you really want to own at a fair price. You may expect such a person to review a wider selection of possible houses, identify problems with a house that need to be repaired, and negotiate the best possible price. If you do choose to hire your own agent, be sure this person does not work for the same agency that listed the property in the first place. There is no reason to believe that such an agent would not communicate any information you provide to the representative of the seller.

If you do complete a real estate transaction using the same agent as the seller, you should realize that this person is not "on your side." Never tell such an agent how much you would be willing to pay for a house or that you are "in love with it." This type of information would almost certainly be used against you and result in your paying more for a house than is necessary.

## How to Bargain

Most Americans are unaccustomed to bargaining because most goods and services are sold at set prices. But buying a house is different. The asking price generally is not the final sale price. If you are unaccustomed to bargaining for a house or feel uncomfortable doing it, you can let a real estate agent do it for you.

You have a right to be shown the price that was paid for any property that has previously been sold. Records of all real estate transactions are public information. When a sale is completed, a record of the transaction is made at an office of your local government, most often the town or city clerk's office. You, or any other interested party, may ask to see this record. In this way, you can learn how much profit the owner would like to make from the sale and how much time has passed since the property was last sold. Although there will be no record for a house built by its owner, the office will have a record of the property's assessed valuation that you may see. Checking this will enable you to compare the house with other properties in the same neighborhood that have similar assessments and have recently been sold. Such comparisons can help you determine if you are being asked to pay more or less than the current market value for homes in an area that interests you. Remember, you have a legal right to be supplied with basic information that can help you determine what a property is worth.

Many times, sellers do not expect to get the prices they are asking on their houses. They set a price that they think may be, say, 5 or 10 percent more than the price they will finally receive. It is up to you to find out how far they will go in discounting that list price. You can start by asking the real estate agent whether he or she thinks the price is "firm." Because the broker's commission

**Question for Thought & Discussion #3:**
Would you pay an agent several thousand dollars to represent you when you buy a house? Could you make a good deal on your own?

is a percentage of the sale price, the higher the price, the more the agent benefits, but not if it means waiting months or years for a sale. The broker's desire to get that commission as soon as possible is an incentive to arrange a mutually agreeable price so that a deal will be made. For a house with a $115,000 list price, for example, you may want to offer $100,000 on condition that you receive the refrigerator, freezer, washer, and dryer that are already in the house. This sort of bargaining happens all the time. You should not accept the list price just because you think you want the house. Before you make an offer, though, certain basic common-sense precautions must be taken.

## Before You Sign Anything

Before you sign anything, make sure you are getting a house that is structurally sound. The list of questions in Figure 12.1 indicates specifically what you want to know. You can pay an expert to go over everything in the house that could cause problems—wiring, frame, plumbing, sewage system, and so

**Figure 12.1**
**Checklist for Inspecting a House**

**Basement**
Has the basement leaked within the past two years? _____
What basement waterproofing repairs have been made within the past two years? _____

**Roof**
What type of roof is it? _____
How old is it? _____
Has the roof leaked within the past two years? _____
What roof repairs have been made within the past two years? _____

**Plumbing**
Has the plumbing system backed up within the past two years? _____
What repairs or servicing has been done within the past two years? _____
Does the house have a septic system, or is it attached to a sewer? _____

**Heating**
How old is the heating system? _____
Is heat provided to all finished rooms? _____
Is it: ❑ Gas  ❑ Electric  ❑ Oil  ❑ Other?  When was it last repaired or serviced? _____

**Air Conditioning**
Is there a central air-conditioning system? _____
If yes, is it provided to all finished rooms? _____
Comments: _____
How old is it? _____ Is it: ❑ Gas  ❑ Electric?
When was the last time it was repaired or serviced? _____

**Utilities**
How much is the average monthly bill for: Electricity $ _____, Gas $ _____, Oil $ _____?
Other than the central heating and air-conditioning systems, are any appliances, fans, motors, pumps, light fixtures, or electrical outlets in need of repair? _____

**Fireplace**
Is the fireplace in working condition? _____
When was it last repaired, serviced, or cleaned? _____
Comments: _____

**Windows**
Are there any storm windows or screens on the premises that are not installed? _____
Specifics: _____

**Figure 12.1**

**Checklist for Inspecting a House, Continued**

**Floors**

What type of floor is under areas covered by wall-to-wall carpeting?
❑ Finished hardwood   ❑ Plywood   ❑ Other
Is flooring material the same for all covered areas? _____
Specifics: _____

**Gutters**

Do gutters and downspouts need any repairs other than routine maintenance? _____
Specifics: _____

**Hot-Water Heater**

How old is the hot-water heater? _____ What is its capacity in gallons? _____
When was it last serviced or repaired? _____

**Rainwater**

Does the property ever have standing water in front, rear, or side yards more than forty-eight hours after a heavy rain? _____
Location: _____

**Electrical**

Do fuses blow or circuit breakers trip when two or more appliances are used at the same time? _____
Specifics: _____

**Locks**

Are all outer door locks in working condition? _____
Will keys be provided for each lock? _____

**Insulation**

Is there insulation in: the ceiling or attic? _____
the walls? _____
other places? _____
Notes: _____
_____
_____
_____
_____
_____
_____
_____
_____
_____
_____

SOURCE: Montgomery County (Md.) Office of Consumer Affairs.

on. The typical cost for an inspection runs from $200 to $500. It is money well invested unless you are an expert at figuring out what can go wrong with the house just by looking at it. Ask your real estate agent about inspectors, or look at the listings under "Building Inspection Service" or "Home Inspection Service" in the Yellow Pages. Again, these companies are selling the same thing a broker is selling—information. Such information can save you hundreds, if not thousands, of dollars in repairs you would later discover had to be made. Often, if a building inspector discovers structural faults in a house, you can have the seller agree to pay for the repairs even after you take over the home. Or, this can be a bargaining point: the price you agreed on can be reduced by the amount of the repair costs. Make sure the purchase contract specifies that the contract is contingent on the results of any inspections you want to have made and that any arrangements you make with the seller concerning necessary repairs are also written into the contract.

**Question for Thought & Discussion #4:**
What qualities would you emphasize if you were selling the building where you live? What would you hope the buyer would fail to notice?

**Earnest Money** Sometimes called a *deposit on a contract* or an *offer* to purchase a house. It is the amount of money you put up to show that you are serious about the offer you are making to buy a house. Generally, you sign an earnest agreement, or a contract that specifies the purchase price you are willing to pay for the house in question. If the owner selling the house signs, then you are committed to purchase the house; if you back down, you could lose the entire earnest money.

**Title** The physical representation of your legal ownership of a house. The title is sometimes called the *deed*.

**Abstract** A short history of title to land; a document listing all records relating to a given parcel of land.

**Title Insurance** Insurance that you pay for when you buy a house so you can be assured that the title, or legal ownership, to the house is free and clear.

# Making the Offer

Generally, when you have decided to buy a house, you make an offer and put it in writing. You also must put up **earnest money,** or deposit binder money. The earnest agreement or binder states in detail your exact offering price for the house and lists any other things that normally are not included with a house but are to be included in this deal, such as a washer or dryer. Within a specified time period, the seller of the house either accepts or rejects the earnest agreement or binder. If the seller accepts and you try to back down, the earnest money you put up, which may be several thousand dollars, is forfeited. But sometimes you can get it back even when you decide against the house after signing the agreement. In any earnest agreement, it is often wise to add an escape clause if you are unsure about getting financing. Put in a statement such as, "This earnest agreement is contingent on the buyer's obtaining financing from a bank for [X thousand] dollars." Remember, the earnest agreement offer is *your* proposal. Put in what *you* want. Let the seller change it; then you review it.

If the earnest or binder agreement is accepted, then a contract of sale is drawn up. Usually, the signing of a contract of sale is accompanied by a deposit, which may be from 2 to 5 percent of the purchase price paid to the seller. Often, the buyer merely adds to the existing earnest money to bring it up to the desired amount.

The deposit may be put into an escrow account or a trusteed savings account that earns interest from the time it is paid to the seller to the time the buyer takes possession of the house. When any substantial sum is involved, it is, of course, advantageous to the buyer to have the deposit put into a trusteed savings account with the interest accruing to the buyer rather than the seller. This is particularly important if there will be a long time between the signing of the conditional sales contract and the actual date of possession of the house.

# Title Examination and Insurance

After the sales contract has been negotiated, the buyer or buyer's attorney (or the broker, escrow agent, or title insurance company) will begin the **title** examination (or "search"), which entails scrutinizing the history of all past transfers and sales of the piece of property in question. Every county has a filing system in which deeds, plats, and other instruments are recorded.

The title examiner will generally obtain an **abstract** from a private abstract company. This document lists all the records relating to a particular parcel of land. After reading the abstract, the examiner will give an opinion as to the validity of the title.

On the basis of this examination, a **title insurance** policy can be drawn up guaranteeing that if any defects arise in the title, the title company itself will defend for the owner and pay all legal expenses involved. Note that this may sound better than it actually is. Title insurance generally does not cover government actions that could restrict use of ownership of the property you just bought. Often, title insurance excludes mechanics' liens not recorded with the proper official agency when the policy was issued. In other words, if work was done on the house and not paid for by the former owner, it is possible, even with the title insurance, that you could end up paying for that work. You should always ask the seller for a copy of paid bills for any obviously recent repairs or additions to the home.

The title search and insurance are important parts of the home-buying process. When you purchase something as expensive as a house, you must be sure that you really own it and that no one with a prior claim can dispute your title to the land and structure.

Another type of insurance many new homeowners either buy or are given is a **home warranty.** This is a policy that promises to pay for repairs or defects in construction in a home over a specific period of time (most often five years). Although such a policy can be of great value if, for example, your new home settles unevenly, causing cracks to appear in its foundation, you should check out the financial condition of the firm that offers the warranty. In the 1990s, a number of these businesses failed, leaving thousands of consumers with no coverage at all. For example, in 1994 the Home Owners Warranty Corporation of Arlington, Virginia, had a capital shortage of $45 million and was placed into receivership by state authorities. It was estimated that as many as 1.7 million homeowners were affected by this firm's failure, leaving them with limited or no protection if their homes developed defects.

# Different Types of Homes

By far the most popular type of residence in the United States (and also the most expensive in terms of energy and maintenance requirements) is the single-family, detached dwelling. But numerous multifamily dwellings are also available to rent or to buy. For home buyers, the three most significant alternatives to the single-family, detached dwelling are condominiums, townhouses, and cooperatives.

## Condominiums

A **condominium** is similar to an apartment, but the apartment dweller has the legal title to the unit that he or she owns. A condominium owner, however:

1. Has a joint ownership interest in the common areas and facilities in the building, such as swimming pools and tennis courts.
2. Must arrange for his or her own mortgage and pay taxes individually on the unit.
3. Must make separate payments for building maintenance and services.
4. Does not accept financial responsibility for other people's units or their share of the overall operating expenses.
5. Votes to elect a board of managers that supervises the property.
6. Has the right to refinance, sell, or remodel his or her own unit.

In many situations, owners who want to sell their condominium apartments are under fewer restrictions than are owners of cooperative units (discussed later). With a condominium, if the owner of a unit defaults on a payment, it affects only the owner's own mortgage. In the case of a cooperative unit, if any owner defaults, the other co-op members must chip in to cover the amount that has been defaulted. Also, condominium owners usually are free to rent or lease their units to others.

## Townhouses

A **townhouse** is a regular house with a yard or garden area but with common side walls. The obvious advantage of a townhouse is economy, for its construction permits savings on the cost of land, insulation, windows, foundation, roof,

**Home Warranty** A form of insurance that protects a home buyer from the cost of major repairs that result from defects in construction.

**Condominium** An apartment house or complex in which each living unit is individually owned. Each owner receives a deed allowing her or him to sell, mortgage, or exchange the unit independent of the owners of the other units in the building. Title to a condominium also gives the purchaser shared ownership rights in common areas.

**Townhouse** A house that shares common side walls with other, similar houses.

A townhouse can be a more economical choice than a single-family house.

and walls. Some townhouses are sold as condominiums. One disadvantage of such housing units may be lack of adequate soundproofing as a result of shared walls and the proximity of neighbors.

## Cooperatives

In a building of **cooperative** apartments, each dweller owns a **pro rata** (proportionate) share of a nonprofit corporation that holds a legal right to that building. In addition, a member of a cooperative:

1. Leases the individual unit she or he occupies.
2. Accepts financial responsibility for her or his own payments and, in addition, accepts responsibility for increases in assessments if other members fail to make their payments.
3. Pays a monthly assessment to cover maintenance, service, taxes, and mortgage for the entire building.
4. Votes to elect a board of directors.
5. Must obtain approval from the corporation before remodeling, renting, or changing her or his unit or selling her or his share in the corporation.

Cooperative housing grew slowly until 1950, when Congress passed legislation that allowed the Federal Housing Administration (FHA) to insure the mortgages of cooperative housing units. Although much cooperative housing has been produced for middle-income families, today co-ops are more popular among higher-income families in large eastern cities, particularly New York, Boston, and Washington, D.C.

Co-ops themselves are nonprofit organizations and are, therefore, owned and operated solely for the benefit of the members. The FHA estimates that the cost of living in a cooperative apartment is about 20 percent less than renting a comparable apartment from a private landlord. In addition, co-op owners reap the various tax advantages of owning a home. Basically, all local taxes and interest on the mortgage for the prorated share in the cooperative are deductible from income before taxes are paid. Because this benefit is not directly available to renters, it is one reason a number of people prefer to join a cooperative instead of renting an apartment, even though an apartment in a cooperative building looks the same and gives the same types of housing services. These tax advantages are also available to owners of condominiums.

## Paying for Your Home

More than 90 percent of all people who buy homes do so with a mortgage loan, which, as mentioned earlier, is a loan that a bank or trust company makes on a house. In some states, you hold the title to the house; in others,

**Cooperative**  An apartment building or complex in which each owner owns a proportionate share of a nonprofit corporation that holds title or a legal right to use the building.

**Pro Rata**  Proportionately; that is, according to some exactly calculable factor.

## THE diverse CONSUMER

### Manufactured Housing

Many Americans who want to own their own homes cannot afford to purchase traditional single-unit housing. For a number of these people, manufactured housing units offer an attractive alternative. At one time most people equated manufactured housing with dilapidated trailer parks. Although this is sometimes still the case, numerous manufactured housing communities are now attractive, well landscaped, and desirable to many people. The great advantage to buying into one of these communities is the relatively low purchase price. In 2002, a double-wide unit with 800 square feet of living space could be purchased for as little as $35,000 in many parts of the United States.

It should come as no surprise that a lot of the people who purchase manufactured housing units have relatively low incomes—many are elderly, some are poorly educated. Unfortunately, some businesses that offer these housing units have exploited unsuspecting consumers in a variety of ways.

The interest rates charged on loans that are offered to people who invest in manufactured housing units are often much higher than those charged on traditional mortgages. Interest rate premiums of 2 to 3 percent are not uncommon.

A more serious problem occurs when artificially low rates seem to be offered. A customer might be told that he or she can borrow $40,000 to buy a manufactured home and pay only 5 percent interest. The problem is that the loan will last only for a certain number of years, after which it must be renegotiated. So after, say, five years, the mortgage is up for review. This time the lending organization demands 10 percent interest. The owner balks at this demand and looks to other lenders for a lower rate. But there is a problem. The property that was purchased for $40,000 is really worth only $30,000. No other bank will lend enough money to pay off the old loan. The original lender becomes the only lender available. The consumer has been victimized, and little can be done about it. This story may sound far-fetched, but it, and other scams, have repeatedly been perpetrated on less affluent Americans.

If you, or anyone you know, ever considers purchasing manufactured housing, be sure to investigate the true value of what is being offered. Most businesses that market these homes are honest. Unfortunately, a few are not.

**Would you consider buying a manufactured housing unit? Why or why not?**

---

the mortgagee does. In several states, a special arrangement is made whereby the borrower (mortgagor) deeds the property to a trustee—a third party—on behalf of the lender (mortgagee). The trustee then deeds the property back to the borrower when the loan is repaid. If the payments are not made, the trustee can deed the property to the lender or dispose of it by auction, depending on state law. As the mortgagor, you make payments on the mortgage until it is paid off.

**Question for Thought & Discussion #5:**
Why are mortgage interest rates usually much lower than the rates charged for other types of consumer loans?

## Sources of Mortgages

Most mortgage loans are made by savings and loan associations, but there are many other sources for mortgages as well. These include mortgage companies, commercial banks, and mutual savings banks, as well as special institutions that, in certain circumstances, may give you a mortgage loan—such as pension funds, mortgage pools, insurance companies, mortgage investment trusts, and state and local credit agencies. Most mortgage lenders maintain Web sites on the Internet.

## Comparison Shopping

Given the variety of sources for mortgage loans, the wise consumer will want to shop around for the best deal available. One way to do this is to check with a mortgage-reporting service. If one is available near you, the local mortgage officer at your bank or at other lending institutions in your

**Discount Points**  Additional charges added to a mortgage that effectively raise the rate of interest you pay.

**Prepayment Privileges**  In a mortgage loan with prepayment privileges, you can prepay your loan before the maturity date and not have to pay a penalty.

**Amortization Schedule**  A table showing the amount of monthly payments due on a long-term loan, such as a mortgage; it indicates the exact amounts going toward interest and toward principal.

community will know of it. The local real estate board or perhaps your broker, if you are using the services of one, may also be able to refer you to this kind of service.

There are also many online mortgage companies. Finding a potential lender is as easy as carrying out an Internet search using the key words *home* and *mortgage.* A review of several Web sites should quickly give you an idea of the rates that are offered, the amount of down payment you would be required to make, and how quickly you may obtain your loan. If nothing else, online mortgage firms can give you a basis for comparing offers made by local lenders. If you can obtain a mortgage at 7 percent online, you might be able to negotiate a better rate with a local bank than the 7.3 percent you have been quoted. A problem with online lending of any type is that you never actually sit down and talk with a representative of the lending firm. Consequently, it is important to investigate any lending organization that you are seriously considering. Check with your state's attorney general's office and with the Better Business Bureau to see if any complaints have been filed. Another way to check out an online lender is to search the Internet using its name. There may well be reports about it online that are less than complimentary. You may even find a chat room where consumers exchange stories about their experiences with different lenders. Remember, anyone can put anything on the Internet. Just because you read something on the Web does not necessarily mean that it is true.

## Mortgage Talk

When you shop for a mortgage, you should know the language of the mortgage trade. Some terms that you will want to understand are *discount points, prepayment privileges, prepayment penalty,* and *amortization schedule.*

**DISCOUNT POINTS**  Sometimes obtaining a mortgage involves paying **discount points.** This is a device to raise the effective interest rate you pay on a mortgage. A point is a charge of 1 percent of the amount of a loan. Basically, discount points amount to prepaid interest. This charge may be assessed against the buyer, the seller, or both. To see how a discount-point system works, suppose that you have to pay four discount points on a $100,000 loan; that means that you get a loan of $100,000 minus 4 percent of $100,000, making the loan only $96,000. You pay interest on the full $100,000, however. Some states have laws that limit discount points, and the Federal Housing Administration and Veterans Administration have restrictions on buyers paying points, so they are charged to the seller. But ultimately they tend to be passed on to the borrower in the form of a higher price.

**PREPAYMENT PRIVILEGES**  Most mortgages today include **prepayment privileges** that allow borrowers to prepay their mortgages before the maturity date without penalty. You might want to do this later on if interest rates in the economy fall below what you are actually paying. You could pay the mortgage off by refinancing it at a lower interest rate. Charging prepayment penalties is against the law in many states.

**AMORTIZATION SCHEDULE**  In calculating the cost of credit for a large loan, such as a mortgage, you will need an **amortization schedule,** which will show you what portions of each monthly payment go to the interest and to the principal on your loan. Lenders usually provide mortgagors with such schedules; if they don't, ask for one. For the first several years of your mortgage loan, you will be paying primarily interest on your mortgage, so the amount you owe on the principal of the loan declines very slowly. But it does decline, and, as it does, the monthly interest is reduced accordingly. By the end of your payment schedule, your payment will be mostly on the principal.

# Different Kinds of Mortgages

Numerous types of mortgages are now available. Although you may not be eligible for all of them, most are available from the same sources: commercial banks, savings banks, mortgage banks, savings and loan associations, and insurance companies.

## Conventional (Fixed-Rate) Mortgage

With a conventional mortgage loan, the money the lender risks is secured only by the value of the mortgaged property and the financial integrity of the borrower. To protect its investment from the start, the lender, such as a savings and loan association, ordinarily requires a down payment of anywhere from 5 to 25 percent of the value of the property, depending on market conditions. Some private insurers will protect lenders against loss on at least a certain portion of the loan. When such extra security is provided, the lender may go to a higher loan figure. The borrower, of course, pays the cost of the insurance.

Most conventional mortgages are given at a fixed rate of interest—which is determined by conditions in the credit market—and repaid over a period of 15, 20, 25, or 30 years. If you expect the interest rate to rise in the future, a fixed-rate mortgage will be advantageous to you because the rate of interest you pay will not change over the period of the loan. If, however, interest rates are high when you are looking for a mortgage, other options may be more favorable to your pocketbook. By making a very large down payment to lower the risk for the lender, you may get a slightly lower interest rate, perhaps a fraction of a percent below the prevailing local rate.

Conventional loans can be arranged on just about any terms satisfactory to both parties. Different lenders favor different arrangements, so it pays to shop around. And because most borrowers pay off their mortgages well before maturity, make sure you will not have to pay a penalty on prepayment; as noted earlier, most mortgages today allow prepayment privileges.

## Adjustable-Rate Mortgage

An adjustable-rate mortgage (ARM) loan carries an interest rate that may change over time. The rate is pegged to interest rate changes in the financial market, as determined by the movement of any public index of interest rates that can be checked by borrowers and that is not under the control of lenders. ARMs allow lenders to keep their flow of funds in step with changing conditions, and this, in turn, could make home loans easier to come by when money is tight. You will get a fractionally lower interest rate at first or be offered other inducements by lenders to make future uncertainties more

> "Conventional loans can be arranged on just about any terms satisfactory to both parties."

**Question for Thought & Discussion #6:**
In 2001, more than 90 percent of new mortgages came with fixed interest rates that would not change over the term of the loan. Why did so many consumers make this choice?

palatable. If you consider obtaining an ARM, you should be sure to ask about its *yearly cap* (how much the interest rate can be increased in a year), *lifetime cap* (the maximum to which the interest rate can ever be raised), and how often the interest rate may be adjusted in each year.

Although ARMs accounted for less than 10 percent of new mortgages in 2001, they made up 60 percent of them in the late 1980s, probably because of the high interest rates that were charged on conventional mortgages at that time.

## Graduated-Payment Mortgage

Although graduated-payment mortgages have become rare, they are still offered by some lenders. These are flexible mortgages that allow borrowers to tailor their monthly payments to their income over time. Payments are arranged so that they start out low and increase later on, perhaps in a series of steps at specified intervals. The term of the loan and the interest rate remain unchanged.

This type of mortgage benefits mainly first-time home buyers who look forward to higher earnings in the future and the ability to afford higher payments at a later date. One disadvantage of the graduated-payment mortgage is the possible "negative amortization" in the early years, which means that, for a time, your debt grows instead of diminishing.

## Rollover Mortgage

In a rollover mortgage, the rate of interest and the size of the monthly payment are fixed, but the whole loan—including principal, rate of interest, and term—is renegotiated, or "rolled over," at stated intervals, usually every three to five years. If interest rates go up, you can expect to be charged more when you renegotiate, but you'll also have the opportunity to adjust other aspects of the loan—such as the term and principal. Or, you can pay off the outstanding balance without penalty. Renegotiation is guaranteed.

Obviously, lenders benefit from rollover mortgages, as they do from ARMs. Benefits to the borrower are the same as those obtained from ARMs, with this plus: periodic renegotiation gives you a chance to rearrange the loan to suit your changing needs without all the expense of refinancing.

## VA Mortgages

Veterans Administration (VA) mortgages (also known as GI loans) are available to qualified veterans who have served in the U.S. armed forces or to their surviving spouses. In general, a person must have served on active duty in the military for two years. Some other special situations also qualify a person for a VA loan; these can be investigated by visiting the Veterans Administration's Web site at **http://www.va.gov**.

The interest rate charged on VA loans is set by the Veterans Administration and does not fluctuate as quickly as other interest rates. A VA loan is not actually a loan from the government. It is a guarantee that if the borrower does not repay the loan, the government will pay at least 60 percent of the amount borrowed. This reduces the risk of lending and makes banks and other financial institutions willing to make loans to people who would not otherwise qualify.

It is possible to take out a VA loan with little or nothing down, and amounts up to $240,000 (in 2003) could be borrowed for as long as thirty years. To buy a house with a VA loan, the house must first be appraised. The VA will not guarantee a loan for more than the appraised value of the home. All VA loans can be prepaid without penalty. Second mortgages and mobile-home loans may also be taken out under the VA program.

## FHA Mortgages

The Federal Housing Administration (FHA) guarantees payment of the entire amount borrowed under an FHA loan. These loans are available to first-time home buyers who meet minimal financial criteria. The added security of this government guarantee causes banks and other financial institutions to extend credit to people who would not otherwise qualify for a loan. In 2003, FHA loans of up to $280,000 could be obtained with as little as 3 percent down.

Before you sign any mortgage agreement be sure you understand what you are signing and that it will require you to make payments that fit within your budget.

Borrowers apply for FHA–insured mortgages in the same way they would apply for a conventional mortgage. They complete a loan application at a financial institution. The bank will then notify the FHA of the application, and the property will be appraised. Like the VA, the FHA will not guarantee a loan for more than the appraised value of the property. FHA interest rates are set by the secretary of the Department of Housing and Urban Development (HUD). Although no rule determines the level at which the rates will be set, they have historically been close to prevailing interest rates charged on conventional mortgages.

The FHA also sponsors a subsidy program for low- and moderate-income families. In this program, down payments can be as low as several hundred dollars, and interest rates can be from 2 to 5 percent. Through the FHA loan guarantee program, millions of lower-income Americans now own homes. You may investigate FHA loans at the Department of Housing and Urban Development's Web site, **http://www.hud.gov**.

# Creative Financing

The term *creative financing* describes loans either provided by a source other than the traditional lenders or with contract features that make them more attractive to mortgagors. Table 12.3 on the next page describes some alternative methods of financing a home.

Because interest rates on existing mortgages are sometimes lower than current interest rates, *mortgage assumptions* can at times be attractive to potential buyers. Note, however, that usually two problems are associated with assuming an existing mortgage:

1. FHA and VA loans can be assumed at the same rate by a new borrower, but conventional loans may contain a *due-on-sale clause,* which states that the balance of the loan must be paid when the home is sold.
2. Even if the existing loan can be assumed, it may represent such a small part of the purchase price that the borrower will have to arrange a second mortgage or come up with a larger down payment.

## TABLE 12.3    Creative Financing Has Its Pitfalls

**Land sales contract.** The buyer and seller form a land sales contract (also called *contract for deed* or *installment sales contract*) under which the buyer promises to make monthly payments to the seller until the property is paid for. The buyer does not receive title to the property until the loan is completely paid.

**Lease-purchase option.** The buyer rents the property until he either moves out or exercises his option to buy it. This can be a profitable arrangement if the option price turns out to be less than the market value of the house when the option is exercised. A disadvantage is that lease payments are not tax deductible. And a lease option is not a way to circumvent a due-on-sale clause because most such clauses say the loan must be paid off if the owner gives anyone a lease that contains an option to purchase the property.

**Ground lease.** The buyer buys the house but leases the ground under it. This arrangement requires a lower down payment for a mortgage because the value of the land is not financed. Usually, the lease contains an option to buy within a few years. A ground lease can help make a mortgage affordable, but lease payments aren't tax deductible, and not owning the land may make the house harder to sell.

**Wraparound mortgage.** Suppose the seller has an assumable mortgage with a below-market interest rate: a 5-year-old, $50,000, 7 percent loan now paid down to $47,926 with 25 years left to go. She sells you the house for $100,000, and you make a 20 percent down payment of $20,000. A lender (who could be the seller) gives you a wraparound mortgage in the amount of $80,000 at 10 percent for 25 years.

You make a monthly payment of $784.09 on the $80,000 loan. The lender, in turn, makes the $379.33 payment for you on the original loan, which you have assumed. The lender pockets the $404.76 difference because, in effect, she lent you $32,074 ($80,000 − $47,926).

**Balloon payment.** In a balloon-payment loan contract, a borrower agrees to make a lump-sum payment of the loan balance at the end of a certain period, typically two to ten years. In the meantime, periodic payments are set up as if the loan were going to run for much longer. This arrangement keeps current payments down and gives the borrower an opportunity to sell the property or refinance the loan before the balloon payment comes due.

Perhaps the easiest and least expensive way to finance the purchase of an existing home is by the use of a mortgage assumption, with the seller holding a *second mortgage* (called a *purchase-money mortgage*) for the balance of the purchase price. The terms and conditions of the arrangement can be tailored to fit the financial requirements of both the buyer and the seller. The main obstacle to this type of arrangement is finding a seller who doesn't immediately need to take his or her *equity* in cash from the sale of the residence. Your real estate agent should be able to help you in your search for this type of financing.

# Mortgage Repayment Terms

Traditionally, a thirty-year payment schedule has been the favored term for repaying a mortgage. In recent years, however, many home buyers have been opting for repayment over a shorter period of time to reduce the amount of interest they must pay for the loan.

## The Fifteen-Year Mortgage

If you pay off your mortgage in fifteen years, the amount you pay each month will be, of course, substantially higher. But you can also reduce the interest paid on the loan. For example, assume that you want to borrow $100,000 toward the purchase of a new home. As indicated in Table 12.4, if you select a thirty-year repayment schedule, your interest rate will be 6 percent, your monthly payments will be $599.55, and the total interest will be $115,838. If, however, you repay the loan over a fifteen-year period instead, your monthly payments will be $843.86 and you will pay only $51,894 in interest. In sum, you will reduce the loan interest by $63,944!

| TABLE 12.4 | Repayment Possibilities on a $100,000 Mortgage with a 6 Percent Interest Rate | | | | | |
|---|---|---|---|---|---|---|
| Type of Repayment | Term to Maturity | Total Interest | Regular Payment | Total Payments | Total Interest Saved | |
| **30-year mortgage** | | | | | | |
| Conventional | 30 years | $115,838 | $599.55 | $215,838 | None | |
| Additional $20 per month toward principal | 28 years, 6 months | 104,373 | 619.55 | 204,373 | $11,465 | |
| Additional $100 per month toward principal | 21 years, 10 months | 75,938 | 699.55 | 175,938 | 39,900 | |
| Extra monthly payment each year toward principal | 25 years, 6 months | 92,197 | 599.55 | 192,197 | 23,641 | |
| **Biweekly mortgage** | 25 years, 6 months | 92,197 | 299.78 | 192,197 | 23,641 | |
| **15-year mortgage** | 15 years | 51,894 | 843.86 | 151,894 | 63,944 | |

# Extra Payments

Another way to save on interest is to take out a conventional, thirty-year fixed-rate mortgage but—in addition to making the required monthly payments—make extra payments periodically toward the principal. As Table 12.4 shows, even $20 extra per month will result in an interest reduction of $11,465—and you will pay off the loan one year and six months sooner.

# Biweekly Payments

Some lending institutions offer yet another alternative—biweekly, instead of monthly, payments. With a biweekly repayment schedule, you pay half of your regular monthly payment every other week. As you can see in Table 12.4, by making biweekly payments you can reduce the interest you pay by $23,641 and pay off the loan in 25 years and 6 months instead of 30 years. In essence, what happens is this: because there are 52 weeks in a year, the 26 payments you make add up to more than 12 monthly payments. In effect, you make the equivalent of an extra monthly payment each year. In addition, the more frequent payments have the effect of reducing the principal on which interest is calculated just enough each month to net you further savings.

**Question for Thought & Discussion #7:**
Why do many consumers worry more about the size of their monthly mortgage payment than about the total amount they pay over time?

# Closing Costs

*Closing*—also called *settlement* or *closing escrow*—is the final step in buying your home. An escrow agent is a neutral third party who keeps your deposit and any pertinent documents until the sale is finalized—which happens at the time of closing. At the close of escrow, the escrow agent will give your deposit and loan funds to the seller and have the deed recorded. You will receive a copy of the deed by mail within about thirty days after it has been recorded.

Several costs must be paid, in cash, at the time of closing; these are fees for services, including those performed by the lender, escrow agent, and title company. These costs can range from several hundred to several thousand dollars, depending on your mortgage and other conditions of purchase, and often home buyers are not aware of them until the time of closing. To be prepared for these out-of-pocket expenses, you should try to obtain in advance an estimate of what they will be. In the initial stages of house hunting, ask your Realtor to give you a rough idea of what the closing costs might be.

Once you've applied for a mortgage, federal law requires the lender to send you an estimate of the closing costs within three days. You may be able to negotiate with the seller to pay for some of the closing fees, with the result of your negotiation being included in your purchase agreement. Under the 1974 Real Estate Settlement Procedures Act (RESPA), the lender must do the following when you borrow money to pay for a house:  .

1. The lender must send you, within three business days after you apply for a mortgage loan, a booklet prepared by the U.S. Department of Housing and Urban Development outlining your rights and explaining settlement procedures and costs.

2. The lender must give you, the applicant, within that three-day period an estimate of most of the settlement costs.

3. The lender must clearly identify individuals or firms that you may be required to use for legal or other services, including title insurance and search.

4. If your loan is approved, the lender must provide you with a truth-in-lending statement showing the annual interest rate on the mortgage loan.

5. Lenders, title insurers, and others involved in the real estate transaction cannot pay kickbacks for referrals.

For further information about the RESPA, visit HUD's Web site at **http://www.hud.gov** and search for *RESPA*.

# Confronting Consumer Issues: Renting a Home

**M**any houses, condominiums, apartments, mobile homes, and other dwelling units are rented by their owners to tenants. Most renters obtain a **lease.** This is a long-term contract that binds both landlord and tenant to specific terms. It is usually for one year and generally requires one or two months' rent to be paid in advance and perhaps one month's rent as a cleaning, breakage, or security deposit.

Alternatively, some housing units are rented on a month-to-month basis, with the rent paid in advance; the renter or tenant automatically has the right to live in the unit for the next month. In this type of agreement, the contract may be terminated on one month's written notice by either party. Given proper one-month notice, the rent can be raised at any time, or the tenant can be asked to leave.

There are advantages and disadvantages to this type of short-term arrangement. Renters are free to move when they want without having to break a lease. At the same time, they face the possibility of having their rent increased or losing their place to live on short notice. Monthly rental rates also tend to be somewhat higher for similar units under short-term agreements. Still, if you know you are going to need a place to live for only three months, this may be the way to go.

## FOUR STEPS FOR POTENTIAL RENTERS

If you have decided to rent a place to live, you face at least four concerns:

1. You must identify and investigate the available rental units in the area where you wish to live.

2. You must choose a rental unit that meets your needs and is within your financial capabilities.

3. You must review the proposed rental agreement to make sure it is appropriate.

4. You must record the condition of the rental unit and keep these records to support your rights if you have a disagreement with the landlord when you move out.

## FINDING AVAILABLE RENTAL UNITS

There are five basic sources of information about available rental units:

1. The Internet

2. Advertisements in local newspapers

3. Property management firms and real estate agencies

4. "For Rent" signs posted on houses or apartment buildings

5. Friends and acquaintances

**Sharing an apartment can be a good alternative for renters who are able to share responsibilities as well as costs.**

### Question for Thought & Discussion #8:
Are you, or people you know who rent, satisfied with the rental unit? Explain your answer.

**The Internet**   The easiest and quickest way to review available rental units is almost always the Internet. Every community of any size has one or more businesses that help landlords and potential renters meet and reach rental agreements. All you need to do is complete a search using the key words *rental housing* and the name of the community where you wish to live. You will probably find listings organized by location, size, price, and type of unit. There is a good chance that photographs of the units will also be posted online. You should be able to quickly "weed out" units that clearly do not meet your needs and concentrate your efforts on ones that do.

Having seen a unit on the Internet, of course, does not relieve you of your duty (to yourself) to visit the unit and check it out. It is easy to make a place that is falling apart look good in a photograph. And you can't tell whether the plumbing works online. Another possible problem with choosing an Internet rental is that the landlord is required to pay to have the unit advertised on the Internet. This cost may be passed on to you in higher rent. Still, the Internet is the logical place to start looking for most people.

*(Continued on next page)*

**Confronting Consumer Issues (Continued)**

### Advertisements in Local Newspapers

Most people who are looking for a home to rent spend at least some time looking through advertisements in local newspapers. Even if they never rent an advertised unit, their time is well spent if they get a better idea of the housing market in a community where they wish to live. Also, advertisements placed by rental businesses can add direction to your Internet search. In large cities so many rentals are available that you may not know where to begin. Seeing a photograph of an attractive development can be very helpful.

### Property Management Firms and Real Estate Agencies

Many people who work long hours or are moving into a new community do not have the time to carry out an exhaustive rental search by themselves. These people often employ a property management firm or real estate agency to do much of their work for them. These businesses are paid by owners to find tenants for their rental units. The services maintain extensive listings of rental units organized by price, size, location, and so on. Renters need to be aware of two potential problems of using these services. First, they will be much more likely to promote rentals that are listed with them rather than others (that may better meet your needs and budget) listed by a competing firm. And, because owners must pay a commission, they are likely to charge tenants higher rents. Nevertheless, if you have little time to find a rental unit for yourself, this may be the best choice for you

### "For Rent" Signs

Many people who know what neighborhood they want to live in simply drive through it looking for "For Rent" signs. At first glance, this would appear to be a very inefficient way to find a rental unit. On closer examination, however, it might not be such a bad idea, particularly if it isn't the only way you search. Canvassing a neighborhood can give you the added benefit of getting a feel for whether you would want to live there. You can locate shopping districts or bus stops that you would want to be near. Or you can decide to avoid living close to a noisy railroad crossing. And you may run across a rental unit that is exactly what you are looking for—just don't be disappointed if it doesn't work out that way.

### Question for Thought & Discussion #9:

How did you, or your friends who rent, find the rental unit? Would you (or they) try a different method in the future?

### Friends and Acquaintances

Never forget to ask your friends and acquaintances about rental units. They might know someone who is about to move out of an apartment that is just what you need. Or some of them may have friends or relatives who are landlords to whom they could direct you. Every year thousands of college students find apartments that used to be rented by seniors who graduated. Again, you should never limit a rental search to word-of-mouth leads, but there is no reason not to use this method, too.

## MAKE SURE THE RENTAL UNIT IS RIGHT FOR YOU

There is no guarantee that a rental unit that gives you a good first impression is really your best choice. There is probably no such thing as the "perfect place to live." Before you sign any rental agreement, you should always take the time to weigh its pros and cons.

### What It Has and Doesn't Have

When inspecting apartments, it is a good idea to have a checklist so that you can compare their qualities and shortcomings. It can be a good idea to take a tape measure as well. You could be disappointed if you rent an apartment only to later find that your wraparound couch is too big to fit in the living room. Figure 12.2 is an example of a typical checklist. You can draw up your own, adding other questions that you feel are important.

### Consider Other Costs

Anytime you move to a new location you are almost certain to encounter unexpected costs. You may have found a wonderful apartment that overlooks a garden in a good neighborhood, but does it have a place to park your car? If you live in a city, there may not be much on-street parking available, and you might not want to leave your car on the street at night even if you could find a parking place. You could end up renting a place to park for $100 per month, and it might be three blocks from your apartment.

Be sure to investigate the cost of utilities. Most rental units require the renters to pay for heat, electricity, and water. A wonderful apartment with high ceilings and big windows in an older building could cost you a fortune to heat. Try to find out what the previous renter paid for utilities. If this information is not available, ask current renters who have apartments in the same building about their utility costs.

Remember to buy insurance for your apartment. You will not need to insure the building itself, but you do need to insure your belongings and to protect yourself if a guest is injured in your apartment by tripping over your loose rug. More is said about renters' insurance in Chapter 17.

**Figure 12.2** A Checklist to Use When Shopping for an Apartment

| | Apartment A | Apartment B | Apartment C | Apartment D |
|---|---|---|---|---|
| Monthly rent (including all expenses that you have to pay directly, such as utilities, recreational fees, parking fees) | | | | |
| Size of security or cleaning deposit | | | | |
| Are pets allowed? | | | | |
| Is a manager or superintendent on the premises at all times? | | | | |
| Garbage-disposal facilities? | | | | |
| Laundry equipment available on the premises? | | | | |
| Is the laundry room safe? | | | | |
| When can the laundry room be used? | | | | |
| Is there a lobby? | | | | |
| Is there a security officer? | | | | |
| Will you have direct access to your unit? | | | | |
| If there is an elevator, what is its condition? | | | | |
| Is the apartment close to public transportation if you need it? | | | | |
| Is it close to food stores? | | | | |
| Is it close to entertainment? | | | | |
| Is it close to other shopping? | | | | |
| Are there sufficient electrical outlets? | | | | |
| Are carpets and drapes included? | | | | |
| Is there enough closet space? | | | | |
| Are the fire exits safe and clearly marked? | | | | |
| Are the tenants around you the ones you want to live near (children, singles, retired, and so on)? | | | | |

## MAKE SURE THE LEASE IS OK

Most standard-form leases seem to favor landlords (who usually provides them). Before you sign a lease, read it carefully, including all the fine print, to make sure you are not signing away your rights as a tenant.

**Clauses to Avoid**   Lease agreements sometimes include a number of clauses that you may ask to have deleted. Try to avoid clauses that do any of the following:

1. Give the landlord the right to arbitrarily cancel the lease because she or he is dissatisfied with your behavior.
2. Forbid you to have overnight guests.
3. Forbid you to sublease your rental unit to another party.
4. Allow the landlord to apply exorbitant charges if you are even one day late paying the rent.
5. Allow the landlord to enter your apartment when you are not there.
6. Make you liable for repairs without your approval.
7. Require you to follow rules that have not yet been written.

**Clauses You May Wish to Add**   There are also a number of clauses that you may ask the landlord to add to your lease:

1. A written listing of all verbal promises that were made when the rental agreement was settled.

*(Continued on next page)*

2. A written statement that extra fees not included in the lease will not be imposed—an extra charge to use the apartment complex swimming pool or weight room could be an example.

3. A statement that promised improvements in the apartment will be made in a specific period of time—your apartment will be painted within two months in a color of your choice, for example.

4. An agreement on early cancellation of the lease. If you expect to be transferred to another city sometime in the next two years, this could be quite important.

5. An agreement that you may take fixtures you install with you when you leave. If you spent $500 for a new dining room chandelier, you would not want to leave it behind.

## BE PREPARED FOR LEAVING YOUR APARTMENT

The best time to prepare to leave your rental unit is before you move in. You do not want to find yourself being charged for scratched walls that were already damaged when you arrived. To try to avoid this type of problem, take these simple steps before you move into an apartment.

1. Go through the apartment with the manager or owner. List every defect you can find and, better yet, take pictures of every room. Have the photographs developed immediately and dated by the developer so there is no question when they were taken.

2. Ask the manager to sign and date the list. Give the manager a copy and keep one for yourself that you store in a safe location. You may want to update the list after you have lived in the apartment for several weeks. By then you may have discovered that the kitchen sink drains slowly or that there is a big chip in the tub you missed earlier.

3. Find out specifically what must be done to have your security deposit returned when you leave. Be sure that this information is included in the lease.

4. Find out where your security deposit will be held. This is important if the building is sold and you have a new landlord to deal with later.

5. Keep a file of all bills you paid to improve your apartment. This will demonstrate that you were a responsible tenant if the need ever arises.

## WHEN YOU HAVE TROUBLE WITH YOUR LANDLORD

Under the laws of most states, landlords are held to a **warranty of habitability.** This means that the landlord implicitly warrants to the tenant that the leased or rented residential premises are in habitable condition—that is, in a condition that is safe and suitable for human living. This warranty is violated whenever a landlord fails to provide essential services that affect a tenant's health and safety.

If you believe that your landlord has violated the warranty of habitability or has in some other way treated you unfairly, there are several steps you can take:

1. Explicitly indicate what your grievance is, such as uncomfortably low (or high) temperatures, a stopped-up sewage system, a continuously leaking toilet, or a refrigerator whose freezing compartment doesn't work.

2. Make several copies of the complaint list. Mail one to the manager (and one to the owner, if two individuals are involved), and keep one for yourself. If there is an organized tenants' group in your area, send a copy to it also.

3. If you receive no response from the manager or owner of your rental unit, contact the agency that administers the housing code in your area and request a visit from a housing inspector, in hopes that he or she will certify the validity of your complaint.

4. If difficulties persist, your best recourse is to contact your state consumer protection bureau. Since landlord-tenant laws vary from state to state, the state consumer protection bureau will be able to advise you on state laws, as well as direct you to personnel who may be able to assist you.

**Withholding Rent**   If your complaints are serious enough, you may, in some states, have the legal right to withhold part or all of your rent until the complaint or complaints are satisfied. Approximately half the states allow the tenant to deduct repairs that are the landlord's responsibility from the rent; they also provide for not paying any rent when the dwelling is uninhabitable. There may be a dollar limit on the amount you can deduct for repairs, however, depending on the state in which you live. Again your state's consumer protection bureau can advise you.

**A Final Alternative**   If your rental situation is truly disagreeable, you might consider looking for another apartment where you won't have similar problems. When one manufacturer's product does not satisfy you, you often turn to a competitor. The same principle could be applied to rental units.

## Key Terms:

**Lease**   A contract by which one conveys real estate for a specified period of time and usually for a specified rent; and the act of such conveyance or the term for which it is made.

**Warranty of Habitability**   An implied warranty made by a landlord to a tenant that leased or rented residential premises are in a condition that is safe and suitable for human habitation.

# Chapter Summary

1. Individuals purchase or rent a house or apartment in order to obtain the flow of services from that particular asset.

2. Renting is an attractive alternative to buying because it offers greater freedom of mobility, no down payment, and no maintenance, and the exact cost can be determined easily.

3. A decision to rent or to buy often rests on whether the added costs of purchasing a house are offset by equity and tax savings.

4. The costs of owning a home include not only initial costs (down payment, closing costs, and so on) but also the costs of repairing and maintaining the home over time.

5. If you are buying a house, you will probably find the services of a real estate agent helpful. Remember, however, that the cost for those services—paid for by the seller—will ultimately be passed on to you in the form of a higher purchase price.

6. A real estate agent essentially brings together the buyers and sellers of homes. He or she is, therefore, a provider of information. Since different agents may be more or less helpful, depending on your preferences in housing and how seriously you are looking, it pays to shop around for the right broker. You might also consider hiring one as a "buyer's agent" to act specifically on your behalf.

7. Buyers and sellers commonly bargain over the price of a house.

8. It is usually advisable to have a building-inspection service inspect the home you wish to buy before you make any offer. The price for such an inspection is small compared with the costs you may incur without one.

9. Cooperatives and condominiums are popular types of ownership arrangements. A co-op is a nonprofit organization that is owned and operated solely for the benefit of its members (the individuals who own residences in the building).

10. The owner of a condominium has title to the unit she or he occupies and faces fewer restrictions than a co-op owner.

11. Numerous types of lending institutions offer many types of mortgages. Shop around for a mortgage just as you shop for anything else. Seek out the best deal in terms of the down payment required, the annual percentage interest rate charged, and prepayment privileges so you won't have to pay a penalty for prepayment if you decide to sell the house in a few years.

12. Before shopping for a mortgage, buyers should understand the terminology commonly used in the mortgage trade, including such phrases as discount points, prepayment privileges, and amortization schedule.

13. Various types of mortgages are available from lending institutions, including conventional (fixed-rate) mortgages, adjustable-rate mortgages (ARMs), graduated-payment mortgages, and rollover mortgages. In addition, government-guaranteed mortgages, such as a VA mortgage or an FHA mortgage, can be obtained.

14. Creative financing of mortgages is frequently arranged through special agreements made between the seller and the buyer.

15. The amount of interest paid for a mortgage loan can be reduced substantially by paying the loan over a shorter period of time and/or by making extra payments toward the principal.

16. Closing costs are out-of-pocket expenses that must be paid at the time of closing. These costs are for services performed by the lender, escrow agent, title company, and other providers.

17. The Real Estate Settlement Procedures Act of 1976 requires that all closing costs and procedures be specifically outlined to you by the lender before you buy a home. The lender must comply with the various requirements of the act within designated time periods.

18. Although some tenants rent dwellings through short-term agreements, most sign a long-term lease. In a lease agreement, the landlord and tenant are bound to certain specified terms.

19. Information about rental units can be obtained online and from newspapers, property management groups, and other organizations that often provide computerized listings of available rental properties. When comparing apartments, consumers should use a standard checklist to evaluate their alternatives to make the best choice.

20. Grievances with landlords can often be handled through a housing agency in your community or an organized tenants' association. You may also wish to check with your state's consumer protection bureau for assistance if you have a rental problem.

## Key Terms

abstract **278**
amortization schedule **282**
condominium **279**
cooperative **280**
discount points **282**

dual agency **274**
earnest money **278**
home warranty **279**
lease **292**
mortgage **271**
prepayment privileges **282**

pro rata **280**
title **278**
title insurance **278**
townhouse **279**
warranty of habitability **292**

## Questions for Thought & Discussion

1. How do you decide how much of your income you will budget for housing?

2. Why are the tax advantages of home ownership less important to most young people than to older people?

3. Would you pay an agent several thousand dollars to represent you when you buy a house? Could you make a good deal on your own?

4. What qualities would you emphasize if you were selling the building where you live? What would you hope the buyer would fail to notice?

5. Why are mortgage interest rates usually much lower than the rates charged for other types of consumer loans?

6. In 2001, more than 90 percent of new mortgages came with fixed interest rates that would not change over the term of the loan. Why did so many consumers make this choice?

7. Why do many consumers worry more about the size of their monthly mortgage payment than about the total amount they pay over time?

8. Are you, or people you know who rent, satisfied with the rental unit? Explain your answer.

9. How did you, or your friends who rent, find the rental unit? Would you (or they) try a different method in the future?

## Things to Do

1. Visit a local real estate agency online and investigate housing prices in your area. What is the average price of a new home versus a previously owned home? How do these prices compare with the national average for both types of homes?

2. Talk with some homeowners you know and ask them about maintenance and repair expenses. How much did they pay in the last year for maintenance and repairs? How many of these expenses were anticipated?

3. Find out how property taxes are determined in your area and how the money is spent.

4. Check to see if there is a tenants' organization in your area. Your county consumer affairs office or local legal aid society may be able to help you. Find out if the organization is attacking some of the problems mentioned in the chapter.

5. Complete a checklist similar to the list in Figure 12.1 for inspecting the building in which you live. Would you be able to sell this property if you owned it? What features would you emphasize to prospective buyers? What subjects would you try to avoid? How does this exercise demonstrate the importance of asking many questions when you consider buying a house?

# Internet Resources

## Finding Consumer Information on the Internet

The following Web sites have been selected for their relevance to topics discussed in this chapter. Search these sites to locate information that can add to your knowledge of government assistance that is provided to home buyers on the Internet. Remember, Web addresses change frequently. If any of these addresses no longer function, find similar sites to investigate using any of the search engines available to you.

1. The Information Center for the Department of Housing and Urban Development's (HUD's) Office of Community Planning and Development (CPD) offers tips for home buyers at its Web site located at **http://www.comcon.org**.

2. FHALoan.com specializes in FHA real estate loans, mortgage rates, home-equity loans, refinances, first-time home buyers, and all other HUD programs. Visit its Web site at **http://www.fhaloan.com**.

3. HUD maintains a Web site where it provides information about HUD homes that are being offered for sale. Its listing of homes is updated daily. Find out what HUD homes are being offered in your community at its Web site at **http://www.usgovinfo.about.com/blhudhome.htm**.

## Shopping on the Internet

The following Web sites have been selected because they offer consumers services similar to those described in this chapter. These are commercial sites that are designed to market products. They do not represent a comprehensive or balanced description of all institutions that offer mortgage loans online. How do the offers made at these Web sites compare with those made by local institutions? Remember, Web addresses change frequently. If any of these addresses no longer function, find similar sites to investigate using any of the search engines available to you.

1. Mortgage Lenders, Online Home Loans arranges for a number of lending institutions to make offers of mortgage loans to potential borrowers. It can be reached at **http://www.secure-a-mortgage.com**.

2. Quicken Loans allows borrowers to apply online for a home mortgage. Its Web address is **http://www.quickenloans.quicken.com**.

3. TCF Bank—Home Mortgages offers to help first-time home buyers find a mortgage loan. Investigate this firm's Web site at **http://www.tcfexpress.com/hm_home.htm**.

## InfoTrac Exercises

Purchasers of new copies of this text are provided with access to the InfoTrac Web site. This Web site links students to thousands of recent articles published in hundreds of periodicals. Use the key words **home, warranty,** or other terms from this chapter to conduct a key-word search. Choose one article that is of particular interest to you and write a brief essay describing what you have learned from the article. Be sure to cite the author and title of the article and the name and date of the publication in which it appeared.

# Selected Readings

Austin, Kenneth T. "Buying a Home: Economics vs. Emotions." *USA Today Magazine,* March 1999, p. 20.

Bill, Peter. "If You Value Your Credibility—Split the Tasks." *Estates Gazette,* February 9, 2002, p. 35.

"Builder Starts Business of Home Inspections." *Sarasota Herald Tribune,* July 8, 2002, p. 8.

"Buying a Home with a Bad Claims History." *Kiplinger's Retirement Report,* December 2002, pp. 8–9.

Carnahan, Ira. "Forget the Spread." *Forbes,* January 22, 2001, p. 142.

DeWesse, Arron. "The Dawn of the High-Tech Era." *National Real Estate Investor,* November 2000, p. 120.

Elswick, Jill. "Benefit Provides Health Insurance for Houses." *Employee Benefit News,* March 1, 2002, p. ITEM02059014.

Gibson, Dale. "Title Search Company Tells an Unhappy Story." *Triangle Business News,* July 14, 2000, p. 1.

*Landlords and Tenants: Your Guide to the Law.* Available from the American Bar Association, 750 North Lake Shore Dr., Chicago, IL 60611.

O'Keefe, Brian. "Buyer Beware—Not Scared." *Fortune,* October 28, 2002, p. 68.

Powel, Michael. "Home Economics." *Money,* March 2000, p. 133.

Razzi, Elizabeth. "The Home Team." *Kiplinger's Personal Finance Magazine,* June 2001, p. 78.

"Take a Walk before You Buy Your New Home." *Business First of Buffalo,* April 16, 2001, p. B-13.

# CHAPTER 13

# Banks Help Consumers Save and Spend

**After reading this chapter, you should be able to answer the following questions:**

- Why should saving be a part of every consumer's financial plan?

- What factors determine how much you can save?

- What are the benefits and costs of different types of saving accounts offered by depository institutions?

- How may consumers save by purchasing government securities?

- What differences are there among checking accounts offered by depository institutions?

- What are the rights and responsibilities of consumers who use banks?

- How may consumers use electronic fund transfer systems to complete financial transactions?

ccording to Benjamin Franklin, "a penny saved is a penny earned." We all know that it's a good idea to put money away for a rainy day, but are you a successful saver? When you try to save, do you set aside $25 this week only to spend it next week on something you don't really need? Do you have a saving goal—something you really want to own that you can buy only if you save? And, if you do have a saving goal, have you made a logical plan that could help you achieve your goal? Finally, if you have a saving plan, do you also have the courage to stick to it? These are questions you should consider as you read this chapter.

> **Saving** The act of not consuming or not spending your money income to obtain current satisfaction.

# Why Save at All?

You probably think of saving as putting money aside so that you can use it to buy something you want in the future. Economists have a different way of defining saving. They will tell you that **saving** is the act of not spending money income to obtain current satisfaction. You might think that this definition is not much different from the one you are used to, but there is an important difference, at least in the way economists look at saving. Instead of thinking of saving in terms of setting money aside, economists think of it as choosing to do without goods or services now in order to be able to purchase other things you value more in the future. From this point of view, saving can be seen as an act of scheduling your consumption over time. It's not money that people save as much as it's purchasing power.

## Americans Save Very Little

No matter what definition you use for saving, the fact is, most Americans don't save very much of their income. Through most of this nation's history, Americans saved between 4 and 8 percent of their after tax income. This percentage declined during the 1990s, reaching levels of less than 1 percent in some months. In fact, near the end of 2002 Americans were saving only 2.1 percent of their after-tax income. This was much less than residents of most other developed countries. Figure 13.1 on the next page shows the total amounts saved by individuals in the United States in the years from 1995 through 2002. These amounts haven't even been adjusted for changes in prices. Consider, too, that some people save a large part of their incomes. This means that others must save nothing at all or are falling into debt (*dissaving*). In which group do you fall? Are you a saver or a dissaver? And, in the long run, what difference will it make in your life?

Most people save because they want to be able to afford special purchases in the future. In the past, you may have saved to help pay for your college education, or to purchase a reliable car. At present, you may save to be able to make a down payment on a house of your own. In the future, you may save to have a comfortable retirement. Many people save to take exotic vacations or to send their children to college. Probably the most important reason to save, however, is not to buy a specific good or a service, but to gain personal or family security.

Few people are fortunate enough to have a steady, uninterrupted flow of income throughout their lives. For most of us, things just don't work out like that. Statistics show that the average worker changes jobs eight times within

**Question for Thought & Discussion #1:**
Do you believe Americans save too little of their income? What problems might current saving rates lead to in the future?

**Figure 13.1** Personal Savings in the United States

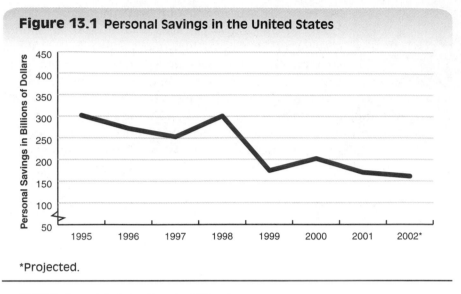

*Projected.

SOURCE: *Economic Indicators*, December 2002, p. 6.

his or her life span. Maybe you will always move from one job to the next without any interruption in your employment or income, but you can't be sure. You could be laid off and out of work for months. In the future you might fall ill or be injured in an accident so you couldn't work. Your home could be destroyed in a flood, or one of your children might need special (and expensive) medical care. No one knows what the future will bring. Having a "nest egg" of savings set aside can make life much easier when unexpected events put a strain on your budget. And even if you never need your "nest egg," you should sleep a little better at night just knowing that it's there.

## The First Step Is Often the Hardest One to Take

Few Americans have more income than they know what to do with. In fact, many people live from paycheck to paycheck. Almost every penny they earn is already allocated to one expense or another before they receive their wages. This is particularly true for students who have books to buy and tuition to pay in addition to their ordinary living expenses. Still, saving is a valuable endeavor, even if you can set aside only a few dollars each week. Getting in the habit of saving now can make it easier to save larger amounts when you earn more income in the future.

Most people find saving easier when they understand why they are saving and have a plan to achieve their saving goals. You know that people, and the things they want from life, are often very different. Your best friend might be saving to buy a new Harley Davidson motorcycle. He may have visions of letting his hair grow, buying a leather jacket, and cruising country roads on his "Hog" all summer long. If that works for him, who are we to say he's wrong? You probably have other goals in your life that saving can help you achieve. You could have plans to marry next year and want to save for a honeymoon or to buy furniture for your new apartment. Whatever is important to you, you should keep it in mind when you make your saving plan and work to follow it. Every time you see something you want but that is not in your budget, you should stop and think. Picture what you're saving for in your mind and remember how much you want it. Then you should find it easier to resist temptation.

## Make a Saving Plan (and Follow It)

Every consumer has the power to control his or her own spending. Equally, then, every consumer has the power to control his or her saving. For most of us, saving doesn't just happen, even if we know why we are planning to save. A saving plan should be given a high priority in every person's or family's budget. You shouldn't be satisfied to save what's left over after you pay your bills and buy other things you want. In fact, you should try to think of saving as a bill that you have to pay to yourself. If you decide to save $25 every week, you should take $25 out of your income at the start of each week and put it somewhere that you won't spend it, such as in a bank. (More will be said about the value of saving in a bank later in this chapter.)

**FOLLOW YOUR SAVING PLAN**   In the eighteenth century, the Scottish poet Robert Burns wrote, "The best laid schemes o' mice an' man gang aft agley." What he meant, of course, is that making plans isn't enough to ensure success. There is the matter of follow-through. Having decided to save $25 every week will do you little good if you don't actually avoid spending that part of your income. Successful savers often make their saving into a kind of ritual. They might take their money to the bank after work on payday and then buy a candy bar, a café latte, or some other small extravagance that won't put too much of a dent in their budgets. The point is to make the act of saving into something memorable so you won't forget.

**DIFFERENT PLANS FOR DIFFERENT PEOPLE**   For some people saving the same amount of income every week is impossible. Servers in restaurants, for example, receive a large part of their income from tips. If the restaurant where they work has a particularly slow week, they may not have any money left over to save after they pay their bills. In other weeks they might serve many customers and do very well. People with irregular incomes often choose to save a set percentage of their earnings over the amount they must use to pay fixed expenses. This isn't the only plan that might be employed, but it works for many people.

When you make your saving plan, you need to find a way to save that fits your personality. People who lack a saving plan, or fail to follow the plan they make, are not likely to be successful savers. And, as a result, they probably will fail to achieve many of their life-span goals.

# Where Should You Stash Your Cash?

So, imagine you have finished your saving plan. Your goal is to save $3,000 in the next year to make the down payment on a used car. You figure you can reach your goal if you set aside $60 from each of your weekly paychecks. Is this enough for your saving plan to succeed? Maybe, but then again, maybe not. Certainly, there are things you can do that will enhance the chances that your saving plan will be successful.

One thing all savers need to decide is where to put the funds they don't spend. It might be easy to keep your savings in the second drawer of your desk under your accounting textbook. But, if you do, you are sure to lose.

How is creating and following a saving plan different from putting a few dollars in a piggy bank when you think of it?

Photodisc/Getty Images

**Question for Thought & Discussion #2:**
What goals have you set in your life that are important enough for you to give up current spending? How successful have you been in saving?

**Depository Institution**  Any organization that accepts deposits and assures depositors that they will be able to withdraw their funds when they need them.

**Fractional Reserve System**  A banking system in which depository institutions are required to keep a minimum share (the *reserve requirement*) of deposits they receive on reserve.

First, when your savings are so easy to get to, you might not be able to resist temptation when a late-night pizza sounds awfully good to you. A greater risk is that someone could find your money in your desk and take it. It would be nice if everyone was totally honest and you could leave anything of value anywhere you want without having to worry that it might not be there when you return to get it. Unfortunately, that's just not how things are. Finally, even if no one takes your money from your drawer, its value (think purchasing power) will decline because of inflation.

Over the last fifty years, consumer prices in the United States have increased at an average rate of roughly 3 percent per year. In some years the rate has been lower, but in others it has been much higher. At the end of the 1970s, for example, prices were going up at a rate of nearly 10 percent per year. When your savings are sitting in your desk, wallet, or purse, their purchasing power is going down. Over a few days this decline won't make much difference, but over several years it can be important. If there is 5 percent inflation in the next year, you might need $3,150 to make your down payment on a used car instead of the $3,000 you intend to save. Banks and other depository institutions can help you solve both the problem of security and the problem that the purchasing power of the money you save might decline.

## What Are Depository Institutions?

A **depository institution** is any organization that accepts deposits and assures depositors that they will be able to withdraw their funds when they need them. At one time most depository institutions in the United States were commercial banks. Only commercial banks were allowed to offer business loans and checking accounts or extend many types of consumer credit. This changed with the passage of the Depository Institutions Deregulation and Monetary Control Act of 1980. Now savings and loan associations, mutual savings banks, credit unions, and even some brokerage firms have joined commercial banks as important depository institutions in our economy. For consumers it can be difficult to distinguish among the services offered by these organizations. They all look pretty much the same. Therefore, even though there are technically a variety of depository institutions in our economy, in this chapter they will all be referred to as "banks" to avoid using the cumbersome term *depository institutions*.

**Question for Thought & Discussion #3:**
What different types of depository institutions exist in your community? What similarities and differences can you identify in the services they offer?

## How Banks Work

The first thing you should realize is that, with the exception of credit unions and some savings and loans, banks are businesses—they exist to earn a profit for their owners. Banks earn profits by selling services to customers for more than it costs them to provide these services. Any bank that fails to earn a profit for very long will cease to exist. In many ways banks are little different from local McDonald's restaurants or Wal-Mart stores.

**THE FRACTIONAL RESERVE SYSTEM**   Banks have essentially two sources of revenue—interest income and the fees they charge for services they provide to their customers. At one time the largest part of banks' income was earned from interest they charged borrowers. Banks in the United States operate on the

**fractional reserve system.** This means that banks are required by law to keep a minimum share (the *reserve requirement*) of deposits they receive on reserve. They may not lend or spend these funds for any purpose. The remaining share may be loaned out to borrowers or used in other ways. Banks pay depositors interest on most deposits and receive interest from those who borrow funds from them. The interest they charge borrowers is significantly higher than the rate they pay depositors. This difference can be as high as 12 percent or more. It allows banks to earn a return on deposited funds.

**BANKS CHARGE MANY FEES** In the past fifty years, fees have become an increasingly important source of bank income. Banks charge fees for many services including checking fees, credit-card fees, safe-deposit box fees, financial advisement fees, estate-planning fees, and so on. One reason banks depend more on fee-generated income today is the strong competition they face in the credit market. There are nearly nine thousand commercial banks in the United States and at least as many other types of depository institutions. The growth of online saving and lending institutions in the 1990s has made banking competition even stronger. Consequently, it is difficult for one bank to pay depositors less or charge borrowers more than other banks and retain its customers. This situation has created a thin profit margin in the credit market that has caused banks to turn to fees as another source of income. Therefore, when consumers buy banking services, they need to consider the fees they are assessed as much as the interest rates they receive for deposits they make or are charged for loans they take out.

## What Banks Offer Savers

Banks offer savers two primary benefits, security and a return on their savings to help them maintain their purchasing power. Banks are well equipped to protect the funds you save. They have heavy vaults with impressive locks, guards, night security systems, professional auditors, and a host of other ways to make sure your money isn't stolen. And, even if funds you deposit are taken by some crook, banks have insurance that will replace the lost money. Further, an agency of the federal government, the Federal Deposit Insurance Corporation (FDIC) for commercial and mutual savings banks, insures deposits in almost every U.S. bank to a maximum amount (currently $100,000). This means that if a bank fails and is unable to repay its depositors, the FDIC would step in and make the payments instead. It is virtually impossible to lose funds deposited in federally insured accounts up to the maximum level.

You have learned how inflation can eat away the purchasing power of savings over time. When you deposit your savings in most bank accounts, you will be paid interest. The rate of this interest will not be high. In 2003 most savings accounts paid less than 1 percent interest. You might think, "Well that's not very much." But the rate of inflation in 2003 was between 2 and 3 percent, so this interest income would have cushioned your savings a little from the ravages of inflation. Remember, though—the interest earned from banks is taxable. After you pay your federal, state, and local income taxes, you may be left with only 0.5 percent interest income that you can keep. Still, even this is better than nothing.

> "Banks offer savers two primary benefits: security and a return on their savings to help them maintain their purchasing power."

**Question for Thought & Discussion #4:**
Compare the current rate of inflation with the rate of interest paid on bank saving accounts. Is this interest rate an important reason why people choose to save?

# How Banks Help Consumers Save

Banks offer consumers a variety of accounts in which they may deposit the money they save. As in most economic situations, both benefits and costs are associated with every saving choice you could make. Different people have different saving goals and needs. Banks, therefore, offer many different types of saving accounts, each designed to meet the needs of various types of savers.

## Day-In-Day-Out Accounts

The simplest and most common saving account is the day-in-day-out account. Different banks may have invented different names for this type of account, but you can recognize one by studying how it works. Day-in-day-out accounts pay interest on funds from the day they are deposited until the day they are withdrawn. Savers have access to their money at any time without having to pay extra fees or penalties for early withdrawal.

**REPLACING PASSBOOK SAVING ACCOUNTS** These accounts were once known as passbook accounts. Each depositor was given a small book in which deposits, interest earned, and withdrawals were entered when transactions took place at a teller's window in the bank. Automation and technology have made these accounts largely obsolete. Now account holders are given a plastic card with a magnetic strip on its back that contains their account information. With the card, and a **personal identification number (PIN),** savers can make deposits and withdrawals at their bank or at thousands of **automated teller machines (ATMs)** that are located in virtually every community throughout the world. It no longer makes any difference whether their bank is open. Savers have access to their funds at any time.

Although the technology of day-in-day-out accounts has changed, they still work much the same way they did fifty years ago. These accounts provide savers with a safe place to store their savings while giving them immediate access to their funds whenever they want. At almost every bank, these accounts are insured by an agency of the federal government up to the $100,000 maximum. Further, day-in-day-out accounts pay a rate of interest that is periodically set by the bank at a specific rate that will last for a set period of time—typically one to three months.

**DAY-IN-DAY-OUT ACCOUNTS DON'T PAY MUCH INTEREST** Unfortunately, day-in-day-out accounts earn a relatively low rate of interest—about 0.75 percent at the beginning of 2003. The fact that depositors may withdraw their savings at any time limits what banks can do with the funds deposited in these accounts. If you deposited $1,000 in such an account, for example, the bank would not know whether it would have the use of your savings for one day, one week, one month, one year, or for even a longer period of time. As a result, it is limited in how it can use your money. Committing funds from these accounts to a long-term loan that has a relatively high rate of interest could cause a problem for the bank if many savers decided to withdraw their deposits. Knowing this, banks tend to use deposits from day-in-day-out accounts to make short-term loans that earn relatively low rates of interest. Banks earn less from deposits in day-in-day-out accounts so they pay depositors lower rates as well.

**Personal Identification Number (PIN)** A number given to the holder of a debit card that allows the card to be used to transfer funds from accounts electronically. Typically, the card will not provide access to an electronic fund transfer system without the number.

**Automated Teller Machine (ATM)** An electronic customer-bank communication terminal that, when activated by an access card and personal identification number, can conduct routine banking transactions.

## Money Market Accounts

**Money market accounts** are an alternative to traditional day-in-day-out accounts. The rate of interest paid by a money market account fluctuates from day to day as prevailing interest rates in the economy change. In most banks, these deposits are also insured to a maximum of $100,000 by a federal agency. As with day-in-day-out accounts, savers have access to their deposits at any time. When you think of opening a money market account, you should consider at least three factors. First, a minimum deposit of $500 or more is required to open such an account. If you can't keep the minimum deposit in your account, you should not open one. Second, the rate of interest paid by money market accounts varies from day to day. There is no guarantee that the rate you are quoted when you open your account will be the one paid next week. When prevailing interest rates change in the economy, banks adjust the rate paid on money market deposits. Still, these accounts usually pay slightly higher rates than typical day-in-day-out accounts. Because banks know they can change the interest rate they pay depositors at any time, they don't worry about paying a little more now. Should prevailing interest rates fall, they can quickly lower the rate they pay their depositors. Banks also know that on average the total deposits by all savers in these accounts are not likely to change very much from day to day. Banks feel more comfortable using these funds to make long-term loan commitments and are able to earn a better return.

## Sweep Accounts

Many banks offer sweep services for larger depositors who have money market accounts. These are sometimes called *sweep accounts*. What happens is that any extra funds that have flowed into any accounts the customer has (such as checking) are automatically deposited into the money market account at the end of each day to earn interest. This service assures customers that all of their deposited funds will earn at least some interest on every day.

## Time Deposits and Certificates of Deposit

You have learned that banks use most of their deposited funds to make loans. As a general rule, the longer the term of the loan they make, the greater the rate of their earnings will be. As a result, banks are willing to pay depositors higher rates of interest if they agree to leave their savings on deposit for extended periods of time. When a depositor places funds in such an account it is called a *time deposit*. The most common type of time deposit is a **certificate of deposit (CD).**

### CDS OFFER ADVANTAGES FOR DEPOSITORS
A certificate of deposit is a time deposit that is committed to the bank for a specific period of time that may range from a few days to as long as ten years. CDs, like other deposits, are insured by an agency of the government to a maximum amount of $100,000 at most banks. CDs are most often issued for periods of six months or for terms of one, two, three, or five years. The longer the term, the greater the rate of interest the CD will pay. CDs normally require a minimum deposit of from $500 to several thousand dollars. Frequently, large CDs (called *jumbos*) of $100,000 or more pay significantly better interest than

**Money Market Accounts** Accounts that pay interest rates that fluctuate from day to day as prevailing interest rates in the economy change.

**Certificate of Deposit (CD)** A deposit that cannot be cashed in before a specified time without paying a penalty in the form of reduced interest.

**Annual Percentage Yield (APY)**
The standard annualized return on a
savings deposit that all savings
institutions must provide to depositors
under the Truth in Savings Act.

smaller CDs with the same term. Again, the reason for this is related to banks' ability to earn income from the funds deposited. In this case, banks can often earn higher returns when loaning large amounts of money. They also have relatively less bookkeeping to do when they work with big deposits. Banks, therefore, are willing to pay more for larger deposits.

**EARLY WITHDRAWALS FROM CDS**  Although depositors agree not to withdraw money deposited in CDs before their terms expire, it is possible to withdraw these funds if necessary; however, the early withdrawal will result in the assessment of an interest penalty. The *early withdrawal penalty* is equal to the interest that would have been earned over a specific number of months, most often three or six. When opening a CD, it is a good idea to find out what the early withdrawal penalty is. You may have no intention of taking your money out early, but you can't foresee the future. You might break your leg next month and need the money in your CD to pay doctor bills. If two CDs offer the same interest but have different early withdrawal penalties, it is preferable to open the one with the lesser penalty.

## How to Evaluate Differences in Bank Accounts

**Question for Thought & Discussion #5:**
Why are young people less likely to deposit their savings in CDs than in day-in-day-out accounts?

Most consumers in the United States could choose to save at many different banks. The Internet has opened up a broad spectrum of saving alternatives as well. So, how can a saver decide which depository institution to use? First, you should be sure that the bank you are considering is a member of the FDIC or a similar agency of the federal government so that your deposits are insured. Second, consider convenience. Most banks are quite competitive in the interest rates they pay depositors. Choosing a bank that is located close to your home or place of work may be a better decision than choosing one that is located halfway across the nation even if it pays slightly more interest.

## APYs Make It Easy

The final task is to compare the rates of interest paid for similar deposits. At one time doing this could be very difficult because banks quoted interest rates in a variety of ways that appeared to be intended more to confuse or mislead customers than to help them make rational decisions. This problem was effectively ended when the federal government passed the Truth-in-Savings Act in 1993. This law requires banks and other depository institutions to advertise a standard interest rate called the **annual percentage yield (APY)** that must be calculated in the same way by all banks. It also prevents depository institutions from advertising their services as "free" unless they really are free. In other words, a bank cannot say its checking accounts are free if depositors are required to maintain a $3,000 balance in their accounts to receive this service. Finally, the Truth-in-Savings Act requires banks to pay interest on at least the average deposit balance for a period (usually one month) instead of the lowest amount.

Since the passage of the Truth-in-Savings Act, it has been easy for savers to compare interest rates paid by different banks. It makes no difference whether the bank pays simple interest once a year or compounds interest twice, four times, monthly, or daily throughout the year (most accounts now pay interest that is compounded daily). All of these methods of payment must be converted to the APY and reported to savers. If Bank A advertises that it pays an APY of 3.11 percent for a one-year CD of $1,000 and Bank B advertises that it pays an APY of 3.17 percent for the same deposit, you know that the deposit placed in Bank B will earn you a larger return.

# Saving by Buying Government Securities

Although the federal and state governments are technically not depository institutions, consumers may use them as a way to insure the security and purchasing power of money they save. Governments borrow funds by selling bonds that are also known as government *securities.* A bond represents a promise to pay a set amount of interest each year to the bond's holder and to repay the face value of the bond at the end of the bond's term. Savers often purchase government bonds instead of depositing their funds in savings accounts. When you purchase a federal government bond, there is essentially no risk of nonpayment. And, unlike FDIC insured deposits, there is no limit to the amount that is guaranteed by the government. If you hold $10 million in federal bonds until their term is finished, you will receive $10 million from the government. For most federal bonds, you will also receive a specified interest payment each year until the bond comes due. In the pages that follow, you will learn that there are several types of government bonds or securities that you may buy.

## U.S. Savings Bonds

There are three types of savings bonds: Series EE, Series HH, and I bonds. The most popular of these are Series EE bonds, although I bonds have begun to catch up in recent years.

Savings bonds differ from most federal securities in a number of ways:

- Series EE or I savings bonds can be purchased through most depository institutions including banks, saving and loans, mutual savings banks, and credit unions, or you may purchase them through a payroll deduction taken by your employer.

- Series EE savings bonds are sold in denominations ranging from $50 to $10,000 each. Their price is equal to half of their face value. In other words, a $50 bond sells for $25. Over time the bond's value will increase to and past its face value if it is held long enough.

- Series EE and I savings bonds do not have a fixed term for which they must be held. They may be cashed anytime after one year from the date they were purchased. If they are cashed within the first five years after they are purchased, they will pay a slightly lower rate of interest.

- Unlike other federal securities, savings bonds cannot be sold to any other person. To get your money from a savings bond, you must cash it at a depository institution that sells savings bonds.

**SERIES EE BONDS** Series EE savings bonds held for five or more years earn interest at a rate equal to 90 percent of the rate paid on five-year U.S. Treasury securities. For example, if the rate of interest paid on five-year Treasury securities is 5.0 percent, Series EE bonds will pay 4.50 percent. This rate is adjusted every six months so there is no way to know what rate a Series

EE bond will pay more than a half-year into the future. Individuals may purchase a limit of $15,000 worth of Series EE bonds in a single year. After thirty years, Series EE bonds stop paying interest and must be cashed or traded for Series HH bonds.

Many savers buy Series EE savings bonds, at least in part, because of the tax advantages they offer. Interest earned on Series EE bonds is not taxed until the bonds are cashed. People may buy Series EE bonds and hold them until retirement. Then, as they cash the bonds to help pay for their retirement, they will pay tax on the interest earned, but probably at a lower rate because of their reduced income. Another important advantage of owning any federal securities is that, although the interest they pay is taxed by the federal government, it cannot be taxed by any state or local government. This is particularly important for people who live in states with relatively high income tax rates. Finally, under the Technical and Miscellaneous Revenue Act of 1986, the interest earned on savings bonds is not taxable by any level of government if they are used to pay for the education of the bondholder's dependent. Many Americans use savings bonds to set money aside for their children's college education. This advantage, however, is phased out for taxpayers with exceptionally high incomes.

**I SAVINGS BONDS** Since the mid-1990s, the federal government has offered consumers the option of buying I savings bonds. The rate of interest paid by I bonds is tied to the rate of inflation as measured by the *consumer price index*. These bonds pay a rate of interest that is set by the Treasury plus the rate of inflation. So, if the set rate is 1.6 percent (the rate on January 1, 2003), and the rate of inflation is 1.8 percent, I bonds will pay 3.4 percent. The great advantage of I bonds is that savers know that the interest rate they receive will always be greater than inflation. This will protect the purchasing power of their savings no matter what happens to prices.

In most ways I savings bonds are the same as Series EE bonds. They have the same redemption rules, denominations, and term. I bonds, however, are sold for their face value, and consumers may buy more of them in a single year. If you believe inflation will increase in the future, I bonds could be a wise way for you to save.

**SERIES HH BONDS** Series HH savings bonds can be obtained only in exchange for Series EE bonds that have matured. They are issued in denominations ranging from $500 to $10,000 and, unlike Series EE bonds, are purchased for their face value. Savers receive their interest from the Treasury by direct deposit to a bank account twice each year. Currently, Series HH bonds have a ten-year maturity, and an interest penalty is levied if they are redeemed before they mature.

As with other saving bonds, interest earned on Series HH bonds cannot be taxed by state or local governments. Further, the interest on the Series EE bonds that were used to purchase Series HH bonds will not be taxed until the Series HH bonds mature. The current interest paid on Series HH bonds, however, will be taxed by the federal government.

If you wish to find additional information about savings bonds or their current interest rates, check with your local banks or other depository institutions. You may also call toll free (800) US-Bonds for bond information, or visit the Bureau of the Public Debt's Web site at **http://www.savingsbonds.gov**.

**Question for Thought & Discussion #6:**
Why do many savers choose to purchase savings bonds instead of depositing their funds in bank accounts? What advantages and disadvantages are associated with this method of saving?

## Treasury Securities

Most of the federal government's massive debt has been financed through the sale of Treasury securities. Treasury securities are also bonds sold by the U.S. government, but the rules that govern their use are very different from those of savings bonds. Treasury securities, for example, will not be redeemed by the Treasury until their terms are complete. People who want to exchange their securities for cash, however, are free to sell their securities to others in the *secondary bond market*. This is a market where previously owned bonds are bought and sold each business day. Typically, sales on the secondary bond market can total several billion dollars in a single day.

In addition to absolute safety for those who hold government securities until their terms expire, these financial instruments pay relatively high interest rates. Their rates are typically higher than those paid on CDs or on savings bonds. And remember, interest earned on any federal security may not be taxed by state or local governments.

**TREASURY BILLS AND TREASURY NOTES** Two groups of Treasury securities are currently sold—*Treasury bills* and *Treasury notes*. Treasury bills are sold in minimum denominations of $10,000 and have varying maturities of three, six, or twelve months. Treasury notes have maturities from two to ten years and are sold in minimum denominations of $1,000. There are still many *Treasury bonds* in existence that were originally issued with thirty-year terms. These bonds are no longer sold by the Treasury but will continue to exist in the secondary market until they mature in the late 2020s.

**HOW TO PURCHASE GOVERNMENT SECURITIES** Individuals may purchase Treasury securities through banks or brokerage firms for a fee or through the Treasury Direct program operated by the Federal Reserve System. You may learn more about Treasury Direct by writing to the Bureau of the Public Debt, Securities Transaction Branch, 13th and C Streets, S.W., Washington, DC 20239, or visiting its Web site at **http://www.treas.gov**.

Why isn't cash always the best way for consumers to spend their money?

# How Banks Help Consumers Spend

Currency can be a weighty commodity. This is not a figurative statement about the moral or ethical uses of money. It is a literal statement of fact. Suppose you have contracted to buy a business for $1 million. The current owner demands that you pay in cash and bring it to her in a brown paper bag. Disregarding laws about reporting cash transactions of this size, what would this mean you would have to do?

A stack of 100 bills of any denomination is half an inch thick, measures 6 inches by 2.5 inches, and weighs 3.5 ounces. The highest-valued bill in circulation today is the $100 bill. So let's do a little math. One million dollars in $100 bills would require 100 packs of currency. They would weigh 350 ounces or almost 22 pounds. They would occupy a volume of 750 cubic inches. That's a space equal to a box that measures 10 inches by 15 inches by 5 inches. And there is the question of security as well as your personal safety. Would you want to carry around a lump of money that big and that valuable to make a payment? Most people wouldn't. Still, Americans make payments that are larger than this every day. They typically use a single piece of paper that

**Question for Thought & Discussion #7:**
What types of financial transactions do most consumers use checks to complete? Why don't they make these payments in cash?

can be as small as a one-dollar bill. What we are talking about here, of course, are checks.

## Demand Deposits

The most frequently used bank accounts in the U.S. economy are checking accounts. Economists call money placed in checking accounts *demand deposits* because once money is deposited in a checking account, customers can demand the use of their funds at any time simply by writing a check. They do not need to contact the bank or any other authority to spend their deposited funds. All they have to do is fill out a few lines on a small piece of paper and it's done.

There are many important advantages to spending with checks rather than with cash.

- Checks may legally be cashed only by the party to whom they are written. If you make out a check to your landlord for $750 to pay your rent, you won't be responsible if he loses it and it is cashed by someone else. The person, business, or institution that cashed the check will bear the loss. Similarly, if someone steals one of your checks, makes it out for $1,000, and forges your signature, you won't be responsible for the money if a bank chooses to honor the check.

- The value of checks is limited only by the amount of funds deposited in the account against which they are written. Remember the example of the $1 million payment above. A single check could have been used to transfer these funds. It would have been small and easy to carry, and no one could have used the check other than the person to whom it was written.

- Checks provide proof of payment. Imagine that you handed your landlord your $750 monthly rent in cash last week. You didn't ask for a receipt—you trusted him and believed he was an honest, responsible person. Now, a week later, your landlord demands your rent payment and threatens to evict you if you don't pay up. When you remind him of your undocumented payment, he denies it ever happened. You don't know if he's getting senile or is just a crook trying to make an easy buck. The fact is, without a receipt or some other proof of payment, you will probably have to make the payment a second time. If you had used a check instead, there would be no problem. Your bank would have a record of when the check was presented for payment and have your landlord's endorsement on the back. You might even have received the actual canceled check from the bank that you could show him.

- Checks help you keep a record of how you spend your income. Some workers cash their paychecks every week and spend their income as time moves on. They may run short on cash as they approach the next paycheck. When asked, they often have no idea how they spent their money. People who make important payments with checks know where their money goes and usually find it easier to keep their spending within their budgets.

## Different Types of Checking Accounts

Banks have invented many different names for checking accounts that are often much the same. They may call them "Golden Access," "Platinum Plus Checking," or "Premium Edge Checking." Whatever a checking account is called, you can probably fit it into one of the following categories if you read the literature about the account that the bank is required to give you.

**ACTIVITY OR "PER CHECK" ACCOUNTS**    This is the basic type of checking account with which most young people start out. Those who don't have enough money to maintain a high balance that would qualify them for special services or lower-cost plans choose activity accounts. Depositors are charged two types of fees under activity plan checking accounts. One is a maintenance fee that must be paid each month regardless of the number of checks written. This fee is typically between $5 and $10 per month. The other is a per check fee that is charged for each check written. This fee is typically between 10 and 30 cents per check. Activity accounts generally pay no interest on funds deposited and have no minimum balance. This is the biggest reason that young people and students often choose them.

**MINIMUM BALANCE ACCOUNTS**    Minimum balance accounts provide depositors with unlimited checking at no charge per check and often with no monthly maintenance fee. Some of these accounts pay a small rate of interest on funds deposited in them. To qualify for a minimum balance account, depositors must maintain some set amount on deposit in their accounts at all times. Some banks will allow depositors to maintain this minimum deposit in another account that pays a better rate of interest. Non-interest-bearing minimum balance accounts typically require deposits of at least $500 to $1,000. Those that pay interest may have minimums as high as $10,000 to $15,000. The fact that depositors are required to maintain these minimum deposits makes these accounts, in the strictest sense, not free.

**FREE CHECKING**    There are very few examples of truly free checking. To be free, a checking account must charge no fees and require no minimum balance. Sometimes banks will provide free checking to specific groups of customers as a marketing tool. Students, for example, are sometimes offered free checking by banks that want to gain their business. The banks reason that once students graduate and start earning good incomes, they will continue to use the same bank and help that institution earn a profit.

**PACKAGE PLANS**    Many banks offer a combination all-in-one package plan. Generally, customers are required to maintain a large minimum balance in their accounts or pay a monthly fee of from $20 to $30 to participate in the plan. In such plans you may expect to find the following services offered, all at no extra charge:

● Unlimited check writing.
● Personalized checks with pictures of a local sports team or a scenic view on them.
● Overdraft protection. (If you write a check for more than you have in your account, funds will either be transferred automatically from a saving account or be charged to a credit-card account to avoid overdraft charges.)
● A small safe-deposit box.
● Free traveler's checks, cashier's checks, and money orders.
● Slightly lower interest rates on consumer loans.
● A small amount of term life-insurance coverage.
● A Visa or MasterCard credit card.

Many of these benefits are offered in very small amounts in the hope that the bank will be able to convince the customer to buy more of the same product to generate a profit for the bank.

**ASSET MANAGEMENT ACCOUNTS**   In the late 1970s, a new type of financial service appeared in the form of *asset management accounts*. Offered first by brokerage firms (such as the Merrill Lynch Cash Management Account, or CMA), and then by banks (including Citibank and Bank of America), major retail stores (such as Sears), and other corporations, these accounts provide a variety of financial services in one financial package. Chief among these are checking services. Asset management accounts also offer consumers brokerage services that allow them to purchase stocks and bonds as well as borrow funds to make these purchases. Typically, a sweep feature is included to move any idle funds automatically into a money market account.

Customers with asset management accounts receive books of checks and a credit card. When a check written by the owner of a CMA, for example, is presented to a bank for payment, the bank contacts Merrill Lynch, which transfers funds from the customer's money market account to the bank. If there aren't enough funds in the money market account to cover the check, the money is automatically charged to the customer's credit-card account. The credit card can be used in a similar way. Charges made to the card can automatically be paid at the end of the account's billing period from the customer's money market account.

At the end of each month, a single report is generated and sent to the owner(s) of each asset management account. The report includes a record of all checking account activity, credit-card use, money market transactions, and any purchases or sales of stocks or bonds. Thus, asset management accounts provide consumers with a comprehensive record of most of their important financial transactions. For many people this is a big help when they evaluate how closely they have stayed to their budgets or prepare their income tax returns.

## Selecting the Right Checking Account

To select the best type of checking account for you, it is important to match your financial needs and situation with the features and requirements of accounts that are offered by depository institutions. If the most you can afford to keep on deposit in a checking account is $100, there is no point in considering one that requires a $1,000 minimum deposit. You may start by completing the following steps:

- Determine the average number of checks you write each month. Light check writers (one to ten) don't need to worry much about per check fees charged by some banks. But, if you are a heavy check writer (thirty or more checks each month), these fees could be an important consideration when you choose the best account for you. A 20-cent per check fee totals $6 for thirty checks each month.

- Determine the average amount of money you could keep in a checking account. Are there months when you know you'll have to pay extra bills and run your balance down? Can you afford to keep several hundred dollars extra as a cushion against unexpected expenses? These are important questions if you are considering a minimum balance account.

**Cyber consumer**

**Learning about CMAs**

Visit the Merrill Lynch Web site at http://www.ml.com. Use the site's search feature to investigate the Cash Management Account (CMA) Merrill Lynch offers consumers. What services do CMAs provide? What fees will be charged for these services? What sort of consumers should consider opening a CMA? Write a short essay that summarizes your findings.

- Review your other financial needs. Do you currently pay fees to have a credit card or keep a safe-deposit box? Could you eliminate some or all of these fees by opening an asset management account? Remember, there is likely to be a monthly or yearly fee for the asset management account as well. Compare the costs of your alternatives before you make your choice.

- Determine whether there are any special deals available to you. Students, for example, may be offered free or reduced-cost checking by some banks. The same can be true for members of specific groups such as unions or public employees.

With answers to these questions, you are prepared to compare checking plans offered by different depository institutions in your community. You may also evaluate services offered on the Internet. Although these offers may appear to be better than those made by local banks at first, you should remember that many of the transactions you intend to make will take place through the mail or at an automated teller machine. This will involve extra time, planning, and, possibly, fees. Most people prefer to choose a local bank for their checking needs.

If you are a heavy check writer but can't afford to keep a large minimum balance, you should look for a bank with a low per check fee. Light check writers should be more interested in low monthly maintenance fees. Those who can afford to keep large balances on deposit can generally find accounts that have no maintenance or per check fees. But they will end up paying for checking services in some way, even if it is only by being required to leave some amount of money in their accounts that they cannot spend now.

Why is it important to keep an accurate record of every check you write?

**MAKE SURE YOUR DEPOSITS ARE INSURED** No matter what depository institution you choose, you must be sure that any funds you deposit are safe. The best way to do this is to select a bank that is a member of the FDIC or some other federally backed agency. It is no longer common to find a bank that is not insured, but there are still a few. It's worth taking the time to find out just to be safe.

**BE SURE TO REVIEW YOUR CHOICE** Remember, when you agree to open a checking account, you have not made a lifetime commitment. The rules and costs of checking change over time. It is a good idea to periodically review services and charges offered by banks in your community. When you find a better deal, don't be afraid to take it. That's what competition is all about.

## Other Types of Checks You May Use

People commonly make some types of financial transactions for which a personal check is not the best choice or, in some cases, will not even be accepted. If, for example, you intend to pay $23,456.12 for a new car, the dealer will not take your personal check and just let you drive away. She will want a check that she is sure will be honored when it is presented for payment at her bank. Special types of checks can be used in this and other unusual situations.

**CERTIFIED CHECKS** A **certified check** is a personal check that has been stamped and signed by an officer of a bank to guarantee that there are sufficient funds on deposit to cover the check when it is presented for payment.

**Certified Check** A check a bank has certified, indicating that sufficient funds are available to cover it when it is cashed.

**Cashier's Check**  A check drawn against the funds of a bank to a designated person or institution. A cashier's check is paid for before it is obtained.

**Traveler's Check**  A guaranteed check, drawn against the funds of a financial institution, that is often used by consumers in place of cash when they travel. Like a cashier's check, it is purchased in advance. A traveler's check can be traced, and if stolen will be replaced by the issuing institution.

When you want to have a check certified, you must go to your bank and write the check in front of a teller or officer of the bank. The bank employee will check the balance in your account and deduct the amount of the check plus a fee (typically, $10 to $20) for the service. The check will then be stamped and signed so that the party who receives it will know that there are funds on deposit to cover the check. Certified checks are not used as often now as they were in the past.

**CASHIER'S OR TELLER'S CHECKS**  A **cashier's check** is actually a bank check. It bears the name of the bank and is drawn against the funds of the bank. Anyone who receives such a check will be paid from the bank's funds rather than those of an individual or business. To obtain a cashier's check, you must go to your bank and have money transferred from your account to cover the amount of the check and a fee (typically, $10 to $20) for the service. Car dealerships and other sellers of expensive consumer goods often require payment in the form of cashier's checks.

**TRAVELER'S CHECKS**  A **traveler's check** is a guaranteed check that is often used by consumers in place of cash when they travel. Most stores refuse to accept personal checks drawn against banks that are not local. The alternative is to use traveler's checks. Traveler's checks, like cashier's checks, are written against the funds of a financial institution—such as American Express or Bank of America. You can purchase traveler's checks at banks as well as at other businesses, including American Express offices and the American Automobile Association (AAA). To do this, you must pay for the checks plus a 1 percent service fee unless they are provided to you with no fee through some account relationship you have with the bank or business. You will be required to sign the checks once in front of the person who sells them to you. You will also be given a record of the checks you have purchased. Then, when you want to spend a check, you will sign it again in front of the person who will receive it. You should expect to be required to present some form of personal identification. You will also enter the check as cashed in your record, which you keep in a safe place apart from the checks. Then, if you lose any of the checks, or if they are stolen, they will be replaced by the issuing firm.

# How to Avoid Checking Problems

Once you open a checking account and start writing checks, a number of problems can arise. This section looks at some of the precautions you can take to protect yourself against potential pitfalls.

## Writing Checks

Writing out a check is, on the surface, very easy. There are, however, certain errors you can make that can cause problems later.

1. *Postdating a check.* In postdating a check, or writing a future date on it, you take a chance that the bank might slip up and cash the check before its postdated date. Thus, if you postdate a check in hopes of having sufficient funds to cover it on that future date, you might end up with an overdraft.

2. *Making checks out to "cash."* Making a check out to "cash," except when you are right in front of a bank teller's window, is not advisable. Anyone

can cash checks made out that way. If you lose such a check, it is the same as losing currency.

3. *Filling out the amount improperly.* It is often easy to be careless and leave spaces before and after the words and numbers indicating the amount of the check. If you do this, you risk an alteration that will increase the apparent amount of the check.

> "Writing out a check is, on the surface, very easy. There are, however, certain errors you can make that can cause problems later."

## Endorsing a Check

Whenever you cash a check that has been made payable to you or attempt to deposit it to your bank account, you have to *endorse* it first. To do this, you sign your name on the back of the check so it can be cleared and the money it represents transferred to you or your bank account. To endorse a check properly, you should sign your name exactly as it is written on the face of the check. If your name has been misspelled or incorrectly written, endorse it with the incorrect version and then write your correct name below it.

You should make your endorsement on the back of the left-hand side of the check. The Federal Reserve System requests that bank customers place their endorsements *only* within the top one and a half inches on the back of the left-hand side of the check. The rest of the space can then be used for bank endorsements as the check is processed through the banking system.

There are three types of endorsements. A *blank endorsement* is simply your name written on the back of the check. A blank endorsement allows anybody in possession of the check thereafter to cash or deposit it, after further endorsing the check with his or her own name, or to transfer it to another.

With a *special endorsement,* you specify a particular individual as the recipient of the funds. "Pay to the order of Annabel Maitland [signed] Sharon Cross" is an example of a special endorsement.

A *restrictive endorsement* indicates, above your signature, "for deposit only." With this type of endorsement, nobody else can cash the check or deposit it into an account other than your own.

## Stale Checks

A check that is more than six months old is considered a *stale check.* Banks have the option of paying or not paying stale checks. Although the bank might, as a courtesy, consult its customer to ask whether the check should be paid, the bank is not required to do so. This means that if you receive a six-month-old check from someone, you may not be able to cash it; similarly, if you receive a check and hold it for six months or more before depositing or cashing it, you may not be able to receive the money that the check represents. In your own accounting, you need to remember not to "forget" about checks you have written, even if they're now a year old; they still could be cashed. To ensure that your bank won't accept a check you wrote that has since became stale, you can submit a *stop-payment order* to your bank, as discussed next.

## Stop-Payment Orders

If you wish to stop a check from being collected after you have given it to someone, you can issue a **stop-payment order.** The bank will then refuse to honor the check. When you think a check has been lost or stolen, or when you realize you have purchased defective merchandise with a check, a stop-payment order is worth the $20 to $30 a bank will charge.

**Stop-Payment Order** An order to one's bank not to honor a particular check when it is presented for payment.

Generally, a phone call will stop payment on a check for only fourteen days; written notice will stop payment for six months, after which time the order can be renewed. Once the stop-payment order has been issued, the tellers in the bank are requested not to pay that particular check. The stop-payment information is also put into the computer so that the check will be rejected if another bank presents it in the bank clearing process. A stop-payment order can be made only on a regular check from a checking account; cashier's and certified checks cannot be stopped. A typical stop-payment order is shown in Figure 13.2.

## Deposit Holds

When you deposit checks into your account, don't assume that you will be able to withdraw the funds immediately—either in cash or by check. Banks are allowed to hold deposited checks for a certain period of time so that they can ascertain whether checks are good before paying them. Hold periods vary from bank to bank, and all banks are now required to disclose their hold policies to customers.

At one time, some banks routinely held deposits without crediting them to consumer accounts for as long as two weeks. Lengthy hold policies benefited banks because checks often cleared before the end of the holding period. Then, the banks could invest these funds and earn a return on them before making them available to depositors. Consumer complaints and public concern about lengthy holds led Congress to pass the Expedited Funds Availability

**Figure 13.2** A Typical Stop-Payment Order

To: [Name of Financial Institution]
I want to stop payment on the following check(s).

ACCOUNT NUMBER: ______ – ______

**SPECIFIC STOP**

*ENTER DOLLAR AMOUNT: ______ *CHECK NUMBER: ______

THE CHECK WAS SIGNED BY: ______

THE CHECK IS PAYABLE TO: ______

THE REASON FOR THIS STOP PAYMENT IS: ______

**STOP RANGE** (Use for lost or stolen check(s) only.)

DOLLAR AMOUNT: 000

*ENTER STARTING CHECK NUMBER: ______ *END CHECK NUMBER: ______

THE REASON FOR THIS STOP PAYMENT IS: ______

**BANK USE ONLY**

**TRANCODE:**

☐ 21—ENTER STOP PAYMENT (SEE OTHER SIDE TO REMOVE)

NON READS: ______

UNPROC. STMT HIST: ______

PRIOR STMT CYCLE: ______

HOLDS ON COOLS: ______

REJECTED CHKS: ______

LARGE ITEMS: ______

FEE COLLECTED: ______

DATE ACCEPTED: ______

TIME ACCEPTED: ______

I agree that this order (1) is effective only if the above check(s) has (have) not yet been cashed or paid against my account, (2) will end six months from the date it is delivered to you unless I renew it in writing, and (3) is not valid if the check(s) was (were) accepted on the strength of my ATM card by a merchant participating in that program. I also agree (1) to notify you immediately to cancel this order if the reason for the stop payment no longer exists or (2) that closing the account on which the check(s) is (are) drawn automatically cancels this order.

IF ANOTHER BRANCH OF THIS BANK OR ANOTHER PERSON OR ENTITY BECOMES A "HOLDER IN DUE COURSE" OF THE ABOVE CHECK, I UNDERSTAND THAT PAYMENT MAY BE ENFORCED AGAINST THE CHECK'S MAKER (SIGNER).

*I CERTIFY THE AMOUNT AND CHECK NUMBER(S) ABOVE ARE CORRECT.

☐ I have written a replacement check (number and date of check).

______

(Optional—please circle one: Mr., Ms., Mrs., Miss) CUSTOMER'S SIGNATURE **X** ______ DATE ______

Act in 1987. Under this act, deposits of government checks, checks on accounts in the same bank, cash deposits, and electronic transfers had to be available for withdrawal by the depositor on the next business day following the deposit. Checks drawn on other local banks could be held no longer than three business days and nonlocal checks for no longer than six business days. Since 1987 this law has been amended to reduce the legal hold periods to one business day for local checks and four business days for nonlocal checks. The act also provided that all bank customers must be allowed to withdraw up to $100 from the total of any checks deposited on the previous business day. Any state laws that require shorter hold periods are unaffected by the federal act.

## Avoiding Bounced Checks

If you write a check and don't have sufficient funds in your account to cover it, the check will "bounce." Your bank will immediately notify you by mail of the problem and charge your account with a fee—which may range from $20 to $30 or even more. The payee—the person to whom you gave the check—will almost certainly also charge a fee for the bad check.

If you are a long-time customer of the bank and have made deposits at regular intervals, the bank might go ahead and pay the check for you and charge your account with an overdraft for the amount of the check. Essentially, if the bank does this, it is temporarily loaning you the amount of the check, and you must reimburse the bank immediately. Frequently, however, the bank will simply return the check to the payee or other person who presented it for payment. When this happens, the payee or other holder can come directly to you and demand payment or redeposit the check in the hope that on the second collection attempt you will have enough money in your account to cover the check.

Bouncing checks can be more than embarrassing; it can also create legal problems for you. If you write a check and are unable to cover it, you could become involved in a civil lawsuit; and, if intent to defraud is involved, you could be criminally prosecuted for writing bad checks.

One way to ensure that you never bounce a check is to arrange with your bank to have *overdraft protection,* which was discussed earlier in this chapter. Another way to avoid bounced checks is to keep good records so that you know at all times how much money you have in your checking account. By entering the amount of each check you write in your check register and comparing your record with the monthly bank statement, you can avoid the problem of bounced checks.

## Reconciling Your Bank Balance

Every month, you will receive a bank statement that includes a list of checks that have been paid. Most banks no longer return the actual canceled checks you wrote. (You should have your deposit slips already.) It is important that you reconcile your bank balance with your checkbook or set of check stubs so that you know exactly how much you have in the bank, can catch any mistakes the bank might have made, and can find out if someone has not cashed a check you wrote.

Since checks written immediately before the closing date on your bank statement will not have been paid by the bank yet, the balance in your checkbook will rarely be exactly the same as the balance on your bank statement.

> **"It is important that you reconcile your bank balance with your checkbook or set of check stubs so that you know exactly how much you have in the bank, can catch any mistakes the bank might have made, and can find out if someone has not cashed a check you wrote."**

Thus, you must reconcile the two by taking account of deposits you made that do not show up on your bank statement and checks you wrote that have not yet been processed. The following is a simple procedure for reconciling your bank statement:

1. Deduct from your checkbook balance any service charges not previously recorded—for new checks, overdraft charges, and so on.
2. Enter your bank statement balance *(T)*. $ _____
3. After adding up all the checks outstanding that are not reflected on your bank statement, subtract the total of these unpaid checks *(U)* from the bank balance entered above and obtain a new balance here. $ _____ $(= T - U)$
4. Add up any deposits you made that did not show on your bank statement and put them here *(D)*. $ _____
5. Now obtain your final balance by adding the unreflected deposit total *(D)* to the new balance you found in Step 4 *(T − U)*. Your final balance is $T - U + D$.

This final balance should be the same as your checkbook balance after the service charge and any other bank charges have been deducted. Figure 13.3 shows the steps to follow in reconciling your bank balance.

Don't destroy your checks and the bank statement after you reconcile your bank balance. For income tax purposes, it's a good idea to keep bank state-

---

**Figure 13.3**
**Reconciling Your Bank Statement**

1. Adjust the balance in your checkbook for the service charge and other bank charges and credits shown on the bank statement but not recorded in your checkbook.

2. See that all deposits you made are properly credited.

3. See that all checks enclosed in your statement are checks issued by you.

4. Check each paid check against your checkbook stubs. List all checks outstanding in the space provided here.

| Checks outstanding not charged to account | | | |
|---|---|---|---|
| No. | $ | | |
| | | | |
| | | | |
| | | | |
| | | | |
| | | | |
| | | | |
| | | | |
| | | | |
| | | | |
| | | | |
| | | | |
| Total | $ | | |

Bank balance $ _____
shown on this statement

ADD +
Deposits not credited $ _____
in this statement $ _____
$ _____

TOTAL $ _____

SUBTRACT −
Checks outstanding $ _____
Adjust for bank fees $ _____
BALANCE $ _____

(Should agree with your checkbook balance)

ments and canceled checks for at least three years. Some people keep them longer in case there is a dispute with the Internal Revenue Service.

# How to Settle a Complaint with a Bank

Because so many of our personal financial transactions filter through the banking system, it is possible that at some point an error will be made and a dispute will arise between you and the bank. Because the error could have a long-lasting effect on your financial status and an impact on others involved in the transaction, it is important to try to settle it quickly.

As a first step, you should contact the bank directly and explain the problem; the bank may not even be aware of it. Many banks have designated employees to deal with these problems. For example, you may have deposited money that was credited to the wrong account; this deposit will fail to appear on your monthly checking account statement. In most cases, errors and misunderstandings can be corrected at this level. If, however, you cannot resolve the dispute to your satisfaction by dealing directly with the bank, you should take your complaint to one of the agencies that regulates your bank. These are listed in Table 13.1.

| TABLE 13.1 | Where to File Unresolved Complaints against Your Bank | |
| --- | --- | --- |
| **Type of Bank** | **Identification Marks** | **Where to Complain** |
| National bank | The word "national" appears in the bank's name, or the initials N.A. appear after the bank's name. | Comptroller of the Currency Customer Assistance Group 1301 McKinney Street, Suite 3710 Houston, TX 77010 (800) 613-6743 |
| State bank, member Federal Reserve, FDIC insured | Look for two signs at the bank: "Member, Federal Reserve System" and "Deposits Insured by Federal Deposit Insurance Corporation." | Division of Consumer and Community Affairs Board of Governors Federal Reserve System 20th and C Streets, N.W., Stop 801 Washington, DC 20551 (202) 452-3693 |
| State nonmember bank or state-chartered mutual savings bank | FDIC sign will be displayed; Federal Reserve sign will not. | Office of Consumer Affairs Federal Deposit Insurance Corporation 550 17th Street, N.W. Washington, DC 20429-9990 (800) 925-4618 or (202) 942-3147 |
| Federal savings and loan association | A sign on the door or in the lobby featuring an eagle surrounded by the words, "Backed by the full faith and credit of the United States Government." | Office of Thrift Supervision Office of Community and Consumer Division 1700 G. Street, N.W. Washington, DC 20552 (800) 842-6929 |
| Federal credit union | Sign will be displayed reading "Member National Credit Union Administration." | National Credit Union Administration Office of Public and Congressional Affairs 1775 Duke Street Alexandria, VA 22314 (703) 518-6330 |

# Confronting Consumer Issues: How to Use Electronic Banking

The last time you wanted cash, how did you get it? Chances are better than fifty-fifty that you visited an automated teller machine to withdraw cash from either your saving or your checking account. Your transaction was easy and quick, and it could have taken place at 11:30 P.M. Banking was not always like this.

Before the 1980s, completing a banking transaction inevitably involved a trip to a bank. There was no alternative. Making a deposit, cashing a check, or transferring funds from one account to another—all these transactions required the customer's personal presence at a teller's window or bank officer's desk. Banking was often slow, inefficient, and frustrating. The proliferation of electronic banking in the past twenty-five years has changed all that. It has also created new challenges for consumers who want to make the best financial decisions possible.

## ELECTRONIC FUND TRANSFER SYSTEMS

Electronic banking is the name commonly used for **electronic fund transfer systems (EFTS)**. EFTS have been developed throughout the financial world by applying computer technology to the banking industry. This development relieved banking institutions and their customers of much of the burden of having to move mountains of paperwork in order to process fund transfers and other financial transactions. An **electronic fund transfer** is a movement of money via an electronic terminal, telephone, computer, or magnetic tape. These transfers often involve the use of the Internet. Automatic payments, direct deposits, transfers of funds from banks or businesses to other banks or businesses, and virtually any other significant movement of funds can now be completed electronically without the use of any cash, checks, or other negotiable financial instruments. Through the use of EFTS, transactions that once took days or weeks to complete can now be carried out in a matter of seconds.

## TYPES OF ELECTRONIC FUND TRANSFER SYSTEMS

Consumers and businesses that directly serve consumers use four primary types of EFTS: (1) automated teller machines, (2) point-of-sale systems, (3) systems that handle direct deposits and withdrawals of funds, and (4) pay-by-telephone or Internet systems.

**Automated Teller Machines** Probably the most obvious application of electronic fund transfers is the use of automated teller machines (ATMs). ATMs are machines that

Digital Vision/Getty Images

**How have electronic fund transfer systems, such as ATMs, changed the way people do their banking?**

allow consumers to access their saving or checking accounts electronically from remote locations. ATMs can be found at many banks and at convenient locations such as supermarkets, drugstores, and shopping malls. Many are placed in small rooms or simply embedded in walls along busy city streets.

To use an ATM, consumers must first open an account with a bank or other depository institution that will issue them an *access card* and personal identification number (PIN). Access cards are **debit cards** that can be used to transfer funds from a consumer's saving or checking account. When used with an ATM, they may also be used to deposit funds. Consumers should always be careful to prevent others from learning their PIN. A PIN's purpose is to prevent others from using a lost or stolen access card to withdraw funds from a consumer's accounts. Your PIN cannot protect your funds if other people have it.

### How to Use an ATM

To complete a transaction at an ATM, you must slide your access card into the machine and wait for it to give you instructions on a screen. It will first ask you to enter your PIN on a key pad and compare your entry with electronic data it has read from a plastic strip on the back of your card. An incorrect entry will cause the machine to reject your card and prevent you (or anyone else) from completing any electronic transfers.

Once your PIN has been accepted, the screen will list a variety of transactions you may complete. All you have to do is choose and wait for instructions for the next thing you should do. If you are making a deposit, the screen will instruct you to place your cash or checks in a special envelope and insert it in a slot in the front of the ATM. Cash withdrawals will be paid from another slot located over a shelf that catches the bills as they are ejected from the ATM. You can withdraw only even dollar amounts in multiples of $10 from ATMs. If you need $53.29, you will have to withdraw at least $60 from the machine. After you have completed your transactions, the machine will return your card and print a record of your transactions. Be sure to remember to take the record with you. A record of your electronic transactions will be included on your next bank statement. You should compare the record printed by the ATM with the record in your monthly statement. If they aren't the same, you need to find out why and correct the mistake.

### Fees for Using ATMs

Fees are often charged for using ATMs. Many, but not all, banks allow their own customers to use their ATMs for free. After all, the banks are saving money because they don't need to pay an employee to complete their side of a transaction. Consumers who use ATMs that belong to banks or institutions where they do not maintain accounts can expect to be charged a fee for the service. This fee can be as large as $4 per transaction. If you complete many ATM transactions when you are out of town on vacation, you may find a substantial charge on your next bank statement that you had not anticipated. ATMs are now required to inform consumers of the fees they charge, either on their video screens or with signs on the front of the machines themselves.

### Point-of-Sale Systems

Because of the growth of electronic fund transfer systems, consumers are able to transfer funds directly to merchants to pay for purchases they make. **Point-of-sale systems** use special terminals located at checkout counters in many stores, including supermarkets, drugstores, and appliance dealers. Instead of paying with cash or a check, a consumer can hand the checkout person his or her debit card. The store's employee inserts the card into a terminal, which reads the data encoded on the card. The computer verifies that the card and

identification codes are valid and that the customer's account contains enough funds to cover the purchase. When the payment is approved, the customer's account is debited for the amount of the purchase, which is automatically transferred to the store's account.

Point-of-sale systems are valuable for consumers because they essentially make it impossible for the consumers to overdraw their accounts when completing this type of transaction. This is an important benefit for people who do not keep accurate records. For stores, these systems guarantee that funds will be transferred to the store's account for purchases that are made. Neither of these advantages is provided by transactions completed with paper checks.

**Question for Thought & Discussion #9:**
Why do some consumers who find it difficult to resist impulse purchases choose to use debit cards instead of credit cards when they shop? Does the use of a debit card protect them from overspending when they shop?

### Preauthorized Direct Deposits and Withdrawals

Most consumers receive periodic payments from their employers or the government that may automatically be deposited electronically in their checking or saving accounts. Direct deposits have many advantages for both the payer (the party that pays) and the payee (the party who receives). For the payer, direct deposit eliminates the need to actually print checks and reduces the possibility that any checks will be lost or misprinted and will have to be reissued. For payees, direct deposit eliminates the possibility that checks might be lost or stolen. Direct deposits are automatically credited to accounts without the need for any action by the payee. This means that directly deposited funds are available when they are paid and will immediately earn interest in many accounts. There is no chance that a check will be forgotten in a consumer's briefcase or purse.

### Making Periodic Payments

In a similar way, most consumers make periodic payments that may automatically be transferred electronically from their accounts to the payee through electronic withdrawals. It is possible, for example, to enforce your saving plan by having funds automatically transferred from your checking account to a saving account on the same day every month. In addition, many consumers pay their telephone, gas, or cable bills with automatic electronic fund transfers.

### Advantages of Automatic Payments

The advantages of making periodic payments automatically are many.

*(Continued on next page )*

**Confronting Consumer Issues (Continued)**

## CONSUMER CLOSE-UP
## How Much Is Convenience Worth?

**N**early 20 million times each day a consumer steps up to an ATM to complete one or more financial transactions. A consumer may withdraw funds, make a deposit, move money from one account to another, or pay a bill. In the majority of cases, consumers pay a fee of $1 to $4 for each transaction they make. Those who use ATMs that are owned by their personal banks may pay only one small fee or possibly none at all. But, if they use a machine that belongs to a different bank or financial institution, they will probably pay two fees—one to the firm that owns the machine and one to their personal banks as well. These fees can be a large percentage of the value of the transactions they make. Often customers are not even aware of how much they will be charged. Until recently, most ATMs provided no information about the fees charged to customers who use them.

Imagine, for example, that while you are shopping one day, you find yourself short of cash and withdraw $50 from your saving account at a nearby ATM that does not belong to your bank. You might be charged $2 by your bank and another $2 by the firm that owns the machine. That's $4, or 8 percent of the value of the funds you withdrew from your account. If you make five transactions like this at ATMs per month, that's $20 per month, or $240 per year. That $240 is a hefty price to pay for convenience.

In recent years many consumers complained that these fees were too high and that it was difficult to know how much they would be charged for using an ATM. In fact, residents of San Francisco voted to outlaw multiple ATM fees in 1999, although this law was overturned by a federal court within a week after it was passed.

The federal government has acted to force ATM owners to provide consumers with the information they need to make rational choices when they use ATMs. On November 12, 1999, then-President Bill Clinton signed the Financial Services Act, which requires ATMs to list fees that will be charged for different types of transactions "prominently" where they can easily be read by customers. This law also required ATMs to display the fees on their screens before transactions are completed. Implementation of this part of the law was postponed because of protests from the banking industry that it is technologically impossible to program fees that are charged by each consumer's bank into every ATM. The General Accounting Office is studying the situation and will determine the technological feasibility of the law's requirements by 2004.

Even if ATM fees are prominently displayed, consumers must choose whether to pay the fees or not use an ATM. When you are short of cash or need to complete banking transactions at night, you may feel you have little choice except to pay these fees.

**Do you believe that charging multiple fees for ATM use is justified? Banks have threatened to remove ATMs if they are not allowed to charge fees for ATM use. Do you believe the banks would really do this? How much are you willing to pay for the financial convenience offered by ATMs?**

For the consumer, it means that bills will be paid on time without any checking fees. This will avoid the possibility of being assessed late charges that can quickly add up for people who forget to pay their bills. Whether they save or pay bills with electronic fund withdrawals, consumers must always remember to keep enough money deposited in their accounts to cover their automatic withdrawals. Consumers who forget are likely to write bad checks that will cause their banks to impose insufficient-funds penalties. These penalties can be as high as $30 per check. For businesses, automatic withdrawals guarantee they will receive payments on time as long as there are sufficient funds in the customer's account. This system reduces their mailing and processing costs.

### Pay-by-Telephone or Internet Systems

In some situations, consumers may wish to make electronic fund transfers, but not automatically. For example, suppose you have agreed to make electronic payments to the

company that supplies your home with natural gas. The amount of your gas bill changes each month. In February it might be as much as $300 or $400 or even more. You might want to find out how much the bill is and then deposit funds in your account before a transfer is made to the gas company. Some banks allow customers to call them on the phone or contact them over the Internet to order the bank to make such payments at a time the customer chooses. This reduces the chance that there won't be enough funds in the customer's account to cover the bill.

### LEGAL PROTECTION FOR ELECTRONIC BANKING

Electronic banking obviously poses some potential problems for consumers. When you write a check, it has your signature on it and leaves a paper trail of evidence as it is cleared through the banking system. If someone forges your signature on a check, there will be paper evidence that can prove you did not write the check. Electronic fund transfers, however, leave no such trail of evidence. Consequently, it can be much more difficult to settle disputes over this type of transaction. For example, what are your rights if someone steals your ATM access card and somehow finds your PIN as well? That person could go to any ATM and withdraw hundreds of dollars from your account. This could go on for days before you find out. If such an event took place, what would your liability be? Would the bank return your funds to your account? How would you prove it wasn't you who made the withdrawals? These are important questions for you to consider when you open electronic banking accounts.

In 1978, Congress addressed these and other issues involved in electronic banking by passing the Electronic Fund Transfer Act (EFTA), which provides a basic framework for "establishing the rights, liabilities, and responsibilities of participants in electronic fund transfers." In addition, the EFTA gave the agency that oversees banking in the United States, the Federal Reserve System, authority to issue rules and regulations to help implement the act. The governing body of the Federal Reserve System is its Board of Governors. This board has set and implemented **Regulation E,** which spells out the respective responsibilities of banks and consumers when disputes arise over electronic fund transfers.

## Disclosure of Terms and Conditions under the EFTA

The EFTA requires that the terms and conditions of electronic fund transfers involving a customer's account be disclosed in writing in readily understandable language at the time the customer opens the account. The required disclosures include the following:

- The customer's liability for unauthorized transfers resulting from the loss or theft of a debit card, code, or other access device.
- The phone number to call to report the theft or loss of a debit card.
- Any fees for using the EFTS.
- The services available and any limits on the dollar amounts or frequency of use.
- The customer's right to be provided with evidence of transactions in writing.
- The procedure for correcting mistakes and resolving disputes.
- The customer's right to stop or prevent transactions.
- The bank's liability to the customer for mistakes it makes.
- Rules concerning the bank's provision of account information to third parties.

When you open an account that provides electronic fund transfer services, you should read the list of disclosures carefully and be sure you understand them. You don't want to find a mysterious $1,000 charge on your bank statement and not know what to do about it.

## Rules to Remember When Using EFTS

The following rules can also help you protect yourself when you use EFTS:

1. *After making an ATM transaction, always wait for the record of your transaction to be printed and take it with you when you leave.* Such records tell you the amount, date, and location, as well as the type of transaction you have completed. This information allows you to verify the bank's monthly statement. Also, if you have a dispute, it helps to have the actual receipt from the machine. Check to make sure the printed record is correct. If there are any mistakes, call your bank immediately. There is often a phone located at the ATM that allows you to contact the bank that owns the machine directly. If there is no phone, a number should be provided on the ATM for you to call.

2. *When you make a mistake that cannot be corrected at an ATM, pick up the customer service phone or contact the bank that owns the machine immediately.* Don't put it off. It is generally easier to correct a mistake now than later. Follow the same procedure if you are ever shortchanged by an ATM.

3. *Keep all EFTS transaction records in a single location.* Keeping your records in several locations is almost a guarantee that some will be lost or misplaced, making it impossible for you to reconcile your statement at the end of the month.

4. *When you make EFTS withdrawals, be sure to record these transactions immediately in your checkbook.* People who don't do this can expect to overdraw their accounts regularly.

5. *Take the time to reconcile your monthly bank statement every month.* Letting this job slide will only make it bigger and more distasteful when you finally decide to do it. You might even decide never to do it. Remember, if you let a mistake go too long, you may not be able to correct it.

6. *Don't share your PIN with anyone.* Avoid using access codes for EFTS transactions that are based on your name, date of birth, address, telephone number, Social Security number, or any other combination of letters or numbers that can easily be associated with you.

7. *When using an ATM, be sure that no one else is able to see you enter your PIN.* If necessary, ask anyone near the machine to step away. Be polite but be firm. It is better to postpone using an ATM if you can't use it with privacy.

8. *Don't use an ATM in a place that you believe might be dangerous.* There is little difference between using an ATM and carrying large amounts of cash. Crooks could force you to

*(Continued on next page)*

**Confronting Consumer Issues (Continued)**

## THE ETHICAL CONSUMER
## Financial Privacy in Cyberspace

Some people are convinced that within a few years virtually all banking will be done electronically and that most financial transactions will be carried out online. Electronic banking offers many advantages for both the banking industry and consumers. The most obvious may be the lower cost. Computers can complete almost any transaction that required the services of a human being only a few years ago. They can transfer funds, record deposits, make payments, and order new checks. Computers make few mistakes, don't get tired, can work twenty-four hours a day, and don't have to be paid overtime on Sundays or holidays.

Another of the great advantages of e-banking, from the banking industry's perspective, is the records of consumers' financial activities that it creates. These records, however, pose an ethical issue for consumers as well as for the banks that serve them.

Electronic banking is a double-edged sword. Although it is convenient and efficient, it also creates the potential for abuse of the individual's right to privacy. Records of electronic transactions build a financial picture of consumers. When you use an ATM, buy a product with a charge or debit card, purchase goods or services online, or simply transfer funds electronically to pay your bills, there is a record of where, how, and for what you used your money. This might not seem important to you, but it may be. Information about your financial transactions could be sold to marketers who might bother you with phone calls or unsolicited e-mails. Furthermore, there are other, more worrisome possibilities.

Suppose you work for your local government and contribute to the political campaign of a mayoral candidate who then loses the election. Would you want the winner to know you had contributed to her or his opponent? If you became ill and made a payment to an oncologist (cancer specialist), would you want your employer to know? What you do with your money can tell others a

great deal about who you are and what you value. It could enable others to take advantage of you. Do you want your financial decisions to be known by others? Are your choices any of their business? Concerns such as these have caused many people to put pressure on the government to pass laws to protect their rights to electronic privacy.

Congress has often discussed writing a comprehensive law to protect the electronic privacy rights of U.S. citizens. As of the beginning of 2003, however, no such law was in place. This does not mean that progress had not been made toward protecting consumers. In October 2001, the chairman of the Federal Trade Commission, Timothy J. Muris, described some of these improvements. He pointed out that only 2 percent of Web sites had privacy notification pages in 1998. According to Muris, by 2001 "virtually all popular sites" had such notifications. This may not have prevented information about your financial transactions from being gathered and passed on, but at least you could find out what a firm's policy was.

Muris explained that one of the most important reasons for Congress's inaction on a comprehensive law was the difficulty in writing one that would work. The Internet is so vast that it is impossible to police every Web site. Further, when a Web site is created that might harm consumer interests, it would be very easy to avoid prosecution or regulation by simply closing down and opening the next day at a different Web address. Muris fears that many suggested laws would harm legitimate, honest businesses that use the Internet but would do little to prevent the abuse of consumers by unscrupulous firms.

**What is your opinion of Muris's point of view? Do you believe laws can be passed to assure consumers that information gathered about their financial choices is used only in ethical ways?**

withdraw funds from the machine almost as easily as they could steal your wallet or purse. As in most situations, a little common sense is in order when you use an ATM.

## DEALING WITH EFTS PROBLEMS

On any given business day, as many as 100 million EFTS transactions may be completed. With that many transactions, it is not surprising that mistakes occasionally take

place. Studies show that the vast majority of EFTS errors are made by consumers who enter transactions incorrectly in their records or make mathematical mistakes when they reconcile their accounts. Still, some of these inaccuracies are made by bank employees. Regardless of who caused a mistake, it is the responsibility of all consumers to investigate and correct errors they believe they discover in their bank statements.

## Steps for Correcting an EFTS Mistake

If you find an error in your bank statement, you should immediately telephone the bank to register your concern. This, however, is only the first step. In general, you are required to inform the bank of the problem in writing within sixty days of receiving the statement, or you may forfeit your right to have the mistake corrected. Your letter must contain the following information:

1. Your name and account number
2. A statement that an error has been made and the amount of the error
3. An explanation of how you think the error occurred

The bank is then required to investigate the problem and report the results of its investigation to you within ten business days. If a full investigation requires more than ten days, than bank can take up to forty-five days—but it must credit your account with the disputed amount until the problem is resolved. After the bank has completed its investigation, it must give you a full, written report of its findings—even if no mistake was made. Banks are held to strict compliance with the EFTA, and if your bank fails to adhere to the letter of the law, it may be held liable for treble damages (three times the amount of provable damages) to you, the consumer.

## How to Stop a Preauthorized Transfer

As the name implies, a preauthorized transfer is one that is approved by you in advance through a special arrangement you make with your bank. In general, the authorization must be in writing, and the bank must give you a copy of the authorization when the arrangement is confirmed. But what if you want to stop a preauthorized payment from being made? Perhaps you want to make a payment in person, or for some other reason you don't want the bank to go ahead with the transfer.

Under the EFTA, you can request the bank to stop the transfer at any time up to three business days before it is scheduled to be made. You can notify the bank orally of your wishes, although if the notification is oral, the bank will probably require written confirmation of the notification within fourteen days.

## How to Deal with a Lost Debit Card

If you lose your debit card, or if it is stolen, you should immediately notify your bank. Under the EFTA, if you notify your bank of the loss or theft within two business days, you will be responsible for only $50 of any unauthorized transfers from your account. If you don't notify the bank until after the second business day, your liability climbs to $500. Your liability may be unlimited if you fail to notify the bank within sixty days of your receipt of a periodic statement that includes an unauthorized transfer.

Electronic fund transfers offer consumers many benefits as long as consumers take their responsibilities seriously. As with most consumer products, the benefits you can receive from electronic fund transfers are directly related to the way you use them.

## Key Terms:

**Electronic Fund Transfer Systems (EFTS)** Systems for transferring funds electronically.

**Electronic Fund Transfer** A transfer of funds via an electronic terminal, telephone, computer, or magnetic tape.

**Debit Card** A card similar to a credit card that allows a consumer to transfer funds from accounts by using a computerized banking system.

**Point-of-Sale System** An electronic communication system that can debit a customer's account when a debit card is used to cover a purchase from a merchant.

**Regulation E** The rules issued by the Federal Reserve Board to protect users of electronic banking services.

# Chapter Summary

1. Economists define saving as the act of not spending current money income. Individuals save in order to achieve personal goals and to provide personal security. Saving allows consumers to continue to spend during periods when their earning capacity falls, such as during sickness or retirement.

2. The amount people are able to save depends on their financial situations and their current flow of income. Almost all people can set some amount of income aside. It can be helpful to think of saving as a payment to yourself. Those with irregular flows of income may save by setting a set proportion of income aside after fixed expenses have been paid.

3. Individuals should place their savings in a secure location that provides a return on their savings. This type of service is provided by most depository institutions. A depository institution is any organization that accepts deposits and assures depositors that they will be able to withdraw their funds when they need them. The most common type of depository institution is the commercial bank. Since the passage of the Depository Institutions Deregulation and Monetary Control Act of 1980, the differences between commercial banks and other depository institutions have become small for consumers.

4. Banks offer consumers a variety of accounts in which they may deposit their savings. These include day-in-day-out accounts that pay a relatively low rate of interest every day funds are on deposit. Day-in-day-out accounts generally have no minimum deposit. Money market accounts also pay interest for each day funds are left on deposit. They require a minimum deposit, and the interest rate they pay may fluctuate from day to day. Savers who make time deposits, most often certificates of deposit (CDs), agree to leave them on deposit for specific periods of time. Although these funds may be withdrawn early, an early withdrawal penalty equal to a specific time period's interest will be charged. Banks pay higher rates of interest for CDs because they are able to earn a better return from these funds.

5. Consumers should consider several factors when they evaluate different banks they could choose. They should be careful to select a bank that belongs to the FDIC or a similar federally sponsored insurance fund. Convenience is an important consideration for many people. Finally, consumers should seek the best interest rate for their savings. At one time it was difficult to compare interest rates offered by different banks. This problem was eliminated by the passage of the Truth-in-Savings Act in 1993. This law requires banks and other depository institutions to advertise a standard interest rate called the

annual percentage yield (APY) that must be calculated in the same way by all banks. Now savers can easily compare interest paid by different accounts by comparing their APYs.

6. Many savers purchase government securities as an alternative to depositing their savings in bank accounts. The federal government offers three types of savings bonds: Series EE bonds, Series HH bonds, and I bonds. EE bonds come in denominations of $50 to $10,000. They may be purchased for half their face value at most depository institutions. EE bonds pay interest that is equal to 90 percent of the rate paid on five-year Treasury securities. I bonds pay a fixed rate of interest plus the rate of inflation as measured by the consumer price index. HH savings bonds may be purchased only by trading in matured EE bonds. HH bonds are purchased for their face value and pay interest each year until they mature. Savings bonds cannot be sold to another person, but they can be cashed in at any time after one year from the day they were originally purchased. An important advantage of owning any federal securities is that the interest they pay may not be taxed by any state or local government.

7. Two other types of federal government securities that consumers may purchase are Treasury bills and notes. These securities are sold in denominations of at least $1,000 and will not be paid off by the Treasury until they come due. They may, however, be sold to others on the secondary market before they mature.

8. Banks help consumers spend by offering checking accounts. Funds deposited in checking accounts are called demand deposits because they may be spent at any time by simply writing a check against them. Consumers may choose from many types of checking accounts. These include activity accounts that may charge maintenance and activity fees, minimum balance accounts that may charge no fees but require that substantial balances be maintained in the account, package plans that charge a monthly fee to provide checking accounts along with a variety of other bank services, and asset management accounts that offer brokerage services as well as most services included in typical package plans. Free checking accounts are rare, but are sometimes offered as a marketing tool to specific groups of consumers.

9. When choosing a checking account, consumers should evaluate their alternatives in terms of their own financial needs and abilities. People who have little money and write few checks should seek an account with a low maintenance fee. Heavy check writers should be more interested in a low per check fee. Those who can maintain a large balance should consider different minimum balance plans.

10. There are a variety of special checks that consumers may use in certain situations. Banks guarantee payment of certified checks by transferring funds from a depositor's account to cover the amount of the check. A cashier's check is drawn by the bank on its own funds and is made payable to a person designated by the purchaser. A traveler's check is purchased from a financial institution and signed by the purchaser at the time of purchase. It can be used only by the purchaser when he or she countersigns the check in front of the person who will receive it.

11. To avoid potential problems with checking accounts, it is important to write and endorse checks properly, to be aware that stale checks may or may not be honored, to know how to issue a stop-payment order on a check, and to realize that banks may hold your deposits for a day or more before you can withdraw the funds by check or in cash.

12. To avoid the embarrassment and potential legal problems that could result from bouncing checks, you can arrange with your bank to have overdraft protection. Good record keeping will help to ensure that you don't write checks when you have insufficient funds in your account to cover them. Monthly reconciliation of your bank statement with your check register allows you to keep informed of the status of your account and your current balance.

13. Electronic fund transfer systems (EFTS) allow vast sums of money to be moved from account to account in sec-onds without mountains of paperwork. The four principal types of EFTS are automated teller machines, point-of-sale systems, preauthorized direct deposits and withdrawals, and pay-by-telephone or Internet systems. The rights and obligations of banks and consumers in EFTS are spelled out by the Electronic Fund Transfer Act of 1978 and the Federal Reserve Board's implementation of Regulation E.

14. Consumers are less likely to be robbed at ATMs when they exercise common sense. If your debit card is lost or stolen, notify your bank of the loss within two business days to keep your potential liability to a maximum of $50. If you report the loss after two business days, your liability could reach $500, and if you take more than sixty days, you could be responsible for all losses resulting from the unauthorized use of your card.

15. If you notice an error on your bank statement and notify your bank within sixty days of receiving the statement, the bank is required to investigate the problem and report its findings to you within ten business days. If a longer time is needed, the bank can take up to forty-five days, but it must credit your account for the disputed amount until the problem is resolved.

16. The EFTA requires that all terms and conditions pertaining to electronic fund transfers, including the bank's and the customer's obligations in EFTS, be disclosed to any customer who contracts for EFT services.

## Key Terms

annual percentage yield
  (APY) **304**
automated teller machine
  (ATM) **302**
cashier's check **312**
certificate of deposit (CD) **303**

certified check **311**
debit card **323**
depository institution **300**
electronic fund transfer **323**
electronic fund transfer systems
  (EFTS) **323**
fractional reserve system **300**
money market accounts **303**

personal identification number
  (PIN) **302**
point-of-sale system **323**
Regulation E **323**
saving **297**
stop-payment order **313**
traveler's check **312**

## Questions for Thought & Discussion

1. Do you believe Americans save too little of their income? What problems might current saving rates lead to in the future?

2. What goals have you set in your life that are important enough for you to give up current spending? How successful have you been in saving?

3. What different types of depository institutions exist in your community? What similarities and differences can you identify in the services they offer?

4. Compare the current rate of inflation with the rate of interest paid on bank saving accounts. Is this interest rate an important reason why people choose to save?

5. Why are young people less likely to deposit their savings in CDs than in day-in-day-out accounts?

6. Why do many savers choose to purchase savings bonds instead of depositing their funds in bank accounts? What advantages and disadvantages are associated with this method of saving?

7. What types of financial transactions do most consumers use checks to complete? Why don't they make these payments in cash?

8. Compare the type of checking account that is appropriate for your current needs and the type that you would expect to use ten years in the future. Why should consumers periodically review the types of accounts they maintain?

9. Why do some consumers who find it difficult to resist impulse purchases choose to use debit cards instead of credit cards when they shop? Does the use of a debit card protect them from overspending when they shop?

## Things to Do

1. Calculate the percentage of your income that you currently save. Identify the goals you have for saving. Determine how much saving you need to accomplish to achieve these goals. Use this knowledge to create a personal saving plan that would better help you reach your life-span goals.

2. Visit three local depository institutions and ask for information about the various types of saving accounts they offer. Assume that you recently inherited $5,000 and want to deposit these funds in an account that offers you the ability to access your money at any time while paying the best interest rate possible. Identify the type of account and depository institution you would choose. Explain your choice in several paragraphs.

3. Visit three local depository institutions and ask for information about the various types of checking accounts they offer. Assume that in a typical month you write twelve checks, that you cannot afford to keep more than $200 on deposit at all times, and that you do not own a car that you can use to drive to a bank. Identify the type of checking account and depository institution you would choose. Explain your choice in several paragraphs.

4. Assume that you deposited your $543.20 paycheck in your checking account three weeks ago. You were given a correct receipt for your deposit, which you have saved. You received your bank statement today and were surprised to find that the bank believes your checking account has only $23.82 left in it. According to your check register, you should have $512.70. On closer inspection, you discover that your deposit of $543.20 was credited to your account as only $54.32. You must write a check for $300 in the next few days for your share of your apartment's rent. Write a letter to the bank that explains the situation and politely asks that the bank make the appropriate corrections to your account.

## Internet Resources

### Finding Consumer Information on the Internet

The following Web sites have been selected for their relevance to topics discussed in this chapter. Search these sites to locate information that can add to your knowledge of saving and checking protection and opportunities available to consumers. Remember, Web addresses change frequently. If any of these addresses no longer function, find similar sites to investigate using any of the search engines available to you.

1. The Federal Reserve System influences money and credit conditions in the United States, supervises and regulates banking, maintains the stability of the financial system, and provides certain financial services. Its Web site is located at **http://www.federalreserve.gov**.

2. The Federal Deposit Insurance Corporation insures deposits and promotes safe and sound banking practices. Visit its Web site at **http://www.fdic.gov**.

3. The Bureau of the Public Debt issues federal government securities that finance the government's borrowing. Investigate its Web site at **http://www.savingsbonds.gov**.

## Shopping on the Internet

The following Web sites have been selected because they offer consumers services similar to those described in this chapter. These are commercial sites that are designed to market products. They do not represent a comprehensive or balanced description of all banking services available online. How do the offers made at these Web sites compare with those made by local lending institutions? Remember, Web addresses change frequently. If any of these addresses no longer function, find similar sites to investigate using any of the search engines available to you.

1. Citigroup advertises that it can help students reach college saving goals. You may visit its Web site at http://www.citigroup.com/citigroup/press/011023a.htm.

2. Wells Fargo offers time accounts, CDs, and repayment options online. You may investigate its products at http://www.wellsfargo.com/per/student/finaid/finaid_overview.jhtml.

3. Bank of America states that it can help you get more from your saving. Its Web site is found at http://www.bankofamerica.com.

## InfoTrac Exercises

Purchasers of new copies of this text are provided with access to the InfoTrac Web site. This Web site links students to thousands of recent articles published in hundreds of periodicals. Use the key words **saving account, checking account,** or other terms from this chapter to conduct a key-word search. Choose one article that is of particular interest to you and write a brief essay describing what you have learned from the article. Be sure to cite the author and title of the article and the name and date of the publication in which it appeared.

## Selected Readings

Allen, Janet. "Money Lessons for Every Age." *Parents Magazine,* March 1999, pp. 104–107.

Boraks, David. "Big Banks Warming to Free Checking." *American Banker,* February 4, 2003, p. 3.

"Bounce and Pay." *Newsweek,* October 18, 1999, p. 86.

"Buy Series I Bonds Online." *Kiplinger's Retirement Report,* December 2002, pp. 8–9.

Clowes, Mike. "Bad Times for Personal Saving Rate." *Investment News,* September 2, 2002, p. 2.

"Court Rules Local Laws Can't Govern ATM Fees." *Banking Wire,* December 9, 2002, p. 10.

"Disclosure of ATM Fees." *Kiplinger's Retirement Report,* August 2001, pp. 10–11.

Henderson, Tim. "More ATM Fees, Free Checking." *American Banker,* June 17, 2002, p. 24.

"Learn How to Shop for Checking Accounts." *Knight Ridder/Tribune Business News,* September 25, 2002, p. ITEM0226811.

Marquis, Milt. "What's behind the Low U.S. Personal Saving Rate?" *Federal Reserve Bank of San Francisco Economic Letter,* March 29, 2002, pp. 1–3.

Stern, Linda. "Money: Stairway to Haven." *Newsweek,* December 23, 2002, p. 72.

"Survey Finds Checking Fees Hit New Highs." *Credit Union Journal,* October 2001, p. 3.

"Top Money Market Yields." *Kiplinger's Personal Finance Magazine,* March 2003, p. 95.

Walden, Mark. "For Savings We're Our Own Worst Enemies." *Triangle Business Journal,* April 30, 1999, p. 46.

Photodisc/Getty Images

# CHAPTER 14

# Using Credit Responsibly

After reading this chapter, you should be able to answer the following questions:

- How much are U.S. consumers in debt, and what is the nature of their debt?
- What are the sources of consumer credit?
- What are the benefits and costs of using consumer credit?
- What legislation protects consumer credit rights?
- How can consumers be responsible in their use of credit?

I n 1786, in the city of Concord, Massachusetts, the scene of one of the first battles of the Revolution, there were three times as many people in debtors' prison as there were in prison for all other crimes combined. In Worcester County, the ratio was even higher—twenty to one. Most of the prisoners were small farmers who could not pay their debts. In August 1786, mobs of musket-bearing farmers seized county courthouses to halt the trials of debtors. Led by Daniel Shays, a captain from the Continental Army, the rebels launched an attack on the Federal Arsenal at Springfield; although they were repulsed, their rebellion continued to grow into the winter. Finally, George Washington wrote to a friend:

> For God's sake, tell me what is the cause of these commotions. Do they proceed from licentiousness, British influence disseminated by the Tories, or real grievances which admit to redress? If the latter, why were they delayed until the public mind had become so agitated? If the former, why are not the powers of government tried at once?

At any given time, the majority of Americans have outstanding *installment debt.* In other words, they have received money from a lender and contracted to pay back what is owed plus interest over a certain period of time—say, monthly for four years. By 2003, the total amount of consumer credit (not including mortgages) was $1.8 trillion—equal to more than one-fifth of personal disposable income. Clearly, we are an indebted society, and buying on credit is part of the American way of life. A number of sources are available to supply that credit for us.

> **"Clearly, we are an indebted society, and buying on credit is part of the American way of life."**

# Sources of Credit

The numerous sources of credit for today's consumer can be divided into two general categories:

1. Sources of credit for consumer loans.
2. Sources of credit for consumer sales.

The first category relates to direct loans that a consumer can obtain. The second category relates to the extension of credit when some item, such as a stereo, is purchased.

## Sources of Consumer Loans

Consumers can turn to a number of sources for direct loans.

**COMMERCIAL BANKS**   The most obvious place to obtain credit is a commercial bank. The personal-loan departments of commercial banks extend about 50 percent of the loans for automobile purchases, as well as about 20 percent of all loans for other consumer goods. In general, these banks are full-service commercial banks that have personal-loan departments.

**CONSUMER FINANCE COMPANIES**   Consumer finance companies make small loans to consumers at relatively high rates of interest. Interest rates are typically very high because consumer finance companies cater to a higher-risk clientele; that is to say, they incur a higher risk of nonpayment on their outstanding loans than do commercial banks. They must be compensated for this higher risk by receiving a higher interest rate. Slightly more than 10 percent of installment credit is financed by these small-loan companies.

**CREDIT UNIONS**   Credit unions are a special type of consumer cooperative agency; only members may borrow from a credit union. Teachers and workers belonging to large unions or companies often have the opportunity to join their own credit unions. Credit unions account for nearly 20 percent of all personal consumer credit.

**LOANS ON YOUR LIFE INSURANCE**   If you have a life-insurance policy with a cash value (see Chapter 18), you may be able to obtain a relatively low-cost loan on that policy. You usually pay something less than 10 percent for a loan on the value of your policy. You cannot be turned down for a loan from your insurance company, and no questions are asked about how you plan to use the money, because it is *your* money. Your credit rating has nothing to do with whether you get the loan. You can take as long as you wish to repay. In fact, whenever the policy becomes payable—either because it matures or the owner of it dies—any outstanding loan is deducted from the amount of the insurance claim that the company must pay. Hence, any loan you take out reduces your insurance protection.

**OTHER**   Approximately 10 percent of consumer loans are extended by other types of lenders. Pursuant to the 1980 Depository Institutions Deregulation and Monetary Control Act, savings and loan associations and saving banks are now allowed to make some personal loans to consumers. Student loans, which are guaranteed by the government but administered through banks, also become installment debt (when they become due). Brokerage houses are another potential source of consumer credit for their customers. In addition to making loans to customers for purposes of purchasing stocks and bonds on credit, brokerage houses allow customers to borrow—by writing checks on their asset management accounts (see Chapter 13)—up to a certain percentage of the value of their investments held by the brokerage firm.

Cash advances on major credit cards, such as MasterCard, Visa, Sears's Discover card, and American Express's Optima card, are another source of installment credit if the loan is repaid over time and draws interest.

## Sources of Consumer Sales Credit

In addition to direct loans, consumers also obtain credit by purchasing products on the installment plan or by using their credit cards.

**AUTOMOBILE DEALERS AND FINANCIAL SERVICES**   Installment loans to pay for the purchase of an automobile are available from car dealers or affiliated financial agencies, such as General Motors Acceptance Corporation. Automobile loans account for approximately 40 percent of all installment debt in the United States.

**CREDIT CARDS**   Today more than 172 million Americans hold at least one credit card. On average, however, cardholders have nine cards each, making a total of over 1.6 billion cards in use. Note, too, that some people must hold more than the average—possibly twenty cards or more.

Two of the most widely known bank credit cards, MasterCard and Visa, account for nearly 500 million of the total—that's almost two bank cards for every adult in the United States. Other credit cards include American Express, Diners Club, and Sears's Discover card. In addition to these major credit cards, most retail outlets offer some form of credit to their customers. Virtually all of

**Question for Thought & Discussion #1:**
What sources of credit do you use? What sources do other members of your family use? Why were these sources chosen?

the major stores, such as Sears, J. C. Penney, Marshall Field's, and Kmart, among others, provide individual credit cards to their customers for making purchases at their stores. Oil companies such as Shell, ChevronTexaco, and Conoco also make it possible for their customers to purchase their products on credit with a credit card.

Credit cards have become an important source for consumer sales credit, as well as for direct loans in the form of cash advances, as already mentioned. Approximately 38 percent of all installment debt is created by credit cards. The average monthly credit-card balance per adult in the United States is now approaching $4,000, and millions of Americans have balances that exceed $8,000.

## Credit Online

Regardless of your purpose or the type of credit you would like to obtain, you can find useful information and identify lending sources online. An Internet search using key words such as *consumer credit, credit cards,* or *home mortgages* will link you to hundreds of institutions that extend credit to consumers online for virtually any reasonable purpose you might imagine. By checking out several Web sites, you can get a good idea of the interest rates, monthly payments, and total cost of borrowing over the Internet. Even if you don't borrow online, an important advantage of online credit for all borrowers is the increased competition it creates. In rural Iowa, for example, there may be only one bank within a fifty-mile area. Because consumers in the area cannot easily get to a different lending institution, this bank could effectively have monopoly power in the local credit market. But the fact that consumers can borrow online may prevent the local bank from charging excessive interest rates. Those who do choose to borrow online should always be sure to check out the reliability and reputation of the lender. Much of this information can easily be found on the Internet.

# Why Borrow?

Why should you ever borrow money? Some of you may answer, "There's no reason you should. Pay cash for everything and never have debt hanging over your head." This is still the prevailing attitude in much of Europe, where many people are reluctant to borrow to purchase goods and services.

## Pay for Future Services with Future Earnings

Many of us might explain our borrowing this way: Suppose you decide that you want to buy an automobile. Remember, you are not buying the automobile *per se;* you're really buying the service of that automobile that you expect to receive over the next few years. In fact, what is likely to be most important to you is the cost of the service flow you will receive over time. Cars, as you realize, are consumer durables. Their value lies in their ability to satisfy our desires or needs over a period of time. So, if you expect to use your new car for four years, why not borrow the money you need to buy it and pay off your debt over the same four years? Does that make sense to you?

In contrast, consider your perspective when you go to a movie. You consume the movie, so to speak, during the two hours you are in the theater—and you pay for your entertainment at the same time you enjoy it. Or, if you go to

Telephone shopping almost always requires the use of a credit card. Do you pay off your credit card debt at the end of each month, or do you carry a balance from month to month?

a restaurant and pay cash for your meal, what are you doing? You are timing your payment to coincide with the value you receive from your spending. But when you buy a large-screen television, just like the automobile, you expect to use the television for many years. You decide to borrow $800 to pay the purchase price. As you use the television, you gradually pay off your debt. What you have done is choose to *synchronize your spending with your consumption.* You might say, "Why should I pay the entire price now for products that I will use over many years? Why not pay for them at the same rate I use and enjoy them?" This argument could be used to justify the purchase of any durable good with borrowed funds. Many people regard this logic as a valid reason for using credit.

## Other Reasons to Borrow

There are, of course, many other reasons why you might choose to borrow money, including:

- Taking advantage of advertised specials when you are short of cash.
- Consolidating bills.
- Having a safeguard in emergency situations.
- Being able to shop or travel without having to carry large amounts of cash.
- Increasing future earning power, such as obtaining a loan to expand your business or to introduce a new line of merchandise.
- Attending school.

Most people don't explicitly think of credit in these terms. Rather, they reason that, because they don't have enough cash to purchase an item they want, such as a car, they must borrow.

# Saving versus Credit Buying

An astute savings and loan association once ran an ad in some national magazines pointing out that if you were to save for 36 months to buy a car for $20,000, you would have to put into the bank only $18,800; the other $1,200 would be made up by the interest you would receive over the three years. If you bought the car immediately and paid for it over 36 months, however, not only would you not receive interest on your savings, but you would have to pay a finance charge on the installment debt. The total price of the car might be $22,000. There is obviously a big difference between $18,800 and $22,000. The conclusion, according to the savings and loan association: it is better to save now and buy later than to buy now and go into debt.

Is anything wrong with the reasoning in that ad? First of all, the interest on your saving account is taxable by the federal government and most state governments. Additionally, the price of the car will rise with inflation during the three-year period. If the interest rate on your saving account does not reflect the inflation rate, the arithmetic will clearly be off. Also, a crucial point was left out: during the three years in which you saved, you would not be enjoying the services of the car or of the other things you could buy. You would be putting off your purchase for three years. Most people do not want to wait that long; they would prefer to have the services of the car immediately and pay the finance charge to do that. After all, the finance charge is merely a payment for using somebody else's money so that you can consume and that other person—the saver who decided not to consume—cannot.

**Question for Thought & Discussion #2:**
How do you decide which products to pay for with cash and which to buy with credit?

# Costs and Benefits of Borrowing to Consume Now

You may therefore decide that the implicit utility you get per service flow of whatever you buy is greater than the interest payments you have to pay your creditor to get the total amount of money to buy the goods right away. No moral judgment need be passed here: it is simply a question of comparing costs and benefits. The benefit of borrowing is having purchasing power today; the cost is whatever future use of your funds that you forgo. Obviously, if the cost were zero, you would borrow as much as you could because you could buy everything you wanted today and pay back whatever you owed at some later date without any penalty. In fact, the ultimate consumer probably would like to die with a giant debt. That way, the person could consume all she or he wanted at everybody else's expense. Of course, when you buy something on credit with no intention of paying back that loan, there is little difference between that action and stealing.

The benefits of borrowing are something that only you can decide. But we should all be aware of the costs of borrowing. There is a very definite relationship between rising prices and high interest rates. But the relationship is not the simple one of cause and effect. Contrary to popular belief, high interest rates do not *cause* inflation. They may, however, result from inflation. When prices are rising, interest rates will have an inflationary premium tacked on to them. A simple example will show you why.

## Inflationary Premiums

Suppose you are a banker who has been loaning out money at 5 percent a year for the last twenty years. Suppose also that, for the last twenty years, there has been no inflation. That 5 percent interest you have been charging is the *real* rate of interest you are receiving. It covers your costs and gives you a normal profit for your lending activities.

Now prices start rising at 5 percent a year, and you expect that they will rise at that rate forever. If someone comes in to borrow money, how much do you think you would want to lend at the 5 percent rate of interest that you have always charged? Think about it. Suppose that a person comes in to borrow $1,000 for a year. At the end of the year, with an inflation rate of 5 percent, the actual purchasing power of that $1,000 paid back to you will be only $950. If you ask for 5 percent, or $50 in interest payments, you will be compensated *only* for the erosive effect of inflation on the value of the money you lend out.

Obviously, you will want to tack on an inflationary premium to the interest rate that you were charging when there was no inflation and none was anticipated. Hence, in periods of inflation, we find the inflationary premium tacked on everywhere. It is not surprising, then, that, during an inflationary period when prices are rising at 5 percent a year, interest rates may be 10 percent (see Table 14.1 on the next page).

## Inflation May Make It Easier to Repay Loans

You, the demander of credit, or the potential debtor, should not be put off by a higher interest rate. After all, you are going to be repaying the loan in cheapened dollars—that is, dollars that have lost part of their purchasing power through inflation. In fact, some interest rates did not react very rapidly to

| TABLE 14.1 | The "Real" Rate of Interest |
|---|---|
| The rate of interest you are paying on a $1,000 loan for one year | 10% |
| The (expected) rate of inflation (loss in value of money) this year | 5% |
| The difference between the rate of interest you are paying and the loss in the value of dollars you will pay back | 5% |
| Therefore, 5 percent is the real rate of interest you pay when you are charged 10 percent on a loan and the rate of inflation is 5 percent. | |

*NOTE:* This example ignores tax deductions, which may reduce real interest rates even more on certain loans, such as home mortgages.

rising inflation in the 1970s. Credit unions, for example, were giving out automobile loans at an interest rate of 8 percent per annum. Since the rate of inflation was 6 percent, those loans cost people only a 2 percent real rate of interest.

# What It Costs to Borrow

Why do you have to pay to borrow? The reason is that somebody else is giving up something. What are they giving up? Purchasing power, or command over goods and services today. For other people to give up command over goods and services today, they have to be compensated, and they usually are compensated with what we call interest. Ask yourself if you would be willing to lend $100 to your friend with the loan to be paid back in ten years, with no interest—just the $100. To do this, you would have to sacrifice what the $100 would have bought, while your friend enjoyed it. Furthermore, you would not be compensated for lost purchasing power caused by inflation. Most people will not make this sacrifice with no reward except as a strictly charitable act (and they also want to be compensated for any risk of not being repaid).

## Interest Rate Determination

Think of the interest rate you pay on a loan as the price you pay the lender for the use of his or her money. What determines that price is no different from what determines the price of anything else in our economy. The various demands and supplies for credit ultimately result in some rate of interest being charged for the different forms of credit.

But we cannot really talk about a single interest rate or a single charge for credit. Interest rates vary according to the length of a loan, the risk involved, whether the debtor has put up something as **collateral** for the loan (that is, secured it), and so on. One rule is fairly certain: the greater the perceived risk, the more the creditor will demand in interest payments from the debtor. Don't be surprised, then, that interest rates in the economy range all the way from relatively low to relatively high. Much of that difference in interest rates has to do with the riskiness of the loan.

## Finance Charges

The cost of credit varies, not only among individuals for the reasons just given, but also among different lending institutions, some of which will charge more than others. The **finance charge** is the total amount you will pay to use credit.

**Collateral** The backing that people often must put up to obtain a loan. Whatever is used as collateral for a loan can be sold to repay that loan if the debtor cannot pay it off as specified in the loan agreement. For example, the collateral for a new-car loan is generally the new car itself. If the finance company does not get paid for its car loan, it can then repossess the car and sell it to recover the amount of the loan.

**Finance Charge** The total costs you pay for credit, including interest charges, possible credit-insurance premium costs and appraisal fees, and other service charges.

This charge includes interest costs, any required credit-insurance premiums or appraisal fees, and other related service charges.

By far the largest part of finance charges is the interest you will pay for your loan. If, for example, you wish to buy a car for $15,000 but have only $3,000 in cash, you will have to borrow the other $12,000.

The **annual percentage rate (APR)** is the cost of credit calculated on a yearly basis. The total amount of interest you pay for your loan is determined by the length and amount of the loan and the interest rate you pay. A $12,000 loan for four years, for example, at an APR of 10 percent would cost $669.43 more in interest than would a three-year loan for the same amount, as can be seen in Table 14.2. Does that mean you shouldn't get a four-year loan? Not necessarily. You will also note that the monthly payment for those four years is substantially less than for a three-year loan; to pay off the loan in three years, you would pay $82.86 more each month. Depending on your personal situation, you might prefer to make lower monthly payments for a longer period of time.

You should also realize that the APR you will be required to pay may differ depending on how you use your borrowed funds. Most lenders, for example, charge a lower APR for loans used to buy new cars than for those used to finance the purchase of a used car. Suppose that you borrow $12,000 to buy a two-year-old car that has been driven 25,000 miles. The bank might charge you a 12 percent APR instead of the 10 percent it would charge if you bought a new car instead. You can see the extra interest cost you would pay in Table 14.2.

# Consumer Credit Legislation

With hundreds of billions of dollars in consumer credit outstanding, the opportunity for abuse is great. Congress has passed a series of laws that are intended to protect consumer rights and prevent the exploitation of borrowers by unscrupulous lenders.

## Truth in Lending

The Truth-in-Lending Act, which is Title I of the Consumer Credit Protection Act of 1968, is essentially a disclosure law. Most kinds of installment debts now have to be properly labeled so that consumers know exactly what they are paying.

In passing the act, Congress's purpose was "to insure a meaningful disclosure of credit terms so that the consumer will be able to compare more readily the various credit terms available to him and avoid the uninformed use of credit." The act, based on the consumer's right to be informed, requires that the various

**Annual Percentage Rate (APR)** The annual interest cost of consumer credit.

**Question for Thought & Discussion #3:**
If you decided to borrow funds, would you be more concerned with your monthly payment or the total you would pay over time? Explain the reasons for your choice.

| TABLE 14.2 | Comparing Credit Terms for a $12,000 Loan | | | |
|---|---|---|---|---|
| APR | Length of Loan | Monthly Payment | Total Finance Charge | Total Cost |
| 10% | 3 years | $387.21 | $1,939.42 | $13,939.42 |
| 10% | 4 years | 304.35 | 2,608.85 | 14,608.85 |
| 12% | 4 years | 316.01 | 3,168.29 | 15,168.29 |

**Right of Rescission** The right to cancel a contract or an agreement that has been signed. For example, if you sign an agreement to buy a set of encyclopedias from a door-to-door salesperson, you have the right to cancel the agreement within a three-day period.

terms used to refer to the dollar cost of credit, such as *interest* and *points,* be described and disclosed under one common label, *finance charge.* Likewise, it prohibits the use of terms that describe the cost of credit in percentage terms, such as *discount rates* and *add-ons,* and prescribes a uniform method of computation of a single rate known as the *annual percentage rate.*

**A RIGHT OF RESCISSION** The Truth-in-Lending Act also grants the consumer-borrower a **right of rescission** (cancellation) for certain credit contracts. (This is also called a *cooling-off period.*) Section 125 of the act gives the consumer three business days to rescind a credit transaction signed with a door-to-door salesperson that results or might result in a lien on the consumer's home or on any real property that is used or expected to be used as her or his principal residence. The right of rescission is designed to allow the person additional time to reconsider using the residence as security for credit. This right of rescission does not, however, apply to first mortgages on homes.

The Truth-in-Lending Act also regulates the advertising of consumer credit. One of the primary purposes of the act's advertising requirements is to eliminate "come-on" credit ads. For example, if any one important credit term is mentioned in an advertisement—*down payment* or *monthly payment*—all other important terms must also be defined.

**TRUTH IN CREDIT-CARD LENDING** A 1970 amendment to the act provides federal regulations on the use of credit cards. This amendment prohibits the unsolicited distribution of new credit cards and also establishes a maximum limit of $50 on liability for the unauthorized use of each such card; that is, the owner of a lost or stolen card that has been used illegally by another person cannot be made liable for more than $50 of illegal purchases. A 1982 amendment to the act required that all installment credit contracts be written in easily readable and understandable English so that consumers can clearly comprehend any credit terms and conditions of the contract.

**A RIGHT TO CREDIT-CARD INFORMATION** The Truth-in-Lending Act does not actually ensure protection, only information. But information can be valuable, because it allows you, the consumer looking for credit, to shop around, to see exactly what you are paying, and to know exactly what you are getting into. The Truth-in-Lending Act requires that an accurate assessment of the annual percentage rate be given, and that's what you should look at when you compare the price of credit from various dealers and companies. In addition, you may want to look at the finance charge, which is the total number of dollars you pay to borrow the money, whether directly or in the form of deferred payments on a purchase. These total finance charges include all the so-called carrying charges that sometimes are tacked onto a retail installment contract plus such things as "setup" charges and mandatory credit life insurance.[1] These all contribute to your costs of having purchasing power today instead of waiting, of having command over goods and services right now, and of taking that command away from somebody else. Expressed as a percentage of the total price, it gives your annual percentage interest rate. In some cases, it may be very high indeed.

---

1. Credit life insurance, which is discussed in Chapter 18, usually isn't a good deal.

## THE ETHICAL CONSUMER
### Getting in over Your Head on Purpose?

According to reports from the federal government, 83 percent of undergraduate students in 2001 had at least one credit card. Thousands of students had two or more. To be sure, there is no reason a student should not have and use a credit card. This type of spending offers many benefits, but it can also present a problem if students get in over their heads and cannot repay their credit-card debt.

Again, at the beginning of 2001, the average U.S. undergraduate had an outstanding credit-card debt of $2,327. Some had accumulated debts on several cards that totaled $7,000 or more. Of all student credit-card accounts, more than 15 percent were one or more payments in arrears in 2001. Roughly 3 percent of these accounts were written off by the issuing banks as bad debts. These payment records and bad debts will haunt thousands of students as they seek to obtain mortgages or other types of credit in the future. They are beginning their adult lives with credit histories that are guaranteed to give them a terrible credit rating. Who do you believe is at fault?

On one side of the issue are the lending institutions that flood campuses with offers of easy credit. These firms want to attract student customers, even if they have no current income. Studies show that once students start to use a particular credit card, there is a good chance that they will continue to use that card after graduation.

Credit-card issuers are willing to risk a few write-offs to get young people "hooked." They promote their cards with statements such as "A spring break in Florida is only a charge away." What they fail to point out, of course, is that a trip to Florida will cost $1,500, which will have to be paid back in the future at an interest rate of, say, 14.5 percent. Many college students are not experienced spenders. Most of what they have was purchased by someone else. For the first time they are on their own, making their own decisions. Often, they just don't understand what falling into debt can do to their futures. Can they be blamed if they are convinced by credit-card companies that charging is a good idea?

In some cases, student borrowers have overcharged intentionally. A student at the State University of New York in Albany applied for, and was granted, more than ten different credit cards. Over a single year, he charged more than $10,000. He was able to make payments on some of the cards by making cash withdrawals with other cards. It took the credit-card companies nearly twelve months to catch up with him. When asked why he did it, he explained that he had enjoyed living the good life for a while and that he had no property he could lose to satisfy the debt. Besides, according to him, the credit-card companies were really at fault because they encouraged him to borrow and never checked to see whether he could pay the debt.

**Do you believe the student discussed above was right? Are his actions ethical? What about the credit-card companies' actions?**

## Regulation of Revolving Credit

Many credit-card companies currently impose an annual membership fee—at least $15, but perhaps as much as $100 or more. So even if you pay your account within the billing period, you still won't receive free credit if you are required to pay a membership fee. In addition, for those people who do not pay off their accounts during the billing period, the creditor will impose a monthly finance charge of generally 1.0 to 1.5 percent. The creditor computes the finance charge by multiplying the monthly rate times the outstanding balance. Creditors use one of the following techniques to compute the outstanding balance and finance charges on revolving credit accounts:

- *Previous-balance method.* The creditor computes a finance charge on the previous month's balance, even if it has been paid.
- *Average-daily-balance method.* The finance charge is applied to the sum of the actual amounts outstanding each day during the billing period

divided by the number of days in that period. Payments are credited on the exact date of payment.

● *Adjusted-balance method.* Finance charges are assessed on the balance after deducting payments and credits.

## Two-Cycle Method

In addition to the methods of calculating finance charges just described, there is another method that you should know about and probably avoid. It is the *two-cycle method.* There is no finance charge under this method if a debt is completely paid off before the end of the grace period. If the balance is not paid in full, however, the grace period is eliminated, and the finance charge is calculated as a percentage of a purchase from the date of the transaction. Furthermore, it is applied retroactively to purchases made in the *previous* billing period from their dates of purchase to the end of that billing period *even if those debts were paid off in full within that time.* This appears to be nothing more than a legal way that some lending institutions have found to circumvent government rules that require lenders to inform borrowers of the APR they will pay. Under certain circumstances, this method of calculating finance charges can result in interest rates that are effectively twice as high as reported APRs.

Under all repayment plans, lenders are required to inform borrowers in writing of how finance charges will be calculated. Make sure you read and understand what your credit contract obliges you to do. Avoid two-cycle lenders unless you are absolutely positive that you will pay off your debts before the end of every billing cycle.

## Know Which Method Will Be Used

It's important to know which method is used in assessing the finance charge you pay, because the various methods can result in quite different finance charges. For example, assume your opening balance or previous monthly balance is $300. You pay $100 on the account, which is credited on the fifteenth day of the month. The monthly interest rate is quoted at 1.5 percent (18 percent annual rate). Table 14.3 shows that the same monthly finance percentage rate can result in a sizable difference in finance charges based on how the creditor computes the outstanding balance. The same monthly finance charge of 1.5 percent results in three different annual rates of interest, depending on which computational method the creditor applies.

The Truth-in-Lending Act requires that all revolving credit contracts and monthly bills state the "nominal annual percentage rate," which equals twelve times the monthly rate. As Table 14.3 illustrates, however, sometimes the annual percentage rate can be misleading when comparing revolving credit accounts. In 1988, Congress addressed this problem by passing the Fair Credit and Charge Card Disclosure Act. Under this act, lenders must reveal to any consumers applying for credit accounts the following information:

1. The annual percentage rate of interest, including the way the rate is calculated if the rate is variable.

2. All fees, including annual fees, minimum finance charges, transaction charges, cash-advance fees, late fees, and over-the-limit fees.

3. The length of the grace period, if any. (A *grace period* is the period of time during which no interest is charged on your account balance.)

**Question for Thought & Discussion #4:**
Do you know how interest charges are calculated for your credit card? Should you know?

| Method | Opening Balance | Outstanding Balance (1) | Monthly Interest Rate (2) | Finance Charge (1) × (2) = (3) |
|---|---|---|---|---|
| A. Previous balance | $300 | $300 | 1.5% | $4.50 |
| B. Adjusted balance | 300 | 200 | 1.5 | 3.00 |
| C. Average daily balance | 300 | 250 | 1.5 | 3.75 |

**TABLE 14.3    Computing Finance Charges by Three Different Methods**

**A. Previous balance.** The interest rate of .015 is multiplied times $300 to obtain the finance charge of $4.50. It is important to note that the lender gives no credit for the $100 payment received on the fifteenth when computing the outstanding balance.

**B. Adjusted balance.** The interest rate of .015 is multiplied times $200 to obtain the finance charge of $3.00. The creditor uses only the ending balance for the period as the outstanding balance.

**C. Average daily balance.** The interest rate of .015 is multiplied times $250 to obtain the finance charge of $3.75. The $250 average daily balance was calculated as follows:

$$
\begin{array}{ccccc}
\text{Number of days} & \times & \text{Balance} & = & \text{Total balance} \\
15 & \times & \$300 & = & \$4,500 \\
15 & \times & \$200 & = & \underline{\$3,000} \\
& & & & \$7,500
\end{array}
$$

$$
\begin{array}{ccccc}
\text{Total balance} & \div & \text{Number of days in month (billing period)} & = & \text{Average daily balance} \\
\$7,500 & \div & 30 & = & \$250
\end{array}
$$

4. The method used to calculate the monthly account balances against which the lender applies interest or finance charges.

# Equal Credit Opportunity

Since October 1975, when the Equal Credit Opportunity Act went into effect, it has been illegal to discriminate on the basis of gender or marital status when granting credit. Regulations pursuant to the act issued by the Federal Reserve Board prohibit:

1. Demanding information on the credit applicant's childbearing intentions or birth-control practices.
2. Requiring cosignatures on loans when such requirements do not apply to all qualified applicants.
3. Discouraging the applicant from applying for credit because of gender or marital status.
4. Terminating or changing the conditions of credit solely on the basis of a change in marital status.
5. Ignoring alimony and child-support payments as regular income in assessing the creditworthiness of the applicant.

The Equal Credit Opportunity Act reaffirms a woman's right to get and keep credit in her own name rather than that of her husband (or of her former husband, if she is divorced). Women who wish to establish their own lines of credit are advised by bankers to do the following:

1. Open checking and saving accounts in your own name.
2. Create a separate credit record, assuming you can afford it.
3. Open a charge account at a retail store. When applying, list only your own salary, not that of your spouse.

4. Apply for a bank credit card.

5. Finally, take out a small loan and repay it on time. Even if you don't need it, this will speed up the establishment of your credit reliability.

Today, a woman has the legal right to credit for her husband's charge cards. In other words, all new accounts automatically give credit references to everyone on the application, including the wife.

# Fair Credit Billing Act

Under the rules of the Fair Credit Billing Act, if you purchase a faulty product with your credit card, you can withhold payment until the dispute over the product is resolved. The purchase must be for more than $50 and must have taken place within your state or a hundred miles of your home. It is up to the credit-card issuer to intervene and attempt a settlement between you and the seller. You do not have unlimited rights to stop payment. You must make a good faith effort to get satisfaction from the seller before you do so. The rules seem to be in the consumer's favor, however. You don't even have to notify the credit-card company that you are cutting off payment (on that item). You just wait for the company to act. It is probably a good idea, however, to let the company know what you are doing. Ultimately, you can be sued by the credit-card company if no agreement is reached.

The Fair Credit Billing Act also established other rules. If you think there is an error in your bill, the credit-card company must investigate, and you can suspend payments until it does so. You simply write to the credit-card company within sixty days of getting the bill, briefly explaining the circumstances and why you think there is an error. It is a good idea to include copies (not the originals) of the sales slips at issue. Under the law, the company must acknowledge your letter within thirty days and resolve the dispute within ninety days. During that period, you don't have to pay the amount in dispute or any minimum payments on the amount in dispute. And, further, your creditor cannot assess finance charges during that period for unpaid balances in dispute. It cannot even close your account; however, if it turns out that there was no error, the creditor can attempt to collect finance charges for the entire period for which payments were not made.

**Question for Thought & Discussion #5:**
Have you, or any persons you know, had a billing dispute with a credit-card company? How was the situation resolved?

# Fair Credit Reporting Act

The Fair Credit Reporting Act (Title VI of the 1968 Consumer Credit Protection Act) was passed in 1970 and went into effect in 1971. Under this law, you have recourse when a credit investigating agency gives you a bad rating. Now when you are refused credit because of a bad credit rating, the company that turned you down must give you the name and address of the credit investigating agency that was used. The same holds true for an insurance company.

The 1971 act was intended to regulate the consumer credit reporting industry to ensure that credit reporting agencies supply information that is equitable and fair to the consumer. In passing the act, Congress was concerned that incorrect, misleading, or incomplete information, as well as one-sided versions of disputed claims, were being reported. A Consumers Union survey of credit bureau reports showed that 50 percent of the files investigated had at least one error. In addition, many people were concerned about the invasion of privacy when reports were distributed to those who did not really have a legitimate business need for them. These reports often contain material about a person's general reputation, personal characteristics or mode of living, and character. The act applies not only to the usual credit bureaus and investigating concerns but also to finance companies and banks that routinely give out credit information other than that developed from their own transactions.

## Gaining Access to Your Credit File

Under the 1971 law, a credit bureau must disclose to you the "nature and substance of all information" that is included under your name in its files. You also have the right to be told the sources of just about all of that information. If you discover that the credit bureau has incomplete, misleading, or false information, the Fair Credit Reporting Act requires that the bureau investigate any disputed information "within a reasonable period of time." Of course, the credit bureau is not necessarily going to do it, but you do have the law on your side, and you can go to court over the issue. In addition, at your request, the credit bureau must send to those companies that received a credit report in the last six months a notice of the elimination of any false information from your credit record. Finally, if the credit bureau does not correct your information, you can file a personal version (limited to one hundred words) of any disputes. In 1991, TRW, which at that time was one of the three largest credit reporting businesses, announced that it would provide free copies of credit reports once a year to people who request them.

Even if you have not been rejected for credit, you still have the right to go to a credit bureau and find out what your file contains. You also have the right to ask the credit bureau to delete, correct, or investigate items you believe to be fallacious or inaccurate. The credit bureau then has the legal right to charge you for the time it spends correcting any mistakes.[2] Also, the Fair Credit Reporting Act specifically forbids credit bureaus from sending out any adverse information that is more than seven years old. There are important exceptions, however. Bankruptcy information can be sent out to your prospective creditors for a full ten years. And there is no time limit on any information for loans or life-insurance policies of $50,000 or more or for a job application with an annual salary of $20,000 or more. That means that adverse information may be kept in your file and used indefinitely for these purposes.

## Credit Scoring

Credit scoring, or more accurately, your FICO score, is a method that businesses often use to decide whether to extend credit to a consumer. FICO scoring is a procedure that was originally developed by Fair Isaac & Company in the 1950s and has been refined since. It evaluates different criteria that are supposed to indicate who is likely to repay a loan and who is not. These criteria include:

● The person's history of making payments on time.
● The length of time credit has been used.

2. In some states, such as California, legislatures have mandated a maximum fee that credit bureaus can charge for such information.

- The amount of credit used relative to the amount approved.
- The length of time a person has lived at his or her current address.
- The person's employment history.
- Negative entries in the person's credit history, such as charge-offs and collections.

FICO scores reported by the three major credit bureaus (Equifax, Experian, and Trans Union) are not necessarily identical, but they should not be dramatically different. Individual creditors use these scores in various ways. From the creditors' point of view, FICO scores simplify their work. All they need to do is to have a cutoff score below which they either won't extend credit or will do so only after careful evaluation. From the borrowers' point of view, low FICO scores can effectively cut them off from being able to borrow funds.

It is possible to increase FICO scores over time. You should start by purchasing copies of your credit history from each of the three major credit bureaus. If you find an important error, you may be able to raise your FICO score by having the mistake corrected. Failing that you should:

- Close any credit accounts that you do not use or use only rarely.
- Pay down any credit accounts that are "maxed" out.
- Use the credit that you have regularly but in limited amounts.
- Pay all of your bills on time.

None of these steps will change your FICO score overnight. But over several years, following these recommendations will increase your ability to borrow additional funds. Essentially, what you need to do is demonstrate that you are a good credit risk.

## Fair Debt Collection Practices Act

In 1977, Congress passed the Fair Debt Collection Practices Act. The purpose of the act is to regulate the debt-collection practices of persons who collect from consumers whose debts arise from purchases that are primarily for personal, family, or household purposes. The act prohibits the following debt-collection practices:

1. Contacting the consumer at his or her place of employment if the employer objects.
2. Contacting the consumer at inconvenient or unusual times, such as three o'clock in the morning, or contacting the consumer at all if she or he is represented by an attorney.
3. Contacting third parties other than parents, spouses, or financial advisers about the payment of a debt unless the court authorizes it.
4. Using harassment and intimidation, such as abusive language, or using false or misleading information, such as posing as a police officer.
5. Communicating with the consumer after receipt of notice that the consumer is refusing to pay the debt, except to advise the consumer of further action to be taken by the collection agency.

The enforcement of this act is the responsibility of the Federal Trade Commission. The act provides for damages and penalties that can be recovered for violation, including attorneys' fees.

**Question for Thought & Discussion #6:**
Do you believe the FICO credit scoring system is fair to consumers? How would you change the system if you could?

# Subprime Lending

There are millions of U.S. consumers who are employed but do not have good credit histories for any number of reasons. They may have been ill and unable to pay bills, out of work, or in a difficult financial situation because of a divorce or death in the family. When these people want to borrow, they are generally turned down by traditional lenders. They often qualify, however, for subprime lending.

Subprime lending refers to loans made to people who have a higher degree of risk of nonpayment than people who have better credit histories. These loans are typically made at significantly higher interest rates than other loans and often require greater initial payments. More than half of the applications for subprime loans are denied. Most of the people who qualify for subprime loans are wage earners with relatively high incomes. A majority of them already have a significant amount of debt accumulated. The vast majority of subprime loans are second mortgages and therefore available only to homeowners.

Subprime lending fills an important function in our economy. It allows millions of people to finance necessary spending. People take out subprime loans to pay for medical treatments, make needed repairs on homes, buy transportation to get to work, and many other legitimate reasons. The fact that a loan is subprime does not necessarily mean that taking out the loan was a consumer mistake.

## The Home Ownership and Equity Protection Act

The federal government passed the Home Ownership and Equity Protection Act (HOEPA) in 1994 to protect consumers who borrow through subprime lending. This law applies to any second mortgage, mortgage refinance, or home-equity loan that has an annual percentage rate that is more than 10 percentage points over the rate paid on U.S. Treasury securities that have similar terms. In other words, if the rate paid by the federal government on its ten-year notes is 4 percent, and you borrow $10,000 through a second mortgage at 14 percent interest, then you are protected by the HOEPA. The HOEPA restricts prepayment penalties, balloon payments, lending without regard to a borrower's ability to repay, and payments made directly from the creditor to a home-improvement contractor. It also requires a lender to carefully and completely explain the implications of the credit agreement that it offers. The HOEPA does not prevent people from taking out loans that they cannot afford to repay. It does require that they be informed of their situations and that their credit rights not be abused. Some lenders, however, have attempted to circumvent the intentions of this law by charging interest rates that are only slightly less than 10 percentage points above federal interest rates.

## The Real Estate Settlement Procedures Act

Another federal law that is intended to help protect borrowers is the Real Estate Settlement Procedures Act (RESPA). This law requires lenders to provide borrowers with a comprehensive list of all the expenses they can reasonably expect to pay when they buy, refinance, or take out a second mortgage or home-equity loan against the value of their home. The RESPA should

**Question for Thought & Discussion #7:**
Do you know anyone who has used subprime lending? Explain why this type of credit was needed and whether the borrower was satisfied with the result.

prevent any unexpected surprises on closing day. If closing costs for a new mortgage will be $2,000, then the consumer will know this beforehand. Again, this does not mean a borrower can't get in over his or her head. It just means that this person will have the information to make a good decision.

# Predatory Lending

Many people confuse predatory lending with subprime lending. Although these types of credit are related, they are not the same. Subprime lenders charge higher interest rates to compensate themselves for taking greater risks. They do not start out with the intention to foreclose on a borrower's property. Predatory lenders, however, plan from the outset to exploit borrowers to make unreasonable profits. There are a number of practices that are common to predatory lenders.

## Flipping

*Flipping* is the practice of convincing borrowers to repeatedly refinance their loans to get a slightly lower interest rate. What is not explained is that many fees are charged for this service and that they far exceed the value of any reduced interest rates. If a person refinances a $20,000 loan to gain a 0.5 percent reduction in interest, she or he will end up paying more when a $4,000 fee is charged to complete the transaction.

## Packing

*Packing* refers to the practice of adding services to a loan that are neither requested nor necessary. These services typically include life and disability insurance, disaster insurance, termite or radon protection, and the like. The premiums charged far exceed the rates that could be obtained from other insurers.

## Stripping

*Stripping* is the practice of forcing a borrower into foreclosure by extending more credit than could possibly be repaid. This allows the firm to have the property sold and charge excessive fees for the sale. The firm might even buy the property itself at a low price and then sell it again to another person who cannot afford to pay.

## Federal Actions to Stop Predatory Lending

The federal government has made efforts to combat predatory lending. In 2001, for example, the Federal Trade Commission filed a complaint that charged the Associates Corporation of North America with "systematic and widespread abusive lending practices, commonly known as 'predatory lending.'" The complaint went on to allege, "The Associates engaged in widespread deceptive practices. They hid essential information from consumers, misrepresented loan terms, flipped loans, and packed optional fees to raise the cost of the loans." The Associates eventually agreed to stop using these tactics when they extend loans. Although government pressure may have ended this firm's practices, many others are still exploiting consumers. The best defense consumers have against predatory lending is to use their good sense and the decision-making process.

# Going into Personal Bankruptcy

The law provides a method for consumers who have become overwhelmed with their debt burden to, in effect, start over again with an almost-clean slate. By filing for personal **bankruptcy,** consumers are able to protect a relatively small amount of their assets while eliminating their debt. The decision to file for personal bankruptcy should never be taken lightly. As you will learn in the following discussion, a number of costs are associated with it that go well beyond financial considerations.

## Why So Many Bankruptcies?

In 1976, the number of Americans who filed for bankruptcy was 193,000. By 1980, this figure had climbed to 409,000. By 2001, the 1980 number had more than quadrupled—to 1.8 million. According to the International Credit Association, 40 million consumers were in financial trouble at that time, and more than 5 million of that group were on the verge of bankruptcy.

Why are so many consumers going into bankruptcy? There seems to be no one answer to this question. After a study of the problem, the Federal Reserve Board concluded that the increase in bankruptcies was related to a parallel increase in consumer debt. Yet the National Association of Credit Unions reports that the number of consumers who are forced to default on their debts has grown only slightly over the past twenty years. The association suggests that the number of bankruptcies is increasing because more people now use bankruptcy as a way of avoiding their debt payments.

Although bankruptcy may carry less of a stigma today than in the past, and obviously more consumers are opting for bankruptcy than previously, it nonetheless can be very costly to an individual. Once you have gone into bankruptcy, it is extremely difficult to reestablish credit as long as the bankruptcy proceeding remains on your credit record (under current laws, for ten years). Because of this black mark on their credit records, some consumers who have resolved their financial problems through bankruptcy have referred to their decision as a "ten-year mistake." The point of this discussion is to emphasize that before you choose bankruptcy as an option, you should very thoroughly explore all other alternatives. Bankruptcy should be considered only as a last resort.

If you do decide to file for bankruptcy protection, however, you have, basically, two choices: Chapter 7 and Chapter 13 of the federal Bankruptcy Code both provide personal relief for debtors. Chapter 7 is often called *straight bankruptcy,* or *ordinary bankruptcy,* whereas Chapter 13 is basically a *debt-repayment plan* worked out by the bankruptcy court between a debtor and his or her creditors.

## Filing for Chapter 7 Bankruptcy

A Chapter 7 filing, the more drastic bankruptcy plan, in effect clears the slate of all obligations except "nondischargeable" debts, such as alimony, child support, taxes, and government-insured loans. The law permits one Chapter 7

**Bankruptcy** The state of having come under the provisions of the law that entitles a person's creditors to have that person's assets administered for their benefit.

**Question for Thought & Discussion #8:**
Do you know anyone who has filed for personal bankruptcy? How did this affect the person's life?

Do you believe that people who file for bankruptcy are more often at fault for their financial problems or just victims of the system?

## Cyber consumer
### Bankruptcy Information Online

In 2003, an organization called Debtworkout.com provided links to information about bankruptcy laws for every state in the Union and the District of Columbia, at **http://www.debtworkout.com/ states2.html**. The most important differences among state laws were in the exemptions to personal property that could be protected. Visit this Web site and investigate how the exemptions in your state compare with those of neighboring states. Are there important differences, or are they pretty much the same? Remember, Web addresses change frequently. If this address no longer functions, find similar sites to investigate using any of the search engines available to you.

Is cutting up your credit cards a good way to reduce your spending?

filing every six years. If you file for straight bankruptcy, you will usually need the services of a lawyer. You must list all your debts and assets and turn control of your assets over to the court. A trustee is named to check on those assets available for distribution to creditors.

**EXEMPTIONS**    An individual debtor is entitled to exempt certain property from the assets that are turned over to the court when Chapter 7 bankruptcy is filed. The debtor may choose either federal or state exemption amounts and classifications. Before the enactment of the federal bankruptcy law, state laws exclusively governed the extent of the exemptions. In some cases, state laws exempted few assets. The federal exemptions may be viewed as a floor, or minimum amount of property that filers are able to protect. When state exemptions are more generous, filers may choose these instead of the federal amounts. The values of exemptions have been adjusted periodically over time. Federal exemptions fall into the following categories:

1. A portion of the equity held in the debtor's residence.
2. A motor vehicle up to a maximum value.
3. Household goods and clothing up to a maximum value.
4. Jewelry up to a maximum value.
5. Other personal property up to a maximum value.
6. Tools required for the debtor to earn an income up to a maximum value.
7. Unmatured life insurance.
8. Dividends earned from life insurance.
9. Necessary health equipment.
10. Social Security and other specified payments that will be received.
11. Certain personal-injury awards made by a court.

### ADVANTAGES AND DISADVANTAGES OF STRAIGHT BANKRUPTCY

The benefit of declaring straight bankruptcy is that most debts are "erased" except for taxes, alimony, support payments, the debts that others have cosigned for you (which now become their debts), and any secured debts.

The disadvantages have already been mentioned: You no longer have control over all your property, and bankruptcy puts a black mark on your credit record. Furthermore, you will incur further debt. After all, you have court and lawyers' costs, which together might run between $500 and $1,000. And remember, you can't file for bankruptcy again for another six years.

## Chapter 13 Plans

Under bankruptcy law, you may choose a Chapter 13 repayment plan. This is basically a debt-consolidation program with legal safeguards that permit you to stretch out your payment of bills. You can develop a plan for full or even partial repayment of debts over an extended period of time. It is a meaningful alternative to full bankruptcy litigation. Under Chapter 13 bankruptcy, there is a maximum amount of debt that may be owed. This ceiling is adjusted periodically. In 2002, this amount was $269,250 in unsecured debt or $807,750 in secured debt. At one time, only a wage earner was eligible for Chapter 13 relief; the person with a small business was excluded. Now the sole proprietor of a small business can also use this plan, as can individuals living on welfare, Social Security, fixed pensions, or investment income.

Under Chapter 13, the debtor is given the exclusive right to propose a repayment plan. No creditor can force a debtor into a plan that she or he does not wish to accept. The plan may provide for repayment over a period of up to five years. It may also provide for payment of claims only out of future income or out of a combination of future income and the liquidation of some of the debtor's currently owned property. Once a Chapter 13 plan has been approved, all creditors must stop collection efforts and suspend interest and late charges on most kinds of debts. Each month, the debtor turns over a specified amount of money to a court trustee, who then dispenses it to the creditors. As long as the plan is working, the debtor may keep all of her or his assets.

One of the major benefits of a Chapter 13 plan is that you may end up paying off your debts at less than 100 cents on the dollar. If you repay at least 70 percent of your debts, you do not have to wait the usual six years until you can file for bankruptcy again.

# Confronting Consumer Issues: Making Credit Decisions

**M**any books available today attempt to give consumers cut-and-dried formulas that tell them when and how much they should borrow. It is not unusual to find a financial adviser who tells consumers that they should never borrow except in emergency situations or for major purchases such as cars or houses. The truth is, because we are all individuals with individual financial situations and values, there is no one rule that can say when each of us should or should not use credit.

We all realize that just about everyone who buys a house automatically assumes that the responsible thing is to borrow; after all, very few of us are in a position to pay out $100,000 or more for the full cost of a house. Because we know the housing services we consume each month represent a very small part of the total price (because houses last so long), we believe the responsible choice is to take out a mortgage. The same holds for cars, especially new ones. A car is such a large expense that very few of us are able to pay for it in cash. After houses and automobiles, though, the reasoning gets pretty fuzzy. Is it responsible to buy a stereo on credit? Or is it alright to buy furniture on credit? Some advisers say yes, some say no. Of course, for clothes and food most financial advisers are adamant about the desirability of paying cash.

When you think about it, the reasoning behind cut-and-dried rules is pretty shaky. Gertrude Stein once wrote that a rose is a rose is a rose; and so, too, a dollar is a dollar is a dollar. Does it matter what you say your dollar is going to buy? You can't earmark it. If you make $300 a week and spend $30 for clothes, $150 for food and lodging, and the rest on entertainment, how do you know which dollar went for "essentials" (food and lodging and clothes) and which dollar for "nonessentials"—entertainment? You don't know because you can't tell one dollar from another. What does it matter if you say you are going to use credit to buy your clothes and pay for your entertainment with cash? It is only important to decide what percentage of your anticipated income you are willing to set aside for fixed payments to repay loans. You should care about the total commitment you have made to creditors. You want to make sure you have not overcommitted yourself. Tables 14.4 and 14.5 may help you determine how much you should borrow.

## SHOPPING FOR CREDIT

Once you have decided that you want to buy some credit— that is, you want to get some goods now and pay for them later—then you should shop around.

Digital Vision/Getty Images

**Why should you always comparison shop** before you apply for credit?

The Truth-in-Lending Act, which requires full disclosure of the annual interest rate charged, makes shopping much easier than it used to be. You can now contact lending institutions and learn within a relatively short period of time which offer the lowest interest rates. To find the lowest interest rates on credit cards, you can contact Bankcard Holders of America and obtain a list of U.S. banks that offer the lowest rates. To obtain a list, write to Bankcard Holders of America, 524 Branch Drive, Salem, VA 24153 or call (540) 389-5445. Many commercial Web sites will link you to credit cards offering low interest rates on the Internet. Creditcardmall at **http://www.creditcardmall.com** was one such site in 2003.

| TABLE 14.4 | Rules for Borrowing |
|---|---|

There are two rules to consider for borrowing:
1. *Never overcommit yourself.*
2. *Limit the times you borrow money but maximize the amounts borrowed.* This way you will pay the lowest interest charges. In other words, don't borrow several small amounts at one-week intervals; wait and borrow the total amount all at once.

| TABLE 14.5 | Determining a Safe Debt Load |
|---|---|

First, you must determine your spendable income per month.
Enter it here.                                              $ _____

Next, determine your monthly debt payments (don't include mortgage payments, if any):

1. Automobile          $ _____
2. Appliances          $ _____
3. Cash loan           $ _____
4. Others              $ _____

                              **Total**    $ _____

Now, determine your monthly payments as a percentage of spendable income. Then, look at the following guidelines.

#### DEBT GUIDELINES: PAYMENTS AS A PERCENTAGE OF SPENDABLE MONTHLY INCOME

| Percentage | Current Monthly Debt Load | Can I Assume Additional Debt? |
|---|---|---|
| Less than 15% | Within safety limits | Yes. |
| 15–20% | Right at limit | Yes, but be careful. |
| 21–30% | Overextended | No, no, no. |
| Above 30% | On the verge of going under | Perish the thought! |

## WHAT IS THE MAXIMUM YOU CAN BORROW?

If you go to a bank or a credit company and ask for a loan, the loan officer probably will require you to fill out a form. On this form, you list your **liabilities** and your **assets** so that the credit officer can estimate your **net worth.** Essentially, you have to list all your assets—whatever you own—and all your liabilities—whatever you owe. The difference is your net worth (see Table 14.6). Obviously, if your net worth is negative, you will have a hard time getting a loan from anybody unless you can show that your expected income in the immediate future will be substantial.

You still do not know what your maximum credit limit is. That, of course, depends on the loan officer's assessment of your financial position. This will be a function of your net worth, your income, your relative indebtedness, and how "regular" your situation is. Regularity means different things to different people, but it generally means the following:

1. You have been working regularly for a long period and, therefore, have been receiving regular income.
2. Your family situation is stable.
3. You have consistently paid your debts on time.

Or your creditworthiness can be measured by the three "C's" that loan officers use as a guide to lending:

1. Capacity to pay back.
2. Character.
3. Capital or collateral that you own.

| TABLE 14.6 | Determining Your Net Worth |
|---|---|

Estimated amounts, end of this year.

**ASSETS**

House (including furniture)—market value                              _____

Cars—resale value                              _____

Life insurance—cash value                              _____

Bonds, securities—market value                              _____

Cash on hand and savings                              _____

Other (for example, jewelry, coin collections, land)                              _____

**TOTAL ASSETS**                              _____

**LIABILITIES**                              _____

Mortgage                              _____

Loans                              _____

Other                              _____

**TOTAL LIABILITIES**                              _____

**NET WORTH**                              _____

An annual net-worth statement may help you and/or your family keep track of financial progress from year to year. Essentially, your net worth is an indication of how much wealth you actually own. We generally find that young people have low net worths—or even negative net worths. In other words, they owe more than they own, but they are anticipating having higher income in the future. As individuals and families get farther down the road, their net worth increases steadily, only to start falling again, usually when retirement age approaches and the income flow slows down, thereby forcing the retired person or couple to draw on accumulated savings. This very simplified statement of family net worth can be filled out easily. Just make sure you include all your assets and all your liabilities. Assets are anything you own, and liabilities are anything you owe.

*(Continued on next page )*

## Confronting Consumer Issues (Continued)

Loan officers may sometimes appear to discriminate against people with unstable living conditions—that is, those who have unstable jobs, unstable family situations, and the like. That may or may not be true, depending on your definition of discrimination. But a loan officer is supposed to make decisions that maximize the profits for his or her company. At the going interest rate, the loan officer may decide to eliminate people who are high risks: loans will be refused to people whose records indicate they will not pay their debts as dependably as people who seem more stable. If you are a credit buyer with an unstable living situation, one way you can persuade loan officers not to refuse you is to discuss your circumstances candidly with them and produce a past record of loan repayments that shows stability in spite of your unstable situation. Or, alternatively, you could offer to pay a higher interest rate.

## WHERE SHOULD YOU GO FOR A LOAN?

For some asset purchases, you immediately know where to go for a loan. If you are buying a house, you obviously don't go to your local small-loan company; you go to a savings and loan association, a commercial bank, or a mortgage trust company, or you sign a contract with the seller of the home. The real estate agent usually helps the buyer of a house secure a loan. If you want to shop around, the easiest thing to do is to call various savings and loan associations and banks to see what interest rates they are charging or visit those that won't reveal that information over the telephone. Chapter 12 discussed in more detail what you should look out for when borrowing money on a house. Remember to check out online lending offers, if only for comparison purposes.

To borrow for a car, again you probably won't go to the small-loan company around the corner. Rather, you should go to a credit union or a commercial bank, where the loan for a car will cost less. As mentioned earlier, the interest rate for a new car is usually lower than for a used car. The reason is that the car becomes the collateral, and a new car is generally easier to sell than a used one. You should also note that if you buy a car that is technically brand new—that is, you are the first owner—but you purchase it after next year's models have entered the showroom, the lending agency may consider it to be a used car and charge you the higher rate of interest.

If you want to borrow money for purchases of smaller items, a credit union might be the cheapest source of credit. The next-best deal would probably be offered by a bank. If you use a credit card to make your purchase, you should be sure to pay off the balance before the end of your billing period, even if this means you must borrow from another

source to do it. This will allow you to avoid paying the high interest rates usually charged on credit-card debt.

The key to purchasing the best credit deal is to treat credit as a good or service and to use the same shopping techniques for purchasing credit as you would to purchase anything else. Your shopping should stop after you've found the best car deal, for example. You may not be getting the best bargain possible if you buy the credit for the car from the dealership or its affiliate. You may do better by going to your local commercial bank. But you can't predict: you have to compare.

If you wish to figure out the cost of borrowing $1,000 for any specified period of time, look at Table 14.7. It shows different specified annual percentage rates for various time periods.

## CONSIDERING REBATES

In the early 1990s, many credit-card issuers began to offer consumers rebates, either in cash or in credits to their accounts, based on their credit-card usage. One of the first was the Sears Discover card, which offered to send customers

| TABLE 14.7 | Cost of Financing $1,000 on an Installment Plan | | | |
|---|---|---|---|---|
| Percentage Rate (Annual) | Length of Loan (Months) | Monthly Payments | Finance Charge | Total Cost of Loan |
| 6% | 6 | $169.60 | $ 17.60 | $1,017.60 |
| | 12 | 86.07 | 32.84 | 1,032.84 |
| | 24 | 44.32 | 63.68 | 1,063.68 |
| | 36 | 30.42 | 95.12 | 1,095.12 |
| 7 | 6 | 170.09 | 20.54 | 1,020.54 |
| | 12 | 86.53 | 38.36 | 1,038.36 |
| | 24 | 44.77 | 74.48 | 1,074.48 |
| | 36 | 30.88 | 111.68 | 1,111.68 |
| 8 | 6 | 170.58 | 23.48 | 1,023.48 |
| | 12 | 86.99 | 43.88 | 1,043.88 |
| | 24 | 45.23 | 85.52 | 1,085.52 |
| | 36 | 31.34 | 128.24 | 1,128.24 |
| 9 | 6 | 171.07 | 26.42 | 1,026.42 |
| | 12 | 87.45 | 49.40 | 1,049.40 |
| | 24 | 45.68 | 96.32 | 1,096.32 |
| | 36 | 31.80 | 144.80 | 1,144.80 |
| 10 | 6 | 171.56 | 29.36 | 1,029.36 |
| | 12 | 87.92 | 55.04 | 1,055.04 |
| | 24 | 46.24 | 109.76 | 1,109.76 |
| | 36 | 32.27 | 161.72 | 1,161.72 |

a check valued at 1 percent of their charges in the previous year. Soon thereafter other organizations jumped on the bandwagon, offering rebates of 2 or even 3 percent. The GM card created a new wrinkle in the rebate competition by offering consumers the opportunity to have 5 percent of their charges (up to $500 a year) credited to an account they could apply to their next purchase of a new General Motors car. Another almost-universal offer is the right to earn "frequent-flyer miles," one per dollar charged, when you use credit cards issued through an airline such as American, Northwest, or United.

Determining the true value of these rebates to an individual consumer can be a real challenge. The first question you need to ask yourself when considering a rebate credit card is whether you want what is being offered. If you know, for example, that you will never buy a General Motors car, there is no reason to use the GM card. But if you intend to buy a GM product, its card is generally recognized as being the best deal. Suppose you charge $5,000 in a typical year. This would give you a $250 rebate on your next new-car purchase. Alternative plans would give you a 1 to 3 percent cash rebate ($50 to $150) or 5,000 frequent-flyer miles (worth an estimated $100 to $150). In this case the value of the GM plan is clearly greater, all else being equal.

When you choose a rebate card, never lose sight of these two important questions: How much does it cost to own? What rate of interest will you pay if you have an unpaid balance? There is little advantage in obtaining a rebate worth several hundred dollars if your credit costs increase many hundreds of dollars.

## THE RULE OF 78

Suppose you take out a loan for twelve months and want to pay it back after five months. The bank or finance company may use what is called *the rule of 78* to calculate what you owe in terms of the percentage of the total year's interest that would have been earned had you carried out the full contractual agreement.

If you pay off the loan after one month, then, based on the rule of 78, you will have to pay your creditor $\frac{12}{78}$ of the year's interest owed, or 15.38 percent. The exact proportional amount of interest that you would have paid on such an installment contract for one month, however, would equal $\frac{1}{12}$ of the year's total interest, or 8.33 percent. Notice the penalty for early repayment: for that one-month loan, you pay almost double the interest that is stated in the installment contract.

If you keep the loan for two months and then repay it, you end up having to pay $(12 + 11)/78 = \frac{23}{78}$ of the total year's interest. Again, comparison shows that $\frac{23}{78} = 29.49$ percent, and $\frac{2}{12} = 16.67$ percent. You pay almost double the effective

amount of interest over what you would have paid had you kept the loan outstanding. In sum, then, using the rule of 78 to calculate early repayment results in the lender's obtaining more than a strictly prorated distribution of interest.

Consider another example. If you are making an early repayment on a two-year loan with twenty-four equal installments, you will use the rule of 300.[3] If you repay the loan after one month, you owe $\frac{23}{300}$ of the total amount of interest that would have been paid over a two-year period (since there are twenty-four monthly installments). That means you will pay 8 percent of the total as opposed to the prorated distribution of interest, which would equal $\frac{1}{24}$, or 4.17 percent.

Lenders do not have to calculate what you owe in this step-by-step manner because they have prepared tables. The method used to prepare the tables, however, is similar to the process just described.

## THINGS TO WATCH FOR

You must be careful when you look at loan agreements because all have various contingency clauses written into them that may or may not affect you.

**Acceleration Clauses**   Some credit agreements include an **acceleration clause**—meaning that the entire debt becomes due immediately if you, the borrower, fail to meet any single payment on the debt. In all probability, you could not pay such a large sum. Obviously, if you could not meet a payment on the debt because you lacked the money, you certainly would be unable to pay off the whole loan at once. The inclusion of an acceleration clause in a credit agreement increases the probability that whatever you bought on credit will be repossessed.[4]

**Add-On Clauses**   You also should be aware of what is called an **add-on clause** in installment contracts, particularly when you go shopping for furniture and appliances. An add-on clause essentially makes earlier purchases security for the more recent purchase. Let's say that you buy furniture for your living room from a particular store on an installment contract. Six months later, you decide that you want new furniture for a bedroom. You return to the same store and also buy the bedroom furniture on an installment contract. If there is an add-on clause and you default on the installment contract for the bedroom furniture, you can lose not only that furniture but all the items you purchased for

*(Continued on next page)*

---

3. Because $(1 + 2 + 3 + 4 + 5 + \cdots + 24) = 300$.
4. Since loans with acceleration clauses usually can be obtained at relatively lower interest rates, they still may be a good deal for people who rarely or never default on loan payments.

## Confronting Consumer Issues (Continued)

the living room, even if you have paid for that furniture after making the second purchase.

### Balloon Clauses
*Balloon clauses* are terms of installment loan contracts that require, after a period of time, specific payments more than twice the normal installment payments. For example, a contract may indicate that $100 a month is due for eleven months; then a single payment of $600 is due in the twelfth month. If you cannot pay the $600, you either have to refinance or, possibly, lose the item purchased on credit. ("Interest-only" loans are balloon loans.)

### Garnishment (Wage Attachment)
It is also possible for a court order to allow a creditor to attach, or seize, part of your property. Your bank account may be attached and used to discharge any debts, or your wages may be subject to **garnishment.** That is, if a judgment is made against you, your employer is required to withhold your wages to pay a creditor.

The Federal Garnishment Law, effective July 1, 1970, is part of the Consumer Credit Protection Act. It limits the portion of an employee's wages that can be garnished. Garnishment can be no more than the lesser of the following:

● Twenty-five percent of take-home pay.
● The amount by which take-home pay is in excess of thirty times the federal minimum hourly wage.

The act also prohibits firms from firing an employee because of a single wage-garnishment proceeding.

### Giving Yourself the Best Chance
How do you give yourself the best chance to get the loan you want? Make sure you plan ahead. Credit experts suggest you do the following:

1. Verify the information about you that credit reporting agencies have in their files. You can do this by calling the local credit bureaus or reporting agencies listed in the Yellow Pages to see if they have a file on you and if any national reporting agency has a file. Then either visit or write to the relevant agency or agencies and request a copy of your records. You will probably have to pay a fee, but it will be worth it. If you find that any information is missing or incorrect, rectify the situation immediately.

2. If you do not have any credit history, start now by opening a checking account and a saving account in your own name. Take out a small loan and repay it on time, and also attempt to get a bank credit card or a store charge account.

3. When filling out a loan application, answer *all* questions.

4. Make a copy of your loan application so that every time you apply for a new loan you can repeat all the information that you have given before. Lenders usually open a new file for every new type of loan.

5. Comparison shopping is useful, but don't try to obtain the same type of loan from several lenders at the same time. Your credit record may then show a streak of applications, which will raise questions among potential lenders.

6. If a lender turns you down, find out why. The rejection may be the result of inaccurate information that was supplied by a credit reporting organization. The Fair Credit Reporting Act of 1970 gives you the right to a copy of this information, and you should take steps to correct it if it is wrong. (See Table 14.8.)

7. Remember, a refusal is not permanent. You may want to try again, even with the same lender.

### Question for Thought & Discussion #9:
Have you examined copies of your credit report for accuracy? If you have, what did you find? If you haven't, why should you obtain copies to examine?

### Reasons for Refusing Credit
Lending institutions that turn down your request for credit are required to explain the reasons for their decision. They typically offer the following reasons for refusing credit to individuals:

● Insufficient length of local residence.
● Insufficient income.
● Inability to verify employment.
● Excessive obligations.
● Insufficient credit history.
● Negative credit history.

### How to Improve a Bad Credit Rating
Once you've gotten into credit difficulties and have a bad credit rating, it may be very difficult to clean the slate and start anew, but it can be done. For one thing, credit reporting agencies are prohibited from giving out negative information that is more than seven years old—except in the special circumstances discussed earlier in this chapter. In the meantime, you should try to rebuild your credit record slowly, starting with applications for department store charge accounts, for which the requirements are usually not as stringent as for major bank cards. Another possibility is to apply for a guaranteed

| TABLE 14.8 | What You Can Do to Protect Yourself against Unfair Credit Reports |
|---|---|

If you are trying to obtain insurance, credit, or a job, you may be subjected to a personal investigation. Under the Fair Credit Reporting Act of 1970:

- The company asking for the report is supposed to let you know that you are being investigated.
- You can demand the name and address of the firm hired to do the investigating.
- You can demand that the investigating company tell you what its report contains—except for medical information used to determine your eligibility for life insurance.
- You cannot require the investigators to reveal the names of neighbors or friends who supplied information.

If the investigation turns up derogatory or inaccurate material, you can:

- Demand a recheck.
- Require the investigators to take out of your file anything that is inaccurate, and send a new report to anyone who has received an incorrect report about your credit history in the past six months.
- Require the investigators to insert your version of the facts, if the facts remain in dispute.
- Sue the investigating firm for damages if negligence on its part resulted in violation of the law and caused you some loss—failure to get a job, loss of credit or insurance, or even great personal embarrassment.
- Require the company to cease reporting adverse information after it is seven years old—with the exception of a bankruptcy, which can remain in the file for ten years.

bank credit card. Certain banks will issue you a MasterCard or other bank credit card, for example, if you deposit a specified amount of cash as collateral for the card. Your credit limit on the card will be equal to (or probably somewhat lower) than the cash on deposit. If you fail to pay your credit-card debt, the bank can draw on the cash deposit to cover the charges made on the card.

Another way to repair a damaged credit rating, if you can afford it, is to open a saving account with a moderate deposit of perhaps $1,000. Leave this money on deposit but borrow against it. The bank will be willing to make such a loan because it has your $1,000 if you fail to make your payments. Make regular payments over the term of the loan to demonstrate your ability to handle money in a responsible way. This will cost you some funds in extra interest (the difference between what the bank pays you for your deposit and what you pay the bank for the loan), but it is probably money well spent.

**Fraud and Your Credit Rating**   Many millions of consumers have black marks on their credit records because of the fraudulent use of lost or stolen credit cards or other types of credit fraud. Identity theft and how to protect yourself against it was discussed in Chapter 4.

**Debt's Danger Signals**   If you observe any of the following, you are in danger of overextending yourself financially:

- You consistently postpone or avoid paying your bills because you lack sufficient income to pay them.
- You begin to hear from your creditors.
- You have no savings or not enough to tide you over a financial upset.

- You have little or no idea what your living expenses are.
- You use a lot of credit, have charge accounts at a number of stores, carry several credit cards in your wallet, and pay only the monthly minimum on each account.
- You don't know how much your debts total.

## HOW TO KEEP FROM GOING UNDER

To prevent yourself and your family from getting into deeper financial trouble, consider the following common-sense actions:

1. Itemize your debts in detail, making sure you note current balances, monthly payments, and dates when payments are due.

2. List the family's total *net* income that can be counted on every month.

3. Subtract your monthly living expenses from your net income. Don't include the payments on debts you already have. The result will be the income you would be able to spend if you had no debts. Now subtract the monthly payments you are committed to making on all your debts. If you come out with a minus figure, you are obviously living beyond your means. If you come out with a very small positive figure, you still may be living beyond your means.

4. If you think that your household is, in fact, living beyond its means, you must inform your family members that the money situation is tight. Tell them that you and every other spender in the family will have to cut expenses, such as those for recreation, food, and transportation.

*(Continued on next page )*

## Confronting Consumer Issues (Continued)

### CREDIT COUNSELING SERVICES

If you still have difficulties making ends meet after you've tried the foregoing suggestions, you might check with a credit counseling service. The best way to locate such a service in your area is to contact the National Foundation for Consumer Credit (NFCC) and ask for a list of nonprofit counseling services that have offices near you. You can write the NFCC at 8611 2nd Avenue, Suite 100, Silver Spring, MD 20910; call (301) 589-5600; or contact it online at **http://www.nfcc.org**. There are some 350 nonprofit credit counseling organizations in the United States, and they are usually better than the for-profit organizations, which charge more for their services. Although the nonprofit organizations charge no fee, they will ask for a contribution, on average about $10, and perhaps small fees for certain debt-repayment services, depending on the assistance they render to you.

According to the NFCC, roughly one-third of the consumers who turn to its credit counselors for help settle their own financial problems satisfactorily after credit counselors advise them how to manage their money so that all bills can be paid. About 40 percent of the clients counseled arrange to make monthly payments to their counselors, who then distribute the funds among creditors under a specially arranged payment schedule. The remainder of those counseled cannot be helped because their situations are too critical, often because of marital problems or drug abuse.

### BE WARY OF DEBT-CONSOLIDATION OFFERS

Although most credit counseling organizations are bona fide consumer groups that will act in your interest, be wary of commercial enterprises that advertise help in debt consolidation or reduction. They might claim they can consolidate all your debts into one fixed monthly payment that will be lower than the total of what you are paying now to all your creditors. Rarely, though, is this claim true; in most instances you won't actually pay a lower interest rate by consolidating your debts instead of paying them separately. Remember, you have already incurred setup charges in taking out the various lines of credit that you want consolidated. Credit companies won't render a service unless they make profit on it. So, if you let a credit company pay off all your existing debts and then lend you the total amount that it paid, you will have to incur the setup charge for that.

**Be Careful of Debt-Reduction Plans**    The Internet is full of offers from firms that promise to reduce your debts through negotiations with your creditors. This practice is supposed to help you avoid bankruptcy that would leave a black mark on your credit record for ten years. Although some of these offers are genuine, many are not. These businesses often demand an up-front fee and offer no guarantee that they will succeed. In some cases, they simply take the fee and disappear without ever contacting your creditors. If you should consider hiring such a firm, agree to pay a fee only after a deal has been negotiated and your debts really have been reduced. In any case, the fact that part of your debts were written off will still appear on your credit record and affect your ability to obtain credit in the future. As a general rule, you should assume that businesses that offer this service are trying to scam you until they prove they aren't.

**Debt Consolidation May Really Help**    There are times when debt consolidation can actually save you money. If, for example, you consolidate all your revolving credit accounts that charge you 18 percent into one 12 percent credit union loan, you will be better off (assuming there are no early-payment penalty charges on the revolving credit accounts).

Furthermore, it may be more *convenient* for you to have all your loans consolidated into one big one. Then you have to write only one check a month instead of many. But this service comes with a price. You may, in fact, have a smaller monthly payment, but it will be for many more months, and you ultimately will pay higher finance charges for the whole consolidation package and thus a higher total payment. If you detest keeping records and writing out lots of checks, you may want to incur this additional cost (and additional debt) by taking a loan consolidation. You should be very careful, however, to keep the result of such a loan in perspective. Many consumers react to a lower total monthly payment by going out and taking on more debt that makes their credit problem even greater. As long as you realize that nobody gives you anything for free, you can make a rational choice, knowing that there are always costs for any benefits you receive. Loan consolidation certainly isn't going to pull you out of dire financial trouble; the only way out is either to make a higher income or to cut back on your current consumption so you can pay off your debts more easily. (You could, of course, sell some of your assets to pay off your debts.)

## Key Terms:

**Liability**   Something for which you are liable or responsible according to law or equity, especially pecuniary debts or obligations.

**Asset**   Something of value that is owned by an individual, a business, or the government.

**Net Worth**   The difference between the value of your assets and your liabilities—that is, what you are actually worth.

**Acceleration Clause**   A clause contained in numerous credit agreements whereby, if one payment is missed, the entire unpaid balance becomes due, or the due date is accelerated to the immediate future.

**Add-On Clause**   A clause in an installment contract that makes your earlier purchases from that source security for the new purchase.

**Garnishment**   A court-ordered withholding of part of wages, the proceeds of which are used to satisfy unpaid debts. Also called *wage attachment*.

# Chapter Summary

1. The majority of Americans have outstanding installment debt at any given time.

2. The sources of credit are many, including commercial banks, consumer finance companies, credit unions, credit-card companies, and retail stores.

3. Individuals borrow in order to obtain the services of expensive durable consumer goods without paying for them all at once. The installment payments can be thought of as matching the service flow from whatever was purchased, such as a house or a car.

4. Interest is the payment for using somebody else's money. As such, it is like any other price. Interest rates must take account of the rate of inflation. Hence, interest rates are relatively high when the rate of inflation is high.

5. The real rate of interest you pay on a loan is the stated rate of interest minus the rate of inflation.

6. The Truth-in-Lending Act requires that both the total finance charge and the annual percentage rate be clearly stated on a loan agreement (except on first mortgages on homes).

7. Lenders use various methods to compute the actual percentage rate you pay on an open-ended credit account: previous balance, average daily balance, or adjusted balance. The resulting finance charge can vary considerably, depending on the method employed. The two-cycle method of calculating interest can result in rates that are much higher than the reported APR. Under the Fair Credit and Charge Card Disclosure Act of 1988, lenders have to disclose to consumers which method of computation is used.

8. Since October 1975, when the Equal Credit Opportunity Act went into effect, it has been illegal to discriminate on the basis of gender or marital status when granting credit.

9. The Fair Credit Billing Act and the Fair Credit Opportunity Act offer protection to the consumer in the areas of disputed charges on credit-card bills and credit reporting methods.

10. In 1977, the Fair Debt Collection Practices Act was passed to protect consumers from unfair debt-collection practices.

11. Although there is no definite way to decide when to rely on credit, you can determine your safe debt load by adding up all your outstanding debt payments and determining what percentage of your after-tax income they represent. As a general rule, you should try to avoid committing more than 20 percent of your after-tax income to debt payments.

12. When you shop for credit, shop as if you were buying any other good or service. Look for the best deal by calling around to get various offers of interest rates and monthly payments. When you actually apply for a loan, remember that the loan officer will look at your capacity to pay back, your character, and the collateral you can produce to back up the loan.

13. Before applying for credit, know where you stand by checking with local credit bureaus to learn what information is in your credit record and to correct any errors that might exist. If you have not established a credit history, it is a good idea to create one by opening checking and saving accounts and applying for loans against your savings.

14. If you are refused credit, the Fair Credit Reporting Act allows you to demand that the firm hired to do the credit check tell you what its report contains, that it do a recheck, and that it insert in its report your version of any facts in dispute.

15. Many consumers with less-than-perfect credit reports borrow through subprime lending. They pay significantly higher interest rates to borrow but are able to obtain credit for important purposes. Predatory lending is related to subprime lending but is not the same. Predatory lending takes place when lenders deliberately set out to exploit consumers through practices such as flipping, packing, and stripping. The federal government has attempted to protect consumers from these practices through the Home Ownership and Equity Protection Act of 1994.

16. Thousands of U.S. consumers file for personal bankruptcy each year. There are two ways to file. Under Chapter 7 bankruptcy, all debts are eliminated when a person's assets are taken by a court to satisfy as much of his or her debts as possible. The individual who files is allowed to retain some exempted assets. Under Chapter 13 bankruptcy, a plan is created to allow the individual to pay off all or part of her or his debts over an extended period of time. Either type of bankruptcy will appear on the filer's credit record for ten years.

17. Signals that you are getting too far into debt include consistently postponing payment of bills because of insufficient income, being contacted by creditors, having no savings or insufficient savings to tide you over a financial crisis, not knowing what your expenses are, relying on credit to make day-to-day purchases because you have no cash, paying only minimum monthly amounts on outstanding balances, and not knowing what your total debts are.

18. Credit counseling services aid many consumers in managing their debt problems. A leading source of credit counseling services is the National Foundation of Consumer Credit. Be wary of debt-consolidation schemes. Although some are genuinely beneficial, they are often overpriced and may tempt you to go further into debt.

## Key Terms

acceleration clause **355**
add-on clause **355**
annual percentage rate (APR) **335**

asset **355**
bankruptcy **345**
collateral **334**
finance charge **334**

garnishment **355**
liability **355**
net worth **355**
right of rescission **336**

## Questions for Thought & Discussion

1. What sources of credit do you use? What sources do other members of your family use? Why were these sources chosen?

2. How do you decide which products to pay for with cash and which to buy with credit?

3. If you decided to borrow funds, would you be more concerned with your monthly payment or the total you would pay over time? Explain the reasons for your choice.

4. Do you know how interest charges are calculated for your credit card? Should you know?

5. Have you, or any persons you know, had a billing dispute with a credit-card company? How was the situation resolved?

6. Do you believe the FICO credit scoring system is fair to consumers? How would you change the system if you could?

7. Do you know anyone who has used subprime lending? Explain why this type of credit was needed and whether the borrower was satisfied with the result.

8. Do you know anyone who has filed for personal bankruptcy? How did this affect the person's life?

9. Have you examined copies of your credit report for accuracy? If you have, what did you find? If you haven't, why should you obtain copies to examine?

10. How will the way you are currently using credit affect your ability to obtain credit in the future?

## Things to Do

1. Select a consumer durable good, such as an automobile or an expensive stereo. Call around or check online to find out where you could get a loan to purchase the product. Are there substantial differences in the APRs charged for the loans? Try to determine why any differences exist.

2. Copy and complete the form in Table 14.5 to determine your personal safe debt load. Are you surprised by what you find? What would you hope your safe debt load would be five years from now? What can you do now that will help you obtain credit in the future?

3. Assume that you are a different person—a loan officer for a bank. Now assume that the real you comes into the bank and asks to borrow $5,000 to buy a used car. Would you give this person (yourself) the loan? Why or why not?

4. Search the Internet using the key words *debt, consolidation,* and *loan.* Visit the Web sites of firms that offer this type of loan. How do they promote their services? Do they make it clear what interest rate you would pay? How much useful information do they provide on their Web sites? Are there significant differences in what they have to offer? If you were really interested in a debt-consolidation loan, would you choose any of these firms? What additional information would you want to have before you concluded an agreement?

# Internet Resources

## Finding Consumer Information on the Internet

 The following Web sites have been selected for their relevance to topics discussed in this chapter. Search these sites to locate information that can add to your knowledge of sources of consumer credit information on the Internet. Remember, Web addresses change frequently. If any of these addresses no longer function, find similar sites to investigate using any of the search engines available to you.

1. The Federal Trade Commission posts a publication titled *Credit and Your Consumer Rights* on the Internet. It can be accessed at **http://www.ftc.gov/bcp/conline/pubs/credit/crdright.htm**.

2. Quicken.com explains your consumer credit rights and offers to send you a copy of your credit report. Investigate Quicken's Web site at **http://www.quicken.com/cms/viewers/article/banking/39643**.

3. The Federal Reserve Board provides consumer information in its *Consumer Credit Handbook to Credit Protection Laws; Personal Finance Building Wealth: A Beginner's Guide to Securing Your Financial Future.* It is located at **http://www.federalreserve.gov/consumers.htm**.

## Shopping on the Internet

 The following Web sites have been selected because they offer consumers services similar to those described in this chapter. These are commercial sites that are designed to market products. They do not represent a comprehensive or balanced description of all firms that offer second mortgages online. How do the loans offered at these Web sites compare with those that are available from local lending institutions? Remember, Web addresses change frequently. If any of these addresses no longer function, find similar sites to investigate using any of the search engines available to you.

1. Lending Tree.com offers homeowners second mortgages that can be used to combine bills into one low-interest payment. Its Web site is located at **http://www.easy-2nd-mortgage.com**.

2. Flagstar Bank offers second mortgages and debt-consolidation loans at its Web site located at **http://www.flagstar.com/lending/mortgage/second**.

3. Second Mortgages Nationwide advertises that it offers the lowest rates for second mortgages. Its Web site can be investigated at **http://www.secondmortgages.com**.

## InfoTrac Exercises

 Purchasers of new copies of this text are provided with access to the InfoTrac Web site. This Web site links students to thousands of recent articles published in hundreds of periodicals. Use the key words **student, credit card,** or other terms from this chapter to conduct a key-word search. Choose one article that is of particular interest to you and write a brief essay describing what you have learned from the article. Be sure to cite the author and title of the article and the name and date of the publication in which it appeared.

# Selected Readings

Bergquest, Erick. "Mandatory Arbitration Latest Subprime Issue in Spotlight." *American Banker,* July 2, 2002, p. 1.

*Building a Better Credit Record: What to Do, What to Avoid.* Available from the Federal Trade Commission, 6th St. and Pennsylvania Ave., N.W., Washington, DC 20580. **http://www.ftc.gov**.

Chatzky, Jean Sherman. "Keeping Score: How Credit Scoring Really Works." *Money,* March 1, 2003, p. 142.

Collins, Brian. "Court Decisions Make Consumer Credit Data a Tougher Sell." *Organization News,* June 2001, p. 3.

*Consumer Handbook to Credit Protection Laws.* Available from the Federal Trade Commission, 6th St. and Pennsylvania Ave., N.W., Washington, DC 20580. **http://www.ftc.gov**.

Jackson, Maya. "The Best of the Web Credit Tools." *Money,* November 1, 2002, p. 139–140.

Lee, Jeanne. "Make Today's Low Rates Pay." *Money,* January 1, 2003, p. 101.

Shilling, Gary A. "The Deadbeat Economy." *Forbes,* December 23, 2002, p. 370.

"The APR Matters." *Essence,* April 2002, p. 56.

"The New Rules of Borrowing." *Consumer Reports,* July 1999, pp. 33–34.

"The Sharks Are Circling: Consumer Credit." *The Economist (US),* March 1, 2003, p. 7.

"What It's Like to Go Bankrupt." *Kiplinger's Personal Finance Magazine,* April 2002, p. 83.

Photodisc/Getty Images

# CHAPTER 15
# Investing for Your Future

After reading this chapter, you should be able to answer the following questions:

- What is the relationship between the risk of an investment and the rate of return on that investment?

- What are the differences among common stocks, preferred stocks, and corporate bonds?

- What are capital gains and losses?

- How do mutual funds work?

- Why is financial planning for retirement important for young people?

- How can consumers create and carry out a personal investment plan?

> 66 **The greater the expected return on an investment, the greater the associated risk will be.** 99

Y ou can do many things with your accumulated savings. You can keep all or part of them in cash, which earns no interest at all and, in fact, loses value at the rate of inflation. You can put them into a saving account that earns a relatively low rate of interest but is extremely secure. You can invest your money in shares of stock of various corporations. You can buy U.S. savings bonds or land. You can purchase consumer durable goods, such as cars, houses, and stereos, that yield a stream of services over their lifetimes. In other words, you can do countless things with your savings. What you *should* do depends on your goals and how much risk you are willing to take.

A basic rule of economics states, "The greater the expected return on an investment, the greater the associated risk will be." When you are offered a "deal" that is guaranteed to earn a 50 percent return in just one year, you can be certain that there is also a good chance that you will lose most or all of the money you invest in this "deal." A little knowledge of the forces of demand and supply explains why this is true. Most investment opportunities are matters of public knowledge—if you find out about one, so will many other people. If a safe investment really offered a 50 percent return, many people would want to take advantage of the opportunity. The supply of funds available to invest would exceed the amount demanded, and the rate of return would fall until the return was justified by the actual risk associated with the investment. This relationship can be demonstrated by a specific example—investing in the stock market. But, first, we need a few facts.

## Some Facts about the Stock Market

**Stock Market** An organized market where shares of ownership in businesses are traded. These shares generally are called stocks. The largest centralized stock market in the United States is the New York Stock Exchange.

The **stock market** is the general term used for all transactions that involve the buying and selling of shares of stocks issued by companies. A share of stock represents a unit of ownership in a corporation that gives its owner the right to a certain portion of the assets of the company issuing the security. Suppose a company wishes to expand its operation. It can obtain the money for expansion by putting up part of the ownership of the company for sale. It does this by offering stock for sale. If a company worth $10 million wants $2 million, it might sell stock. Suppose you alone own the company, and you arbitrarily state that there are one million shares of stock, all of which you own; you then put out on the market 200,000 shares of additional stock, which you sell at $10 a share. You get the $2 million for expansion, and the people who paid the money receive the 200,000 shares of stock. They now have claim to one-sixth of whatever the company owns or earns because their stock represents one-sixth of the firm's new total value of $12 million (your original $10 million plus their $2 million).

There are many different submarkets within the stock market. Probably the best known are centralized markets such as the New York Stock Exchange and the American Stock Exchange. With this type of exchange, all transactions between stock buyers and sellers are completed in one central location, such as New York City. There are smaller centralized regional stock exchanges in many large cities, including Boston, Chicago, San Francisco, and Philadelphia. In addition to centralized stock exchanges, there are a variety of over-the-counter (OTC) markets that carry out transactions via telephone and computer linkups throughout the nation. Until the late 1980s, most stocks traded on OTC markets were for relatively small corporations. Although this is still

Photodisc/Getty Images

Remember, when you buy stock, there is no guarantee you will earn a profit.

true of many stocks traded on OTC markets, an organization called the National Association of Securities Dealers Automatic Quotation (NASDAQ) system now trades stocks for large corporations, including Intel, Oracle, and Cisco Systems. The NASDAQ is a computerized system that links stockbrokers throughout the nation so that thousands of listed stocks can be exchanged electronically without the need for the transactions to take place in a single, centralized location.

## Common Stock

Most stock that is owned and traded in the United States is **common stock.** Common stock has **equity** because it represents a share of ownership in a corporation. Owners of common stock have a vote in choosing the management of the firm. They also may receive a share of the firm's profit in dividends, an amount that often varies with the firm's success. If the firm does well, the dividend could be large; if it does poorly, no dividend may be paid at all. For this reason, more risk is associated with owning common stock than with many other types of business investments.

## Preferred Stock

**Preferred stock** also represents ownership, or equity, in a corporation. It is different from common stock in at least two important ways. Owners of preferred stock have no vote in choosing the firm's board of directors. They must rely on the owners of common stock to decide who will run the firm. However, they have a right to receive the profits and assets of the firm before the owners of common stock.

Another important difference is that, unlike common stock, preferred stock pays a fixed dividend that does not change over time, regardless of the amount of a firm's profit. Although it is possible for a corporation's board of directors to choose not to pay preferred-stock dividends if the firm is losing money (as Chrysler Corporation did in 1979), these dividends generally must be made up before any common dividends can be paid in successful years. Also, if the firm fails, assets that remain after its other debts are satisfied will be paid to preferred stockholders before common stockholders receive anything. For these reasons, less risk is associated with buying and holding preferred stock than with owning common stock.

## Corporate Bonds

A *corporate bond* is basically an IOU, or a promissory note of a corporation, usually issued in multiples of $1,000. A bond is evidence of a debt in which the issuing company promises to pay the bondholder a specific amount of interest for a specified length of time and then to repay the loan on the expiration date. In every case, a bond represents debt of the issuing corporation. Its holder is a creditor of the corporation and not a part owner.

Interest on a bond and its principal must be paid on time, or the firm is in default and may be forced into bankruptcy by the bondholder. In such cases, bondholders have a right to the firm's assets before either the common or the preferred stockholders. See Table 15.1 on the next page for a summary comparison of common stocks and bonds.

**Common Stock** A unit of ownership that has a legal claim to the profits of a company. For each share owned, the common-stock owner generally has the right to one vote on such questions as merging with another company or electing a new board of directors.

**Equity** A legal claim to the profits of a company. This is another name for stock, generally called common stock.

**Preferred Stock** A unit of ownership in a corporation; each share entitles the owner to a fixed dividend that the corporation must pay before it pays any dividends to common stockholders; owners have a claim on the firm's assets before common stockholders if the firm fails, but have no vote in choosing the firm's board of directors.

### Question for Thought & Discussion #1:
Do you or a member of your family own shares of corporate stock? How could investing in stock fit into your life-span plan?

**Capital Loss**  The difference between the buying price and the selling price of something you own when the selling price is lower than the buying price.

| TABLE 15.1 | Comparison of Common Stocks and Bonds | |
|---|---|---|
| **Common Stocks** | | **Bonds** |
| 1. Represent ownership. | | 1. Represent owed debt. |
| 2. Have no fixed dividend rate. | | 2. Require interest to be paid, whether or not any profit is earned. |
| 3. Allow holders to elect a board of directors, which, in turn, controls the corporation. | | 3. Usually entail no voice in or control over management. |
| 4. Have no maturity date; the corporation usually does not repay the stockholder. | | 4. Have a maturity date when the holder is to be repaid the face value. |
| 5. Allow holders to have a claim against the property and income of the corporation after all creditors' claims have been met. | | 5. Give bondholders a prior claim against the property and income of the corporation that must be met before the claims of stockholders. |

Some types of preferred stock and corporate bonds are *convertible*. This means the owners of these investments may choose to trade them for common stock at a specific rate, often within a specific period of time. The advantage of owning convertible preferred stock or bonds is the opportunity they offer investors to take advantage of a firm's success, which could push the value of its common stock up, while assuring them the security of the preferred stock or bond if the firm is less profitable.

## Capital Gains and Losses

Stock can go up and down in price. If you buy a stock at $10 and sell it at $15, you make a *capital gain* equal to $5 for every share you bought and then sold at the higher price. That is called an *appreciation* in the price of your stock, which you realized as a capital gain when you sold it. If the value of your stock falls and you sell it at a loss, you have suffered a **capital loss** because of the *depreciation* in the market value of your stock. Normally, when you buy a stock that has never paid a dividend, you expect to make money on your investment by an increase in the value of the stock. That is, if the company is making profits but not giving out dividends, it must be reinvesting those profits. A reinvestment in itself could pay off in the future through higher profits. The value of the stock would then be bid up in the market. Your profit would be in the form of a capital gain rather than dividend payments.

## What Affects the Price of a Stock?

Some observers believe that individual psychological or subjective feelings are all that affect the price of a stock. If people think a stock will be worth more in the future, they will bid the price up; if they think it will be worth less in the future, the price will fall. That is not a very satisfactory theory, however. Usually, psychological feelings are based on the expected stream of profits that the company will make in the future. Past profits may be important in formulating a prediction of future profits. Past profits are bygones, however, and bygones are forever bygones. A company could lose money for ten years and then make profits for the next fifteen.

The fact that a stock has done well in the past is not a guarantee that it will also do well in the future.

If a company with a poor performance hires a new management team that has a reputation for turning losing companies into winning ones, investors may expect profits to turn around and rise. If a company has a record number of sales orders for future months, one also might expect profits to go up. Whenever profits are expected to rise, the value of the stock also typically rises, as people bid up the price of the stock. Any information about future profits, therefore, should be valuable in assessing how a stock's price will react.

# Making Money in the Stock Market

You may have read about the famous nineteenth-century investor J. P. Morgan, who supposedly made his fortune by manipulating the stock market. You may also have heard of Ken Lay, who was the chief executive officer (CEO) of Enron Corporation when its stock value fell from over $60 to less than $1 per share in only a month. Lay advised Enron's employees and other investors to buy Enron stock at the same time that he was selling millions of dollars' worth of the stock before its value fell. Although Lay made enormous profits on Enron stock, his employees and many other investors lost almost everything they owned when Enron filed for bankruptcy.

Unfortunately, Lay was not alone. In the first years of the twenty-first century, executives of many large corporations, including Tyco, WorldCom, and Adelphia, were accused of manipulating stock prices or of using company funds for their personal benefit. These executives were ultimately dismissed by their boards of directors, and several were indicted for criminal acts. The point to remember from this is that most individuals who have made large amounts of money quickly in the stock market have done so at the expense of other people. Although it is possible to invest wisely in stocks and bonds, it is also possible to be "taken" by unscrupulous persons. Consumers should never invest their money until they understand what they are investing in. And they should seek expert assistance when they realize that they are not qualified to make investment decisions on their own. For many investors, their primary source of information is the broker they employ to complete their market transactions.

## Should You Employ a Full-Service Broker?

Unless you know a current stock owner who is willing to sell his or her stock to you directly, you cannot purchase stock without paying for the services of a brokerage firm. A brokerage firm earns its income by buying or selling stocks or bonds for other people, businesses, or institutions. Brokerage firms have purchased memberships on various stock exchanges that give them the right to trade securities through these organizations. People who represent these firms to the public have historically been called *brokers,* although more recently they have preferred to be called *account executives.* Brokers can basically be divided into two categories: full-service brokers and discount brokers.

Full-service brokers provide consumers with investment research and advice as well as buying or selling securities for them. For these services individual investors are charged a relatively high fee for each trade (typically between $100 and $200) and may also pay a small annual fee for maintaining an account. Discount brokers charge lower fees (typically $50 or less) but offer no, or very limited, information and advice about investments. Another

**Question for Thought & Discussion #2:**
Most stocks lost value in the years between 2000 and 2003. Should this loss cause people to reevaluate their stock investments? Explain why or why not.

alternative is to use an online brokerage service. These services do not offer advice, but they will complete trades for as little as $7.99. Consumers who use discount brokers or online services must decide for themselves how to invest their funds with information they have gathered from other sources.

**CHOOSING A FULL-SERVICE STOCKBROKER**   Many consumers assume that the only safe way to invest in stocks is first to seek the advice of a professional stockbroker. They believe that people employed by major firms such as Merrill Lynch and Paine Webber must be able to offer valuable advice because of their access to in-depth research, knowledge of economic and financial events, and general stock market savvy. Besides, they provide personalized service tailored to the needs and goals of each customer. Right? Well, maybe.

Consumers should recognize that in many ways stockbrokers are little different from any other salespeople. They have a product to sell (information, stocks, and bonds), and their personal income (commission) depends on how much of that product they convince consumers (you) to buy. If you enter into a relationship with a stockbroker with this understanding, you are more likely to be satisfied with your results. The final responsibility to buy or not is yours alone. A stockbroker cannot take advantage of any consumer without that person's cooperation.

**ASK YOUR FRIENDS AND RELATIVES**   Probably the best way to choose a broker is to ask your friends or relatives to recommend someone they have consulted and trust. This not only puts you in contact with a broker who has satisfied customers, but it increases the probability that this person will work to give you the best advice, because to do otherwise could jeopardize an existing relationship (one that earns them money) with your friend or relative.

When you first discuss your investment plans with a broker, he or she should inquire about your financial situation as well as your personal goals and objectives. Your broker should always understand your tolerance for risk. Remember that any investment that offers extraordinary returns is also likely to involve substantial risk. If you are not willing to accept a substantial risk to your investment principal, your broker should be aware of that. If you do have money you are willing to risk, however, your broker may be able to direct you to investments that bring in equally substantial returns. Consumers should realize that all investments involve some risk (even a government bond's value may be diminished by an unexpected high rate of inflation), and with few exceptions there are no absolute guarantees that consumers will not lose money.

**DOES YOUR BROKER WORK FOR YOU?**   Some people argue that brokers are not the people to ask for investment advice because their livelihood depends on encouraging you to buy or sell stock, whether or not this is in your best interest. In some cases stockbrokers have provided false or misleading information to investors. It has also been suggested that brokers are no more likely to be right in picking stocks than you are. The logic behind this view is that stockbrokers do not have access to unique information. All other stockbrokers and professional investors for banks, stock funds, insurance companies, and other large financial organizations will probably know the same facts passed on to you by your broker, and long before you learn of them. They are likely to have bid the price of recommended stocks up or down before you have a chance to buy at a low price or sell at a high price.

The stock and bond markets are extremely efficient. Any new information affects market prices almost immediately. Your chances of getting in on the ground floor are really quite remote. Even if your broker has **inside information,** it is unlikely to do you any good. After all, why would your broker give you this information for you to turn a quick profit rather than earning a profit for herself?

For most consumers, the best way to invest in the stock market is to buy the stocks of a variety of established and promising firms at regular intervals and to hold them over the long term. History suggests that this policy is likely to earn an average 10 percent return on investments.

Photodisc/Getty Images

## Investing Online

In the late 1990s, the investment world was shaken by an explosion of online investment brokerage firms that offered services to investors. By 1999, hundreds of Internet brokerage firms were completing trades for fees as low as $7.99. These businesses accept orders twenty-four hours a day, seven days a week. They have help desks that consumers can call if they have problems completing trades or accessing their accounts. Table 15.2 lists a few of the large online firms and the fees they charged in 2003 for basic transactions. The different fees reflect various levels of service.

As the Internet brokerages expanded, traditional full-service brokerage firms saw their sales and market share decline. They have reacted by offering a combination of services and a choice of fees to their customers. Many now will complete transactions online for free or for low fees for consumers who maintain various levels of assets in their accounts. Consumers may also have the choice of paying a separate fee to receive investment advice and research. As the variety of options has increased, it has become progressively more difficult to determine exactly how much trading in stocks or bonds will cost— it all depends on what else you want. The once simple choice between a full-service and a discount broker has become a confusing range of services and fees. Under these circumstances, you are likely to appreciate your ability to make rational choices by using the decision-making process.

Making your own investment decisions will save you fees, but it requires you to spend more time and effort evaluating your alternatives.

**Question for Thought & Discussion #3:**
Who would you trust to give you good investment advice? Why would you trust this person?

| TABLE 15.2 | Leading Online Brokerage Firms in 2003 | |
| --- | --- | --- |
| American Express | http://www.accutrade.com | $29.95 for up to 4,000 shares |
| Ameritrade | http://www.ameritrade.com | $10.99 per trade |
| Datek | http://www.datek.com | $9.95 per trade |
| E*TRADE | http://www.etrade.com | $9.95 for up to 5,000 shares |
| Fidelity | http://www.fidelity.com | $29.95 for up to 1,000 shares |
| Charles Schwab | http://www.schwab.com | $29.95 for up to 1,000 shares |
| T. D. Waterhouse | http://www.waterhouse.com | $17.95 for up to 5,000 shares |

**Inside Information**  Information about a company's financial situation that is obtained before the public acquires it. True inside information is usually known only by corporate officials or other insiders.

**Mutual Fund** A fund that purchases the stocks of other companies. If you buy a share in a mutual fund, you are, in essence, buying shares in all the companies in which the mutual fund invests. The only business of a mutual fund is buying other companies' stocks.

## What about Investment Advisers?

If you wish to have help in making your investment decisions, you can hire an investment counselor or adviser. Generally, when you hire an investment adviser, you sign a contract that gives the adviser discretion over your investment funds. The adviser will then buy and sell securities for you without further authorization. In the past, investment counselors generally were not interested in handling investment accounts of less than $100,000, but owing to competition many of them now offer investment counseling services for much smaller accounts, even accounts of $10,000. Investment counselors are numerous, and many investors—particularly, active investors—turn to them for investment assistance.

In deciding whether you should hire an investment adviser, you need to keep in mind that whatever profits you may realize on your portfolio because of the counselor's advice do not include the fees for managing your account (typically, ½ to 2 percent of the assets managed, depending on the size of the account) or brokers' commissions that must be paid whenever stock is traded. For example, a typical advertisement for an investment counselor might claim that his or her stock portfolio returns 15 percent a year, rather than the average 10 percent rate of return for all stocks listed on the New York Stock Exchange. But the 15 percent rate does not take into account either the counseling fees or the trading costs. Trading costs (brokers' commissions) can be expensive because investment services usually do a lot of trading: they go in and out of the market—buying today, selling tomorrow. Each time a stock is bought or sold, a commission is paid to the broker. Thus, the more trading your investment counselor does for your account, the more trading costs you incur. In fact, investments made through counselors generally do no better than the general market averages because any special profits they make are eaten up by brokerage fees and their own counseling charges.

## Mutual Funds

**Mutual funds** take the money of many investors and buy and sell large blocks of stock; the investors get dividends or appreciation on their shares of the mutual fund. The mutual fund, then, is a company that invests in other companies but does not sell any physical product of its own. You can buy shares in mutual funds just as you can buy shares in General Motors. A study of mutual funds concluded that mutuals that did the least amount of trading made the highest profits, an expected result if you understand the competitive nature of the stock market. Buying into mutual funds provides investors with two important advantages. It diversifies their investments so that they are not too dependent on any one type of stock or bond. Also, mutual funds employ people who will constantly keep track of the fund's investments and, it is hoped, buy and sell investments at the most advantageous times. This is something that most individual investors lack the time to do.

### The Two Types of Mutual Funds

Most mutual funds or investment trusts can be classified as either closed-end funds or open-end funds. Shares in *closed-end* investment trusts (mutual funds), some of which are listed on the New York Stock Exchange, are readily transferable in the open market and are bought and sold like other shares.

*Open-end* funds sell their own new shares to investors, stand ready to buy back their old shares, and are not listed on the stock exchange. Open-end funds are so called because they issue more shares as people want them.

The only commission you pay to buy closed-end mutual funds is the standard commission you would pay on the purchase of any stock. There are two types of open-end mutual funds, no-load and load funds. A *no-load* mutual fund charges no setup, or loading, charge for you to get into the fund, whereas a *load* mutual fund charges you about 8 percent to get into the fund. Both may charge a yearly management fee. The salesperson or stockbroker who sells you an open-end mutual fund with a loading charge usually keeps most of that charge as a commission. Mutual-fund experts divide open-end and closed-end funds into many categories. Among these are the following:

- *Income funds.* These funds attempt to achieve high yields by concentrating on high-dividend common stocks or bonds or a combination of these.
- *Balance funds.* To minimize risk, these funds hold common stocks and a certain proportion of bonds and preferred stocks.
- *Maximum capital gains funds—dividend income incidental.* These are often aggressively managed and take higher-than-average risks by buying into little-known companies.
- *Long-term growth funds—income secondary.* Fund managers go after large, seasoned, high-quality growth stocks that do not generate dividends.
- *Specialized funds.* These funds restrict themselves to certain types of securities, such as gold-mining stocks.
- *Global funds.* These funds purchase stock issued by firms that are located in other countries or whose business is primarily in international trade.
- *Environmental funds.* Funds in this category purchase stock issued by firms whose corporate policies are theoretically beneficial to the world's environment.
- *Index funds.* Index funds purchase stocks that are intended to mirror an index such as the Standard & Poor's 500.

In addition, there are the following types of closed-end funds:

- *Real estate funds.* Otherwise known as REITs, or *real estate investment trusts,* these are of two types—*mortgage trusts,* which borrow money from banks and relend it at a higher rate to builders and developers, and *equity trusts,* which own income-producing property.
- *Dual-purpose funds.* These types of closed-end funds sell two classes of stock—*income shares* and *capital shares.* The first group of purchasers receive all the fund's net income; the second group participates only in capital gains.

There is, of course, no guarantee that a mutual fund will be a good investment. Study the data in Table 15.3 on the following page to see mutual funds with good and bad records in recent years.

## Money Market Mutual Funds

Perhaps the best-known type of mutual fund today is a *money market mutual fund.* We have already mentioned such funds when we talked about checking accounts and saving instruments. Banks and thrifts offer money market

**Question for Thought & Discussion #4:**
How concerned should investors be about the fees mutual funds charge? Explain your answer.

| TABLE 15.3 | Returns of Selected Mutual Funds, January 2003 | | |
| --- | --- | --- | --- |
| | **Total Return over Previous:** | | |
| **Name and Objective of Fund** | **1 Year** | **3 Years** | **5 Years** |
| American Express Global Growth (global) | −22.9% | −22.8% | −4.4% |
| Citizens Global Equity (global) | −21.4% | −23.4% | +0.7% |
| Fidelity Select Environment (environment) | −16.4% | −0.7% | −8.9% |
| Fidelity Utilities (utilities) | −26.6% | −20.4% | −4.2% |
| Janus Venture (growth) | −27.2% | −29.7% | +0.6% |
| John Hancock (balanced) | −18.2% | −8.7% | −2.1% |
| Merrill Eurofund (Europe) | −12.7% | −7.8% | +2.3% |
| U.S. Global Gold Shares (gold) | +81.4% | +12.2% | −1.6% |

SOURCE: *The Wall Street Journal*, January 6, 2003, pp. R-22–R-33.

## Cyber consumer
### Checking Out Mutual Funds Online

There are thousands of mutual funds you could choose for your investment dollars. Virtually all of them maintain Internet Web sites that provide information about the funds they operate and the types of investments they make. At the beginning of 2003, Fidelity held a greater value of assets in its family of nearly a hundred mutual funds than any other investment organization. Access the Fidelity Web site at **http://www.fid-inv.com**. Investigate the types of mutual funds it offers and how you could establish a periodic investment plan for your future.

accounts insured up to $100,000. Alternatively, there are literally hundreds of uninsured but very safe money market mutual funds available to investors. The term *money market* refers to the fact that all proceeds are invested in relatively short-term money market instruments, such as Treasury bills, commercial paper sold by reputable corporations, and other short-term debt instruments. There are actually three types of money market funds:

● *Government funds.* These invest only in U.S. Treasury obligations and those of other federal agencies.

● *General-purpose funds.* These invest in banks and corporations, as well as in government obligations.

● *Tax-exempt funds.* These invest in obligations of state and municipal governments whose interest payments are not taxable by the federal government. Typically, investors earn smaller dividends, but the tax-free advantage of these funds is appealing to those investors in higher tax brackets.

Most money market mutual funds offer check-writing privileges and telephone transfers. There are restrictions, however. For example, checks may have to be for $500 or more in certain funds. The advantage of money market mutual funds is that they often offer higher interest rates than those offered by commercial banks and thrift institutions. As with all investments, those money market mutual funds that offer higher interest rates than others do carry a slightly higher risk. For certain mutual funds to make relatively higher rates of interest, they must invest in slightly riskier assets, such as debt obligations of longer maturity and debt obligations issued by corporations that do not have the highest possible security ratings.

For the small saver, a disadvantage of money market mutual funds is that they are not insured by an agency of the federal government. Although the uninsured money market funds have a high safety rating, some care should be taken in selecting them. To check the rating of a money market fund, you can look it up in Morningstar, a publication that evaluates all major mutual funds four times each year. Morningstar can be found at most public libraries, or you can pay to access it over the Internet at **http://www.morningstar.com**.

# Is There No Way to Get Rich Quick?

The general conclusion to be reached from our analysis of the stock market is that the investing schemes everybody talks about are really quite useless for getting rich quickly. That does not mean, of course, that some people will not get rich by using them. Luck has much to do with making money in the stock market—just as it does with winning at poker or craps. If you do make money with your particular scheme, it does not mean you are extrasmart, a better investor, or a prophet. You may be lucky. Or you may make more than a normal rate of return on your invested capital if you spend a tremendous amount of time searching out areas of unknown profit potential. But then you are spending resources—your own time, for example. Your above-average profits can be considered as payment for the time you spent—the value of your opportunity cost—analyzing the stock market and different companies.

The question still remains: How can you make money? Experience suggests that by investing in a variety of stocks, people will earn a return of 6 percent to 10 percent over many years. Attempts to earn higher rates of return almost always require investors to accept a greater risk of losing their original investment.

The key to making money in the stock market is to think in long-run terms. As Figure 15.1 shows, although in some years stock investment returns may be lower than 10 percent, these lower rates of return will be compensated for by a higher-than-10-percent return in other years. Although you may not be able to "get rich quick" in the stock market, a long-run rate of return of 10 percent, even when adjusted for inflation, can help you create a substantial nest egg to aid you in achieving your long-term financial goals.

> " The key to making money in the stock market is to think in long-run terms. "

## Question for Thought & Discussion #5:

Do you know someone who has made a lot of money in the stock market? Do you think this person's good fortune was the result of luck or good planning?

**Figure 15.1** Nominal and Real Values of the Dow Jones Industrial Average, 1995–2003

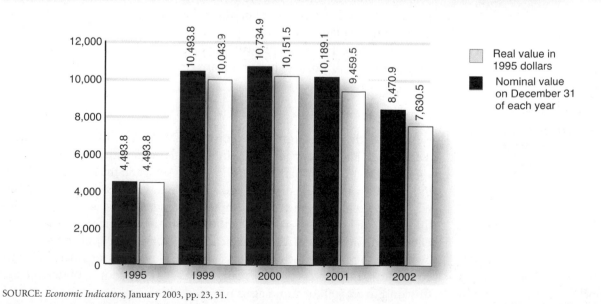

Real value in 1995 dollars

Nominal value on December 31 of each year

SOURCE: *Economic Indicators,* January 2003, pp. 23, 31.

# Watch for Those Sure-Fire Schemes

By now, you ought to be quite suspicious of any special investment deals about which you hear. Because so much competition exists in the investment markets and because you, as a single consumer, are not likely to be smarter than any of the experts around, you should consider every single investment as a trade-off between risk and rate of return (and also liquidity). The higher the potential rate of return, the higher the risk. There is no reason why you should expect to do better than average unless you have some special information.

An example is real estate. Will Rogers once said, "It's easy to make money: just figure out where people are going and then buy the land before they get there." Obviously, if Will Rogers was aware of this truism, all the experts know it, too. What do you think happens when the experts figure out where people will be going? They use that information to bid up the price of land in those areas. Only if you think you have such information ahead of everyone else can you expect to make a higher-than-normal rate of return in any type of land investment.

So don't be taken in by a statement such as "Land is always a safe investment." The value of land can fall like the value of anything else. Even when the overall price of land has been going up for a long time, that does not mean that you will make more than you could make investing money in something else. Although, on average, you might make more in land, you also take a greater risk because land deals frequently fall through completely.

You can think of a thousand and one other investment opportunities to which the same logic applies. Just remember that you do not get something for nothing; any time you do something with your savings, you are going to take a risk, however small.

# Pensions and Retirement Funds

For many people, one of the smartest ways to save is to set up a personal pension plan. Depending on your income level and whether you have an employer-provided pension plan, it can be beneficial to save via an individual retirement plan that protects your savings from taxes until you take them out of the plan. There are basically two plans available to individuals whose employers do not provide pension plans or who are self-employed.

## Individual Retirement Accounts

When **Individual Retirement Accounts (IRAs)** were introduced under the federal Tax Reform Act of 1981, individuals who did not have employer-sponsored retirement plans were allowed to set aside up to $2,000 each year in an IRA. In 1986, IRAs were made available to most taxpayers regardless of whether they had employer-sponsored retirement plans, although this benefit was phased out for taxpayers with large incomes. Since the passage of the Economic Growth and Tax Reconciliation Act of 2001, all taxpayers have been able to take advantage of IRAs regardless of their incomes. The annual contribution limit was also increased to $3,000 for people under the age of fifty and to $3,500 for those aged fifty and older.

IRAs offer tax benefits in two ways. First, the annual contributions are not taxed as current income, and income earned by the funds in the account is

**Individual Retirement Account (IRA)**  An investment account on which the earnings are not taxed until funds are withdrawn from the account, usually at retirement. IRA contributions may also be tax deductible, depending on one's income level.

## THE ETHICAL CONSUMER

# Business Leaders Should Be Ethical, Too

It seems to be a law of economics that in any given year some people will be accused of illegal financial activities. But, by anyone's standards, 2002 was a special year. Consider this partial list of disturbing events in 2002 involving top leaders of some of our nation's largest corporations.

| | |
|---|---|
| January 11 | Accounting firm Arthur Andersen announces that its employees destroyed thousands of documents related to its audits of Enron. It is later charged with obstruction of justice on March 15, 2002. |
| January 29 | Global Crossing files for Chapter 11 bankruptcy, blaming "aggressive accounting" and its former executive officers. |
| February 12 | Nortel's chief financial officer (CFO) resigns after it is revealed that he sold stock in the firm before it announced poor financial results. |
| April 2 | Xerox announces that it will restate its earnings for 1998 through 2001 and pay a $10 million fine to settle Securities and Exchange Commission (SEC) charges that it engaged in fraudulent accounting. |
| April 30 | WorldCom's CEO quits under pressure from the firm's directors over its sinking stock price and misleading accounting practices. |
| April 30 | Sotheby's former CEO is sentenced to three years' probation for price fixing. |
| May 11 | Merrill Lynch agrees to pay $100 million to settle accusations that its analysts deliberately misled investors. |
| June 13 | Imclone's former CEO Samuel Waksal is arrested on charges of insider trading. He pled guilty to these charges on October 16, 2002. |
| June 18 | Enron announces that it made $745 million in special payments to its |

| | |
|---|---|
| | executives in the year preceding its filing for bankruptcy. |
| June 24 | Three Rite Aid executives are charged with illegal accounting practices. |
| June 26 | WorldCom announces it has found accounting mistakes that total $3.8 billion. |
| June 27 | Former Tyco CEO Dennis Kozlowski is charged with tax evasion. |
| July 11 | WorldCom files for bankruptcy, listing $41 billion in debt. |
| August 16 | Salomon Smith Barney analyst Jack Grubman resigns after being accused of promoting stock to help the firm obtain investment banking business. |
| September 24 | Adelphia Communications founder and former CEO is indicted on charges of looting the firm of hundreds of millions of dollars. |
| October 11 | Two former WorldCom officials plead guilty to accounting fraud involving $7.2 billion in expenses that were not properly reported. |
| November 1 | Former Enron CFO Andrew Fastow is indicted for fraud. |
| November 6 | WorldCom announces that it reported $9 billion in profits that it did not earn. |
| December 23 | The SEC and ten stock analyst firms reach an agreement that requires the firms to pay $1.4 billion in fines and restitution to investors who were misled by false reports. |

**What steps can ordinary investors take to protect their investments from dishonest corporate executives and stock analysts? What should the government do to reduce the chance that there will be another year like 2002?**

not taxed until the funds are withdrawn. Second, although the contributor must pay tax on withdrawals from the IRA as they are made, presumably the tax rate will be lower because the individual will be retired and have a lower income. Funds must be left in an IRA until the contributor has reached at least 59½ years of age, and withdrawals must begin no later than when the individual is 70½ years of age. Individuals who withdraw their funds from an IRA before they reach 59½ years of age are taxed on their withdrawals and charged a 10 percent early-withdrawal penalty.

IRA funds must be deposited in a government-approved account. The account may invest the funds in a simple savings account, certificates of deposit (CDs), or stocks and/or bonds. Funds invested in one account may be transferred to different accounts in the same or different financial institutions. The investor, however, must be careful never to take possession of the funds when they are transferred. Doing so could make the investor liable for the taxes on the entire amount transferred. All banks and other financial institutions that maintain IRAs can make these transfers in ways that will avoid this possibility. As the owner of an IRA, you should never withdraw your funds from one IRA and try to reinvest them in a different IRA on your own. Always let the bank or broker do it for you.

## Roth IRAs

In 1998, a new type of Individual Retirement Account became available to U.S. consumers—the Roth IRA. Unlike ordinary IRAs, you invest after-tax dollars in these accounts. The advantage is that the proceeds are not taxed when they are withdrawn, *including* the income that has been earned and reinvested in the account over the years. Suppose that you place $1,000 each year in a Roth IRA that invests your funds in stocks. Further, assume that the stocks increase in value at a rate of 10 percent each year. At the end of twenty years, you will have invested $20,000 of your after-tax income, but the account will be worth $59,886. You will not pay a penny of tax on either your $20,000 investment or the $39,886 return your investment earned. Obviously, the longer you can leave your funds in a Roth IRA, the greater the tax advantage you will enjoy. For people who are near retirement, a Roth IRA is not as advantageous, particularly if they expect to be in a lower tax bracket after they retire.

## Investing through 401(k)s

Since 1978, federal tax law has allowed people who work for corporations to save and invest through what are called **401(k) plans.** These plans enable roughly 45 million U.S. workers to contribute part of their earnings to a retirement fund. The plans offer several advantages to workers. First, the money they contribute is not taxed as current income (although they must pay taxes on it when it is withdrawn from the program after they retire). Another benefit for more than half of the participants in 401(k) plans is that their employers also contribute to the plans, resulting in even more rapid growth in their value. The amount that could be contributed to 401(k) plans was limited to $30,000 in 2000, but this limit is adjusted upward for inflation each year. Another advantage of some 401(k) plans is that contributors can borrow against the value of their savings at low interest rates.

**THE ENRON EXAMPLE**    Although most 401(k) plans are well run, some are not. In recent years, there have been a number of 401(k) disasters. The 2001 financial collapse of Enron is possibly the best example of a 401(k) plan gone bad. Thousands of Enron employees belonged to the firm's 401(k) retirement plan. For years the firm's policy had been to match employee contributions to its plan up to a maximum amount. When Enron's stock price increased to more than $90 per share in 2000, employees believed that they had hit the "jackpot." Many ordinary employees had accounts valued in the millions of dollars. Even when the price of the stock began to fall in 2001, cor-

**401(k) Plan**    A form of retirement plan that can be used by corporations to set aside part of employee earnings for retirement savings before taxes are withheld.

porate executives advised employees to keep all of their retirement savings invested in Enron stock. Unfortunately, while these executives told their employees to keep buying, they were selling thousands of shares of Enron stock themselves.

In the fall of 2001, accusations of accounting fraud caused the market for Enron stock to collapse. In less than one month, Enron's stock declined from about $60 per share to less than $1. Unfortunately for Enron's employees, under the firm's rules they could not sell the stock that had been purchased with corporate matching funds until they were at least fifty years old. Even then, they had to give thirty days' notice to sell their stock whether it had been purchased with their money or the firm's. By the time employees could sell their stock, it had lost nearly 99 percent of its value. People who had thought they were millionaires ended up with only a few thousand dollars in their retirement funds. When Enron filed for bankruptcy, many of these same workers were laid off. And during this time, many Enron executives had not only been able to sell their Enron stock, but had also been given bonuses worth a total of $745 million. Is it any wonder that many Enron employees were angry and felt cheated?

**NOT ALL 401(k)S ARE EQUAL** Because there is no one universal contract for 401(k) plans, you should investigate your firm's plan carefully before investing in it. The events of 2001 and 2002 did lead the government to impose restrictions on how 401(k) funds may be invested and to strengthen reporting requirements. Still, there are no universal rules on how these plans must be run. Most plans give employees some discretion as to how funds they contribute are invested. It is not uncommon, however, for a firm to require that the funds it contributes be invested in the firm's own stock for at least a period of time. In some cases, stock purchased with the firm's contribution must be held for a set number of years, such as two, five, or ten. In other situations, employees are not allowed to sell their stock until they reach a specific age. In any case, employees may take a number of steps to reduce the chance that an Enron-type disaster will ruin their retirements.

- Consult a financial expert who is not associated with your employer about the 401(k) plan before joining.
- Avoid any 401(k) plan that requires you to invest your money in the stock of the firm for which you work.
- Avoid putting more than 10 percent of your investments in any single stock.
- When a firm requires that the money it contributes be invested in its own stock, don't leave the funds there longer than necessary, particularly if they are a large part of your retirement savings.
- Take an active role. If you are not happy with what the fund's managers are doing, inform them of your feelings. Encourage other workers to do the same.

## Keogh Plans

**Keogh plans** are named for Eugene J. Keogh, the congressional representative from New York who sponsored the Keogh Act of 1962. The Keogh Act was passed to allow self-employed individuals to set up their own pension plans. Keoghs are similar to IRAs in that they allow you to defer tax payments

**Keogh Plan** A retirement program designed for self-employed persons by which a certain percentage of their income can be sheltered from taxation. As with IRAs, interest earnings are not taxed until withdrawal.

on interest earned in the plan until the time of withdrawal at the age of 59½ or later; as with IRAs, if you withdraw money from your Keogh plan before that age, you face a 10 percent penalty on the amount withdrawn. Also, you must begin to withdraw from the account by the age of 70½. Unlike IRAs, you can continue contributing to a Keogh plan after retiring from one job as long as you earn self-employment income from another job.

Even if you are covered by a pension plan at work or have an IRA, you can have a Keogh plan as well—as long as it covers only the income you earn from self-employment. With just a small sideline business, you can set up a Keogh account and save tax dollars while contributing to your retirement fund.

There are a variety of distinct Keogh plans to choose from, each with different rules and limitations. If you believe you could benefit from having a Keogh plan, it probably would be wise to find a reputable retirement adviser who can provide you with more specific information.

# Confronting Consumer Issues: Making Personal Investment Plans

As we have noted, schemes to get rich quick in the stock market are useless, because the stock market is so competitive. That is also true for just about every other investment opportunity into which you could put your savings.

Nonetheless, you need to invest your savings someplace. And you are more likely to be successful in reaching your investment goals if you develop and follow a personal investment plan.

## SET YOUR INVESTMENT GOALS FIRST

Before consumers put even one penny in an investment, they should decide why they are investing. Not only is saving to make investments usually easier when consumers have goals they are trying to achieve, but knowing their goals will usually help them decide what types of investments to make. If, for example, you want to have money to make a down payment on a new home next year, it could be a mistake to invest your money in a bond that will not mature for ten years. Similarly, it probably would be unwise to invest your savings for your children's college education in stock in a gold mine that could either make a great profit or fail in the next year. Once you have decided what your investment goals are, you should find it easier to create the rest of your investment plan and carry it through successfully.

## HOW TO START INVESTING

There are various ways to start investing, and your choice will depend largely on (1) how much time you want to devote to the process and (2) the extent to which you want—or feel you need—the assistance of professionals.

**Investment Clubs**   Investment clubs are a popular resource for many who want to learn about the stock market and earn money on their savings at the same time. There are an estimated 30,000 investment clubs across the country. Typically, they have a membership of fifteen investors and holdings of around $60,000 or so—meaning that, on average, each member contributes about $4,000. Some clubs consist of no more than two friends who subscribe to investment newsletters; others are large-scale partnerships of experienced investors who do their own research. Since it can cost quite a bit to join an established club—if you are required to match the other members' previous contributions—you might want to consider forming your own club with a friend or group of friends.

Before you talk to an investment advisor you should determine your financial goals and tolerance to risk.

The National Association of Investment Clubs (NAIC) helps new investment clubs get started. You can contact the NAIC at 1515 East Eleven Mile Road, Royal Oak, MI 48067 or at **http://www.better-investing.org**. On request, it will send you a pamphlet on how to organize an investment club and an investor's manual. Annual dues for membership in the NAIC were $40 per club and $14 for each member in 2002.

In addition to advising you on how to launch your club, the NAIC will send you a model portfolio and worksheets to help you analyze stocks on your own. Generally, the NAIC advises that you aim for 10 percent annual growth.

**Asset Management Accounts**   Asset management accounts, discussed previously in Chapter 13, are offered by a number of major brokerage firms, banks, and other financial organizations. For many who wish to invest in the stock market, as well as have a checking account and a credit card, these one-stop financial services are certainly something to consider. Combination money market and brokerage accounts let you earn high interest on your cash, buy and sell securities, and borrow on your investments, in addition to offering check-writing services and credit cards.

*(Continued on next page)*

**Confronting Consumer Issues (Continued)**

If you want to open an asset management account (and have the $10,000 or so that may be required to do so), you need to decide which type of institution best suits your needs. If you want investment advice, your best bet is to open an account with a brokerage firm. Discount brokerage service, with commissions that can be 90 percent less than the fees charged by full-time brokers, is available at banks and through mutual funds.

One of the great benefits of an asset management account is the "sweep" feature. Any extra cash in your account above a certain minimum, including dividends and bond income, is automatically reinvested, typically in a money market fund, so your money is constantly earning interest.

**Financial Planners**    A growing number of consumers are turning to financial planners for help in investment planning and asset management. Most financial planners work independently and charge anywhere from $500 to $5,000 for their services, depending on their client's net worth, the planner's reputation, and whether the planner's fee is based on commission. Nationwide financial services—such as Merrill Lynch and Prudential, among others—also have financial planners on their staffs. Financial plans from nationwide services are much less expensive and are often offered to customers free of charge.

Unlike investment counselors, who are experts in the stock market and recommend specific stocks and bonds to purchase, financial planners help consumers make overall budgeting, saving, and investing decisions and recommend allocation of funds among different *types* of investments—in order to get the most favorable return for the investors' money.

---

### Question for Thought & Discussion #7:

At what age do you think a person should create a financial plan? What is likely to happen to a person who never creates a financial plan?

---

**How Helpful Are Financial Planners?**    Many consumers have slept better at night and improved their financial situations considerably with the help of financial planners. But the quality of plans and planners varies widely. Consumers Union (CU) has investigated such services and found that their cost and quality are inconsistent and often difficult for novice investors to evaluate. CU sent a man and woman posing as a married couple to ten advertised financial planners. The couples asked each planner the same questions and provided the same information about their financial situation and goals.

Although CU reached no clear conclusions, its research did lead to a number of general observations. First, most financial planners have "favorite" types of investments toward which they steer consumers. In many cases, they have a financial relationship with the investments or firms that they recommend. Financial planners are not likely to promote innovative methods of investment that consumers can carry out for themselves, such as buying stocks, bonds, or mutual funds on the Internet. The quality of the advice given is often, but not always, related to the size of the fee that is charged. The lowest-cost planners charged about $500 for a basic analysis of a family's financial situation and elementary investment recommendations. They also gave the least-complete advice. CU was particularly distressed to find that many planners ignored the specific requests and financial goals identified by their clients. The only recommendation made by all ten planners was that the couple prepare wills.

**Finding a Good Financial Planner**    The fact that some financial planners provide advice and services of limited or questionable quality doesn't necessarily mean that you should avoid purchasing financial-planning assistance. It does, however, mean that you should be cautious and exercise your rational decision-making powers.

Thousands of cases of financial-planning fraud are reported to the government every year. Victims lose the funds they paid for the service as well as a large part of the money they invested. According to estimates, U.S. consumers lose as much as $5 billion to fraudulent investment advisers each year. To avoid losing your money to a con artist, quiz potential planners carefully. What are their credentials? What about their past performance? Are the planners registered with the National Association of Securities Dealers, the Securities and Exchange Commission, or another professional group? Also, ask for references from several current customers. By studying Table 15.4, you can learn what the various titles often used by financial planners really mean. For further details on how to evaluate a financial planner, contact the Federal Trade Commission on the Internet at **http://www.ftc.gov** or at FTC, 6th St. and Pennsylvania Ave., N.W., Washington, DC 20580 and ask for a copy of its publication *Money Matters*. This brochure contains a lengthy list of questions to ask financial planners to ensure that you won't be taken in by a bogus planner.

**Elements of a Good Financial Plan**    A good financial plan will include the nine elements listed in Table 15.5. It will also reflect a planning process that usually involves the following steps:

1. *Gathering data.* A good planner will request that you bring all relevant financial information—including bank

## TABLE 15.4 Financial Planner Terms and Definitions

**Certified financial planner (CFP)** A financial planner who has passed an examination accredited by the certified Financial Planner Board of Standards that tests for knowledge and understanding that will enable the applicant to organize a client's estate, insurance, investments, and tax affairs.

**Certified public accountant (CPA)** An accountant who has passed a state-administered examination that tests for knowledge and understanding of accounting practices and tax issues but not for knowledge of investments.

**Chartered financial analyst (CFA)** A title earned by people from the Institute of Chartered Financial Analysts in Charlottesville, Virginia. To receive this designation, analysts must pass exams in economics, financial accounting, portfolio management, securities analysis, and standards of conduct.

**Chartered financial consultant (ChFC)** A title awarded by the American College in Bryn Mawr, Pennsylvania, to people who complete ten courses and twenty hours of examinations covering economics, insurance, taxation, estate planning, and other financial areas. These people work primarily in the insurance industry.

**Registered investment adviser** This title, granted by the Securities and Exchange Commission, allows holders to offer financial advice to clients for a fee. To obtain it, one need only pay a $150 fee to the government. It in no way indicates that the title holder has any training or other qualifications to give financial advice.

**Registered representative** A designation to indicate that an employee of a stockbrokerage firm has passed a series of tests administered by the National Association of Securities Dealers and is qualified to represent that firm in financial transactions.

SOURCE: Ken Sheets, "Money Managers Are Everywhere," *Kiplinger's Personal Finance Magazine*, February 1995, p. 64.

## TABLE 15.5 Elements of a Good Financial Plan

**1.** It is clearly written, in language the intended user can understand.

**2.** Recommendations are clear and unambiguous.

**3.** It contains a *cash-flow analysis*, a sort of glorified budget, that shows your income from all sources, minus all your expenses.

**4.** It includes a *net-worth statement* (a snapshot of your assets and liabilities) and examines your current debts to see if any should be consolidated, paid off from other available funds, or refinanced.

**5.** It includes an examination of your current insurance and recommends ways to bolster your coverage, if necessary, and to save on premiums, if possible.

**6.** It examines your current investment portfolio and makes recommendations for restructuring your investments if appropriate.

**7.** It includes a tax analysis and tax-saving suggestions.

**8.** It touches on retirement planning and estate planning.

**9.** It includes a statement of your goals, objectives, and tolerance for investment risk.

SOURCE: *Consumer Reports*, January 1986, p. 38.

certificates, insurance policies, tax returns, wills, and other documents—to your first meeting with the planner.

2. *Identifying goals and objectives.* Your goals and objectives will be the foremost concern of your planner. Essentially, this step involves learning why you want financial assistance.

3. *Identifying financial trouble spots.* A good financial planner will be able to identify any trouble spots—if you are underinsured or overinsured, for example, or paying more taxes than necessary.

4. *Following the plan.* A good planner will follow through on recommendations and coordinate the implementation of the plans with others—such as a lawyer—if necessary.

5. *Reviewing the plan periodically.* Usually, a good planner will review your plan with you annually to make sure that it is still appropriate to your planning goals.

If you want to know more about the ins and outs of the stock market, there are hundreds of books to consult. Some

of the most recent publications are included in the "Selected Readings" section of this chapter. Stock market and investment information can also be found in such publications as *Fortune, Forbes,* the *Wall Street Journal,* and consumer-oriented magazines such as *Kiplinger's Personal Finance Magazine* and *Money.*

## READING QUOTATIONS ON THE NEW YORK STOCK EXCHANGE

By far the best-known public market is the New York Stock Exchange. As with all other stock exchanges, at the beginning of each trading day shares of stock often open at the same price at which they closed the day before. At the end of each day, each stock has a closing price. This is the information that newspapers report. Table 15.6 on the next page shows information on some of the stocks traded on the New York Stock Exchange as it would appear in the financial pages of a typical newspaper. American Stock Exchange quotations are often given in most newspapers, too. Major newspapers throughout the country also carry regional or local stock exchange listings.

In a listing, each company's name is printed in an abbreviated form. For example, in Table 15.6, Am Ele. Pw. means American Electric Power. Often, other letters will appear next to the name of the company. For example, the letters *Pf* mean that preferred stock is quoted in that row. Prices are listed in dollars and cents.

*(Continued on next page)*

**Confronting Consumer Issues (Continued)**

| TABLE 15.6 | | | Partial List of New York Stock Exchange Composite Transactions, February 27, 2003 | | | | | | | | |
|---|---|---|---|---|---|---|---|---|---|---|
| **A** YTD % Change | **B** High | **C** Low | **D** Stock | **E** Symbol | **F** Div. | **G** % Yld | **H** P-E Ratio | **I** Vol. 100s | **J** Close | **K** Net Change |
| −13.2 | 14.00 | 2.92 | AAR | AIR | 0.0 | 0.0 | 0.0 | 446 | 4.47 | −0.03 |
| −18.0 | 27.97 | 8.70 | AOL Time | AOL | 0.0 | 0.0 | 0.0 | 200,823 | 10.74 | +0.47 |
| 1.4 | 41.97 | 17.62 | Alcoa | AA | .60 | 2.9 | 45 | 26,191 | 28.52 | +0.38 |
| −6.4 | 48.80 | 15.10 | Am. Ele. Pw. | AEP | 2.40 | 11.5 | 10 | 115,682 | 20.95 | +0.62 |
| −2.5 | 57.10 | 43.49 | Avon Pds. | AVP | .84 | 1.6 | 24 | 12,345 | 52.51 | +0.21 |

| | | |
|---|---|---|
| A | YTD % Change | The percentage change in the stock's value since the start of the current calendar year. |
| B | High | The highest price the stock has traded for in the past 365 days. |
| C | Low | The lowest price the stock has traded for in the past 365 days. |
| D | Stock | The name of the firm, usually abbreviated. |
| E | Symbol | The ticker symbol for the firm. |
| F | Div. | The amount of dividend paid by the firm in the past 365 days. |
| G | % Yield | The dividend of the stock as a percentage of its price. |
| H | P-E Ratio | The price of the stock divided by its earnings in the past year. |
| I | Vol. 100s | The number of 100-share lots sold on the day. |
| J | Close | The price of the last shares sold on the day. |
| K | Net change | The difference between yesterday's close and today's close. |

SOURCE: *The Wall Street Journal,* February 28, 2003, p. C3.

## QUOTATIONS ON THE NASDAQ

Only large, well-known corporations have their stocks listed on formal stock exchanges such as the New York or American Stock Exchanges. Most securities are traded on the over-the-counter market (OTC). Dealers in the OTC market are able to buy or sell stocks by contacting firms that "make the market" for specified corporations. Until the late 1980s, most of these stocks were listed at their *bid* (the amount people were willing to pay for the stock) and *asked* (the amount at which current owners were willing to sell) prices. This changed for many OTC stocks in the late 1980s when the NASDAQ was created. This system allows OTC trades to be completed electronically. Volume of these trades has increased to the point that it now makes sense to list a stock's high, low, and closing price for a trading day. As a result, quotations for stocks traded on the NASDAQ are essentially the same as for stocks listed on formal exchanges.

## MUTUAL FUND QUARTERLY REVIEWS

One way to check out mutual funds is to study the quarterly reviews that appear in the *Wall Street Journal, Barron's, Money,* and many other financial publications. Although the reports do not all contain exactly the same information, you can be reasonably sure that you will find the minimum initial investment, sales charge, annual expense, dividends, and performance over recent periods of time. Table 15.7 shows

part of a January 2003 quarterly mutual fund review. Remember that past success is no guarantee of future performance. In other words, evaluate a fund's future prospects before you invest.

## WHAT ABOUT BONDS?

Many people invest in bonds as an alternative to stocks. Unanticipated inflation can make bonds a bad deal, however. This is particularly true, of course, for low-interest U.S. savings bonds, but it may also be true for any other type of bond—federal, state or municipal government, plus corporate—that is *long term* and has an interest rate that fails to reflect fully the decreasing purchasing power of the dollars you loaned the people who gave you the bond.

In effect, most bonds are fixed income–bearing types of investments. You buy a bond, and it yields a specific annual return in dollars that can be translated into an interest yield. In other words, if you buy a bond that yields $100 a year and it cost $1,000, you receive a 10 percent rate of return; if it cost only $800, you get a 12.5 percent rate of return. Generally, as with all investments, the higher the rate of return, the higher the risk that the issuer of the bond will not be able to pay interest—or will not be able to pay at all.

If you decide to buy bonds, make sure you go through a broker who knows the bond market. Tell the broker how much risk you are willing to take and when you want the

## TABLE 15.7 — Partial List of Mutual Fund Quarterly Review, January 6, 2003

| A Fund Name | B Min Inv. | C Assets (mil $) | D Sales Charge | | E Annual Exp. | F NAV $ | G Dividends | | H Performance | | | |
|---|---|---|---|---|---|---|---|---|---|---|---|---|
| | | | Init. | Exit | | | Income | Cap. Dist. | Fourth Quarter | 1 Year | 3 Years | 5 Years |
| Smith Barney Aggressive | $1,000 | 1,689 | 0% | 5% | 1.99 | 57.73 | 0.0 | 0.0 | +9.6% | −33.3% | −9.9% | +10.1% |
| Strong Overseas | $2,500 | 84 | 0 | 0 | 1.90 | 8.69 | 0.0 | 0.0 | +4.8% | −20.0% | −24.5% | NA |
| USAA Income | $3,000 | 1,699 | 0 | 0 | .55 | 12.40 | .67 | 0.0 | +1.1% | +8.7% | +9.8% | +6.8% |
| Vanguard Health Care | $3,000 | 14,115 | 0 | 0 | .31 | 96.16 | 0.96 | 6.53 | +5.3% | −11.4% | +9.8% | +14.8% |

A   Name of the fund.
B   Smallest initial investment.
C   Value of fund's assets in millions of dollars.
D   Sales charge as a percentage of price. Init. = when buying shares. Exit = when selling shares.
E   Annual expense as a percentage of the value of each account.
F   Price of each share at the time of the report.
G   Dividends per share: Income = from dividends received by the fund. Cap Dist. = from capital gains earned per share.
H   Percentage return on investment for last three months, year, three years, and five years.
SOURCE: *The Wall Street Journal*, January 6, 2003, pp. R-22 R-33.

bonds to mature. You can buy bonds that mature in six months or thirty years, if you want.

**Zero-Coupon Bonds**   Corporate and government zero-coupon bonds pay no interest until the bond matures, yet you must pay taxes annually on the *implicit* interest earned. The attraction of these bonds lies in the profits you can accumulate by investing in them, for they are often sold at deep discounts. Zero-coupon bonds may make a good long-term investment for, say, your child's education fund or for your IRA, on which you pay no taxes until later anyway. If you had put $5,000 into a U.S. Treasury zero-coupon bond in mid-1990, for example, by the year 2010 you would receive back $40,000.

**Reading Bond Market Quotations**   Bonds normally have a face value of $1,000, but they can sell for more or less than that amount on the secondary market. In other words, they sell at a premium or a discount from their face value. Prices for bonds are listed as a percentage of their face value. A figure of 79¼, for example, means that a $1,000 bond is selling for $792.50.

Table 15.8 shows a listing of sample bond quotes from the New York Bond Exchange. Actually, the majority of bonds, including all tax-exempt bonds, are traded in the OTC market. Listings of OTC bond transactions include a bid price and an ask price. Often, there will be more than one listing of bonds for a particular company; this simply means that the company has different bond issues, each maturing at a different date or having different characteristics.

## TABLE 15.8 — Reading Bond Quotations

| A Bond | B Int. Rate & Dt. of Mat. | C Current Yield | D Volume | E Close | F Net Change |
|---|---|---|---|---|---|
| AT&T | 6¼ 04 | 6.2 | 110 | 102.88 | −0.88 |
| Bell So. | 7½ 33 | 7.1 | 10 | 105.50 | −0.38 |
| Duke En. | 7¼ 13 | 8.0 | 205 | 100.50 | . . . |
| Ford | 6½ 08 | 6.5 | 51 | 100.00 | +0.25 |
| TVA | 7¼ 43 | 6.8 | 25 | 106.88 | . . . |

A   The name of the firm that issued the bond.
B   The interest paid by the bond as a percentage of its face value and the date of maturity.
C   The interest paid by the bond as a percentage of its current price.
D   The value of bonds sold in thousands of dollars.
E.  The price of the last bond sold on the day per $100 of original face value.
F   The difference between yesterday's closing price and today's.
SOURCE: *The Wall Street Journal*, February 28, 2003, p. C11.

### Question for Thought & Discussion #8:
What sort of person is most likely to invest in stocks or bonds through a mutual fund? Are you this sort of person? Why or why not?

*(Continued on next page)*

## Confronting Consumer Issues (Continued)

**Tax-Exempt Bonds**   Municipal bonds generally are tax exempt; that is, the interest you earn on those bonds is not taxed by the federal government, and, in some cases, it is not taxed by state governments either. Nevertheless, these bonds are not always a good deal. To decide whether a tax-exempt bond is worth buying, you first determine how the yield compares with the rate you can earn on another investment that is not tax exempt—and this will depend, in large part, on your tax rate. Generally, you must determine whether the *after-tax* income you can receive from an alternative investment is greater or less than the (nontaxable) yield on a municipal bond.

Tax-exempt bonds are usually available in both $1,000 and $5,000 denominations. Unfortunately, most of the newer bonds are being issued in $5,000 denominations, so small investors cannot purchase them directly. You can, however, buy shares in tax-exempt bond mutual funds instead of buying the bonds themselves. Table 15.9 shows a sample of tax-free mutual funds and the average annual return from a family of Vanguard funds from December 1997 through December 2002. Interest rates fell significantly during those years. As a result, the return on municipal bond funds may be lower now than these averages. If you would like to receive more information about Vanguard Funds, call (800) 622-7447 or visit **http://www.vanguard.com**.

Tax-exempt bonds are classified not only according to the organizations that issue them—states, territories, cities, towns, villages, counties, local public housing authorities, port authorities, water districts, school districts—but also according to the sources of funds that the issuing organizations can utilize to pay interest and principal. As an example, *general obligation* tax-exempt bonds are backed by the full credit, and ordinarily by the full taxing power, of the state or municipality. *Revenue* tax-exempt bonds are backed only by revenues from a specific activity, such as a water-supply sys-

tem or a toll road. In addition, bonds are rated according to their riskiness, ranging from very risky to not at all risky. In the 1970s, owners of bonds issued by New York City found out all too painfully that tax-exempt municipals may not be such a good deal. You have to be wary about them, as you do about all investments.

You can check several sources to find out about the financial stability of the issuing organization for any bonds you wish to buy. Specifically, you can go to a library to look at *Moody's Bond Record*, where bonds are rated from the highest grade Aaa to the lowest grade C (speculative).

## REAL ESTATE DEALS

By now you should be suspicious of any investment that promises unusually high yields. Many real estate deals do exactly this, and they should be avoided because they involve large risks. You should be particularly suspicious of telephone solicitations or those that come to you in the mail. Most of the investments offered in this way involve high commissions for the salesperson, but the property is usually *illiquid* (hard to sell). The fact that many real estate deals border on fraud does not mean that investing in real estate is a bad choice—only that you should be selective about it.

Throughout history, real estate has increased in value at a rate greater than the rate of inflation most of the time. Owning buildings that can be rented generates current income as well. Of course, this type of investment often requires a personal commitment to maintaining property and to dealing with tenants, but it can be financially rewarding. Generally, real estate investments should be looked at as long-term investments. Few people become wealthy in real estate quickly.

## ALL THAT IS GOLD DOES NOT GLITTER

What about investing in precious metals? Isn't it a sure thing? The answer is unequivocally no. People who bought gold in 1980 when the price was over $1,000 an ounce probably weren't too happy when it fell to $500 an ounce a few months later. To be sure, the price of gold goes up from time to time. But that does not mean that you are guaranteed a high rate of return forever. In February 2003, gold was selling at less than $360 an ounce. The price of silver also fluctuates. In 1977, it was selling for $4.41 an ounce, and, during the first couple of months of 1980, it went up to $40 an ounce, but then the price dropped back down to $10 an ounce. Some people made a killing; some people lost a fortune. In 2003, silver was selling at around $4.50 an ounce. The price of platinum was $155 an ounce in 1977 and reached $850 an ounce in 1980; by 1985, it was down to $388.

| TABLE 15.9 | Tax-Free Mutual Funds |
|---|---|
| Fund | Average Return December 1997–December 2002 |
| Vanguard Muni Bond— Short Term | 4.0% |
| Vanguard Muni Bond— Intermediate Term | 5.4% |
| Vanguard Muni Bond— Long Term | 5.9% |
| Vanguard Muni Bond— Insured Long Term | 6.1% |

SOURCE: *The Wall Street Journal,* January 6, 2003, p. R-33.

You should not construe these examples to mean that investment in precious metals is never a good choice. Sometimes it is. A well-diversified investor is the best investor around.

If you wish to diversify your investments, you could hold a certain percentage of your investments in precious metals, such as Canadian gold maple-leaf coins, U.S. gold eagle coins, pre-1965 U.S. silver coins, and a variety of other precious metals. But do not assume that you are guaranteed a high rate of return. Nothing is intrinsically a sound investment at all times for all people.

## Question for Thought & Discussion #9:
Do you think most homeowners think of their homes as an investment or just a place to live? Why do you think they feel the way they do?

## VARIETY IS THE SPICE OF INVESTMENT

It is generally advisable to seek variety in your savings plans for several reasons. First, not all your savings should be in liquid assets. If you unexpectedly require money quickly, you would like to have some cash or saving account reserve that you can take out immediately without losing anything. Remember, however, that you pay a cost for keeping cash—the cost of the rate of inflation that shrinks the purchasing power of those dollars.

The second reason to vary your savings is that you can reduce your overall risk by having a large variety of different investment assets. There are no fixed rules to follow, although many investment counselors have their own. They might suggest you keep a certain fraction of your assets in cash, a certain fraction in a CD, and so on. But there is no scientific rule or reason behind such advice. You must decide yourself how much *liquidity* you want, how much risk you wish to take, how many long-term investments you desire, and how many short-term investments you prefer.

Remember, as you increase the variety of risks you have in your investment portfolio, you lower the overall risk in that whole portfolio but you also lower the overall rate of return you will receive. You may want to gamble as part of your investment program. You may want to buy, for example, penny stocks that sometimes jump tremendously in value. You may want to buy stocks on OTC markets that have a high variability and sometimes really hit. But you certainly should not put all your eggs in this basket, because if you lose, you will have nothing. At the other end of the spectrum, you could be absolutely safe by keeping everything in government bonds; but because you would be unable to make a higher rate of return, you probably don't want to do that, either. Remember, successful investment does not mean making a killing. It means avoiding losses that deprive you of retirement savings and, at the same time, earning a reasonable rate of return on those savings. Any other goals you choose may be costly.

# Chapter Summary

1. In the stock market, shares of U.S. businesses are bought and sold just about every weekday throughout the year. When you buy and sell stocks, you may either sell them for more than you paid and experience a capital gain or sell them for less than you paid and experience a capital loss.

2. When you buy a share of stock in a corporation, there is no guarantee that you will receive dividends or that you will make any particular rate of return on your investment. If you loan money to the company—that is, buy one of its bonds—you are guaranteed, as long as the company does not go bankrupt, a specified dollar interest payment every year and a specified principal payment when the bond matures.

3. In many ways stockbrokers are merely salespeople. They sell information about stocks and bonds. The more transactions they are able to convince you to complete, the greater their income will be. Investors should realize this before they employ a stockbroker to advise them.

4. Full-service brokers provide investors with advice about possible investments. The fees they charge for transactions will be more than those charged by brokers who do not provide investment advice. Investors may also complete stock and bond trades through discount brokers and online services. These services generally offer no investment advice but charge much lower fees. In 2003, it was possible to trade stock online for as little as $7.99 per transaction.

5. Do not be taken in by investment counselors who guarantee you a higher-than-normal rate of return in the stock market. Generally, investment counselors receive any above-normal rates of return in the fees they charge you or the commissions you must pay to buy and sell stocks often.

6. Mutual funds may be an easy answer to your investing problems, for they purchase a wide variety of stocks. It is generally advisable to buy into a no-load mutual fund that has no sales charges. Money market mutual funds provide a relatively safe investment vehicle, although they are not insured by the federal government the way money market deposit accounts are.

7. All schemes to make you rich should be investigated thoroughly, for, on average, they rarely guarantee you a higher-than-normal rate of return unless you accept a greater amount of risk. For example, even though the amount of land is fixed and the population is growing, real estate is not always a good investment.

8. For many people, a good way to save is to set up a pension fund. Three plans available to individuals are Individual Retirement Accounts (IRAs), 401(k) plans, and Keogh plans (for the self-employed).

9. Before consumers invest their savings, they should develop a personal investment plan that starts by identifying their investment goals. Many new investors join clubs, hire financial advisers, or seek assistance from full-service stockbrokers.

10. Although financial planners may be helpful to consumers who lack knowledge of financial markets and tax laws, they are expensive and should be investigated carefully. Not all financial planners are qualified to sell advice, and a few are dishonest.

11. Whenever consumers invest their money, they should keep in mind the relationship between high promised returns and high risk of financial loss. A diversified investment portfolio can help decrease the overall risk to any investor. In general, investing regularly in a varied selection of financial vehicles offers the best opportunity for long-term gain.

# Key Terms

401(k) plan **372**
capital loss **362**
common stock **361**

equity **361**
Individual Retirement Account (IRA) **370**
inside information **365**
Keogh plan **373**

mutual fund **366**
preferred stock **361**
stock market **360**

# Questions for Thought & Discussion

1. Do you or a member of your family own shares of corporate stock? How could investing in stock fit into your life-span plan?

2. Most stocks lost value in the years between 2000 and 2003. Should this loss cause people to reevaluate their stock investments? Explain why or why not.

3. Who would you trust to give you good investment advice? Why would you trust this person?

4. How concerned should investors be about the fees mutual funds charge? Explain your answer.

5. Do you know someone who has made a lot of money in the stock market? Do you think this person's good fortune was the result of luck or good planning?

6. Do you know anyone who has invested in an IRA or 401(k) plan? Is this person satisfied with the results of his or her retirement plan? Why or why not?

7. At what age do you think a person should create a financial plan? What is likely to happen to a person who never creates a financial plan?

8. What sort of person is most likely to invest in stocks or bonds through a mutual fund? Are you this sort of person? Why or why not?

9. Do you think most homeowners think of their homes as an investment or just a place to live? Why do you think they feel the way they do?

---

# Things to Do

1. Visit several stores that sell rare coins or stamps either in person or online. Find out what these businesses have to say about the investment value of the items they sell. Do you believe they are telling the truth? Are rare coins or stamps a good method of investing for most people? Why or why not?

2. Visit the New York Stock Exchange Web site at **http://www.nyse.com**. Investigate the services that this organization offers to consumers. Write an essay that summarizes what you find.

3. Identify a stock that you might consider buying. Why are you interested in owning this stock? Graph its price over the remainder of this semester. If you had purchased 100 shares of the stock, would you have made a profit or a loss? Assume you paid $7.99 to buy the stock online and the same to sell it later.

4. Identify and investigate three different mutual funds that invest in small firms they hope will grow. Compare their results over the past three years. How much difference was there in their returns? Why did some do better than others?

---

# Internet Resources

## Finding Consumer Information on the Internet

The following Web sites have been selected for their relevance to topics discussed in this chapter. Search these sites to locate information that can add to your knowledge about the regulations placed on brokerage firms. Remember, Web addresses change frequently. If any of these addresses no longer function, find similar sites to investigate using any of the search engines available to you.

1. The Securities and Exchange Commission (SEC) provides information about regulations imposed on stockbrokers. It also provides links to other organizations that oversee broker activities. You can access the SEC's Web site at **http://www.sec.gov/investor/brokers.htm**.

2. The National Association of Securities Dealers (NASD) has a public disclosure program that provides information to the public about specific stockbrokers and brokerage firms. Its Web site is located at **http://www.nasdr.com/2000.asp**.

3. The North American Securities Administrators Association (NASAA) maintains a Web site with links to authorities in each state that deal with stockbrokers and brokerage firms. Search this Web site to find out what services are offered in your state. It is located at **http://www.nasaa.org/nasaa/abtnasaa/find_regulator.asp**.

## Shopping on the Internet

The following Web sites have been selected because they offer consumers services similar to those described in this chapter. These are commercial sites that are designed to market products. They do not represent a comprehensive or balanced description of all firms that offer financial planning online. How do the services at these Web sites compare with those that are available from local financial planners? Remember, Web addresses change frequently. If any of these addresses no longer function, find similar sites to investigate using any of the search engines available to you.

1. Lifetime Financial Planning offers residents of Virginia fee-only planning that is advertised as being able to help clients reach their lifetime goals. Visit this firm's Web site at **http://www.lifetimefp.net** to see what it has to offer.

2. Compass Financial Planners, a firm based in Atlanta, Georgia, offers fee-only financial planning that is based on "integrity, objectivity, and service." Find out more about this firm by visiting its Web site at **http://www. feeonlyplanning.com**.

3. E. F. Moody, which offers financial planning to California residents, advertises that "knowledge makes obsolete the inequities that ignorance and prejudice justify." Investigate this firm's Web site to see if it offers a service you would consider employing. It is located at **http://www. efmoody.com**.

## InfoTrac Exercises

 Purchasers of new copies of this text are provided with access to the InfoTrac Web site. This Web site links students to thousands of recent articles published in hundreds of periodicals. Use the key words **SEC regulations** or other terms from this chapter to conduct a key-word search. Choose one article that is of particular interest to you and write a brief essay describing what you have learned from the article. Be sure to cite the author and title of the article and the name and date of the publication in which it appeared.

## Selected Readings

April, Carolyn A. "Investigation: SEC Charges 33 with Internet-Based Stock Fraud." *InfoWorld*, September 11, 2000, p. 22.

Berman, Phyllis. "Loaded Questions." *Forbes*, March 18, 2002, p. 181.

Birnbaum, Jeffery H. "The Last Word on Roth IRAs." *Fortune*, August 16, 1999, p. 122.

Clash, James M. "Focus on the Downside." *Forbes*, February 22, 1999, p. 162.

Goldwasser, Joan. "Getting Online Savvy." *Kiplinger's Personal Finance Magazine*, February 2002, p. 52.

"Mutual Funds: Pieces of the Action." *Consumer Reports*, March 2000, pp. 27–35.

*New York Times. Financial Planning Guide.* Published annually in November.

Norcera, Joseph. "Portfolio Analysis." *Money*, October 1, 2001, p. 63.

"Online Help for Fund Investors." *Consumer Reports*, March 2000, pp. 36–38.

Resendez, Alex. "Choosing a Broker." *Futures*, March 2001, p. 423.

Scott, Matheys. "Investing on a Budget." *Black Enterprise*, September 2001, p. 42.

Solomon, Deborah, and Aaron Llucchetti. "House Panel Grills Mutual Funds." *The Wall Street Journal*, March 13, 2003, pp. C1, C6.

"Taking Stock of Equity Mutual Fund Expenses." *Investment News*, September 25, 2000, p. 19.

*What Is a Mutual Fund?* Available from the Consumer Information Center, Pueblo, CO 81002. **http://www.pueblo.gsa.gov**.

Yasin, Rutrell. "Bubble Bursts for Online Brokers." *Investing News*, December 17, 2001, p. 28.

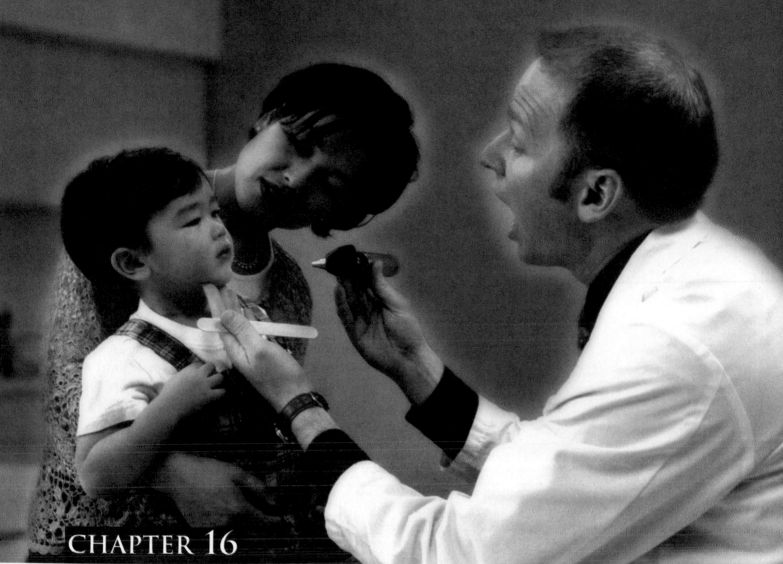

Rubberball Productions/Getty Images

# CHAPTER 16
# The Health-Care Dilemma

**After reading this chapter, you should be able to answer the following questions:**

- How does insurance enable consumers to control financial risks?

- Why are medical costs so high?

- What role does the government play in seeing that health care is provided for Americans?

- What types of health insurance are available to consumers?

- What steps have been taken to control health-insurance costs?

- What benefits and costs might result from a national program of health insurance paid for by the government?

- How can consumers make the best medical-care decisions?

**Insurable Interest** Something of value that is to be insured.

**Insurable Risk** An insurer's understanding of the risk of insuring a particular object or condition.

 e are all exposed to unknown risks every time we get out of bed. Some risks are small, and even if we do suffer a loss, such as breaking a breakfast plate, it makes little difference to our style of life. We should, however, take steps to avoid losses that are greater than we can afford to suffer. To do this, most U.S. consumers purchase insurance policies of various types.

## The Foundation of Insurance— Risk Pooling

All types of insurance are based on the principle of shared risk, or *risk pooling.* No matter how careful we are, it is impossible to eliminate every chance we have of suffering a loss to our property or our health, or of causing injuries to others. If risk is spread among many people, however, by having each individual pay money to an insurance company that will compensate those who suffer losses, no individual need suffer a loss that is beyond his or her ability to bear. In effect, each insured party is trading the cost of the insurance for an assurance that he or she will be paid for any covered losses.

In order to provide this service, insurance companies must be able to predict the probability and value of losses that may be suffered by insured parties. The companies accomplish this by studying past events and projecting them into the future. Two terms that describe these concepts are *insurable interests* and *insurable risks.*

An **insurable interest** is the property or condition that is insured against loss. There must be some way to determine the value of this interest. An **insurable risk** is the insurance company's understanding of the probability that the insurable interest will suffer a loss and the estimated value of any loss that may take place. Given enough time to collect and evaluate data concerning past losses, insurance companies can predict with great accuracy the probability of future losses being suffered by different groups of people. They use this information to set premiums that will allow them to pay claims by insured parties and earn a fair return on their investments. Insurance companies are not really in the business of gambling. They are able to calculate the amount of risk they take when they issue insurance policies. This allows them to set prices that almost always yield a profit. The only real dangers an insurance company faces are that an enormous disaster will cause many of its customers to suffer large losses or that it will issue so few policies that the law of probability will not protect it from the chance of a too-large proportion of its customers suffering losses.

## Our Need for Medical Insurance

It probably comes as little surprise that U.S. spending on medical care has been growing in money values and as a share of the value of our total production. Medical expenditures increased dramatically during the last century—from $4 billion in 1929 (roughly 4 percent of the value of production in that year) to over $1.4 trillion in 1999 (nearly 14 percent of the value of production). As Figure 16.1 shows, in 2000, medical costs accounted for about 14 percent of national expenditures. Medical costs consistently outpaced the rate of inflation during most years in the past decade. In other

**Question for Thought & Discussion #1:**
What is the insurable interest for medical-care insurance?

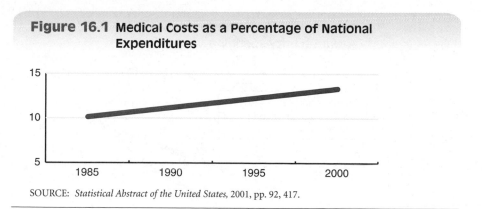

**Figure 16.1** Medical Costs as a Percentage of National Expenditures

SOURCE: *Statistical Abstract of the United States,* 2001, pp. 92, 417.

words, the relative price of health care is higher than it once was. Still, the growth in demand for medical care has resulted in an increase in the amount of care that is delivered to the U.S. population regardless of higher prices.

Although few people expect medical care to be free, many wonder why the cost for health-care services has gone up so much faster than other costs. There are several reasons for this, and the first has to do with increases in demand brought about by government programs.

# The Government Steps In

Before the introduction of *Medicare*—government-subsidized medical care for the aged—congressional estimates of what that program would cost were only a fraction of what the actual expense turned out to be. In the mid-1960s when Medicare was instituted, the actual cost of health-care services to the elderly and others covered under Medicare was drastically lowered, because the government paid the greatest portion of their medical bills. The quantity demanded of health-care services therefore rose so much that the available supply of medical-care services was taxed beyond capacity. The only thing that could give was the price, and it gave. Since then, the Medicare budget has continued to grow to meet ever-increasing prices for medical care. Through most of the 1990s, the Medicare budget increased by nearly 12 percent each year, more than doubling between 1990 and 1999. This rate of growth was nearly three times the overall rate of inflation. During the same period, the number of Medicare beneficiaries increased by only a little more than 10 percent. These data indicate that the amount expended for each Medicare beneficiary grew by about 80 percent during the decade of the 1990s.

In the future, the aging of our population will cause Medicare costs to continue to rise more rapidly than most other prices in our economy. According to predictions (see Figure 16.2 on the next page), government spending for Medicare will exceed spending on Social Security benefits shortly before the year 2015.

# The Insurance Framework

Another reason for the increased demand for (and consequent higher price of) medical services has to do with the insurance framework. More than 190 million Americans are covered by some kind of private health insurance—either through a group insurance plan at their place of employment or through individual health policies. Herein lies the problem. Insurance rarely covers as many

**Figure 16.2** Growth in Medicare Costs Relative to Defense and Social Security Spending

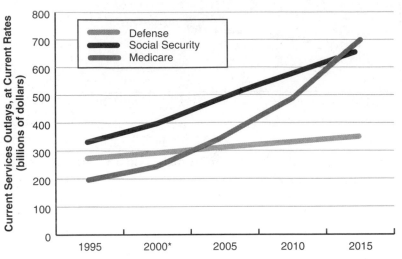

*Values for 2000 and after are projections.
SOURCE: *Statistical Abstract of the United States*, 2000, pp. 121, 348, 391.

of the costs incurred for **outpatient services** as it does for **inpatient services.**[1] Individuals covered by insurance, therefore, have an incentive to go to the hospital to be taken care of by their private doctors.

Hospitals also have an incentive to administer exotic tests because the fees for those tests help pay for the hospitals' investments in the sophisticated technology needed to perform them. At one time, insurance plans generally exercised little control over the number of tests and examinations that were performed on patients. In recent years, however, many insurers have instituted cost-management systems that have somewhat reduced the number of tests administered.

The lure of private insurance dollars and government assistance with capital costs for Medicare providers probably has encouraged the tendency to build too many hospitals. As a result, for several years hospitals have been operating at only 60 to 65 percent capacity. To recoup their overhead costs, these hospitals have had to charge higher prices than they would if they were fully utilized. In addition, to attract doctors—and thus patients—to their services, hospitals feel it necessary to acquire the latest scanners, lasers, magnetic resonance imaging machines, and other high-tech items, even though the same equipment may be available at another hospital only a few miles away. Such investments are very costly, and they ultimately result in higher charges to patients who use hospital services. Every year, the daily cost of hospitalization rises significantly. By 2003, the cost of spending one day in the hospital averaged nearly $1,300.

The problem is that patients covered by health insurance do not pay the *direct* costs of the medical care that they receive. Because of this, they demand more medical services than they otherwise would. This increase in quantity demanded causes medical expenses to rise, if all other things are held constant.

**Outpatient Services**  The services of doctors and/or hospitals that do not require the individual to remain as a registered patient in the hospital.

**Inpatient Services**  Services rendered to an individual by doctors and/or hospital staff while the patient remains in the hospital for at least one night.

---

1. A few insurance companies are now reversing this situation by requiring outpatient treatment for certain procedures.

Consider an example. When Anna Brown, a courtroom interpreter in Los Angeles, was charged $400 for blood tests while she was hospitalized for surgery, her husband, Jack, was not at all concerned about the bill until he learned that the insurance company deemed the tests unnecessary and refused to pay for them. At that point, Jack became concerned—so concerned, in fact, that he demanded that the doctor either justify the charges or drop them from the bill. The result? The charges were dropped.

Many consumers have had similar, if less dramatic, experiences. Often, if your doctor learns that your insurance won't cover certain expenses, she or he won't bill you for them. This courtesy to you and your pocketbook is rarely extended to insurance companies, however; on the contrary, insurers must almost universally pay at least the going rate for tests and procedures covered under their policies and—depending on the ethics of the physicians involved—sometimes far more than that.

# The Supply of Medical Care

Medical care includes but is not limited to the services of physicians, nurses, and hospital staff; hospital facilities; maintenance of the facilities; and medications and drugs. What determines the supply of the most important item—physicians' services—in the total medical-care package?

## Restricted Entry

Generally, only a small percentage of those who take the Medical College Admissions Test (MCAT) are accepted into medical schools. The more prestigious schools, such as Harvard Medical School, accept as few as 5 percent of their applicants. Some students apply to as many as ten different medical schools and, when turned down, reapply two or three times. Moreover, probably two or three times as many students do not bother to apply because they know that the odds are against them. Why is there such a large discrepancy between those who want to go to medical school and those who are accepted? If you compare the number of students who wish to attend law school with the number of students who actually go, the discrepancy is much smaller than for medical school. Obviously, the number of medical schools in the United States is severely restricted, as is the number of entrants into those schools.

Over the last fifty years, the limited number of seats in medical schools combined with an increasing demand for medical care caused doctors to be in short supply. As a result, physicians rarely lacked work or the ability to dictate the terms of their practices and fee schedules. Their incomes rose accordingly, as can be seen in Figure 16.3 on the following page; by 1999, the average income for a physician was over $194,000 (although it had declined slightly from 1997 when it had approached $200,000).

The era of the demand for physicians exceeding the supply may be drawing to a close, however. As Figure 16.4 on the next page shows, the growth in the number of doctors in the United States has exceeded our nation's population growth in the last decade. In 1950, there were roughly 1,000 physicians for every 1 million people. By 1999, this ratio had changed to 2,925 doctors per 1 million people. Although advances in medical science have created a greater demand for physicians with specific specialties, the time is approaching when the supply of doctors may exceed the demand. At that time the forces of demand and supply are likely to cause doctors' average earnings to level off or even fall.

> "Over the last fifty years, the limited number of seats in medical schools combined with an increasing demand for medical care caused doctors to be in short supply."

**Question for Thought & Discussion #2:**
How has the push by medical-care insurance providers to shorten the length of hospital stays affected hospitals' finances?

**Figure 16.3** The Rise in Physicians' Incomes, 1950–1997

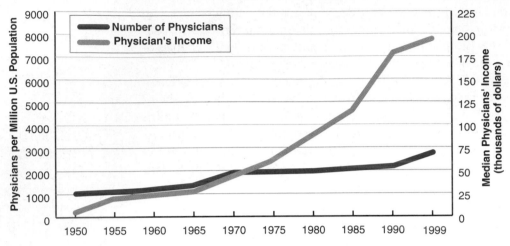

SOURCE: *Statistical Abstract of the United States*, 2001, pp. 8, 106, 108.

## A Surplus of Physicians?

What has caused the turnaround in the supply of doctors? For one thing, the shortage of doctors in the 1980s and the relative high incomes in that profession drew the maximum number to the medical field. Notwithstanding the limited number of medical schools and the high costs of attending medical school, there has been a gradual increase in the number of graduates in recent years. In addition, foreign-trained doctors have immigrated to the United States to obtain high-paying jobs.

Although the increasing supply of physicians may eventually lead to lower fees, this cost is only one of the concerns many people have regarding the U.S.

**Question for Thought & Discussion #3:**
Why hasn't the growing supply of doctors brought about a significant reduction in the fees they charge?

**Figure 16.4** The Increasing Supply of Physicians Relative to Population Growth

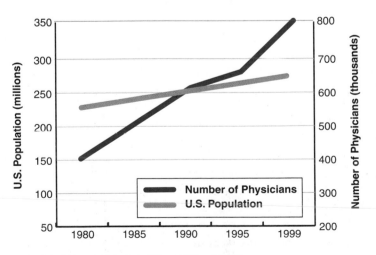

SOURCE: *Statistical Abstract of the United States*, 2001, pp. 8, 106.

health-care system. For many years, demographic experts have predicted financial problems for the Medicare program because of the increasing cost of medical care and the growing number of older people who will need care in the future. Specifically, between 2000 and 2010, 15 million people will join the Medicare system. At the same time, another 20 million workers will join the labor force to pay the costs of Medicare benefits for these people. As a result, there will only be about five people working to pay the cost of each Medicare beneficiary by 2010. Given these facts, many people believed that the Medicare system would become insolvent soon after 2005.

This problem has been postponed but not eliminated by imposing Medicare taxes on all earned income instead of only a limited amount since 1994. Although the Medicare trust fund is currently solvent, some still predict that it will run out of funds around 2030 unless benefits are cut or taxes are again increased. How much would you be willing to pay to assure the solvency of the Medicare program?

> **"If aspirin had to meet present requirements for FDA certification, it would probably fail to pass muster because we do not know why it works, and it can have potentially dangerous side effects when taken by children or in large quantities."**

# Drug Regulation and Its Costs

Consumers spend billions of dollars each year for drugs, about half of which are sold by prescription only. Since the Kefauver-Harris Amendment to the 1938 Food, Drug and Cosmetic Act was passed in 1962, the Food and Drug Administration (FDA) has required drug companies to follow quite detailed, extremely lengthy procedures before it will approve a new drug for the market. To test and evaluate most new drugs currently takes between six and ten years.

The benefits of this legislation are obvious—Americans can be fairly sure that the drugs they purchase are safe and effective. On the cost side is the suffering of those who are deprived of the benefits of new drugs during the lengthy testing process (except for some drugs for terminally ill patients, for which the FDA requires less testing). Some drugs never reach the market because the FDA is uncertain about their effectiveness and side effects. If aspirin had to meet present requirements for FDA certification, it would probably fail to pass muster because we do not know why it works, and it can have potentially dangerous side effects when taken by children or in large quantities.

The costs that pharmaceutical companies pay for complicated testing procedures for new drugs are factored into the price that consumers pay for prescriptions for brand-name drugs. The increasing availability of generic drugs, however, is helping to control this part of our medical bills. Once the patent on a brand-name drug expires, the drug can be approved by the FDA for marketing under its generic name, and this is frequently done. Fewer regulatory costs are associated with generic drugs because they require less time for FDA approval—since brand-name equivalents have already been tested. We discuss the cost-saving aspects of generic drugs later in this chapter.

# High-Tech Medicine

The application of technology to the field of medicine has resulted in an astonishing array of sophisticated medical equipment and testing procedures. Life-saving artificial organs, organ transplants, wonder biotech drugs, new cancer treatments, *in vitro* fertilization techniques—these are but a few of the options available today in the medical marketplace. With the aid of

> "The number of malpractice claims filed each year jumped by almost 100 percent during the 1990s."

technology, diagnostic tests are now available for almost every disorder. It is now possible, for example, to run 900 different types of tests on blood alone.

But as those who benefit—or wish to benefit—from these services are learning, high-tech medical care is very costly. The $80 X-ray has now been joined by the $600 CAT (computed tomography) scan and the $1,500 magnetic imaging resonance test. Costs for a kidney transplant range from $40,000 to $50,000, and for a liver transplant you can pay anywhere from $180,000 to $600,000. If you have to wait in a hospital intensive-care unit until a donor organ becomes available, your costs could mount to many times that of the procedure itself. Increasingly, the lives of premature infants are being saved, but, again, at a cost that may range from $250,000 to more than $1 million, depending on the medical attention required.

Ironically, researchers and health-care officials initially felt that technological innovations would reduce medical costs as a result of more efficient diagnostic and treatment procedures. It is now becoming clear, however, that these developments have contributed enormously to the high cost of health care in this country. According to some experts, more than 50 percent of the annual inflation in medical costs is due to high-tech medical care.

# The Malpractice System

The need of doctors and hospitals to defend themselves against potential malpractice lawsuits also contributes to the high cost of health care. The number of malpractice claims filed each year jumped by almost 100 percent during the 1990s, as did the amount of damages that juries awarded to claimants—today, frequently in excess of $1 million. As a result, doctors and other health-care providers have been forced to pay higher premiums for malpractice insurance to protect themselves against the possibility of a malpractice claim.

The average malpractice-insurance premium among family doctors (who do not perform surgery or other risky procedures) was just over $12,000 by 2003. But for surgeons, obstetricians, gynecologists, and other physicians with high-risk practices, premiums often exceeded $80,000 per year. Doctors who were successfully or frequently sued for malpractice could have their premiums increased to $200,000 or more. To be sure, no one wants a careless or unqualified doctor to treat patients, but it is unreasonable to expect a physician who treats thousands of patients over a lengthy career never to make a mistake. Such large malpractice premiums can effectively end a doctor's career.

The cost of malpractice premiums is passed on to patients and their insurance providers in higher fees. A typical obstetrician, for example, will treat about 1,200 women in a year and deliver 100 babies. When an $80,000 malpractice premium is divided among these patients, it adds about $68 to each of their annual bills. This amount may not seem unreasonable, but it is a significant cost of obtaining medical care.

## Related Costs of Malpractice Claims

Increased malpractice-insurance premiums are not the only cost of malpractice claims. The fear of being sued has caused many physicians to practice what has been called "defensive medicine"—that is, they order tests and other evaluative procedures and consultations not because they feel the procedures are needed, but to protect themselves from possible malpractice suits. All such procedures entail more costs for insurance providers as well as for the physicians who must spend additional time and money for record keeping in case their treatments are ever questioned in a court of law. Some analysts believe that defensive medicine adds as much as $150 billion to our annual health-care costs.

## A Possible Solution

A number of states have taken action to try to limit the cost of malpractice insurance. The most common method has been to cap the amounts that can be awarded for pain and suffering that result from medical mistakes. In states that have limited these awards to amounts of typically $250,000, the costs of malpractice insurance have increased at a slower rate. In 2002, for example, in states with economic caps on pain and suffering, premiums for obstetricians ranged from $17,786 to $55,084. In states without these caps, such as Michigan, the premiums approached $90,000. In his 2003 State of the Union address, President George W. Bush suggested imposing a national cap on malpractice awards.

Differences in malpractice-insurance premiums have caused some physicians to move their practices to lower-cost states, leaving a shortage of qualified doctors in states with high premiums. In 2002, doctors in West Virginia staged a two-day work stoppage to protest their malpractice premiums. Although they did treat emergency cases, elective surgeries were postponed to make their point. In 2003, many state legislatures recognized that something needed to be done to control malpractice-insurance costs. Until they act, malpractice premiums will remain an important expense and problem for our medical-care system.

> **Question for Thought & Discussion #4:**
> Would you support capping awards for pain and suffering at $250,000 to reduce malpractice-insurance premiums? Why or why not?

## Who Can Afford the Medical Bill?

The government, insurance companies, employers, and private individuals must foot the medical bill, and they are all gradually realizing that they can no longer afford to do so. If present cost trends continue, the government (and taxpayers) cannot continue to pay for Medicare without making totally unsatisfactory trade-offs—such as cutting national defense or Social Security spending. Insurance companies, which pay the direct costs of medical care for another substantial percentage of Americans, cannot absorb increasing costs without seeing their profit margins shrink. Therefore, insurers pass on these costs to employers and individuals in the form of higher premiums. In the last few years, health-insurance premiums have risen dramatically—on average by about 10 percent a year, but in some cases up to 35 percent.

For employers who offer group health-insurance coverage to their employees, the cost per employee of this coverage is mounting daily; it now averages almost $12,000 a year. Some corporations pay even more, depending on their benefits package. For many firms, health insurance has become the highest cost outlay after wages.

Because of increased insurance costs, some employers have chosen to reduce insurance coverage or require employees to share some (or more) of the expense through larger contributions to the plan or higher deductibles, or both. Many

CONSUMER CLOSE-UP
## Should You Have a Right to Know?

**W**ould you want to know if your doctor was about to be sued for malpractice by one of his or her patients? Imagine that you were rushed to a hospital emergency room one day because you were having severe chest pains. A surgeon examines you and recommends that you immediately undergo bypass surgery. You should always obtain a second opinion if there is time, but suppose you are told that you need the operation NOW! What would you do? Would it make a difference if you knew that the surgeon was about to be sued by someone else for performing a similar operation in which the patient died? Pretty scary, isn't it?

Now, consider the same situation from the surgeon's point of view. You have been educated and trained for many years at great personal expense. After successfully completing your training and performing hundreds of bypass operations, you operate on a patient who is in very bad health, and he dies. You tried your best, but nothing you could do made any difference. The patient's family refuses to face reality, hires a lawyer, and threatens to sue you and the hospital for malpractice (remember that anyone can sue anyone for any reason—a lawsuit doesn't mean that the plaintiff is right, or that he or she will win). You are convinced that you will be exonerated if the case ever comes to trial. In the meantime, should your patients be told that you may be sued? What would this knowledge do to your practice and your ability to earn a living?

The Florida Health Department faced this question in 1999. Al Sunshine, a reporter for a local Florida television station, sued the department to force it to provide a list of doctors whose patients had filed papers indicating their intention to sue. A state judge ruled that the information did not have to be released under Florida law until ten days after a suit was actually filed, but after that time it was to be public knowledge.

**Do you feel this was a proper decision? How does this type of litigation add to the cost of medical care?**

SOURCE: Rebecca Lentz, "Shine a Little Light: Release of Potential-Lawsuit Records Riles Florida Docs," *Modern Physician*, November 1999, p. 13.

small businesses have been forced to drop group insurance for their employees entirely, simply as a survival measure. The result of these developments is that employees are now paying higher prices for fewer insurance benefits, and some workers cannot afford health-insurance coverage at all.

# Health-Care Dilemma—or Crisis?

Recent surveys conducted in Canada, England, and the United States found that the U.S. health-care system is the most expensive, least liked, least fair, and, in some ways, least efficient of the three countries' medical systems. The extent of consumer dissatisfaction came as a surprise to many: not just a majority, but a full 89 percent of the Americans surveyed believe that a fundamental change in our health-care system is necessary.

Although we pay more for medical care than any other nation on earth, Americans apparently feel they are not getting their money's worth. Despite our technological capabilities, we live, on average, no longer than individuals in other Western countries. Our infant mortality rate remains higher than that of other developed countries—and even that of many developing countries. Our Medicare program does not cover the single greatest expense of our elderly population: long-term nursing care. Millions of our poor are not receiving adequate health care, and an increasing number of not-so-poor Americans cannot afford the costly high-tech procedures now available. More and more frequently, medical care seems to be allocated to those who can afford it; those who can't, do without. Media exposure of horror stories, such as patients being "dumped" from for-profit emergency rooms because they can't afford to pay for medical services, underscores the plight of the poor under our current system of health care. The disturbing ethical implications of these consequences of high medical costs have led many to conclude that what was perceived as a health-care dilemma in the past has now become a health-care crisis.

Physicians, employers, insurance providers, and government officials are all seeking solutions to the high cost of health care and its troubling results. The following sections examine both the measures that are currently being undertaken to bring medical costs under control and those that have been proposed for possible future implementation.

# Tackling Waste and Inefficiency

According to a *Consumer Reports* article published in the 1990s, as much as 25 percent of Americans' expenditures on medical care are wasted on unnecessary treatments, procedures, tests, or hospitalization. This would have amounted to more than $300 billion being wasted in 2003. Several possible explanations can be suggested for this waste of time, money, and resources. One possible explanation is the profit motive. The more medical services provided, the greater the income for members of the medical-care industry. Another possible reason is the desire (previously discussed) of doctors and hospitals to avoid being sued by patients for undiagnosed or improperly treated sickness or injury. Finally, as pointed out earlier, because most U.S. patients do not pay directly for medical care, they probably are less interested in keeping costs down. They may demand treatments that are not appropriate or necessary. In any case, we clearly are not making the most efficient use of our medical-care resources. In recent years a variety of medical-insurance organizations have instituted policies intended to control or even reduce the spiraling cost of medical care.

## Health Maintenance Organizations

For decades, **health maintenance organizations (HMOs)** have offered an attractive alternative to individuals and employers facing high health-insurance premiums. One of the oldest and largest HMOs, Kaiser Permanente, began in 1933. It was followed by other, similar programs around the country. Today more than 65 million Americans are insured by HMOs.

Unlike the fee-for-service standard that has been traditional in private medical practices, HMOs operate on a prepaid plan. Members of an HMO pay a certain, preestablished premium for all medical-care services, which are provided by physicians and hospitals participating in the HMO. Doctors are usually paid a set fee per patient that is not related to the patient's income. In other words, it is not possible to price-discriminate in HMOs as it is in fee-for-service practices.

**HMO LEGISLATION** In 1973, employees were given the opportunity to switch from standard group health plans to HMOs by a provision in the Health Maintenance Organization Act of that year. This law required all companies that had twenty-five or more workers and some kind of group health plan to offer their workers the alternative of participating in an HMO—provided that a qualified HMO was in the area and that the employer was first contacted by the HMO organizer. The legislation also required HMOs to accept individual members as well as groups of employees.

**Health Maintenance Organization (HMO)** A type of insurance plan in which members pay a flat fee in return for all medical services, provided they are administered by participating doctors and hospitals.

**Question for Thought & Discussion #5:**
Why do many physicians choose not to become members of HMO or PPO systems?

Later amendments to the 1973 law, however, abolished the requirement that employers must offer HMO services if available. Since 1995, that choice has been left to the employer. Under the amendments, employers are also free to negotiate rates based on their claims histories rather than paying the same premiums as everyone else. These changes are causing employers to look at HMOs with renewed interest as business firms try to cope with rising insurance premiums.

**THE PROS AND CONS OF HMOS**   From a patient's point of view, the major criticism of HMOs is that their members cannot choose just any doctor but only one from the list of qualified HMO physicians.[2] There is also concern that, given the profit incentive, doctors operating for a flat fee will not deliver high-quality medical services but will skimp on time spent with, and tests run for, each patient. Advocates of HMOs say that, because the worry of paying for each visit to the doctor is removed, people tend to seek treatment earlier, reducing the chances of a serious health crisis. Whatever the reason, studies have found that HMO members are sent to hospitals 40 percent less often than nonmembers.

Consumers should also be concerned about the financial stability of HMOs. During the past ten years, many HMOs lost money due to a combination of factors including increases in costs of medical care, reduced reimbursements from Medicare and Medicaid, limits on the premium increases allowed by state regulatory agencies, and, in some cases, their own inefficiency. Doctors Health Plan of North Carolina, for example, lost $8.5 million in 2000. In Detroit, Blue Care Network lost $14.7 million in the first three months of 2001. PacificCare of Colorado lost $24.1 million in 2000. And, in a worst-case scenario, the California Department of Managed Care placed UHP Healthcare of Los Angeles into receivership in 2001 because its net worth had fallen below $20 million.

Although some financially weak HMOs have merged with larger and more successful organizations, many are still in difficult financial situations. Consumers need to be aware of the financial strength of their insurance providers. If an insurer appears to be having financial difficulties, consumers should consider looking for another insurance provider or encourage their employers to do so if their medical insurance is furnished as a benefit of their jobs.

## Preferred Provider Organizations

Another cost-cutting insurance plan, which is somewhat similar to an HMO, is the **preferred provider organization (PPO).** Here an employer contracts for low-cost insurance coverage in return for having the employees use the services of particular physicians and hospitals that agree to a specific, set schedule of fees. PPOs offer insured employees more flexibility than HMOs in terms of choice because, for a higher deductible and copayment, the employee can see a non–PPO physician. The ability of employers to negotiate insurance bargains through PPOs—which has not been possible with HMOs—has been one of the major reasons for the rapid growth in PPOs in the last few years.

**Preferred Provider Organization (PPO)**   A type of insurance plan similar to an HMO but more flexible. In a PPO, members are allowed to choose the services of non–PPO medical providers in return for a higher copayment or deductible.

---

2. Under the amendments to the HMO Act, HMOs are allowed, but not required, to permit their members to use non–HMO physicians in some circumstances and be entitled to reimbursement of the cost.

Because of the cost savings available through HMOs and PPOs, more than 85 percent of insured employees are now covered by these or similar insurance plans. By contrast, in 1984, 85 percent of employees were covered by traditional group policies, under which employees could see any doctor they wished or receive the services of any hospital they chose. As Joseph Califano, Jr., former secretary of the U.S. Department of Health, Education, and Welfare (now Health and Human Services), observed, the era of free choice in medical care is largely over. Such choice has simply become too expensive.

## Cost-Management Systems

The runaway cost of medical insurance has caused many insurers to institute a variety of cost-management systems. Although these systems have the goal of preventing medical premiums from growing at unreasonable rates, they have reduced the medical-care choices of many consumers and increased the amount of red tape with which health-care providers must deal.

**PREAUTHORIZATION FOR MEDICAL TREATMENTS**   For some time, a number of insurance companies have required individuals to obtain their preauthorization before hospitalization. If your doctor recommends that you undergo coronary angiography, for example, you (or perhaps the doctor) must first call the insurer for approval. Coronary angiography, a diagnostic test used to visualize arteries to the heart, is a very expensive procedure, and the insurance company may request that you get a second opinion to see if it is really indicated.

If your doctor wanted to give you a test to check your health, would you want your insurance company to decide whether or not it was an appropriate procedure?

In the past, preauthorization for physician-recommended procedures was frequently given as a matter of course. As insurance companies fight increasing losses, however, they are devoting more resources to preauthorization and precertification procedures as a way to curb the costs of unnecessary surgeries and expensive diagnostic tests. Increasingly, insurers are consulting computer databases to evaluate whether a patient's symptoms warrant the recommended treatment. A variety of computer programs indicate appropriate medical procedures that can be carried out for patients with specific diseases or symptoms. Here is the way these systems typically work: Before hospitalization, you or your doctor must call your insurance carrier and describe your symptoms and the recommended treatment. The nurse receiving your call asks you a number of questions, which are determined by the computer program, and feeds your answers into the computer. The computer then analyzes your answers to see if the procedure is medically justified. If not, a doctor under contract to the insurance company will check with your physician to make sure that your circumstances warrant the procedure.

**THE AMA JOINS IN**   Such computer-screening programs have not been greeted with enthusiasm by doctors, who resent being second-guessed by computers. Nevertheless, some physicians' organizations have recognized the inevitability—and the potential benefits—of such programs and have established their own precertification databases. Even the American Medical Association (AMA) is working closely with Blue Cross/Blue Shield and the Health Insurance Association of America to devise comprehensive guidelines for precertification.

**Question for Thought & Discussion #6:**
How do you feel about your doctor's decisions regarding your health care being evaluated by a computer program?

Obviously, computer technology is affecting the medical world just as it is all other aspects of life. The role of computers in medical diagnoses and treatment will likely expand in the future. At least one medical visionary, Dr. Paul Ellwood, who pioneered the HMO concept, foresees a national computer network linking all doctors' offices in the country. The computer system would analyze patient data and treatment outcomes fed into it by doctors and provide constant feedback on the results. Such a system would reduce the number of misdiagnosed illnesses and inappropriate treatment procedures—and therefore costs.

**WHEN YOU CONSIDER MEDICAL INSURANCE**  A number of terms you are likely to encounter when you investigate medical insurance are defined in Table 16.1. A list of types of insurance you should avoid appears in Table 16.2.

| TABLE 16.1 | Terms Consumers May Encounter When They Shop for Medical Insurance |
| --- | --- |

**125 or flex plan** Federal law allows employers to establish this type of plan through which employees may set aside part of their earnings, up to a maximum of $2,000 a year *before* taxes are deducted. This money may be used to pay medical expenses the employees would otherwise have to pay out of pocket. Any unspent money in such a plan is forfeited by the employee at the end of the plan's financial year.

**Cafeteria plan** A form of insurance offered by some employers that allows employees to spend a fixed amount of money on any of a selection of approved medical-insurance plans.

**Coordination-of-benefits provision** A contractual part of many insurance policies that requires the insurance carrier to work with other organizations that also cover an insured party.

**Deductible** An amount that an insured party must pay before an insurance carrier will pay any part of a claim.

**Disability-income insurance** A type of insurance that pays insured parties a specific amount of money to cover living expenses when they are unable to work due to a debilitating injury or illness.

**Disability insurance** A type of insurance that pays for medical equipment and/or treatment that is necessary because of a permanent or temporary mental or physical condition that prevents the insured party from leading a normal life.

**Duration period** A maximum time that a particular benefit will be paid. This is typically associated with hospital or nursing-home services, which are frequently limited to a maximum of one to two years.

**Experimental treatment** A new medical procedure that has not been tested or approved by most insurance companies for payment.

**Internal limits** A limitation on payments an insurance policy will make for a specific purpose even if the limits of the policy are not reached. Typically, there is a limit on payments for mental-health or cosmetic-surgery benefits, for example.

**Major medical insurance** A supplemental insurance plan that pays for a share (most often 80 percent) of medical expenses beyond those covered by a primary medical-insurance plan.

**Medical IRA or savings account** An insurance plan that is available from some employers that provides an amount of money to employees that may either go to pay for a deductible on a basic medical-insurance policy if they become ill or are injured or be deposited in a savings account (either IRA or normal) if the employees use none or only part of the money. These plans are intended to encourage workers to use medical insurance only when it is really needed.

**Participating doctor** A doctor who has joined a preferred provider organization and has agreed to accept its payments as full satisfaction for patient obligations for medical care.

**Rehabilitation insurance** A type of insurance included in many basic medical policies that covers all or part of the cost of rehabilitation treatments that are carried out as the result of disease, injury, or mental illness.

**Renewability** A medical-insurance policy may have guaranteed renewability that ensures the policyholder that he or she will continue to be able to purchase insurance in the future, although possibly at an increased premium. If a policy has optional renewability, the insurance carrier may choose to allow or not to allow the insured party to continue to purchase insurance.

**Preexisting condition** A physical or mental situation that exists before the inception of an insurance policy. Until 1996 legislation, insurers could refuse to cover such conditions under a policy for a specified period of time, most frequently two years.

**Second-opinion requirement** A clause in many medical-insurance policies that requires insured parties to obtain a second medical opinion before specific medical procedures (generally elective surgery) are carried out.

**Self-insurance** A form of insurance in which a business or government agency sets aside money to pay all or part of the medical expenses for employees, to avoid the cost of buying insurance from other organizations.

**Stop-loss provision** A limitation on the amount an insured party will be required to pay (typically $10,000) for a particular illness or within a particular year as his or her share of a major medical plan claim.

**Waiting period** A period of time (typically from two to ten weeks) between the date a party signs an individual medical-insurance contract and the date the contract goes into effect.

| TABLE 16.2 | Types of Insurance to Avoid |
|---|---|

Although there are differences among these types of insurance, they all suffer from relatively high costs and limited benefits that bear little relationship to the costs incurred by the beneficiary. Some of these policies have been found to pay less than 20 percent of their premium income to insured parties in benefits.

**Accident policies**  A type of insurance that pays a specific amount for a specific accidental injury. These policies tend to be relatively expensive, and benefits bear no relationship to the actual cost associated with an injury.

**Loss-of-member policies**  A type of insurance that pays a specific amount for the loss of the use of a particular part of the body, such as a hand or a foot, or the loss of sight. Such policies tend to be relatively expensive, and benefits bear no relationship to the actual cost associated with a loss of use of a body member.

**Sickness or dread-disease policies**  A type of insurance that pays only if an insured party suffers from a particular disease. These policies tend to be relatively expensive, and benefits often bear no relationship to the actual cost associated with a sickness.

**Income-replacement policies**  A type of insurance that pays a specific amount per day for a set period of time when a person is hospitalized. Most of these policies pay nothing while a patient is recuperating out of a hospital, and the amount paid bears no relationship to the amount of the individual's lost income.

## Controlling Medicare Costs

In recent decades, the government has been concerned with reducing the high cost of the Medicare program. A major step in this direction was taken in 1983 by the institution of a far-reaching change in the way Medicare costs were reimbursed. Instead of reimbursing hospitals for actual costs, the government established a set fee according to 467 *diagnostic-related groups*, or DRGs. By testing and using other diagnostic procedures, physicians determine which DRG applies to a particular patient before hospital admission. The government pays the hospital the specific fee relating to the DRG, regardless of how long the patient is in the hospital or how expensive the care may be.

**REDUCING THE LENGTH OF HOSPITAL STAYS**  This "prospective-payment" plan was designed to curb Medicare costs by reducing the length of hospital stays and the number of unnecessary procedures and services. In this, it has succeeded. In the two years following the institution of the prospective-payment system, the average hospital stay for Medicare patients decreased from ten to eight days, and by 2000 the average stay was five days. Overall, it is estimated that, since its introduction in 1983, the prospective-payment plan has saved the government an estimated $320 billion in Medicare costs.

The government has found it difficult, however, to deal with the twin needs of cutting medical costs for the Medicare program while also providing necessary care for the increasing number of older Americans. In 1988, an attempt was made to provide "catastrophic-care" coverage for long-term, serious ("catastrophic") illnesses in Medicare patients. This plan was generally opposed because it would have been supported largely by increasing Medicare supplement payments by people enrolled in the system. It also failed to cover the costs of nursing-home care—essential for many older people who suffer catastrophic illness. The law was repealed in 1989, roughly three months after it was put into effect.

**Capitation** A plan for medical care in which providers are given a specific amount of money for each patient regardless of what treatment is provided.

**CAPITATION**    Today, a common method of controlling Medicare costs is **capitation.** Under this plan, Medicare providers are given a specific amount of money for each patient regardless of what treatment is provided. On patients who receive no treatment, the providers get to keep the entire payment, which covers their basic costs and provides them with a substantial profit. Those patients who receive many or expensive treatments may create greater costs for the doctor, clinic, or hospital than the amount of the payment. This obviously gives Medicare providers a strong incentive not to provide unnecessary treatments. At the same time, however, necessary treatments may be forgone, resulting in substandard care. Desirable or not, it appears that capitation will be more the rule than the exception in the coming years.

**PRESIDENT BUSH MAKES A PROPOSAL**    In March 2003, President George W. Bush proposed giving Medicare recipients a choice of three different forms of Medicare benefits that would provide some type of prescription drug coverage. Under his plan, Medicare beneficiaries could (1) remain in the traditional plan and receive discounts of 10 to 25 percent on their prescription drugs; (2) choose "Enhanced Medicare," which would allow them to obtain coverage from preferred provider organizations but with significant copayments for both drugs and medical care; or (3) join "Medicare Advantage," under which they would enroll in an HMO subsidized by the government and receive their prescription drugs with small copayments. Under the last option, of course, members would have to choose health-care providers from a limited number of physicians who were members of the HMO they joined.

All three plans were designed to save money for the government in the long run. Democrats in Congress responded with proposals that called for all Medicare recipients to receive prescription drug coverage without being required to change plans. The cost of the president's proposal was estimated at $400 billion over five years, and the cost of the Democrats' plan was put at $900 billion over the same period. Soon after Bush's plan was announced, it was generally recognized that it would not be adopted in its original form. Even Republicans argued that it did not provide adequate prescription drug coverage. Nevertheless, many expected that something would be done to help our nation's elderly with their drug costs because 2004 was an election year.

# Universal Health-Insurance Coverage

The idea of universal health-insurance coverage provided by the government is nothing new. Such insurance has existed in most European countries and Canada for more than fifty years. Although the possibility of having such a system in the United States has often been discussed, it has never come close to being implemented for various reasons—some are economic, some are political, and some relate to our desire to maintain consumers' freedom of choice. A little background information about universal health coverage will be useful before we address the actual programs proposed for the United States.

## Potential Benefits of Universal Health-Insurance Coverage

Obviously, the greatest benefit of universal medical insurance would be a decline in the sickness and suffering experienced by the roughly 48 million Americans who are currently uninsured. With government-provided medical insurance, people would no longer be reluctant to seek medical treatment in nonemergency situations. As things are, many uninsured people choose not to seek medical care because they cannot afford to pay its costs. As a result, their medical conditions become chronic and are more expensive to treat when these individuals finally are forced to see a physician or are rushed to an emergency room. When this happens, many of the uninsured are not able to pay for their emergency treatment. The cost is then borne by the hospital and passed on to other patients or to the government.

Universal medical insurance would probably result in other benefits as well. Most of our nation's uninsured are employed at low-paying jobs. When these people become seriously ill or are injured, they lose pay and their employers lose their services. At this point, because these uninsured individuals no longer have any income, they often qualify for the government's Medicaid program. In other words, many uninsured people cannot obtain medical insurance unless they become seriously ill and have little or no income. Only at that point will the government pay for their medical treatment. With universal coverage, we would no longer have the ironic situation in which people must lose their jobs to qualify for medical insurance. Another problem is that many of our nation's uninsured go to work even when they are ill. Thus, they are likely to spread any contagious diseases they have to other workers, who then will also lose work time.

## The Costs of Universal Medical Coverage

No one can say with certainty how much universal medical coverage would cost. The cost would depend on many factors including how the plan would be administered, the types and amounts of coverage, and how often people would use these services. Estimates range from $600 billion to more than $1 trillion a year in addition to what the government already spends for Medicare and Medicaid. Of course, there would be savings as well. Individuals and businesses would no longer have to pay private insurers for medical insurance. Although taxes would have to increase to pay for universal coverage, some people suggest that a single government program might cost less to administer than the hundreds of private plans that now exist.

**INCREASED USAGE** One concern about universal coverage is that people might use more services. It is reasonable to believe that if medical treatments were free or available at very low cost, individuals would demand more of them. This is a simple application of the law of demand. Many people who now have a cold do not visit a doctor because they know they will have to pay some or all of the cost. With universal coverage, however, they might choose to visit a physician even though all they need to do is take two aspirin and get bed rest. Studies have shown that a copayment as small as $5 can reduce doctor visits by as much as 25 percent.

Another concern is that physicians might have an incentive to increase the number of tests and treatments they prescribe. Any universal plan would probably impose caps on the fees that could be charged for specific procedures. If a physician could charge only $50 for an office visit, he or she might order unnecessary blood tests that could add another $90 to the fees for which the government is billed. There are ways to get around most rules that are intended to limit a program's costs.

**LONGER WAITS** One of the greatest problems of universal health programs in other nations is the length of time it takes to obtain medical treatments. This

**Question for Thought & Discussion #7:**
Have you or someone you know ever chosen not to seek medical care to avoid the cost? What were the final results of the decision? Was it a rational choice?

goes far beyond merely waiting in an office for many hours to see a physician. With limits on the fees physicians and other health-care providers can charge, there is a reduced supply available. Again the laws of economics apply—in this case, the law of supply. In Canada, for example, patients who suffer from cancer have been required to wait a month or more to begin chemotherapy or radiation treatments because Canadian hospitals were overcrowded. In fact, many Canadians who are able to afford the price have come to the United States to obtain immediate treatments that they pay for themselves.

**MORE BUREAUCRACY**    Although some people believe that a government-run universal insurance program would be more efficient than our current system, many others disagree. The government is not well known for efficiency or good organization. Consider your own interactions with government agencies. Have they always gone smoothly? Furthermore, what would happen to the private insurers that currently exist? Would they simply be put out of business? Would the government compensate their owners for their loss? Would their workers be laid off? Remember, more than 200,000 people are currently employed by private medical-insurance providers. What would happen to them? Could they be retrained and employed by the government to do the same types of jobs they do now? There are no easy answers.

## Proposed Systems

Over the years many universal health-insurance programs have beeen proposed for the United States. Some believe that the only fair system would be similar to the Canadian and European programs that provide all medical care on demand for free. This obviously would be the most expensive type of coverage for the government, and therefore for the U.S. taxpayers. Such a system has never been seriously considered by our federal government.

**THE CLINTON PROPOSAL**    In 1993, President Bill Clinton organized a task force to investigate the U.S. health-care system and suggest a plan to provide basic medical coverage for all Americans. In 1994, this task force recommended a plan that would have required all employers to either provide medical insurance for their employees or contribute to a government program that would help individuals pay for basic medical coverage that they would purchase from private insurance companies. The government would also have helped those who were self-employed or unemployed purchase insurance by providing them with vouchers that could be used to pay part of the cost of buying medical coverage. The Clinton plan was defeated in Congress for a number of reasons.

Chief among the reasons were the projected costs of the Clinton proposal and its potential impact on the economy. Estimates of its costs ran as high as $240 billion per year for only the most basic coverage. Further, many small businesses argued that they could not afford to pay for medical insurance for their employees or to contribute to a government program. Such businesses often operate on a small profit margin and are very competitive. Their owners explained that they could not afford to pay for medical insurance because they could not pass the increased costs on to their customers. Whether these contentions were correct or not, the Clinton plan was soundly defeated in Congress, and no other proposals for universal health coverage were considered during the remaining years of the Clinton administration.

**PRESIDENT BUSH'S SUGGESTION** After President George W. Bush was inaugurated in 2001, he suggested alternative paths to reduce the number of uninsured Americans. His idea was to have the federal government help create organizations that individuals could join that would bargain with health-care providers for lower prices. According to President Bush, individuals pay more than groups for health insurance and health care because they do not have a strong bargaining position. As members of a large group, they could achieve much lower medical-care costs. He also proposed that individual taxpayers be given increased tax deductions or credits for medical-insurance costs on their federal income tax.

There were a number of problems with the Bush plan. In addition, the U.S. Senate, which was controlled by the Democrats, refused to pass the proposal. In any case, although the plan might have reduced the high cost of medical insurance somewhat, it would not have eliminated these costs. For many of our nation's uninsured, any medical-insurance premium is too much. Furthermore, providing a tax credit or exemption to help cover the cost of medical-insurance premiums only helps people who pay federal income taxes. Many of our nation's uninsured earn so little that they pay no federal income taxes at all even without a special medical-insurance deduction.

After the elections of 2002, the Republican Party controlled both houses of Congress. It is possible that with these majorities President Bush will be able to get some sort of insurance program enacted that will help our nation's uninsured obtain medical coverage. But with the other items on the president's agenda, this might not happen.

## Health Care as a Consumer Product

That we have not created a national program for health care does not change the fact that many Americans do not have medical insurance or that we all pay a cost directly or indirectly for the sicknesses they suffer, which often go untreated. At the heart of the debate over the role of the government in providing health-care services is a fundamental question concerning the nature of medical care: Is health care just another consumer product, or is it a constitutional right of Americans? Traditionally, opposition to government involvement in health care has been based, implicitly or explicitly, on the assumption that health care is a consumer product and, as such, should be subject to the laws of the marketplace just as other goods and services are. In this view, government intervention in the medical marketplace leads to inefficiency because it interferes with the working of supply and demand. If the government would stay out of the medical-care market, supply and demand would regulate a price that the market could bear. Cost-conscious consumers would demand less medical care, thus lowering prices. Physicians and medical-service providers who performed badly would see their revenues decline, and those who performed well would reap rewards in the form of profits.

To be sure, not everyone would be able to afford the same degree of medical care under such a system, but more people would receive greater, and higher-quality, health care than they would under a government-regulated system. There would be no way around the problem that expensive medical treatments would be rationed to those who could afford them, just as there is no way around the problem that not everybody can own a Rolls Royce or a country mansion or a vacation home on the lake.

> **"Is health care just another consumer product, or is it a constitutional right of Americans?"**

**Question for Thought & Discussion #8:**
Would you support a program to provide universal health insurance to all U.S. citizens? How could such a program be paid for?

## The Ethical Consumer

### What Do You Do When Your Insurer Says No?

Although every business wants to save money, when it comes to medical-insurance providers, saying no in some cases is almost the same as a sentence of death. Take the case of Richard Cohen, a fifty-three-year-old resident of Seaford, New York, who was diagnosed as having myeloma, a type of cancer of the bone marrow. The accepted method of fighting this disease is to extract bone marrow, remove the cancerous cells from the marrow, and return it to the patient, a procedure that costs $100,000 or more. Twice Cohen's operation was canceled because his insurance company was still "reviewing the case." A problem with such delays is that either chronically ill patients may die while awaiting approval or their disease may progress to such an extent that their bodies are too weak to withstand the stress of the operation. Only after Cohen hired a lawyer who threatened the insurance company with litigation did the company finally agree to pay for the treatment.

As a basic rule, insurance companies refuse to pay for treatments that are "experimental." According to attorney Richard Carter, a specialist in medical litigation, however, many companies read the word *expensive* as *experimental*. They don't stay up with the latest treatments, so payment for tested and accepted treatments is often refused because they will cost the insurance company a substantial amount of money. According to Carter, the insurance companies are often being short-sighted when they refuse to pay for expensive new procedures. He argues that traditional care that drags on for many months or years may end up being more costly in the long run than newer methods that actually cure patients. He believes that the most important first step insured parties can take when their request for coverage is refused is to not take no for an answer. By continuing to fight, patients make it clear to the insurance company that they are not going to go away and that it may be better for the company to deal with them. Carter points out that it is not even necessary to hire a personal lawyer in many situations because a number of consumer advocacy groups offer help for free or at a small charge. Among these groups are:

- AIDS Clinical Trial Information Service at <u>http://www.actis.org</u>.
- Can Help Cancer Treatment at <u>http://www.canhelp.com</u>.
- Children's Organ Transplant Association at <u>http://www.cota.org</u>.
- National Children's Cancer Society at <u>http://www.children-cancer.com</u>.
- National Organization for Rare Disorders at <u>http://www.rarediseases.org</u>.

Those who view medical care as a consumer product are not heartless or lacking in compassion. On the contrary, most of them would agree that, ideally, everyone *should* have access to adequate health care. Their position is simply that, from a realistic point of view, this ideal may not be attainable within the framework of our present economic system—which, for all its faults, has succeeded in creating a higher quality of life than any other yet devised. Rather than tossing out our present system, they feel, we should build on its strengths by further cost-cutting measures and by placing more emphasis on the role of charity.

## Health Care as a Constitutional Right

At the other end of the debate spectrum are those who maintain that medical care is a right and not a privilege. The first sentence of the U.S. Constitution states that one of the aims of our government is "to promote the general welfare" of its citizens. Since health care is essential to our general welfare, shouldn't the government bear the ultimate responsibility for seeing that our health needs are met? A majority of Americans seem to think so. In a survey, nine out of ten Americans polled responded that they felt every American had a right to medical care "as good as a millionaire gets." Only recently, however, have a significant number of Americans wished for more direct involvement of government in health care as a means of securing that right. Although at no time has the average American received medical services comparable to those obtained by millionaires, until recently most Americans could at least be assured of receiving medical treatment during emergencies—and possibly of receiving emergency treatment as good as that a millionaire would receive. If they couldn't pay the charges, the hospital absorbed the costs and, normally, passed them on to insurance companies through higher billings. Increasingly, that is no longer the case in some of our cities.

That even a minority of our nation's physicians—who in the past opposed government intervention in the medical-care field—would promote a government-sponsored health program signals that a change is taking place in this country. Old arguments seem to be giving way in the face of new and urgent needs. Whatever the outcome of the current debate over national health care, at the present time, you, as a consumer, face the problem of dealing with high medical costs.

**"Old arguments seem to be giving way in the face of new and urgent needs."**

# Cyber
## consumer
### Finding Medical Insurance Online

Virtually every firm that offers medical insurance to consumers maintains a Web site where it promotes its products. CIGNA Health Care, one of the nation's largest HMOs, provides such a site. Visit this firm's Web site at http://www.cigna.com, and investigate the types of information it provides to its subscribers. How helpful is this Web site? Would it encourage you to become a CIGNA subscriber?

# Confronting Consumer Issues: Purchasing Medical Care

**H**ave you ever gone to work or to school only to have someone say, "You look terrible! Why don't you go see a doctor?" Although you may hope to visit a doctor's office only for routine physical examinations, the probability is that sooner or later you will need the services of a doctor, or even a hospital, for illness or injury. In 1999, Americans consulted with or were treated by doctors nearly 3.5 billion times. These physicians were paid more than $290 billion for their services. Another $415 billion was spent on hospital care and almost $117 billion for medicine and drugs. Other expenses brought the total spent on medical care to more than $1.3 trillion. The cost of medical care is the most rapidly growing part of the typical U.S. consumer's annual spending.

Photodisc/Getty Images

**What personal and professional qualities** would you look for when you choose a physician?

## CHOOSING A DOCTOR

Relatively few doctors are **general practitioners (GPs)** or *family-practice physicians.* Of the nearly 800,000 physicians in the United States in 1999, only 75,000, or about 9 percent, had this sort of practice. The others all specialized in something. Although GPs earn relatively less than most specialists, they are your first line of defense in protecting your health. Most of the medical care you will need in your life can be provided by a GP. If you do require the services of a specialist, your family-practice physician can refer you to one.

### Choose Your Doctor When You're Healthy
It is vital that you have a working relationship with your doctor that is based on mutual respect and trust. If you do not already have an established relationship with a doctor, the sooner you develop one, the better. To make accurate diagnoses of ailments, a doctor needs to be familiar with you and your medical history. Without a regular doctor, you may find yourself needing a doctor quickly and calling numbers at random from the phone book until you find one who is willing to see you; or you may have to go to an emergency room when your medical problem is not a true emergency, so you will have to wait hours for treatment as others are tended to ahead of you. You should choose a doctor when you are healthy and can take the time to investigate your alternatives.

### Pay Attention to Recommendations
One of the best ways to find a good doctor is to ask your friends and acquaintances for recommendations. There is, of course, no guarantee that you will relate well to a doctor just because someone else you know does. Nevertheless, you can at least

eliminate those who are unable to get along with any of their patients. If you have moved from another community where you had a doctor, you might ask him or her for a recommendation or referral. You might also call a local hospital. Hospitals sometimes recommend physicians who are popular with other patients. You could also call the local office of the American Medical Association (AMA). Do not expect the AMA to evaluate doctors; it will be very hesitant to say anything bad about any doctor. But the AMA may at least be able to identify doctors who are accepting new patients. Deciding among physicians who are identified is then up to you.

> ### Question for Thought & Discussion #9:
> How did you choose the physician you see when you become ill? Are you satisfied with your choice?

### Consider a Doctor's Age and Gender
When you look for a personal physician, you should try to choose one you believe you will be able to have a relationship with over an extended period of time. Choosing someone who is experienced can be valuable, but there is such a thing as going too far. If you are in your twenties, choosing a doctor in his or her fifties guarantees that you will need to find another physician in the future.

Some people are not comfortable discussing personal health problems with a doctor of the opposite sex. If you think you might not be as forthright in discussing your health problems with a doctor of the opposite sex, you should respect your feelings and select a physician of the same gender.

### Consider a Physician's Association

In most communities, some hospitals have better reputations than others. Some may also be more conveniently located to where you live. When selecting a doctor, you should inquire with which hospitals she or he is affiliated. Remember, a doctor who does not work out of hospitals with the best reputations in your community may not be a good choice.

Consider whether the doctor has an individual practice or is part of a group. If you choose one who is in practice alone, ask who would provide treatment for you if your doctor is out of town when you become ill. When a doctor is part of a group, you should be sure you understand how the group works. Will you always see the same doctor, or will you see whoever happens to be working when you make an appointment? If there is a particular doctor in the group whom you would prefer to avoid, can you request that someone else treat you? Can you always be treated by a male or female physician, if that is your preference? Patients need to be sure they understand what they are agreeing to when they choose a doctor who is part of a group.

## WHAT IS THE DOCTOR'S ATTITUDE TOWARD YOU?

Although doctors have special knowledge and skills, they are still just people. If your doctor talks down to you, or refuses to answer your questions, or seems uninterested in what you have to say, then you probably need to look for someone else. The attitude of the staff in a doctor's office is often a good gauge of the doctor's character. If they are rude and demanding, then you may be sure, if nothing else, that the doctor is willing to tolerate employees who are rude to patients. This implies a degree of insensitivity that should concern you.

Keep one fact in mind, however: that a doctor, or anyone else who provides a service, is friendly and personable does not prove that he or she is competent. Always look beyond the doctor's personality to the quality of the treatment being provided.

## LOOK FOR AN INTEREST IN PREVENTIVE MEDICINE

Good doctors are just as interested in **preventive medicine** as in curing a disease you already have. If you are seeing a doctor for the first time, and she or he expresses concern about your weight, your smoking, your high cholesterol, or other similar conditions, it may be one sign that the doctor is truly interested in your health. You should also expect a doctor to have basic diagnostic procedures carried out after your first visit to establish a "baseline" for your physical condition—data against which to measure future changes.

## DO YOU REALLY NEED AN OPERATION?

One of the most feared statements any doctor can make is "You need an operation." The immediate reaction of many consumers is to panic. What they need to do is calmly consider their alternatives and remember that they are in charge of their own fates.

No reputable doctor will suggest surgery without having a good reason. You should ask for, and expect to receive, an explanation you can understand for why your doctor feels an operation is necessary. Don't be intimidated. It is your body and your life. You have a right to know. If your doctor refuses to provide you with an acceptable explanation, you should look for a new doctor.

### General Practitioners Can't Do It All

Most general practitioners do not perform surgery other than some minor outpatient procedures. If you need an operation, you will probably be referred to a specialist. Again, you should ask the specialist to explain what procedures will be used and why they are necessary. There is no benefit in being rude, but there is value in asking questions that are to the point. If you want to know how much risk is involved, you should say so. Asking a question such as "Is this really the best solution?" is not as likely to elicit helpful answers as asking, "What are the risks of this operation?" or "What is likely to happen if I choose not to have it done?"

Although many insurance companies require a second opinion before they will pay for surgery, you should seek another doctor's opinion whether it is required or not. People who have studied medical care in the United States estimate that between 20 and 25 percent of all surgery performed in this country is not necessary. It is usually a good idea not to ask for a second opinion from another doctor who is closely associated with your original doctor. Doctors who work together are not likely to dispute each other's professional opinions. Many Internet sources provide guidance for people who are seeking a second opinion prior to surgery. Most large hospitals maintain Web sites that list information and offer specific instructions to people who live in their areas. Search the Internet for hospitals or other organizations that provide this type of assistance. Use the key words *second-opinion surgery* and the names of large hospitals in or near your community.

### A Surgeon Is Only Part of a Team

Many consumers mistakenly believe that surgery is sort of a one-person show. In fact, surgeons operate with the assistance of other medical professionals who are part of a team. Patients should investigate all members of their surgical teams. For example, it is very important to know the qualifications of

*(Continued on next page)*

**Confronting Consumer Issues (Continued)**

your **anesthesiologist.** You should find out if a certified anesthesiologist will personally administer the anesthesia, or if this person will direct a nurse or a resident doctor to complete the procedure. If someone else will administer the anesthesia, you should find out what this person's qualifications are. In addition, someone on the team should ask you about any allergies you have, your medical problems, if you are taking medication, how you have reacted to anesthesia in the past, or if you have a history of abnormal bleeding. If no one asks for this type of information, you should find out why. The anesthesiologist and other members of the team need to know these facts to provide you with proper care and protect your life.

## WHAT MEDICATION IS RIGHT FOR YOU?

Most drugs prescribed by doctors represent no risk, or very little, to most patients. Broad-spectrum antibiotics, antihistamines, and other common drugs are routinely prescribed to help cure many common sicknesses. If you are not familiar with a drug that is prescribed to you, ask your doctor what it is and what it does. If you feel that you need a more extensive explanation, many drug source books are available in almost every public library. A common one is the *Physicians' Desk Reference,* published periodically by Ballantine Books. This book lists most common drugs by their scientific and trade names. It describes their purposes and possible side effects and identifies other drugs that should not be taken at the same time.

### Beware of Drug Interactions   Many drugs can
cause severe and even life-threatening **drug interactions** when taken together. This is less of a danger when all medications you are taking are prescribed by one doctor. It is much more likely to become a problem when you are seeing two or more doctors who do not communicate with each other about your treatment. If you have different drugs prescribed by more than one doctor, you should identify them to your doctors and ask if the drugs are likely to react adversely with each other. In recent years many pharmacies have purchased computers that will flag drugs prescribed to a customer that are likely to interact in a dangerous way. This is one reason to fill all of your prescriptions in the same location.

### Taking Too Much of a Good Thing   Consumers
should realize that the fact that one tablet makes them feel better does not mean that two tablets will make them feel wonderful. Patients should never take it on themselves to change the dosage prescribed by their doctors. It is also important to take medication for the full length of time

prescribed. Suppose an antibiotic is prescribed for seven days to fight an infection. If the symptoms abate after three days, a person might be tempted to stop taking the medication. This would be a mistake. The infection might not be completely gone. It could redevelop and could even become resistant to the antibiotic, making it more difficult to cure in the future.

As a general rule, you should be careful of taking too many antidepressant drugs, painkillers, sleeping pills, or other drugs that affect the level of activity in your body. Never take more of these drugs than have been prescribed by your doctor, and question your doctor about the need for a drug that is prescribed over an extended period of time. Taking any medication that can alter the rate at which your body functions has the potential for harming, or even killing, you.

### What about Generic Drugs?   It is often possible
to purchase non–brand-name **generic drugs** that are much less expensive than their brand-name counterparts. You should ask your doctor or pharmacist if there is a generic alternative for any prescribed drug, even if your insurance pays for the drug. The extra cost of brand-name drugs may force an increase in insurance premiums for your prescription plan and reduce other forms of compensation your employer can afford to provide to you.

### Buying Prescription Drugs Online   Many
Web sites offer to sell prescription drugs at reduced prices. In recent years, sites located in Canada have offered U.S. consumers the best deals. The prices of drugs sold in Canada are regulated by the Canadian government. As a result, Canadian consumers can often purchase the same drugs as U.S. consumers for 30 to 50 percent of the U.S. price. By shopping for these drugs online and paying for shipping, U.S. consumers have been able to benefit from huge savings.

Technically, this practice is against the law, but the federal government has done little to stop U.S. consumers from shopping for drugs in Canada online. The drug industry, however, may not be so permissive. Prescriptions filled in Canada represent a loss of income to U.S. pharmaceutical firms. In 2003, GlaxoSmithKline stopped shipping its drugs to Canadian businesses that sold their drugs to U.S. residents. If other firms follow GlaxoSmithKline's lead, this source of low-cost prescription drugs may be eliminated for U.S. consumers.

## A VISIT TO THE EMERGENCY ROOM

At this moment, approximately 12,000 people are in the process of being admitted to emergency rooms in the United States. Most of them are not seriously injured—only roughly 15 percent are. Whether or not your visit involves a life-threatening situation, you should know your rights and responsibilities in an emergency room.

## When an Emergency Is an Emergency

Most people who visit emergency rooms are not truly in an emergency situation. Examples of situations that doctors identify as true emergencies include severe bleeding, choking, poisoning, severe chest pain, gunshots or other wounds that have penetrated internal organs, severe burns, and the like. If you visit an emergency room with a true emergency condition, you will probably receive emergency treatment. If your condition does not demand immediate treatment, you will not be served quickly in most cases. Other people will probably receive treatment before you, even if you arrived earlier.

## When Your Need Is Urgent but Not Immediate

If you have a fever of 103 degrees, have sprained your ankle, have a cut that you have stopped from bleeding, or have been unable to keep food down for twelve hours and have cramps, you may need care soon, but there is time to consider alternatives. Your first step may be to call your doctor. Many group practices have "emergency" numbers. Others provide their home numbers for emergencies. Of the situations just described, the only one that would have warranted an immediate call is the last, which could indicate a serious internal problem.

If you do call your doctor, she or he will tell you whether you need emergency care and, if so, where you should go for it. If you do not succeed in reaching your doctor, you are left with the responsibility of making the decision yourself. In such cases, consumers should realize that not all emergency rooms are equally equipped or able to provide treatment. If you visit an emergency room in a large urban hospital, it may or may not have the most up-to-date equipment and qualified doctors, but it certainly will have many patients with true emergencies. You may wait a very long time to be treated.

Smaller or rural hospitals may have no emergency room at all and may have to send you on to a larger facility.

The best choice, if it is available, is likely to be a medium-sized hospital in a relatively affluent area. It is likely to have a well-equipped emergency room and qualified doctors present or at least on call. An important step you can take now is to identify the best emergency room facility near your home.

### Question for Thought & Discussion #10:

What experiences have you had with hospital emergency rooms? Were you satisfied with the treatment you received? Why or why not?

This is, of course, no use to you if you become ill when you are away from home. If you need an emergency room when you are on a trip or visiting another city, try to seek recommendations from people who live in the area. Wherever you go, be sure to have the necessary insurance and other medical information with you if you possibly can. No one who has medical insurance should leave home without such information, because people never know when a medical emergency will arise. Not having such information can slow or even prevent necessary treatment.

## When You Arrive at the Emergency Room

If you have not come to the emergency room in an ambulance, it is often better to stay in the car until contact has been made with the emergency room's officials. When you enter the emergency room, you will first talk to a trained receptionist or a nurse, whose job is to determine how serious the problem is and how quickly you need treatment. At this point your answers should be brief and to the point. Don't try to tell the person what you think he or she needs to know. You will probably save time by just answering questions and saving lengthy explanations for doctors later on.

If the problem is a major one, you may wish to have your own doctor or a specialist called in. Many teaching hospitals have interns and residents who are looking for patients with certain conditions about which they are interested in learning. Although this is reasonable from their point of view, you might prefer an established doctor with more experience.

## DEALING WITH A HOSPITAL

No one likes to be in a hospital, but if it is necessary, there are a number of steps you can take to make the best of it. First of all, though, make sure you really need to be hospitalized. Hospitals are intended for people who require medical care that cannot be provided in other settings. Unfortunately, today many people are in hospitals who really shouldn't be there. They may have been admitted because they demanded it, because their doctors found it easier (and possibly more lucrative) to treat them there, because the hospital had empty beds it wanted to fill, because the hospital and/or the doctor wanted to avoid the possibility of being sued for lack of appropriate treatment, or for any number of other reasons.

One of the best reasons for avoiding the hospital when you can is that hospitals are full of sick people. No matter how hard the staff tries to prevent the spread of disease or infections among patients, the potential is always there. A hospital is often a good place to get sick.

*(Continued on next page)*

**Confronting Consumer Issues (Continued)**

Furthermore, hospitals, particularly large ones, are often run much like factories. They are administered by bureaucracies that are separated from those who provide and receive care. In such situations people may be treated more like numbers than as individual patients. Regrettably, mistakes do take place. Patients are sometimes given inappropriate treatments, and harm is done.

## When Hospitals Are Necessary

If you do have to enter a hospital, probably the most important thing to remember is to ask questions. Yes, the hospital staff is probably overworked and has a thousand other things to do. But that is exactly why you should ask questions. It may be an old saying, but the squeaky wheel does usually get the grease. If you are admitted to a hospital, you should politely ask about anything you don't understand, or that you feel is not being done properly. You might be able to prevent a mistake in your treatment this way.

Whenever you stay in a hospital, you may be poked and prodded at regular intervals. This may seem to happen most often late at night. If you find it happening for no apparent reason, you should ask your doctor what tests are being done and what their purposes are. It is possible for unnecessary tests to inflate your hospital bill in addition to disturbing your sleep. Even if your insurance company pays the bill, you and other consumers pay indirectly through higher premiums.

To find information about a specific hospital or nursing home, you can write to the following:

American Hospital Association
840 North Lake Shore Drive
Chicago, IL 60611
**http://www.aha.org**

## When Hospitals May Be Avoided

Many medical procedures that were once done in a hospital are now administered on an outpatient basis. This saves money and time. If your doctor has scheduled you for relatively minor surgery, you should inquire whether it could be done on an outpatient basis. Remember, however, that many insurance policies will not cover or will limit their payments for outpatient services.

Another case where it may be better to refuse hospital treatment is when a patient is clearly dying. **Hospice** organizations throughout the country give assistance to the families of people who are terminally ill. Many people who realize they will die soon prefer to spend their final days in the warmth of their homes and with the people they love, rather than in the often cold and sterile environment of a hospital.

If you ever find yourself dealing with such a situation, consider contacting your local hospice organization for help and guidance.

## CHOOSING A NURSING HOME

Although you may not need the services of a nursing home now, medical records indicate that every American has a better than 50 percent chance of spending at least some time in an intermediate-care facility, or nursing home, before he or she dies. Even if you never enter a nursing home yourself, you almost certainly have a close relative who will.

Nursing homes are particularly worrisome for most families because of their reputation of being little more than holding centers for the elderly and infirm, and because most medical-insurance policies will not pay for the costs of the services they provide. In 2003 the average cost of nursing-home care was between $130 and $190 a day, depending on the state of residency. This amounts to a yearly cost of $43,800 to $65,700 before any special treatments or necessary medications are paid for. It is not uncommon for the cost of keeping a person in a nursing home to exceed $100,000 a year. This kind of expense can quickly eat up savings and leave other family members in poverty.

### Is a Nursing Home Necessary?

There are several alternatives to placing a person in a nursing home. One of the most promising is an arrangement that has been called a "nursing home without walls." There are programs that, for example, send their staff to the homes of the clients and help them get up in the morning and prepare for a trip to the program's day-care center, where they receive both treatment and stimulation. At the end of the day, clients are returned home with a meal and are helped into bed if necessary. Because there are no sleeping facilities to maintain, the cost of this program is somewhat lower than a typical nursing home, and it allows clients to stay in their own homes.

Home health care is another alternative for people who need some help but not twenty-four-hour-a-day assistance. Many businesses and some not-for-profit organizations will send people to help the elderly in their homes. The cost of this service averaged $26 an hour in 1999. Therefore, if assistance is necessary for only an hour or two per day, it can save money. If more time is required, however, this alternative will be more expensive than a nursing home.

Many communities have adult day-care services run either by businesses or by not-for-profit organizations. When these programs are appropriate for the needs of clients, they often provide quality care at the lowest possible cost. These programs, however, do not relieve family members of the responsibility of providing care for patients in the morning, at night, or on weekends and holidays. They do offer people an escape from the need to provide full-time care, but they do not take these responsibilities for more than a few hours at a time.

There is a growing number of what might be called halfway houses in the nursing-home industry. These are rental apartments that offer a range of services at different costs for elderly people who are able to care for themselves at least partially. Residents are able to opt for various levels of care as their needs change. They may choose to have their meals prepared for them, cleaning and laundry done, and the assistance of a staff member to help them up in the morning or to bed at night. When they need full-time care, they can be transferred from their apartments to a nursing facility that is in the same building complex. For people who can afford it, this is often the best choice.

## What to Look for in a Nursing Home

If you believe a member of your family may require nursing-home care, the time to investigate services in your community is before the need is immediate. The best nursing homes are most often full. You should try to avoid being forced to accept a substandard home just because it is the only one available. Put your relative's name on the waiting lists of your first and second choices and wait for a space.

Visit nursing-home facilities you are considering and talk to many different people, both staff members and residents who seem to be mentally alert. You do not need to cross-examine residents. Simply ask how they feel about the facility or how they spend their time there, and you will often be able to tell what kind of care they are receiving. Observe whether the facility is clean, staff members are friendly and appropriately dressed, the residents are clean and well groomed, the food is well prepared, activities for residents are provided, therapy equipment is available, and so on.

### Question for Thought & Discussion #11:

Has a member of your family ever been admitted to a nursing home? How was the nursing home chosen? Is your family satisfied with the services provided by this facility? Why or why not?

Always ask about arrangements that need to be made for medical care. You should find out if the nursing home will expect to have drugs delivered with instructions from the patient's doctor. Other questions you should ask include the following:

1. How many registered nurses are available at different times of day?

2. Does the nursing home have its own doctor, or is the patient's doctor expected to provide necessary care?

3. What procedures will be followed if a patient becomes ill and requires hospitalization?

4. What is the ratio of staff to patients?

5. Can physical therapy be carried out at the nursing home, or must patients be transported to other locations?

Try to get a feel for the organization's ability to help patients recover their health rather than simply house them.

## Paying for a Nursing Home

If your family has the financial resources to choose any nursing facility regardless of its cost, you are indeed fortunate. For most people who need nursing care, cost is a primary consideration. The rules of when state or federal programs will help pay for nursing-home care, and the specific services that are covered, change so often that it is almost pointless to try to explain them here. As a general rule, Medicare will pay for nursing-home care only when a patient has been discharged from a hospital and is not yet able to return home. In 2003, Medicare did not pay for long-term nursing home care. Many state-administered Medicaid programs will pay for this type of care when an individual's wealth has been exhausted. Some states try to require the patient's children to pay for all or part of the expense. When states pay for nursing-home care, the patient and relatives often have little choice in what facility is used.

Most private medical-insurance plans will not pay for nursing-home care except for brief periods of time after a patient has been discharged from a hospital. It is possible to buy long-term care insurance, but it is very expensive. Premiums for older people can be as high as $6,000 a year or more. Even these policies have limitations. They most often will not pay for nursing-home stays made necessary by pre-existing conditions (illnesses the patient already had when the policy was taken out). Many financial advisers suggest depositing money that would have been used to pay for such insurance in a special account. Putting $6,000 aside each year for ten years will result in a balance of over $70,000, which would cover the cost of most stays in a nursing home. In fact, more than half the people who are in nursing homes pay the costs from their own wealth or are supported by their relatives.

Selecting a nursing home that delivers quality care for a reasonable cost can be difficult. The Medicare Administration maintains Web pages devoted to making such a choice at its Web site located at **http://www.medicare. gov/Nursing/Overview.asp**. This site also contains links to many other Web sites that can give you specific information

*(Continued on next page)*

## Confronting Consumer Issues (Continued)

about nursing homes that exist in your community. A publication, titled the *National Nursing Home Survey: 1999 Summary,* is available from the Government Printing Office (GPO) at http://bookstore.gpo.gov. Although the most recent edition in 2003 was four years out of date, a revised edition was in the process of being prepared. Even if this publication is not brand new, the quality of individual nursing homes does not often change quickly. Therefore, it should still be useful. You can also check local Medicare and Medicaid offices, welfare agencies, and public libraries that may have evaluations of nursing homes in your area.

## ALWAYS BE WILLING TO ASK FOR HELP

It is impossible to keep up with all the changes in medical care, and most of us lack the expertise to make medical choices without help. We should never be reluctant to ask for such assistance. In virtually every community, there are organizations that provide it. A good place to begin is the closest office of your state health department. People there are trained either to help you with your problem or to direct you to those who can. The following are some of the best sources of information and help about specific diseases:

Centers for Disease Control
Health Education Information
1600 Clifton Road, MS-K13
Atlanta, GA 30333
http://www.cdc.gov

Public Inquiry
(404) 639-3536
AIDS Clearinghouse
(800) 458-5231

Whenever you need to make a choice about medical care, however, you should remember that the final decision is yours. You can and should ask for information and advice, but it is your responsibility to evaluate what others say. Your life is your own, and you should make the important decisions concerning it.

## Key Terms:

**General Practitioner (GP)**   A doctor who has a family practice rather than a specialized practice.

**Preventive Medicine**   Medical procedures carried out with the intention of preventing people from becoming ill.

**Anesthesiologist**   A doctor who specializes in administering anesthesia to patients before surgery.

**Drug Interactions**   Reactions among drugs that can sometimes be life threatening.

**Generic Drugs**   Non–brand-name medicines that are often sold at much lower prices than the same brand-name products.

**Hospice**   An organization that helps the families of terminally ill patients.

# Chapter Summary

1. Insurance is based on the concept of shared risk. Insurers are able to provide protection from various types of loss by studying past data to determine the probability of a loss occurring to an insurable interest. This information is used to determine how much an insured party should pay for protection to cover the losses that are claimed and allow the insurance company to earn a profit.

2. In 2003 the cost of medical care in this country was an estimated $1.4 trillion and accounted for nearly 14 percent of our national expenditures. Not only are medical costs rising dramatically, but the relative price of health care has also gone up consistently in recent years.

3. The institution of the Medicare program during the 1960s caused medical costs to rise because of the subsequent increased demand for medical services. The costs of Medicare continue to rise and are expected to be more than those for Social Security shortly before 2015.

4. The private insurance framework has increased the cost of medical care because medical expenses covered by insurance, particularly hospital expenses, are not paid directly by the individuals receiving the services. This creates a lack of incentive on the part of doctors and patients to control costs.

5. Another reason medical costs have risen is the limited supply of doctors due to restricted entry into the profession. Although the supply of doctors is increasing, and we may have a surplus of doctors soon, doctors' charges for medical services still make up a major portion of our overall costs of health care.

6. Government regulation of the drug industry and drug-certification requirements have increased the prices that consumers must pay for prescription drugs. The marketing of generic drugs, which require far less testing than new brand-name drugs, has helped to control growth in prices for prescription drugs.

7. High-tech medical equipment and procedures have brought many (sometimes life-saving) benefits to consumers, but they have also contributed enormously to our health-care costs.

8. The increased number of medical-malpractice lawsuits and higher jury awards have meant, in turn, higher malpractice-insurance premiums for doctors, especially those in high-risk areas such as obstetrics and neurology. To guard against malpractice suits, doctors engage in defensive medicine—extensive testing, consulting, and record-keeping procedures—that may add as much as $150 billion a year to our national medical bill.

9. Rising insurance premiums are one of the consequences of rising medical costs. Individuals and employers have faced dramatic premium increases in recent years, and some small business firms have been forced to drop insurance coverage for their employees as a result. The growing number of uninsured individuals has become a source of national concern.

10. To curb costs due to waste and inefficiency, employers are turning increasingly from traditional group insurance plans to lower-cost health maintenance organizations and preferred provider organizations. Today, 85 percent of insured employees are covered under such plans.

11. Some insurers are instituting cost-management programs that use computer databases to determine whether certain medical treatments are medically justifiable. Insurance carriers using cost-management programs require that individuals seeking medical treatment obtain precertification from the insurance company before treatment.

12. To control Medicare costs, the government instituted a prospective-payment plan in 1983. Under this plan, the government pays a preestablished fee, determined by the nature of the patient's illness, to hospitals for treatment given to Medicare patients.

13. A growing number of Americans, including, significantly, some physicians, support the idea of a national health plan. Issues currently being debated include whether the government has a responsibility to ensure adequate health care for all citizens and whether further government involvement in health care would improve or only worsen the situation.

14. Consumers should choose a doctor before one is desperately needed. Friends, relatives, and/or other doctors can make recommendations. A doctor's age, gender, associations, and attitude are all factors consumers should consider.

15. It is important for consumers to investigate any significant medical procedure, whether it involves prescription of drugs or major surgery. Patients should always ask for explanations of what is being suggested, the likely results, and if there are any alternatives that should be considered. When in doubt, consumers should seek a second opinion.

16. Consumers should avoid being admitted to a hospital unless they have a clear need for the services only a hospital can provide. In the hospital, patients and their relatives should be sure to ask for explanations of any care they do not understand. If a patient's illness is clearly terminal, it may be better for that patient to be cared for at home or at a hospice organization.

**17.** About half of all Americans will stay in a nursing home before they die. Before having a person admitted to a nursing home, people may want to investigate alternative methods of care. It is better to begin the investigation before such care is absolutely needed. Nursing homes are very expensive, and most medical insurance does not pay for them. More than half the people in nursing homes today are being supported by private funds.

## Key Terms

anesthesiologist **412**

capitation **400**

drug interactions **412**

general practitioner (GP) **412**

generic drugs **412**

health maintenance organization (HMO) **395**

hospice **412**

inpatient services **388**

insurable interest **386**

insurable risk **386**

outpatient services **388**

preferred provider organization (PPO) **396**

preventive medicine **412**

## Questions for Thought & Discussion

1. What is the insurable interest for medical-care insurance?

2. How has the push by medical-care insurance providers to shorten the length of hospital stays affected hospitals' finances?

3. Why hasn't the growing supply of doctors brought about a significant reduction in the fees they charge?

4. Would you support capping awards for pain and suffering at $250,000 to reduce malpractice-insurance premiums? Why or why not?

5. Why do many physicians choose not to become members of HMO or PPO systems?

6. How do you feel about your doctor's decisions regarding your health care being evaluated by a computer program?

7. Have you or someone you know ever chosen not to seek medical care to avoid the cost? What were the final results of the decision? Was it a rational choice?

8. Would you support a program to provide universal health insurance to all U.S. citizens? How could such a program be paid for?

9. How did you choose the physician you see when you become ill? Are you satisfied with your choice?

10. What experiences have you had with hospital emergency rooms? Were you satisfied with the treatment you received? Why or why not?

11. Has a member of your family ever been admitted to a nursing home? How was the nursing home chosen? Is your family satisfied with the services provided by this facility? Why or why not?

## Things to Do

1. Survey your friends and relatives about their attitudes toward our health-care system. Do they feel it provides adequate care, is efficiently run, or is in need of an overhaul? What changes would they make in the system if they could? Use the results of your survey to write an essay about how people regard the U.S. health-care system.

2. Discuss growing medical-care costs with someone who is employed in the health-care system. How does this person believe medical costs could best be contained? In what ways does this "insider's" point of view differ from your own?

3. Interview a person who has had a serious illness or been severely injured. Ask this person to describe his or her experiences with the U.S. health-care system. Is this person satisfied with the treatment that was provided? Were the costs of care appropriate? What changes would this person suggest for the system?

4. Visit a local nursing home. Evaluate the apparent quality of care that is provided. Would you want a relative of yours to stay in this nursing home? Explain why or why not.

5. Investigate what the federal government has done to extend prescription drug coverage to Medicare recipients since March of 2003. Has any action been taken? How successful has the government been in helping the elderly pay for their prescription drugs?

## Internet Resources

### Finding Consumer Information on the Internet

The following Web sites have been selected for their relevance to topics discussed in this chapter. Search these sites to locate information that can add to your knowledge of online services that provide medical-care advice. Remember, Web addresses change frequently. If any of these addresses no longer function, find similar sites to investigate using any of the search engines available to you.

1. The American Academy of Family Physicians maintains a Web site that provides extensive information to consumers about drugs, basic health care, and choosing a family physician. Its Web site is located at **http://www. familydoctor.org**.

2. The Achoo Medical Network provides links to many different health-care Web sites. Visit its Web site at **http:// www.achoo.com**.

3. The American Academy of Pediatrics provides a wide variety of information directed to the medical care of children and young adults. Investigate its Web site at **http://www.aap.org**.

### Shopping on the Internet

The following Web sites have been selected because they offer consumers services similar to those described in this chapter. These are commercial sites that are designed to market products. They do not represent a comprehensive or balanced description of all firms that offer prescription drugs online. How do the services at these Web sites compare with those that are available from local pharmacies? Remember, Web addresses change frequently. If any of these addresses no longer function, find similar sites to investigate using any of the search engines available to you.

1. Get Drugs Online is a service that links consumers with a variety of firms that sell drugs online. Its Web site is located at **http://www.getdrugsonline.com**.

2. Pharmacy One Online advertises that it can offer people "the convenience of not having to visit a doctor's office or pharmacy to get the prescription drugs" they need. Evaluate its offer at **http://www.pharmacy-one.com**.

3. Med Pharmacy.com guarantees that it can supply low-cost drugs within twenty-four hours to any location in the United States. Visit its Web site at **http://www. medpharmacy.com**.

### InfoTrac Exercises

Purchasers of new copies of this text are provided with access to the InfoTrac Web site. This Web site links students to thousands of recent articles published in hundreds of periodicals. Use the key words **Medicare costs** or other terms from this chapter to conduct a key-word search. Choose one article that is of particular interest to you and write a brief essay describing what you have learned from the article. Be sure to cite the author and title of the article and the name and date of the publication in which it appeared.

## Selected Readings

"Advancements Revolutionize Disease Diagnosis and Treatment." *Medical Devices and Surgical Technology Week*, February 25, 2001, pp. 2–3.

Albiniak, Paul. "Drug Money." *Broadcasting and Cable*, March 11, 2002, p. 25.

Feldman, Judy. "Is More Medical Choice Better?" *Money*, December 1, 2002, p. 193.

"Get Results Using the BBB." *Black Enterprise*, May 2000, p. 167.

Gleckman, Howard. "Bush's Medicare Reform." *Business Week*, February 3, 2003, p. 49.

"The Health Care Struggle Today." *Dollars and Sense*, May 2001, p. 25.

Henkel, John. "Help for Choosing a Nursing Home." *FDA Consumer*, November–December 2002, p. 37.

Kinney, Eleanor D. *Protecting American Health Care Consumers*. Durham, NC: Duke University Press, 2002.

Landers, Susan. "People Happy with HMOs until They're Sick." *American Medical News*, June 26, 2000, p. 6.

Miceli, Thomas J., and Dennis Huffly. "Do HMOs Encourage Prevention?" *Contemporary Economic Policy*, October 2002, pp. 429–440.

"Most Emergency Room Visits Are Not for Emergencies." *Health and Medicine Weekly*, September 9, 2002, p. 4.

Perlstein, Steve. "Public Health Needs to Focus on Prevention." *Family Practice News*, November 15, 2002, p. 43.

Rice, Berkeley. "Malpractice Premiums Soaring Again." *Medical Economics*, December 9, 2002, pp. 51–52.

Rose, Joan R. "For Seniors Fewer Choices—Higher Costs." *Medical Economics*, January 10, 2003, p. 15.

Photodisc/Getty Images

# CHAPTER 17

# Insuring Your Home and Your Automobile

**After reading this chapter, you should be able to answer the following questions:**

- What types of insurance does a homeowners' policy provide?
- Why do renters need insurance protection for their homes?
- How can you determine the types and amounts of insurance that are right for you?
- What is no-fault insurance?
- What determines the cost of automobile insurance?
- What steps can you take to control the cost of insurance?
- What are your rights if you are refused insurance coverage?
- How can consumers choose to purchase legal advice?

Consumers purchase homeowners' and automobile insurance for two reasons. One reason is, of course, to protect their investments in case of loss. For example, if your house burns down, you're in serious trouble unless you have insurance coverage. You lose not only your personal possessions but also your investment (if your house is paid for). If you have a mortgage on your home, you will end up paying for a house that no longer exists or be forced into bankruptcy.

The second reason is that consumers need to protect themselves against possible liability claims if they are the cause of another's harm. If, for example, you are at fault in an auto accident that results in injury or death to another, you may face a lawsuit for damages—possibly in excess of $1 million. Similarly, if someone is injured on your property, you could be sued for damages, even though your only "fault" might be owning the property on which the injury occurred.

People protect themselves against both of these types of potential hazards—loss of property and liability claims—by purchasing homeowners' and automobile insurance.

# Insuring Your Home

Part of the process of buying a home is taking out a homeowners' insurance policy. Depending on where you live, the value of your home, and the type of coverage you buy, you can expect to pay anywhere from $300 to $1,400 (or more) a year for homeowners' insurance coverage. You will probably need to arrange for insurance coverage for any house that you pay for through a mortgage before you close on the property. This is because most mortgage holders require the *mortgagor*—that is, the homeowner—to prepay taxes and insurance as part of the monthly house payments. Part of each mortgage payment is placed in a special reserve account (called an *escrow account*) that is usually set up within the lending institution. Money for property taxes and insurance accumulates in the account, and the lending institution then makes these payments directly from the account. In this way, the *mortgagee*—the lending institution—does not have to worry about foreclosure on the house because of unpaid taxes or problems if the house is destroyed in a storm or a fire. Escrow accounts, however, pay mortgagors very low rates of interest. If the lending institution will allow borrowers to be responsible for their own taxes and insurance, they will do better financially to take care of these costs themselves by keeping their money in accounts that pay a better return.

Could you afford to repair or replace your home or your car if they were damaged or destroyed in a storm?

## Types of Homeowners' Coverage

You can purchase two basic types of insurance coverage—property coverage and liability coverage. *Property coverage* includes the garage, house, and other private buildings on your lot; personal possessions and property whether you

**Question for Thought & Discussion #1:**
How much liability coverage—both for your home and for your car—do you think is a "reasonable" amount? Explain why you chose these amounts.

**Basic Form Policy**   A homeowners' insurance policy that covers eleven risks.

**Broad Form Policy**   A homeowners' policy that covers eighteen risks.

**Comprehensive Form Policy**   A homeowners' policy that covers all risks except (usually) flood, war, and nuclear attack.

are at home or are traveling or at work; and additional living expenses paid to you if you could not live in your home because of a fire or flood.

There are three types of *liability coverage:* (1) personal liability in case someone is injured on your property or you damage someone else's property and are at fault; (2) medical payments for injury to others who are on your property; and (3) coverage for the property of others that you or a member of your family damages.

There are two types of insurance policies you might buy for your home: one provides just property insurance; the other provides both property and liability insurance.

A standard *fire-insurance policy* protects the homeowner against fire and lightning, plus damage from smoke or water caused by the fire and the fire department. If you pay a little bit more, the coverage can be extended to protect you against damage caused by hail, windstorms, explosions, ice, snow, and so on. It is important to include hazards that might be likely in your area, even if the coverage may cost a little more.

A *homeowners' policy* provides protection against a number of risks under a single policy, allowing you to save over what you would pay if you bought each policy separately. It covers both the house and its contents.

A number of forms of homeowners' policies are available. They differ in the number of risks that they cover. Figure 17.1 shows what each is like. As you can see, the **basic form policy** covers eleven risks, the **broad form policy** covers eighteen risks, and the **comprehensive form policy** covers those risks and all other perils except those listed at the bottom of the chart.

**ADDING A PERSONAL-ARTICLES FLOATER POLICY**   Under the basic homeowners' policy, the contents of your home are insured up to 50 percent of the value of your house, but only for losses resulting from stated perils. You may wish to pay a slightly higher premium to insure specific personal articles such as cameras, musical instruments, works of art, jewelry, or your personal home computer. Under the basic homeowners' policy, for example, there is usually a jewelry cap of a specified sum, such as $1,000. If you apply for a personal-articles floater addition to your homeowners' policy, you will be asked to submit a list of the specific items you want to have covered under it and an affidavit from an appraiser giving their current market value. When you insure under a floater, you have all-risk insurance and can omit the covered property from your fire and theft policies.

**PERSONAL-EFFECTS FLOATER POLICY**   You also can take out a personal-effects floater policy to cover personal items when you are traveling. In most cases, a personal-effects floater isn't necessary because your regular homeowners' insurance covers you. Because a personal-effects floater covers only the articles taken off your property, you still need insurance for them when they are on your property. The policy does not cover theft from an unattended automobile unless there is evidence of a forced entry.

**EXCESS-LIABILITY POLICY**   The growing number of million-dollar personal-liability lawsuits has caused many Americans to pay between $200 and $500 a year for personal-liability policies that protect them for $1 million above what they carry in their homeowners' (or automobile) policies. Because this kind of policy covers such a broad range of potential liability claims, whether from injuries on your property or from a car accident, it is

**Question for Thought & Discussion #2:**
Why have umbrella policies become much more common in recent years?

### Figure 17.1 Standard Forms of Homeowners' Insurance

| Basic | Broad | Comprehensive | Perils against Which Properties Are Insured |
|---|---|---|---|
| ▓ | ▓ | ▓ | 1. Fire or lightning |
| ▓ | ▓ | ▓ | 2. Loss of property removed from premises, endangered by fire or other perils |
| ▓ | ▓ | ▓ | 3. Windstorm or hail |
| ▓ | ▓ | ▓ | 4. Explosion |
| ▓ | ▓ | ▓ | 5. Riot or civil commotion |
| ▓ | ▓ | ▓ | 6. Aircraft |
| ▓ | ▓ | ▓ | 7. Vehicles |
| ▓ | ▓ | ▓ | 8. Smoke |
| ▓ | ▓ | ▓ | 9. Vandalism and malicious mischief |
| ▓ | ▓ | ▓ | 10. Theft |
| ▓ | ▓ | ▓ | 11. Breakage of glass constituting a part of the building |
|  | ▓ | ▓ | 12. Falling objects |
|  | ▓ | ▓ | 13. Weight of ice, snow, sleet |
|  | ▓ | ▓ | 14. Collapse of building(s) or any part thereof |
|  | ▓ | ▓ | 15. Sudden and accidental tearing asunder, cracking, burning, or bulging of a steam or hot-water heating system or of appliances for heating water |
|  | ▓ | ▓ | 16. Accidental discharge, leakage, or overflow of water or steam from within a plumbing, heating, or air-conditioning system or domestic appliance |
|  | ▓ | ▓ | 17. Freezing of plumbing, heating, and air-conditioning systems and domestic appliances |
|  | ▓ | ▓ | 18. Sudden and accidental injury from artificially generated currents to electrical appliances, devices, fixtures, and wiring (TV and radio tubes not included) |
|  |  | ▓ | All perils except flood, earthquake, war, nuclear accident, acts of terrorism, and others specified in your policy. Check your policy for a complete listing of perils excluded. |

SOURCE: Insurance Information Institute.

called an **umbrella policy.** Typically, umbrella policies cover slander and libel defenses, too.

**FLOOD INSURANCE**  You will notice that even a comprehensive home-owners' insurance policy does not cover floods. If you live in an area that may experience flooding, it is advisable to purchase federally subsidized (that is, by all federal taxpayers) flood insurance. You must live in an area designated eligible by the Federal Insurance Administrator of the U.S. Department of

**Umbrella Policy**  A type of supplemental insurance policy that can extend normal liability limits to $1 million or more for a relatively small premium.

> **"Although Americans spend billions a year to insure their homes, many still have inadequate insurance coverage."**

Housing and Urban Development. Your insurance agent will be able to tell you if you are eligible.

## How Much Insurance Should You Have?

You should have 80 percent of the total value of your property insured—that is, 80 percent of its replacement value. You may wish, or be required by the lender, to use the services of a professional appraiser to get an accurate idea of the replacement value of your house. You can find appraisers in the Yellow Pages; or your insurance agent, a local home builder, or your banker may be able to help you.

You need not insure your house for more than 80 percent of the replacement value for two reasons. First, the land has a value that would not be destroyed in a fire or flood, and even if the house were totally burned down, the foundation, sidewalks, driveway, and so on would remain. Second, if you have coverage for 80 percent of the replacement value, you can collect the full replacement cost, not the depreciated value, of any damaged property (up to the limits of the policy). For example, say, your ten-year-old roof is damaged in a fire, and replacing it costs $3,500. If you have at least 80 percent coverage on your house, your insurance company must pay you the full amount of the roof damage, whereas if your house is covered for less than 80 percent of replacement, you will get less. Specifically, you will be paid only that portion of the loss equal to the amount of insurance in force divided by 80 percent of the replacement cost of the entire house times the loss on the roof. If your house would cost $120,000 to replace and you have only $60,000 of insurance, then, on your roof damage of $3,500, you will be paid:

$$\frac{\$60,000}{\$96,000} \times \$3,500 = \$2,188$$

A little explanation of this formula is in order. The fraction consists of the actual amount of insurance that you have on your structure ($60,000) divided by the amount of insurance you would need to cover 80 percent of the replacement value, or $.8 \times \$120,000 = \$96,000$.

Homeowners' insurance policies generally insure the contents of a home at their current market value. This means that a dining room table and chairs that cost $3,000 four years ago might be insured for only $1,500 if that is their value as used furniture. Yet it might cost $4,000—that is, even more than the original purchase price—to replace them today because of inflation. It is possible to buy replacement-value insurance for the contents of your home for only a few dollars more each year than an ordinary policy costs.

**PERIODIC REVIEW OF COVERAGE**   Although Americans spend billions of dollars a year to insure their homes, many still have inadequate insurance coverage. If you already have a homeowners' policy, you should periodically review its provisions to make sure that the policy reflects changes in values brought about by inflation, improvements to your home, or property appreciation. If you live in a house that you bought many years ago, for example, the cost of replacement may be much more than you think. Take that into account when you renew a homeowners' policy. You may want to arrange with your insurance company to have the value of your insurance increased by a certain percentage every year or two to keep pace with current price levels.

**Question for Thought & Discussion #3:**
Would you choose to purchase periodic review coverage for your home? Why or why not?

Photodisc/Getty Images

Do you own any valuable items that would not be protected by a standard homeowner's policy? Should you purchase extra coverage to protect them?

**HOW TO SAVE ON HOMEOWNERS' INSURANCE**    There are a number of ways you may be able to reduce the insurance premium you pay without sacrificing coverage. By increasing the deductible, for instance, you can get a much lower premium. In most states, the premium is around 30 percent less for a $500 deductible than for a $100 deductible and about 10 percent less for a $250 deductible than for a $100 deductible. If you install a burglar alarm, smoke detectors, deadbolt locks, and fire extinguishers, you can also save between 2 and 15 percent on your insurance bill. You might also consider moving to a low-crime area, where theft insurance will cost less. Sometimes paying your premiums on a three-year basis rather than a one-year basis can save you money, as can purchasing a package policy rather than separate policies for different perils.

## Homeowners' Warranties

Over the past thirty years, many contractors and some real estate firms offered home warranties as an inducement for consumers to purchase new or previously owned houses. Contractors who were members of the National Association of Home Builders, for example, commonly paid a premium to an insurance fund that provided homeowners' warranties called "HOWs" to individuals who purchased homes the members had constructed. These warranties typically were promoted as providing homeowners with protection from losses caused by major structural defects in their homes for as long as ten years after construction. Although the idea was to encourage consumers to purchase homes protected by HOWs, there is some doubt about the value of such programs.

**HOW MUCH ARE HOWS WORTH?**    Some home warranty contracts are filled with exclusions that may render them valueless in many situations. Policies issued by the Residential Warranty Program, for example, excluded defects that included problems in "non–load bearing walls, tile, plaster, drywall, subflooring, bricks, stucco, siding, and roof shingles." Furthermore, several of the firms that offered this type of protection filed for bankruptcy in the 1990s, leaving insured homeowners with no protection. Still, many contractors continue to offer HOWs with the homes they sell. In 2001, the California Home Warranty Association reported that more than 400,000 new houses were sold with home warranties in California in 2000. If you are offered a HOW with any house you purchase, it would be worth your time and effort to investigate what is really being proposed.

**HOME SERVICE CONTRACTS**    As an alternative, consumers can purchase service contracts from insurance providers that cover houses (new or old) against structural defects and unexpected repairs of electrical, plumbing, or heating systems. Some of these policies also cover the breakdown of major appliances such as refrigerators or central air-conditioning units. The largest provider of this type of coverage in 2000 was American Home Shield, which marketed basic service contracts for most houses at an annual premium of $300 to $500. When a loss occurred, consumers were required to pay a deductible that typically ranged from $35 to several hundred dollars, depending on the type of loss and the specific contract. Although service contracts may give homeowners a degree of security, many experts believe that their cost exceeds their value.

## Cyber consumer
### Searching for Home Protection Online

American Home Shield is the largest provider of home service contracts in the United States. Its policies offer homeowners protection from defects and breakage of most home appliances and major systems such as heating, cooling, plumbing, and electrical systems. Although this type of insurance is not cheap, it can prevent homeowners from being forced to pay an unexpected and budget-breaking bill when their hot-water heaters spring a leak or their freezers quit and hundreds of pounds of meat spoil. Investigate what AHS has to offer at its Web site located at http://www.homeshield.com.

**Residence Contents Broad Form
Policy**   A renters' insurance policy that
covers possessions against eighteen risks.
It includes additional living expenses and
liability coverage in case someone is
injured in the apartment or house you
are renting.

**Question for Thought
& Discussion #4:**
How many of your friends who
rent an apartment buy renters'
insurance? If they don't, what
risks are they taking?

> **"According to the
> Insurance Information
> Institute, fewer than
> one-third of all renters in
> the United States carry
> renters' insurance."**

# What About Renters' Insurance?

According to the Insurance Information Institute, fewer than one-third of all renters in the United States carry renters' insurance. This is partly because many renters assume that their landlords' homeowners' policies will cover any damages to their possessions caused by fire or theft. Unfortunately for many renters, this isn't the case. Although the landlord or owner of your building is liable for damage to the building and for injuries occurring in common areas, such as the lobby or hallway, you are responsible for protecting the inside of your dwelling, and you are liable for accidents that occur there. Renters' insurance, called a **residence contents broad form policy,** is a policy that covers personal possessions. It includes additional living expenses and liability coverage.

Before you look for a renters' policy, make a detailed inventory of your possessions. Decide on the coverage you want and then do some comparison shopping.

Note that household possessions are sometimes insured for half their value. Everything, including linens and plants, should be listed in your inventory. The inventory list must be kept in a safe place away from the dwelling, and a copy should be given to your insurance agent.

Typically, a renters' policy can be obtained from most property-insurance companies for between $200 and $400 a year. If you have especially valuable equipment or possessions for which you want extra protection, a personal-articles floater can be added—just as in a homeowners' policy—for an additional premium.

# Automobile Insurance

When you buy a car, one of the first things you must think about is insuring yourself against theft, fire, liability, medical expenses, and damage related to the automobile. It is foolish to risk driving without automobile insurance. Of the many kinds of coverage you can purchase, the most important is liability insurance.

## Liability Insurance

Liability insurance pays when the policyholder is at fault for injury to another person. It covers bodily injury liability and property damage. Liability limits are usually described by a series of three numbers, such as 100/300/50, which means that the policy will pay a maximum of $100,000 for bodily injury to one person, a maximum of $300,000 for bodily injury to more than one person, and a maximum of $50,000 for property damage in one occurrence. Liability insurance also pays for the cost of defending a policyholder should a covered loss result in litigation. Table 17.1 shows the minimum bodily injury limits and property damage liability coverage required by law in each state.

Because the cost of additional liability coverage is relatively small, it is wise to consider taking out a much larger limit than you would ordinarily expect to need because awards in personal-injury suits against automobile drivers who are proved negligent are sometimes astronomical. Some dependents of automobile accident victims have been successful in suing for $1 million or more.

People who aren't satisfied with the maximum liability limits offered by regular automobile-insurance coverage can purchase a separate amount of

| TABLE 17.1 | Automobile Financial Responsibility/ Compulsory Limits, 2003 |
| --- | --- |

There may have been changes in some states since these data were published. For the latest information, check your own state department of motor vehicles or your auto-insurance agent.

| State | Liability Limits* | State | Liability Limits* |
| --- | --- | --- | --- |
| Alabama | 20/40/10 | Montana | 25/50/10 |
| Alaska | 50/100/25 | Nebraska | 25/50/25 |
| Arizona | 15/30/10 | Nevada | 15/30/10 |
| Arkansas | 25/50/15 | New Hampshire | 25/50/25 |
| California | 15/30/5 | New Jersey | 15/30/5 |
| Colorado | 25/50/15 | New Mexico | 25/50/10 |
| Connecticut | 20/40/10 | New York | 25/50/10 |
| Delaware | 15/30/5 | North Carolina | 30/60/25 |
| District of Columbia | 25/50/10 | North Dakota | 25/50/25 |
| Florida | 25/50/25 | Ohio | 25/50/25 |
| Georgia | 20/40/10 | Oklahoma | 10/20/10 |
| Hawaii | 25/10/25 | Oregon | 25/50/10 |
| Idaho | 25/50/15 | Pennsylvania | 15/30/5 |
| Illinois | 20/40/15 | Rhode Island | 25/50/25 |
| Indiana | 25/50/10 | South Carolina | 15/30/10 |
| Iowa | 20/40/15 | South Dakota | 25/50/25 |
| Kansas | 25/50/10 | Tennessee | 25/50/10 |
| Kentucky | 25/50/10 | Texas | 20/40/15 |
| Louisiana | 10/20/10 | Utah | 25/50/15 |
| Maine | 50/100/25 | Vermont | 25/50/10 |
| Maryland | 20/40/10 | Virginia | 25/50/20 |
| Massachusetts | 20/40/5 | Washington | 25/50/10 |
| Michigan | 20/40/10 | West Virginia | 20/40/10 |
| Minnesota | 30/60/10 | Wisconsin | 25/50/10 |
| Mississippi | 10/20/5 | Wyoming | 25/50/20 |
| Missouri | 25/50/10 | | |

*The first two figures refer to bodily injury liability limits and the third figure to property damage liability. For example, 10/20/5 means coverage up to $20,000 for all persons injured in an accident, subject to a limit of $10,000 for one individual, and $5,000 coverage for property damage.

SOURCE: Insurance Information Institute Web site, http://www.iii.org, 2003.

coverage under an *umbrella policy,* which sometimes goes as high as $5 million in coverage.

## Medical Payment Insurance

Medical payments on an auto-insurance policy will cover hospital and medical bills and, sometimes, funeral expenses (for those in your car). Medical payment insurance pays regardless of who is at fault in an accident. Some policies allow you to buy medical payment insurance for your passengers. You may not want to buy medical payment insurance for your personal coverage through your auto-insurance policy, however, if you have sufficient medical coverage through an individual medical-insurance policy or through a group policy at your place of employment.

**Zero Deductible**  In the collision part of an automobile-insurance policy, the provision that the insured pays nothing for any repair for damage to the car due to an accident that is the fault of the insured. Zero deductible is, of course, more expensive than a $250 or $500 deductible policy.

## Collision Insurance

Collision insurance covers damage to your car in any type of collision not covered by another insured driver at fault. It is usually not advisable to purchase full coverage (otherwise known as **zero deductible**) on collision. The price per year is quite high because it is likely that, in any one year, small repair jobs will be required and will be costly. Most people take out $250 or $500 deductible coverage, which costs about one-quarter the price of zero deductible.

## Comprehensive Insurance

Comprehensive auto insurance covers loss, damage, or anything on your car destroyed by fire, hurricane, hail, or just about all other causes, including vandalism. It is separate from collision insurance. Full comprehensive insurance is quite expensive. Again, a $250 or $500 deductible is usually preferable.

## Uninsured Motorist Insurance

Uninsured motorist coverage insures the driver and passengers against injury by any driver who has no insurance at all or by a hit-and-run driver. Many states require that all insurance policies sold to drivers include this coverage. The risk is small, so the premium is relatively small.

## Accidental Death Benefits

Sometimes called *double indemnity*, accidental death coverage provides a lump sum to named beneficiaries if you happen to die in an automobile accident. Although it generally costs very little, you may not want it if you feel you have already purchased a sufficient amount of life insurance.

## What Determines the Price of Auto Insurance?

It is not uncommon for different prices to be charged for identical insurance coverage, so it pays to shop around before purchasing a policy. But prices can be misleading because different insurance companies offer varying qualities of service. One company may be less willing to pay off claims than another. One may have an insurance adjuster at your house immediately if you have a small accident; another may never send one out, leaving you to do the adjusting yourself. Although insurance costs are not consistently related to quality of service, they are consistently differentiated among various classes of drivers.

Why? The reason is simply that the probability that an accident will occur is different for different classes. Competition among insurance companies has forced them all to find out which classes of drivers are safer than others and to offer those classes lower rates. For example, statistically, single males from the ages of sixteen to twenty-five have the highest accident rate of all drivers. According to the National Safety Council, one-third of the drivers within this group have auto accidents. Some states have a history of higher rates of insurance claim settlements, both in and out of court, which increase

Being involved in an accident can have a dramatic effect on your insurance premium.

Photodisc/Getty Images

the cost of providing auto insurance. For this reason, drivers in such states pay a much higher price for auto insurance—as can be seen in Table 17.2. Statistics also indicate that female drivers have fewer accidents overall than male drivers; therefore, women often pay lower insurance rates for cars that they use exclusively. Gender-based insurance rates have been under attack in recent years because this method of rate determination places more emphasis on gender than on safe-driving records, health practices, social practices, and so on.

Insurance rates also depend on the following variables:

- The car you drive.
- Where you drive (see Table 17.2).
- What the car is used for.
- Marital status.
- Occupation.
- Safety record.

| TABLE 17.2 | Average Auto-Insurance Expenditures by State, 2000 | | | | |
|---|---|---|---|---|---|
| Rank | State | Average Expenditure | Rank | State | Average Expenditure |
| 1 | District of Columbia | $996.39 | 27 | Illinois | 651.60 |
| 2 | New Jersey | 977.07 | 28 | Oregon | 625.37 |
| 3 | Massachusetts | 945.61 | 29 | Utah | 620.05 |
| 4 | New York | 935.64 | 30 | Kentucky | 615.69 |
| 5 | Connecticut | 871.20 | 31 | South Carolina | 612.07 |
| 6 | Delaware | 848.51 | 32 | Missouri | 611.73 |
| 7 | Nevada | 829.28 | 33 | Arkansas | 606.05 |
| 8 | Rhode Island | 825.44 | 34 | Oklahoma | 602.72 |
| 9 | Louisiana | 806.01 | 35 | Alabama | 593.65 |
| 10 | Arizona | 791.99 | 36 | Tennessee | 592.33 |
| 11 | Alaska | 770.11 | 37 | Ohio | 579.27 |
| 12 | Maryland | 757.41 | 38 | Virginia | 576.08 |
| 13 | Colorado | 754.88 | 39 | Indiana | 570.12 |
| 14 | Florida | 746.29 | 40 | Vermont | 568.39 |
| 15 | Washington | 722.08 | 41 | North Carolina | 563.66 |
| 16 | Michigan | 702.32 | 42 | Kansas | 558.27 |
| 17 | Hawaii | 700.09 | 43 | Wisconsin | 551.33 |
| 18 | Pennsylvania | 698.56 | 44 | Nebraska | 532.74 |
| 19 | Minnesota | 695.55 | 45 | Montana | 530.43 |
| 20 | West Virginia | 680.09 | 46 | Maine | 528.08 |
| 21 | Texas | 677.83 | 47 | Idaho | 505.16 |
| 22 | New Mexico | 674.27 | 48 | Wyoming | 495.60 |
| 23 | Georgia | 674.12 | 49 | Iowa | 484.08 |
| 24 | New Hampshire | 665.47 | 50 | South Dakota | 478.75 |
| 25 | California | 658.32 | 51 | North Dakota | 477.28 |
| 26 | Mississippi | 654.16 | | USA | 686.71 |

SOURCE: Insurance Information Institute Web site, http://www.iii.org, 2003.

**No-Fault Auto Insurance** A system of auto insurance whereby, no matter who is at fault, the individual is paid by his or her insurance company for a certain amount of medical costs and for the damage to the car.

> "Many early proponents had expected no-fault insurance programs to result in lower insurance premiums, but this has not been the case."

**Question for Thought & Discussion #5:**
Do you believe no-fault insurance is a good idea? Why or why not?

# No-Fault Insurance, Pros and Cons

The original proponents of **no-fault auto insurance** in the 1960s believed that all states would soon adopt the system. To date, however, only fourteen states and the District of Columbia have "pure" no-fault insurance. Ten other states have a modified form of no-fault, in which no-fault benefits have been simply superimposed on an unchanged liability-based system.

**WHAT IS NO-FAULT?**   No-fault auto insurance is basically an extension of the principle of workers' compensation insurance to auto accidents. Under workers' compensation, for example, an injured worker is compensated for the injuries sustained regardless of whether the employer or the worker was at fault. Likewise, under a no-fault auto insurance program, your insurance company does not have to decide whose fault the accident was before paying you for medical expenses arising from injuries sustained in that accident. In a traditional, liability-based fault system, a determination must first be made as to who caused an accident. The insurer of the party deemed to be "at fault" then pays the other party's bills—medical bills, auto-repair bills, lost earnings, and, in some cases, pain and suffering.

In many no-fault states, a certain dollar threshold—varying between $500 and $4,000—must be met before you are allowed to sue for pain and suffering compensation. In these states, however, a "descriptive" threshold is required. In Michigan, New York, and Florida, accident victims must meet a serious-injury requirement instead of a dollar requirement before they can sue for compensation.

**NO-FAULT VERSUS LIABILITY—WHICH IS BETTER?**   In those states that combine high-benefits programs with a restriction on lawsuits to those with serious injuries, no-fault works well. This is the case in New York, Michigan, and Florida, where the descriptive threshold limits insurance payments to those who are seriously injured and who need compensation.

A study by the Department of Transportation revealed that twice as many victims in states with no-fault coverage were being compensated than was the case in states with a traditional liability system. And payments were made more quickly. Most no-fault laws require that payment be made within thirty to sixty days after the submission of the proof of claim. By contrast, in liability states, victims sometimes have to wait months—even years—to be compensated. Furthermore, less money is involved in litigation costs under no-fault, and more of the insurance dollar goes toward benefits; overall, 10 to 15 cents more of each premium dollar goes toward compensation instead of litigation costs.

Many early proponents had expected no-fault insurance programs to result in lower insurance premiums, but this has not been the case. Premiums in no-fault states are about as high as in liability states. The reason is that more victims are being compensated for medical care, and medical-care costs have risen dramatically in the past three decades. Since the mid-1970s, no additional states (other than the District of Columbia in 1983) have converted to no-fault plans.

# How to Shop for Automobile Insurance

Shopping for automobile insurance is usually easier than shopping for a car. You may want to look first to your local credit union or to some special insurance source available to you if you are a member of certain organizations.

### CONSUMER CLOSE-UP
## Who Should Fix Your Car?

**O**kay—sooner or later it happens to almost everyone—you've been involved in an automobile accident, and you need to have your car's damaged right front fender replaced. You're insured, so after you pay the $250 deductible, you should be able to have your car repaired wherever you want. Right? Well, maybe not.

In recent years, many insurance companies have worked to reduce the cost of accident claims by trying to require customers to use their direct repair programs (DRPs). A DRP works like this: The insurance company makes deals with automobile repair businesses throughout your state. In exchange for the company's directing insured repairs (like yours) to these firms, the businesses agree to provide repairs at reduced costs. Unfortunately, to be able to charge lower prices, many of these businesses cut corners on the quality of the repairs they make. For example, they often install remanufactured parts instead of new parts, or they use "secondary market" parts, which are produced at low cost in developing nations and frequently lack the quality of original

manufacturer parts. Many consumers have complained that cars repaired through DRPs are not restored to their original conditions and therefore lose resale value.

Some insurers have told consumers that they must use a DRP repair facility if they expect to be compensated for their losses. Other consumers have been told that they must pay the difference between what a DRP would charge and the cost of the repair at a facility they choose. Many consumers have complained about these practices to the insurance industry and to their state governments. Although the industry argues that the use of DRPs lowers costs and therefore premiums, some state legislatures have not accepted this position. In 2001, North Carolina passed a law that requires insurance companies to inform their policyholders that they are not required to choose a repair shop that participates in the firms' DRPs and that they may not be charged part of the cost of repairs if they use a repair facility that does not participate in a DRP. Although the North Carolina law requires insurance companies to provide consumers with this information, it does not prevent these firms from continuing to urge consumers to use their DRPs.

**If you were involved in an accident, would you want to take your car to a repair facility of your choice, or would you agree to take it to a DRP participant? Do you believe a law like North Carolina's should be passed for the entire nation?**

Sometimes companies get special rates for their employees. If you are a government employee, you can often obtain special types of automobile insurance from a government employees' insurance company. When comparing insurance companies, remember to look at the service they provide. You can shop for insurance by figuring out the exact policy you prefer, including liability, uninsured motorist, medical, collision, comprehensive, and perhaps towing, with the specific limits you want; then get a written statement from several insurance companies' agents. Insurance premiums can vary by 10 to 30 percent, depending on what company you select.

The insurance agent you work with is also important. If one in your area has a reputation for being fair and knowledgeable, you may want to take suggestions from that person. Again, you are being sold information as part of the package. (You may also be buying "clout" if you are dealing with a company agent rather than a broker for many different companies.)

Table 17.3 on the next page will help you compare insurance policies. When calling around to get insurance, you can fill in the chart and compare policies.

The Insurance Information Institute has a hot line you can call for further information on auto insurance. If you have questions about insurance, call its toll-free number, (800) 331-9146, or visit its Web site at **http://www.iii.org**.

## How to Lower Your Insurance Rates

Following are some tips on how to lower your automobile-insurance rates:

1. Don't buy coverage that you don't need, such as collision and comprehensive insurance on an older car. For example, if you have a five-year-old car

| TABLE 17.3 | Comparing Auto-Insurance Companies | | | |
|---|---|---|---|---|
| | | Company | | |
| Type of Coverage | Limits Desired | A | B | C |
| 1. Liability: | | | | |
| Bodily injury | $ _____/person, $ _____/accident | _____ | _____ | _____ |
| Property damage | $ _____/accident | _____ | _____ | _____ |
| 2. Physical damage: | | | | |
| Compensation for total loss | Only wholesale price available | _____ | _____ | _____ |
| Collision | $ _____/deductible | _____ | _____ | _____ |
| 3. Medical payments | $ _____/person | _____ | _____ | _____ |
| 4. Uninsured motorist | $ _____/person, $ _____/accident | _____ | _____ | _____ |
| 5. Accidental death benefits | $ _____ | _____ | _____ | _____ |
| 6. Towing | $ _____ | _____ | _____ | _____ |
| 7. Comprehensive | $ _____/deductible _____ | _____ | _____ | |
| 8. Other | $ _____ | _____ | _____ | _____ |
| | ANNUAL TOTAL: | _____ | _____ | _____ |

**Question for Thought & Discussion #6:**
Did members of your family take insurance costs into consideration when they chose the cars they drive? Should they have done this?

whose Kelley Blue Book value is relatively low, you may not want to bother with collision insurance because you will never collect more than Blue Book value (and the damage may be more than the car is worth).

2. Avoid high-performance or expensive cars for which auto insurance is much higher.

3. Take a higher deductible on collision and comprehensive insurance. Remember, the higher the deductible, the lower the premium.

4. See if you qualify for a discount for not smoking, not drinking, belonging to a car-pool, having an accident-free record for the past three years or more, or keeping your mileage low each year. If you are a student, ask about discounts for good grades and for driver's education courses.

5. Don't use your car for work if you can obtain other transportation.

6. Don't duplicate insurance. If you have a comprehensive health- and accident-insurance policy, then you don't need medical coverage in your automobile-insurance plan.

7. Any time your situation changes, notify your company. Do this when your estimated yearly mileage drops, when you join a car-pool, when a driver of your car moves away from home, and so on.

## The Special Problem of Insurance for Rental Cars

For many years, some car rental firms have pressured customers to purchase *insurance waivers* (special policies that cover damage to rented cars) that cost from a few dollars to as much as $15 a day. These waivers can almost double the cost of renting a car if they are accepted. Many states (New York, for example) now require insurance companies that do business in these states to provide insurance for rental cars automatically. Before you agree to pay for such a waiver, be sure to check whether you are already covered.

# When You Are Refused Insurance

Sometimes, because of a bad driving record, you will be refused liability coverage by an automobile-insurance company. When this happens, you become an **assigned risk.** To get insurance as an assigned risk, you must first certify that you have attempted within the past sixty days to obtain insurance in the state where you live. A pool of insurance companies (or sometimes the state) will then assign you to a specific company in the pool for a period of three years. At the end of three years, you can apply for reassignment, provided you are still unable to purchase insurance outside the pool.

If you are an assigned risk, you can purchase only the legal minimum amount of insurance in your state. In most cases, you will pay a much higher premium for the same amount of coverage than someone who is not an assigned risk.

**Assigned Risk** A person seeking automobile insurance who has been refused coverage. That person is assigned to an insurance company that is a member of the assigned-risk pool in the person's state.

# Confronting Consumer Issues: Choosing a Lawyer

When consumers' homes are damaged or when they are involved in automobile accidents, they may need to hire people with specialized legal knowledge to help them protect their rights. Often, such individuals are lawyers who charge for the legal knowledge they have and the services they provide.

Most consumers have heard stories about lawyers who charge $200 an hour or more and drag cases out for twice as long as necessary to inflate their bills. Rumors of lawyers receiving enormous bonuses and maintaining opulent lifestyles are common. There may be individual cases where such stories are true, but most lawyers earn their fees by providing honest value and a necessary service.

In many situations, consumers do need the services of a lawyer to protect their rights. There are other circumstances, however, that involve the law but do not require the services of a lawyer. Consumers can often receive legal information and guidance from other sources at a fraction of the cost of hiring a lawyer.

## OBTAINING HELP FROM NONLEGAL PROFESSIONALS

Accountants are probably the most common **nonlegal professionals** who serve as a source of information for consumers who need legal advice but not a lawyer. Certified public accountants and professional financial advisers are often qualified to provide information about tax laws and accounting methods that can solve many financial and legal questions consumers have. Most accounting firms charge roughly half as much as qualified tax attorneys for similar services. This does not mean consumers should never consider using a lawyer when they have financial problems. It does mean they can save money by using other professionals for help that does not require a lawyer.

Real estate agents are not lawyers but are often able to explain the legal requirements of real estate transactions. They are generally qualified to provide basic information about title searches, zoning ordinances, easements, leases, or registering deeds that can save prospective buyers hundreds or even thousands of dollars in legal fees. In many cases, when a consumer buys or sells a home, a lawyer will probably be needed to finalize the deal, but much of the initial work can be performed by nonlegal professionals.

Bankers are also qualified to provide certain types of financial information. Most banks maintain entire departments that are devoted to trust and estate planning. Although banks will charge a fee for their services and may

What qualities would you look for when you choose a lawyer to defend your legal rights?

require legal consultations in some matters, the cost of estate planning through banks is likely to be less than the cost of using only a lawyer. For consumers who have specialized needs, however, the bank may not have sufficiently qualified personnel. In such cases, consumers need to seek help from a tax lawyer. There is little sense in saving a few hundred dollars in legal fees if it results in paying the government thousands more in taxes.

### Question for Thought & Discussion #7:
Why are lawyers a necessary part of any effective legal system?

## SOURCES OF LOW-COST LEGAL ADVICE

Several sources of legal information are available that consumers may use at little or no cost. When you feel, for example, that your civil rights have been violated, you can ask for assistance from the American Civil Liberties Union, the National Organization for Women, the National Association for the Advancement of Colored People, or the federal government's Equal Employment Opportunity Commission. Most states also maintain departments that can help people protect their civil rights. There is, of course, no guarantee that these organizations will choose to support a particular case, but you have little to lose from asking.

When you believe you have been harmed by a fraudulent business practice or by a defective and dangerous product, you can contact either the Federal Trade Commission or the

Consumer Product Safety Commission. Again, there is no guarantee that these agencies will help a particular individual obtain satisfaction, but you may as well ask.

Free legal aid may also be available from state family service agencies, insurance commissions, divisions of employment, or health and human services departments. The point to remember is that consumers may not need to hire a lawyer when they have a legal problem. It often pays to investigate other alternatives.

## WHEN YOU NEED A LAWYER

The best time to choose a lawyer is before you require one. It is better to avoid being placed in the situation of needing to choose a lawyer today to represent you in court tomorrow. Consumers should use the same process to select lawyers that they would use to select any other important purchase, for they are buying a service that may be expensive both in cost and in the impact it may have on their lives and financial well-being. No one is likely to make the best possible choice hiring a lawyer in a hurry.

A good way to start is to ask yourself what type of services you expect an attorney to provide. People who lead a simple life with few legal complications may need to employ only a **generalist** who can draw up a will or help complete the purchase or sale of a home. People who are wealthy, run a complex business, intend to sue someone, or are involved in complicated disputes with the IRS, however, need to hire specialists.

## CHOOSING A SPECIFIC LAWYER OR LAW FIRM

The actual selection of a specific lawyer or legal firm may be accomplished through a series of steps. It is often best to begin by asking for recommendations from friends or relatives whom you respect. This will provide insight as to a particular lawyer's ability and character, and knowing she or he was recommended may cause the lawyer to take a greater interest in you. Few lawyers would jeopardize an existing relationship with a customer by providing unsatisfactory service to a friend or relative of that person.

Other professionals you employ can also be a good source of advice. If you have an established relationship with an accountant, he or she might be able to recommend a lawyer who can provide specialized legal advice for your financial situation. Real estate agents are often able to recommend lawyers. Remember, however, that just because a lawyer has been recommended to you by someone you trust doesn't mean you shouldn't evaluate him or her and the services offered for yourself.

Directories that list lawyers and their qualifications may be helpful. The best known is the *Martindale-Hubbell Law*

*Directory*, http://www.martindale.com, available at many large libraries. This directory rates lawyers based on information solicited from other lawyers and provides information about each lawyer's law school degree, professional associations, fields of practice, and honors. It probably is not wise to accept blindly what this or any other directory has to say about individual lawyers, but it is a good place to begin an investigation.

When choosing a lawyer, consider the size of the legal firm and the length of time the firm or an individual lawyer has been in practice. New lawyers who are out on their own for the first time are not necessarily bad lawyers. Everyone was new once; however, new lawyers will not have had time to develop the relationships and understanding of the legal system that can make the difference between winning and losing a case. Large firms are more likely to have lawyers who can share their expertise to solve more complicated legal matters. At the same time, large, established legal firms are likely to charge higher fees for simple legal services that could be provided equally well by smaller organizations. Large firms may also act more slowly or be less sensitive to the wishes of a client with limited needs who will generate smaller rather than larger fees.

## MEETING YOUR PROSPECTIVE LAWYER

When you have identified a firm and a specific lawyer you believe you might want to employ, your next step is to make an appointment to meet this person to discuss your legal needs. The purpose of this meeting is often more to develop a personal relationship with the lawyer than to resolve a specific legal problem. It is important for people to feel that lawyers they hire will responsibly protect their interests. A woman involved in a divorce, for example, would probably not want to employ a male lawyer if she believed he had sexist attitudes.

### Prepare a List of What You Want to Know
Consumers should ask a number of questions at the initial meeting with a lawyer:

1. *What are your areas of expertise?* Assure yourself that your prospective lawyer is qualified to provide the services you are likely to need. There is no reason to hire a lawyer who specializes in divorce if you need help running a business.

2. *What is your availability?* Good lawyers tend to be busy. A lawyer who is already overcommitted is not a good choice if you expect to need frequent and lengthy consultations.

3. *What fees will be charged for standard services?* There is, of course, no way to foretell the future; however, a lawyer should be willing to explain how billing will be computed for standard services.

*(Continued on next page)*

**Confronting Consumer Issues (Continued)**

**4.** *Will you sign an agreement or provide a written statement that outlines your fees and the services they will purchase?* Lawyers are often hesitant to put such agreements in writing, but without them clients are often left to the lawyer's mercy.

Once a service has been provided and billed, clients must pay that amount, even if it seems unreasonable, unless they have a contractual agreement on which they can base their refusal or are willing to hire another lawyer and go to court—where they may lose.

## What to Bring with You to the Meeting

Consumers can save time and money by anticipating information and documents that their lawyers will need. Whether you need help with your business, an estate, or an adoption, bringing documents the lawyer is likely to want to see to your initial meeting should speed the legal process. Moreover, having such information with you can reduce the number of meetings you will need to have, the amount of time these meetings will take, and consequently the cost of your legal consultation.

In addition to information and documents, consumers who have a specific legal objective for meeting with a lawyer will benefit from preparing questions they need answered. List these questions in a logical sequence and leave space under each question to summarize the lawyer's responses. Your personal notes are important because although you may expect a lawyer to provide you with a written letter or brief that includes specific information you have requested, it will not include every comment made during an interview or your impression of what was said.

## WHAT DETERMINES A LAWYER'S FEES?

Many factors contribute to a lawyer's or law firm's determination of fees that are charged for services rendered. Obviously, the complexity of a case and the amount of time it will require are primary considerations. Related factors include the lawyer's existing workload and the length of time that is available before the work must be completed. The more pressure there is to finish a case quickly, the more a customer can expect to be charged.

An attorney's reputation and experience also have a direct bearing on the fee. A consumer who hires a nationally famous lawyer should expect to pay more than one who hires a recent law school graduate. Lawyers are more likely to charge lower fees when they believe a lasting relationship has been established that will generate future income. In some cases a client's financial situation will affect what a lawyer charges as well. A final, and often important, consideration is the lawyer's overhead, or cost of doing business. If a law firm

maintains a large, professionally decorated office in an expensive high-rise building, it will need to earn the money from its clients to pay for the office. There is no reason to be put off by a lawyer who has a well-furnished office, but there is no reason to be impressed, either. The important question to consider when choosing a lawyer is "What will this lawyer be able to do for me?" not "Is his or her office impressive?" In the early 2000s, the least a consumer could expect to pay for a qualified lawyer's services in most parts of the United States was roughly $120 an hour.

The following are common methods of billing used by lawyers and legal firms:

- *A flat fee, in which a specific amount is charged for each particular service.* This allows clients to know how much they will be charged before a service is rendered, but gives lawyers little incentive to "go the distance" to provide the extra work that may win a case or successfully resolve a problem.

- *An hourly rate, in which a specific amount is charged for each hour a lawyer works on a legal problem.* This is a common method of setting fees for nonstandard services when a lawyer cannot be sure how long a case will take to complete. Consumers who agree to this type of arrangement should ask for a written cap, or a limit, on the number of hours a lawyer may put in on a case without client approval. Without such a cap, it is possible for legal fees to grow rapidly to an amount that is far more than a client expected to pay.

- *A contingency fee, meaning a share of any settlement that is paid to the lawyer.* In accident cases, many lawyers charge one-third of any judgment regardless of the size of that judgment. Lawyers will agree to this type of fee only when they are reasonably sure that a significant judgment will be made. It obviously can be used only in cases where a financial settlement is expected.

- *A fee-plus-cost agreement that allows lawyers to bill clients for the costs they pay for researching and preparing a case in addition to their regular legal fee.* Consumers should be very careful when they agree to this type of arrangement. They need to know how the costs of a case will be deter-

### Question for Thought & Discussion #8:
If you were permanently disabled in an accident, would you accept a contingency agreement that would pay your lawyer one-third of any settlement you received? What would the costs and benefits of this type of agreement be?

mined. There have been situations in which lawyers have charged $50 an hour for the services of a legal assistant who was paid $10 to $20 an hour for her or his work. This arrangement provides lawyers with an incentive to have unnecessary work done by assistants to increase the size of their own income.

If consumers feel a lawyer's fees are excessive, they can try to negotiate lower fees. They may also choose to employ a different attorney who charges less. Whatever they do, they must do it before services have been rendered. The time to discuss a lawyer's bill is before a fee has been charged.

## Key Terms:

**Nonlegal Professional**    A person who is not a lawyer but is qualified to provide legal advice about specific situations.

**Generalist**    A lawyer who does not practice in a specialized field of law but is qualified to provide legal advice in most normal situations.

## Chapter Summary

1. Consumers purchase homeowners' and automobile insurance for two reasons—to protect their investments in case of loss and to protect themselves against possible liability claims if their property is the cause of another's harm.

2. There are basically two types of home-insurance policies—fire-insurance policies and homeowners' policies. Property coverage and liability coverage are both available under homeowners' policies.

3. You should make sure that you obtain sufficient insurance on your house so that at least 80 percent of its replacement value is covered.

4. Special types of insurance can also be added to your basic homeowners' coverage to protect particularly valuable items in your home or to provide extra liability limits.

5. Homeowners' warranties (HOWs) may be purchased or are sometimes provided by contractors to protect against structural defects on newly built homes.

6. Renters can obtain insurance coverage for their personal possessions through a renters' insurance policy.

7. Various forms of coverage are available under automobile-insurance policies, including liability coverage, medical payments, collision and comprehensive insurance, and uninsured motorist coverage. Accidental death benefits are also obtainable.

8. Auto-insurance rates are determined by many factors, including age, gender, the kind of car you drive, where you drive, what your car is used for, marital status, occupation, and safety record.

9. In states that have adopted no-fault auto insurance, the liability-based system is eliminated. Essentially, if you have no-fault insurance, your insurance company pays you in case of an accident, no matter who was at fault.

10. To get the best deal on auto insurance, you need to comparison-shop, because rates and claims-handling services can vary considerably for the same kind of insurance coverage. You can lower your insurance premium considerably by taking a higher deductible, searching for discounts for which you might be eligible owing to safe-driving habits and other precautions, reducing the miles driven on your car, not duplicating insurance coverage, and letting your insurance company know when changes occur that might reduce your premium.

11. When an individual is refused auto-insurance coverage because of a bad driving record, he or she becomes an assigned risk. Special insurance programs allow minimum insurance coverage for assigned risks, but the premiums charged for assigned-risk coverage are very high.

12. Although there are many situations in which consumers need to employ a lawyer, they should recognize that legal advice is often available from other, less expensive sources.

13. Lawyers should be chosen before a consumer has the immediate need for a lawyer's services. A lawyer or law firm can be selected through a series of steps that includes asking friends or relatives for recommendations, making inquiries with professionals, and consulting directories of lawyers. Consumers should also consider the size and reputation of the firm for which a lawyer works.

14. The fees lawyers charge depend on the complexity of the cases being considered, their workload, their experience and reputation, and their overhead expenses for office space and assistants.

## Key Terms

assigned risk **429**
basic form policy **418**
broad form policy **418**

comprehensive form policy **418**
generalist **433**
no-fault auto insurance **426**
nonlegal professional **433**

residence contents broad form
   policy **422**
umbrella policy **419**
zero deductible **424**

## Questions for Thought & Discussion

1. How much liability coverage—both for your home and for your car—do you think is a "reasonable" amount? Explain why you chose these amounts.

2. Why have umbrella policies become much more common in recent years?

3. Would you choose to purchase periodic review coverage for your home? Why or why not?

4. How many of your friends who rent an apartment buy renters' insurance? If they don't, what risks are they taking?

5. Do you believe no-fault insurance is a good idea? Why or why not?

6. Did members of your family take insurance costs into consideration when they chose the cars they drive? Should they have done this?

7. Why are lawyers a necessary part of any effective legal system?

8. If you were permanently disabled in an accident, would you accept a contingency agreement that would pay your lawyer one-third of any settlement you received? What would the costs and benefits of this type of agreement be?

# Things to Do

1. Visit the office of a local insurance agent to determine the cost of obtaining basic coverage for an automobile you own or would like to own. Do the same by completing an Internet search of insurance providers. How much difference is there between the cost of insurance furnished by your local agent and the cost of insurance that can be purchased on the Internet? Which would you choose? Why would you choose it?

2. Survey your friends and relatives to find out the amount of automobile liability insurance they carry. How many carry the minimum required by law? Do you believe people who carry the minimum are making a rational choice? Why or why not?

3. Look through your local telephone book's Yellow Pages for advertisements by lawyers or legal firms. Evaluate these advertisements in terms of the quality of the information they provide to potential customers. What do they emphasize? What important types of information do they fail to provide? Do you believe it is a good idea to allow lawyers to advertise their services? Why or why not?

4. Talk to a person who has recently employed a lawyer to help with a legal procedure. How was the lawyer chosen? Was the person satisfied with the services the lawyer provided? Would this person employ the same lawyer again in the future?

# Internet Resources

## Finding Consumer Information on the Internet

The following Web sites have been selected for their relevance to topics discussed in this chapter. Search these sites to locate information about consumer services provided by state governments to consumers who purchase insurance. Search for your state's department of insurance Web site and compare what is offered on its site with the ones identified below. Remember, Web addresses change frequently. If any of these addresses no longer function, find similar sites to investigate using any of the search engines available to you.

1. The New York State Department of Insurance maintains a Web site that provides New York residents with basic information about their insurance rights and provides them with an e-mail link to report problems. Its site is located at **http://www.ins.state.ny.us**.

2. The Texas Department of Insurance maintains a Web site that provides Texas residents with basic information

about their insurance rights and provides them with an e-mail link to report problems. Its site is located at **http://www.tdi.state.tx.us**.

3. The office of the Washington State Insurance Commissioner maintains a Web site that provides Washington residents with basic information about their insurance rights and provides them with an e-mail link to report problems. Its site is located at **http://www.insurance.wa.gov**.

## Shopping on the Internet

The following Web sites have been selected because they offer consumers services similar to those described in this chapter. These are commercial sites that are designed to market products. They do not represent a comprehensive or balanced description of all firms that offer renters'-insurance protection online. Visit these Web sites and compare the products they offer. Remember, Web addresses change frequently. If any of these addresses no longer function, find similar sites to investigate using any of the search engines available to you.

1. Geico Direct offers renters'-insurance protection online. It advertises that its policies are the "sensible alternative." Its Web site can be investigated at **http://homeowners. geico.com**.

2. State Farm Insurance is the largest supplier of consumer insurance policies in the nation. It offers renters'-insurance protection at its Web site that is located at **http://www. statefarm.com**.

3. Fidelity Investments maintains a Web site that links consumers to insurance companies that offer renters' insurance in every state. Visit its Web site at **http://www. insurance.com**.

### InfoTrac Exercises

 Purchasers of new copies of this text are provided with access to the InfoTrac Web site. This Web site links students to thousands of recent articles published in hundreds of periodicals. Use the key words **homeowners' insurance** or other terms from this chapter to conduct a key-word search. Choose one article that is of particular interest to you and write a brief essay describing what you have learned from the article. Be sure to cite the author and title of the article and the name and date of the publication in which it appeared.

## Selected Readings

Coolidge, Carrie. "How Much Insurance?" *Forbes,* June 14, 1999, p. 266.

"Counsel Heed Thyself." *New Jersey Law Journal,* May 21, 2001, p. 26.

"Dogs and Teen Drivers Could Leave You Broke." *Business Week,* September 16, 2002, p. 86.

Gagoi, Pallavi. "Quandary at the Rental Counter." *Business Week,* July 15, 2002, p. 140.

Geddes, Dwight. "No-Fault Auto Insurance: Views from the Front Line." *Claims,* March 2002, pp. 52–53.

"How to Save on Auto Insurance." *Consumer Reports,* September 1999, pp. 12–18.

"It's a Dangerous World, Don't Forget Your Umbrella." *Business Week,* August 13, 2001, p. 5.

Nagel, Jennifer, and Ernie Condon. "Choosing a Lawyer." *Consumers' Research Magazine,* September 2002, p. 2.

Ottolenghi, Hugo H. "Coverage Crisis." *Daily Business Review,* January 9, 2003, pp. AA2–5.

Pennisi, Albert F. "New Home Warranties Offer Owners Protection." *Real Estate Weekly,* January 2, 2002, p. 16.

"Policies for Peace of Mind: Does a Renter Need Renters' Insurance?" *Consumers' Research Magazine,* December 2002, pp. 23–27.

Stern, Linda. "Home Coverage Hike." *Newsweek,* December 16, 2002, p. 86.

"Teen Drinking and Driving May Be on the Rise." *Alcoholism and Drug Abuse Weekly,* May 10, 1999, p. 118.

Thinkstock/Getty Images

# CHAPTER 18

# Life Insurance and Social Security

After reading this chapter, you should be able to answer the following questions:

- What are the benefits and costs of buying life insurance?

- What different types of life insurance can consumers buy?

- How much life insurance is enough?

- How can consumers investigate the financial strength of life-insurance companies?

- What benefits does Social Security provide?

- How can consumers make the best choices for their own life-insurance needs?

e have discussed homeowners' insurance, liability insurance, automobile insurance, and medical insurance at some length. These are basically forms of protection or income security. Now we'll discuss two additional forms of protection—life insurance and Social Security.

# Life Insurance and Our Need for Security

A sense of security is important for most families. In fact, psychologists contend that the average American wants a sense of security more than just about anything else in life. The following are some of the major hazards to *financial* security, along with the ways Americans protect against them:

- *Illness.* Health and medical insurance, savings account for emergencies, Medicare, Medicaid.
- *Accident.* Accident insurance, savings account, state workers' compensation, Social Security, aid to the disabled, veterans' benefits.
- *Unemployment.* Savings fund for such an emergency, unemployment compensation.
- *Old age.* Private retirement pension plans, savings account, annuities, Social Security.
- *Premature death.* Survivors' insurance under the Social Security Act, life insurance, workers' compensation, savings, investments.
- *Desertion, divorce.* Savings, investments, Temporary Assistance for Needy Families (TANF).
- *Unexpected, catastrophic expenses.* Health insurance, property insurance, liability insurance, Medicare.

The responsibility for providing economic security for a family may be assumed by the family members themselves, their relatives, charitable institutions, employers, or the government. Seventy years ago, financial security was provided primarily by the first three, but now employers and the government are assuming some of the responsibility.

## Premature Deaths

As in most parts of the world, the mortality rate in the United States has been declining for many years. People are suffering from fewer fatal diseases and are living longer, as can be seen in Figure 18.1. Nonetheless, every year there are more than 300,000 premature deaths in the United States. In many cases, particularly when a wage earner dies, these premature deaths cause financial hardship for other members of the family. The main purpose of life insurance is to provide financial security in this event.

## History of Life Insurance

The first recorded life-insurance policy was written in June 1536 in London's Old Drury Ale House. A group of marine **underwriters** agreed to insure the life of William Gybbons for the grand sum of $2,000. This coverage was obtained for an $80 **premium.** (Unfortunately for the underwriters, Gybbons died a few

**Underwriter** The company that stands behind the face value of any insurance policy. The underwriter signs its name to an insurance policy, thereby becoming answerable for a designated loss or damage on consideration of receiving a premium payment.

**Premium** The payment that must be made to the insurance company to keep an insurance policy in effect. Premiums usually are paid quarterly, semiannually, or annually.

**Figure 18.1** Life Expectancy at Birth, 1900–2010

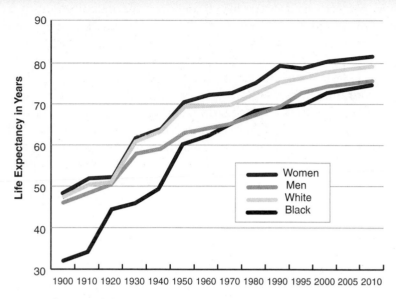

*Note:* Figures for 2000 and after are projections.

SOURCES: *Historic Statistics of the United States to 1957*; and *Statistical Abstract of the United States, 2001,* p. 73.

days before the policy was to run out.) And so it was that life insurance became a sideline for marine underwriters. Then in 1692, the Society for the Equitable Assurance of Lives and Survivorship began issuing policies covering a person for his or her lifetime. Old Equitable, as it became known, still exists today. In North America, the first corporation to insure lives was the Presbyterian Ministers' Fund, started in Philadelphia in 1759. In 1800, there were only 160 life-insurance policies in force in the United States. Only after the Civil War did the industry begin to flourish; since the early twentieth century, it has grown rapidly.

Today, roughly 83 percent of all households headed by a married couple, widow, or widower own at least some life insurance. In 2003, the average total coverage for all insured people in each household was about $190,000. On average, a premium of $1,780 was paid for this coverage (roughly 25 percent of this amount was paid by employers as a fringe benefit for their employees).

# The Basic Types of Life Insurance

Although the basic principles of life insurance are simple to grasp, the various insurance programs you can purchase are many and complex. In this chapter, we outline the basic types of life-insurance policies. We also offer some ideas to help you determine your own life-insurance needs and some recommendations about the appropriate type of insurance for you.

Basically, there are two types of life insurance—term and whole life. **Whole life insurance** combines protection with a cash value, whereas **term insurance** offers only protection. Whole life is also called *straight line, ordinary life,* or *cash-value insurance;* these are merely different names for the same type of life insurance.

**Whole Life Insurance**  Insurance that provides both death and living benefits; that is, part of the premium is put into a type of savings account.

**Term Insurance**  Life insurance that is for a specified term (period of time) and provides only a death benefit; it is a form of pure insurance with no saving aspect.

**Uniform Decreasing Term Insurance** A term-insurance policy on which the premiums are uniform throughout the life of the policy but its face value declines.

## Term Insurance

Premiums for term insurance, unlike those for whole life, commonly increase at the end of each term, such as every five years, if you wish to keep the same face value on your insurance policy. The increased premium reflects the rising probability of death as age increases. Thus, term life insurance will cost you relatively little when you are twenty-five years old, but by the time you are sixty, your premiums will have risen dramatically. By that time, however, you probably won't want as much term insurance because your children will be well on their way to financial independence, and you will have built up other forms of financial resources for any dependents you still have. That means you can reduce the premium burden by reducing the amount of insurance carried to protect your family. An important point to note here is that term insurance is one way for a young, growing family to have more insurance at an *affordable price* than it could have with the more expensive types of insurance. At this time in the life cycle, adequate insurance is needed most, but family income is normally relatively low.

Families often choose **uniform decreasing term insurance,** which has a level premium but a decreasing face value. A similar type of policy is called a *home-protection plan.* It is decreasing term insurance that decreases at *approximately* the same rate that the outstanding amount of money owed on a house declines as payments are made on the mortgage. Thus, when a home-protection policy is taken out for a face value equal to the amount of the mortgage on the home, the home can be paid off (or the mortgage payments can be paid) with the insurance benefits if the breadwinner dies anytime during the life of the mortgage.

Consumers can easily obtain data that will help them review their alternatives for buying term and other types of life insurance by contacting insurance-quotation services, such as MasterQuote (**http://www.masterquote.com**), Quotesmith.com (**http://www.quotesmith.com**), and Top Term Rates (**http://www.toptermrates.com**). These and other similar organizations will provide consumers with free comprehensive listings of policies that are available from a variety of insurance companies, with their prices for individuals of various ages and both genders. They also sell life-insurance policies over the Internet or the phone. Although this type of service may make insurance buying easy and convenient, it does not tell the consumer about the strength or reliability of the firms that are listed. There is little advantage in buying insurance at a good price if the firm is not financially sound. The importance of checking out a firm's financial stability is discussed later in this chapter.

**RENEWABILITY** Standard term insurance is often labeled one-year term or five-year term because those are the intervals, or terms, between premium increases. Other periods are also available. A term policy is *renewable* if the coverage can be continued at the end of each period merely by payment of the increased premium without the need for a medical examination. The renewability feature must, of necessity, add to the cost of the policy, but if you wish to preserve your insurability despite any changes in your health, you certainly would want to pay the extra costs for this feature. Term policies are commonly renewable until the policyholder reaches some age of retirement, such as sixty-five or seventy. All coverage then stops.

In one sense, the premiums for any term policy are constant for the life of the policy, but since most term policies are written with a five-year or ten-year "life," the constancy of premium is not too meaningful. The premium is truly constant throughout a long period of time only with decreasing term insurance, in which the face value falls every year.

**CONVERTIBILITY**  Often, **riders** can be attached to term policies that give you the privilege of converting them to other than pure insurance without the necessity of a medical examination. You pay for this additional feature, however. If you have a convertible term policy, you can convert it to whole life without any problems. The main reason you might want to convert is to continue your coverage after you pass the age of sixty-five or seventy. After converting the policy, you would pay whole life premiums based on your age at the time of conversion. Most insurance experts believe that these two features—convertibility and renewability—should be purchased. They give you much flexibility at an appropriate additional cost.

Table 18.1 shows the costs of $50,000 worth of one-year–renewable term insurance for a thirty-five-year-old man. If this man keeps $50,000 worth of term insurance until age sixty-five, he will pay total premiums of $17,893. He will have no cash value in the policy, as he would in a whole life policy.

## Whole Life Insurance

Whole life insurance accounts for a little less than half the total value of all life insurance in force in the United States. Life-insurance salespersons will almost always try to sell you a whole life policy because it is more profitable for them and the company. (It has been estimated that a salesperson earns about nine times more selling the same amount of whole life as she or he does selling term insurance.)

**PREMIUMS**  Whole life premiums generally remain the same throughout the life of a policy. As a result, the policyholder pays more in each of the early years than is necessary to cover the company's risk in later years. Table 18.2 on the next page gives an example of a $10,000 ordinary life-insurance policy with an annual level premium of $222.70 for a thirty-five-year-old man. In the first year, $205.50 of the $222.70 goes to the insurance company to

**Rider**  A written attachment to an insurance policy that alters the policy to meet certain conditions, such as convertibility, double indemnity, and so on.

| TABLE 18.1 | A Typical $50,000 Yearly Renewable Term Policy for a 35-Year-Old Male | | |
|---|---|---|---|
| Year | Annual Premium | Year | Annual Premium |
| 1 | $165.50 | 11 | $312.50 |
| 2 | 172.50 | 12 | 339.00 |
| 3 | 181.00 | 13 | 368.00 |
| 4 | 192.00 | 14 | 400.00 |
| 5 | 204.50 | 15 | 435.00 |
| 6 | 219.00 | 16 | 473.50 |
| 7 | 235.00 | 17 | 515.50 |
| 8 | 252.00 | 18 | 560.50 |
| 9 | 270.50 | 19 | 609.50 |
| 10 | 290.50 | 20 | 642.50 |
| 20th year total | $6,838.50 | Total at age 65: | $17,893.00 |

| TABLE 18.2 | Composition of Cash Value and Operating Charges in 20-Year Ordinary Life Premiums | |
|---|---|---|

**$10,000 Ordinary Life**
**Dividends\* to Purchase Paid-Up Additions**
**Annual Premium: $222.70     Male: Age 35**

| Year | Deposit to Cash Value | Deposit to Insurance | Total Cash Value |
|---|---|---|---|
| 1 | $ 17.20 | $205.50 | $    17.20 |
| 2 | 179.71 | 42.99 | 196.91 |
| 3 | 190.43 | 32.27 | 387.34 |
| 4 | 201.97 | 20.73 | 589.31 |
| 5 | 213.47 | 9.23 | 802.78 |
| 6 | 225.43 | −2.73 | 1,028.21 |
| 7 | 237.14 | −14.44 | 1,265.35 |
| 8 | 250.35 | −27.65 | 1,515.70 |
| 9 | 262.61 | −39.91 | 1,778.31 |
| 10 | 275.17 | −52.47 | 2,053.48 |
| 11 | 270.17 | −47.47 | 2,323.65 |
| 12 | 282.60 | −59.90 | 2,606.25 |
| 13 | 294.64 | −71.94 | 2,900.89 |
| 14 | 306.82 | −84.12 | 3,207.71 |
| 15 | 320.64 | −97.94 | 3,528.35 |
| 16 | 333.21 | −110.51 | 3,861.56 |
| 17 | 346.11 | −123.41 | 4,207.67 |
| 18 | 360.95 | −138.25 | 4,568.62 |
| 19 | 376.12 | −153.42 | 4,944.74 |
| 20 | 391.60 | −168.90 | 5,336.34 |

| | Summary 20th Year | At Age 65 |
|---|---|---|
| Total cash value | $5,608.97† | $10,566.83† |
| Total deposits | 4,454.00 | 6,681.00 |
| Net gain | 1,154.97 | 3,885.83 |

\*Dividends are neither estimates nor guarantees but are based on the current dividend scale.
†Includes terminal dividend.

cover the insurance costs, and $17.20 goes to the **cash value** of the policy for the purchaser. By the sixth year, the deposit to cash value is greater than the level annual premium and stays greater throughout the life of this particular policy. As you can see by the twentieth year—that is, when the policyholder is fifty-five years old—after $4,454 has been paid in, there is a total cash value of $5,608.97 in the policy.

The cash value of a whole life policy certainly is not the same as a savings account. Insurance industry people often promote whole life as an insurance policy combined with a saving plan, but it is definitely not. The cash reserve is not given to your named **beneficiary** as a separate payment; rather, it is included in the face amount of the policy. Thus, looking at Table 18.2 again, let us assume you have paid in for ten years. Your total cash value is $2,053.48. What if you die at the end of ten years? You have a $10,000 ordinary life policy, and your named beneficiary gets $10,000, not $10,000 plus your cash value of $2,053.48.

**Cash Value**  Applied to whole life policies only, it represents the amount of "savings" built up in the policy and available to the living policyholder, either to borrow against or to receive if the policy is canceled.

**Beneficiary**  The designated person or persons for any insurance policy. In a life-insurance policy, the beneficiary is the person who receives the benefits when the insured dies.

Owners of whole life policies often take comfort from the fact that their premiums are level and, therefore, represent one of the few costs that do not go up with inflation. (The real value of the policy, however, as well as the real value of premiums, declines as the buying power of a dollar falls.) True, the cost is relatively high to begin with, but it gets no higher. The exact level of premiums that you would pay for a $10,000 ordinary life-insurance policy, such as the one in Table 18.2, depends on your age when you buy the policy; the younger you are, the lower the premium will be, because the company expects to collect premiums from you for many years. The older you are, the higher it is.

**Question for Thought & Discussion #3:**
Why do you believe the rate of growth in sales of whole life insurance has declined in recent years?

**WHOLE LIFE INSURANCE OFFERS LIVING BENEFITS**   As we will see when we compare whole life with term insurance, whole life is relatively expensive because it is a form of financial investment as well as insurance protection. The investment feature is known as its "cash value." In Table 18.2, the cash value at the end of twenty years was in excess of $5,000; at age sixty-five, it was actually in excess of the death benefit of the policy. You can, of course, cancel a whole life policy at any time you choose and be paid the amount of cash value it has built up. Individuals sometimes "cash in" a whole life policy at the time of their retirement when the cash value can be taken out either as a lump sum or in installments called annuities. These are the so-called **living benefits** of a whole life policy—the opposite of death benefits. The death benefit of a life-insurance policy is obviously the insurance you have purchased that is payable to your beneficiary on your death. In contrast, the living benefit includes the possibility of converting an ordinary policy into some sort of lump-sum payment or retirement income. In any one year, up to 60 percent of all insurance company payments are in the form of living benefits.

Note that the level premium for a whole life policy is paid throughout the life of the policyholder—unless you reach the ripe old age of, say, ninety-five or a hundred.

**BORROWING ON YOUR CASH VALUE**   One feature of a whole life policy is that you can borrow on its cash value any time you want. The interest rate on such loans is relatively favorable. If you should die while the loan is outstanding, however, the sum paid to your beneficiary is reduced by the amount of the loan. In any event, the borrowing power given to you by the cash value of a whole life policy can be considered a cushion against financial emergencies. Note that if you ever have to drop a whole life policy because you are unable to pay the premiums or because you need its cash value, you most certainly will take a loss. And, of course, you will give up the insurance protection.

**WHEN YOU REACH RETIREMENT AGE**   When you reach retirement age, you may maintain all or part of your death benefit by:

1. Choosing protection for the rest of your life, but at a lower value.
2. Choosing full protection, but for a limited number of years in the future.

Or you may accept a living benefit by:

3. Choosing a cash settlement that gives back whatever savings and dividends have not been used to pay the insurance company for the excessive costs it has incurred for your particular age group.
4. Choosing to convert the whole life policy into an annuity that provides a specified amount of income to you each year for a certain number of years.

**Living Benefits**   Benefits paid on a whole life policy while the person is living. Living benefits include fixed and variable annuities.

"The latest research
suggests that whole life can
be a sensible long-term
investment for those who
could otherwise expect their
own investments to earn
only about 4 percent
after taxes."

**DEATH BENEFITS**   In most life-insurance policies, you specify a beneficiary who receives the death benefits of that policy. If you buy a $10,000 ordinary life policy and do not borrow any money on it, your beneficiary will receive $10,000 when you die. You may also have other options for settling a life-insurance policy, however. Before you purchase any insurance policy, you should discuss the particular settlement terms that are available with the underwriter of that insurance. Generally, four optional settlement plans are available:

● *Plan 1.* The beneficiary receives a lump-sum payment.
● *Plan 2.* The face value (principal) of the insurance policy is retained by the insurance company, but a small interest payment is made to the beneficiary for a certain number of years or for life. Then the principal is paid to the children or according to the terms in the contract.
● *Plan 3.* The face value is paid to the beneficiary in the form of installments, either annually, semiannually, quarterly, or monthly. The company makes regular payments of equal amounts until the fund is used up. In the meantime, the company pays interest on the money remaining to be paid out. There are two options here. Under one option, each payment is for a specific amount, and the length of time during which the payments will be made depends on the amount of the payment, the face value of the policy, and the rate of interest guaranteed on the policy. Under the other option, the payments are made for a given time period. In this case, the amount of each payment depends on the number of years the income is to be paid, the face value of the policy, and the rate of interest guaranteed on the policy.
● *Plan 4.* Regular life income is paid to the beneficiary. The insurance company guarantees a specified number of payments that will total the face value of the policy. If the beneficiary dies before the guaranteed payments have been made, the remainder goes to the estate of the beneficiary or as directed in the contract. This is sometimes called an *annuity plan.*

**INSURANCE PLUS FORCED SAVINGS**   In sum, whole, straight, or ordinary life insurance gives you pure insurance plus forced savings and, hence, the possibility of retirement income, as can be seen in Figure 18.2. You can instead buy pure insurance—that is, term insurance—at a lower cost than whole life. You can invest the difference in your own saving and retirement plans and perhaps be better off (or, at least, no worse off) if you can get a higher rate of return on your savings than the insurance company offers. Let's use an example to demonstrate this idea. Suppose Fred buys a $100,000 whole life policy at the age of forty. He pays annual premiums of $532 for twenty years, or a total of $10,640. When he reaches sixty, the cash value of his policy will be just over $10,000. If Fred could have bought a $100,000 level term policy for an average premium of $232 a year, and had invested the remaining $300 each year at a 5 percent tax-free return, he would have accumulated more than $10,400. He would be roughly $400 better off than he is with the whole life policy.

The latest research suggests that whole life can be a sensible long-term investment for those who could otherwise expect their own investments to earn about 4 percent *after* taxes. But if, on your investments, you can make 5 percent or more after taxes, whole life may not be the type of policy for you.

**Figure 18.2** How Whole Life Insurance Works to Provide Both Savings and Protection

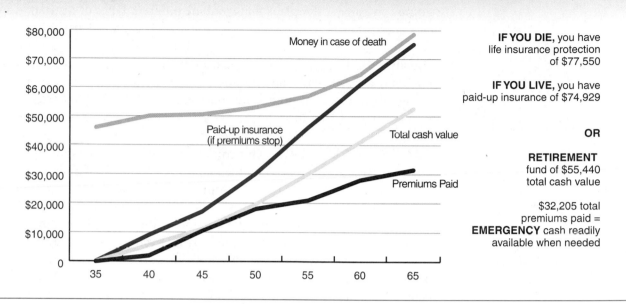

**IF YOU DIE,** you have life insurance protection of $77,550

**IF YOU LIVE,** you have paid-up insurance of $74,929

**OR**

**RETIREMENT** fund of $55,440 total cash value

$32,205 total premiums paid = **EMERGENCY** cash readily available when needed

## Universal Life Insurance

Another type of insurance policy, which combines some aspects of term insurance and some aspects of whole life insurance, is called *universal life*. Every payment, usually called a *contribution*, reflects two deductions made by the issuing life-insurance company. The first one is a charge for term insurance protection; the second is for company expenses and profit. The money that remains after these deductions earns interest for the policyholder at a rate determined by the company. The interest-earning money in the policy is called the policy's *cash value*, but that term does not mean the same thing as it does for a traditional whole life policy. With a universal life policy, the cash value grows at a variable interest rate rather than at a predetermined rate.

Universal life insurance offers two major advantages over whole life insurance. The first is a complete disclosure of the fees that the insurance companies take out for managing the policy. The other is that the interest rate earned on the cash value, at least for now, seems to be higher than for traditional whole life policies. Simply stated, universal life insurance is a term-insurance package with an investment fund. As with any package arrangement, you have to ask yourself whether you can get a better deal by purchasing the components separately. In other words, you must decide whether you should buy the best term policy you can find and then find the best investment.

**Question for Thought & Discussion #5:**
Would you be likely to choose to purchase universal life insurance coverage in the next five years? Why or why not?

## Some Specialized Insurance Policies

Various companies offer a number of special life policies. They include combination plans and variable life-insurance policies. Every year, new plans are added and old ones modified. A qualified insurance agent can describe them for you.

## Combination Plans

Some companies offer plans that combine different types of insurance, such as a family plan, family income plan, or extra-protection policy.

**FAMILY PLAN**   This insurance plan combines some features of term insurance and some of whole life insurance. Under a family plan, every member of the family has some insurance; newborns are automatically covered a certain number of days after birth.

**FAMILY INCOME PLAN**   This combination of term and whole life insurance is designed to provide supplemental income to the family should the breadwinner die prematurely. In a typical twenty-year family income policy plan, if the policyholder dies, his or her beneficiary might receive $10 a month for each $1,000 of the term portion of the policy during the balance of the twenty years. Then, at the end of the twentieth year, the beneficiary would receive the face value of the whole life portion of the policy, either in a lump sum or in monthly installments. Under a variation of this policy called the *family maintenance plan*, the monthly payments continue for a full twenty-year period *after* the insured dies.

**EXTRA-PROTECTION POLICY**   This policy also combines term and whole life insurance in double, triple, and even quadruple amounts. A triple-protection policy, for example, provides $2,000 of term insurance for each $1,000 of whole life insurance. The term insurance usually continues until age sixty or sixty-five and then expires; the whole life portion of the policy remains in force. Insurance experts point out that these policies give less protection per extra premium dollar than the family policies previously mentioned, but the extra protection continues for a longer time.

## Modified Life Policies

Modified life plans generally are sold to newly married couples and young, single professionals just starting out. For the first three to five years, the policy is term insurance. It then converts automatically to whole life protection at a higher premium. In the trade, these plans are called *Mod 3* and *Mod 5*.

## Adjustable Life Insurance

One of the newest insurance policies available, adjustable life, presumably offers insurance plans adjusted to each customer's needs and budget. Adjustable life differs from conventional life policies in two ways: you can switch back and forth between whole life and term coverage, and you can change the amount of insurance protection. Basically, you can increase your coverage as much as you please if you can pass a medical examination or present evidence of insurability. If you keep the same policy, you can adjust for inflation once every three years by adjusting your coverage and the premium. For an extra cost, you can also buy a guaranteed-insurability rider that assures you the right to buy more life insurance in the future regardless of your health.

## Other Types of Life-Insurance Policies

In addition to life insurance that you buy as an individual, you also may be eligible for certain other types of life-insurance policies that generally are offered at more attractive rates.

**GROUP INSURANCE**  Group insurance is usually term insurance written under a master policy that is issued to either a sponsoring association or an employer. Some types of group insurance are currently available through employers and various fraternal and professional organizations. Per $1,000 of protection, the cost of group insurance is generally lower than individually obtained insurance for many reasons, but primarily because of the lower selling and bookkeeping costs. The selling costs are lower because the employer or sponsoring group does all the selling; there is no commission to be paid to a selling agent. And the bookkeeping costs are lower because, again, the employer or the association may do all the bookkeeping. Generally, no medical examination is required for members of the group unless they want to take out an unusually large amount of group insurance.

**CREDIT LIFE INSURANCE**  If you take out a loan, in many cases you may be required to buy life insurance in the amount of the loan. The reason is simple: without such insurance, if you die with part of the loan outstanding, the creditor may have trouble collecting it. But if the creditor is named the beneficiary in the life-insurance policy you are required to take out as part of the loan, the creditor is assured payment of any amount due. The average amount per policy is small, perhaps $2,000. Credit life insurance may seem inexpensive, but it isn't. It typically makes sense only for a person fifty years old or more who lives in a state with a low maximum rate, and then only if an existing insurance program is inadequate. It might, for example, be better than nothing for a person with a health problem who cannot otherwise be insured. Most consumers, however, are better off simply upgrading their basic insurance portfolio. Thus, be careful: your creditor may be abusing his or her right to require life insurance on your loan by demanding an overly high premium. Check to see that the rate you are actually paying is commensurate with other group policy rates. If it is not, then the difference you pay should be added to the total finance charge so you can figure out the true percentage rate of interest you are paying on the loan. Although lenders often require borrowers to buy insurance, they might not require you to purchase it through them. If their rates are higher than alternative policies, find out if you are allowed to buy your insurance from a different source.

No matter what type of life insurance you buy, you will be required to complete an application. If you fail to provide complete and accurate information on that application, you may not be entitled to benefits under the policy, as illustrated by this chapter's *Consumer Close-Up* feature on the next page.

## Reading a Life-Insurance Contract

As with any major investment, you should carefully review the life-insurance coverage that is provided in your contract with the insurance company (your insurance policy). A plethora of clauses and options can be included in a life-insurance policy; the following are some of the major ones:

> **"If you fail to provide complete and accurate information on [your insurance application], you may not be entitled to benefits."**

**Question for Thought & Discussion #6:**
What advantages are there to purchasing whole life insurance when you are young?

## CONSUMER CLOSE-UP
# Honesty Is the Best Policy

**A** few years ago, Michael Berthiaume applied for mortgage life insurance with the Minnesota Mutual Life Insurance Company. Berthiaume wanted to make sure that if he should die, the balance due on his mortgage (about $44,000) would be covered. The insurance company did not require Berthiaume to take a physical examination for the policy, but he did have to answer questions concerning his health status on the application. One of the questions was whether the applicant had ever been treated for, or diagnosed as having, high blood pressure. Although four months earlier Berthiaume had been advised by his doctor that he had hypertension, he nonetheless answered "no" to the question. In a word, he lied.

Eight months later, Berthiaume died. When his widow submitted a claim for the mortgage insurance, however, she was denied payment. The company stated that it had no obligations under the policy because while it was investigating the claim, it discovered that Berthiaume had misrepresented his health status on the application. Under Minnesota law (and in most other states), an insurer can avoid obligations under a policy if an applicant misstates necessary information or misleads the insurance company into accepting a risk of which it is not aware. Hypertensive individuals find it difficult to obtain insurance coverage because they present a higher risk of death. When they do obtain insurance, they usually have to pay twice what other individuals do for the coverage.

Obviously, insurance companies, like other businesses, have an eye toward profits. The fewer claims they have to pay, the higher those profits, and therefore it is in their interest to investigate insurance claims carefully to see if they might uncover a problem that would allow them to avoid payment. What this means for you, the consumer, is that you must be careful not to misstate facts on an insurance application. Honesty is normally always the best policy, but especially so when it comes to insurance applications.

**If you had a serious health condition, would you report it fully and honestly when you applied for medical or life insurance? How might your choice affect your ability to purchase insurance?**

- *Guaranteed-insurability option.* This option, sold with whole life policies, allows the policyholder to purchase additional insurance at specified ages and amounts without having to meet medical qualifications.

- *Automatic premium loan option.* With this provision, the insurer will automatically pay any premium that is not paid when due. The premium then becomes a loan against the cash value of the policy. This option will continue until the total of the automatic loans is equal to the cash value; then the policy is terminated.

- *Convertibility.* This is a clause or option applied to term-insurance policies that allows you to switch the policy to whole life or endowment at standard premium rates regardless of any change in your health.

- *Accidental death or double indemnity.* With this option, an additional sum is paid to your beneficiary if you die as the result of an accident. Because this payment usually doubles the face amount of the policy, it is called *double indemnity.* Most people who have evaluated insurance believe double-indemnity insurance costs more than it is worth.

- *Incontestability.* Most policies have a clause that prohibits the company from challenging statements made in your application after two years if you should die; thus, even if you made false statements, the company cannot nullify the policy after a stated period.

- *Guaranteed renewability.* This type of clause typically applies to renewable term insurance. It requires that the insurance company renew the term policy for a specified number of term periods, even if there has been a significant change in the health of the insured.

- *Settlement options.* This portion of your policy details the methods by which the death benefits can be paid to your beneficiary.

- *Grace period.* This is the period of time that your life insurance will remain in force if your premium is overdue.
- *Suicide.* In this clause, the insurance company stipulates what will happen should you commit suicide. The clause will limit the insurance company's liability in such a case.
- *Misstatement of age.* This provision states that if you have given an incorrect age on your policy application, the death benefit will be adjusted to reflect your correct age.
- *Reinstatement.* This clause describes the way in which your policy, should it lapse for any reason, can be reinstated.

# Choosing a Sound Life-Insurance Company

How would you feel if you had paid thousands of dollars into an insurance policy only to have the company fail just before you were ready to retire? Unfortunately, this is exactly what has happened with increasing frequency in recent years. The best way to avoid this problem is by carefully checking out the financial condition of insurance companies before you purchase a policy.

Money "saved" in life-insurance policies is not insured by any federal government agency like the Federal Deposit Insurance Corporation. All states now have "guaranty funds" that pay off customers of failed insurance companies for amounts up to, but not over, $100,000. When life-insurance companies fail, policyholders will generally receive their money, but they may have to wait years while courts review the situation and decide how to proceed.

Although most life-insurance companies are financially strong, it is worth your time to be sure that you don't choose one that isn't. The following are some of the firms that rate the strength of insurance companies:

*Best's Insurance Reports*
Ambest Rd.
Oldwick, NJ 08858
(908) 439-2200
http://www.ambest.com

*Moody's Investors Services*
1350 Treat Blvd., Suite 400
Walnut Creek, CA 94596
(925) 945-1005
http://www.moodys.com

*Fitch Services*
One State Street Plaza
New York, NY 10004
(212) 908-0621
http://www.fitchratings.com

# Involuntary Benefit Programs— The Case of Social Security

During the depths of the Great Depression, the nation realized that many people had not provided for themselves in case of emergencies. It also realized that a large percentage of the elderly population, unable to rely on their children for support, had become destitute. In an effort to prevent a recurrence of so much pain and suffering by elderly people, Congress passed the Social Security Act of 1935. In January 1940, when the first monthly benefit was paid, only 22,000 people received payments. Today, however, well over 90

> "How would you feel if you had paid thousands of dollars into an insurance policy only to have the company fail just before you were ready to retire? The best way to avoid this problem is by carefully checking out the financial condition of insurance companies before you purchase a policy."

**Question for Thought & Discussion #7:**
Do you believe it would be a good idea to allow people to invest part of their Social Security taxes in stocks, as President George W. Bush has suggested? Why or why not?

**OASDHI** Old-Age, Survivors, Disability, and Hospital Insurance—the government name for Social Security insurance.

**Income Transfer** A transfer of income from some individuals in the economy to other individuals. This is generally done by the government. It is a transfer in the sense that no current services are rendered by the recipients. Unemployment insurance, for example, is an income transfer to unemployed individuals.

percent of people aged sixty-five or older are receiving Social Security benefits. If U.S. population growth continues to slow, the average age of the population will continue to rise. Hence, the total number of people eligible for and receiving Social Security will increase as a percentage of the total population. Figure 18.3 shows the past and projected percentages of the population aged sixty-five and over.

We have called this section "Involuntary Benefit Programs" because you and I, with few exceptions, have no choice. If we work, we must participate in the Social Security program. Self-employed people were at one time able to avoid it, but today they must pay self-employment Social Security taxes if they do not work for someone else. If you work for someone else, your employer must file Social Security taxes for you if you earn over $1,000 in any year. Of the people earning money in the United States, fully 95 percent contribute to Social Security.

The Social Security Act provides benefits for old-age, survivors, disability, and hospital insurance. It is therefore sometimes called **OASDHI.** It is essentially an **income transfer** program, financed out of compulsory payroll taxes levied on both employers and employees: those who are employed transfer income to those who are no longer employed. You pay for Social Security while working and receive the old-age benefits after retirement. The benefit payments usually are made to those reaching retirement age. Also, when an insured worker dies, benefits accrue to his or her survivors. Special benefits

**Figure 18.3** U.S. Population over Age 65

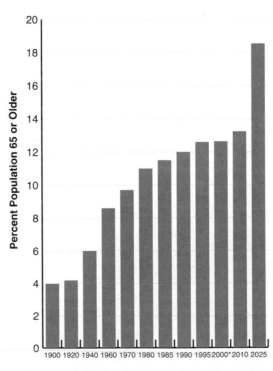

*Note:* Percentages for 2000 and after are projections.
SOURCE: *Statistical Abstract of the United States, 2001,* p. 15.

provide for disabled workers. Additionally, Social Security now provides for Medicare, which was discussed in Chapter 16.

The Social Security Act of 1935 also established an unemployment-insurance system. Unemployment insurance is not really a federally operated program. Rather, it is left to the states to establish and operate such plans. Although all fifty states have these plans, they vary widely in the extent and the amount of payments made. Programs are basically financed by taxes on employers; these taxes average about 2 percent of a business's total payroll. A worker who becomes unemployed may be eligible for benefit payments. The size of these payments and the number of weeks they can be received vary from state to state.

## Basic Benefits of Social Security

Later in this chapter, you will learn where to get information with which to determine, tentatively, the benefits you are allowed under Social Security. (The predictions must be tentative because Congress frequently changes the benefits.) Essentially, Social Security is a form of life insurance. Every time you have a child, the maximum life-insurance benefit of Social Security is automatically restored, and its term is automatically increased to a potential eighteen years. Here is what you can expect from Social Security:

1. Medicare payments in the future.
2. If you should die, payments to your beneficiary.
3. Payments to you or your dependents if you are totally disabled and unable to work.
4. A retirement annuity—that is, a payment of a certain amount of money every month after you retire until you die. (This payment, however, is legislated by Congress and can be changed by Congress.)
5. If you die, a modest lump-sum payment, presumably to take care of burial expenses.

Whenever you figure out your insurance needs, you must consult the basic coverage that you have under your Social Security. (You will learn how to do that later in the chapter.)

> **Question for Thought & Discussion #8:**
> Do you know a retired person who is living on Social Security benefits alone? How is this person doing? What does this show you about your need to save for your own retirement?

## How Social Security Is Paid

In theory, Social Security is supported by a tax on the employee's income that is matched by the employer. The Social Security tax is indicated on your payroll receipt as FICA—Federal Insurance Contributions Act. When this tax was first levied, it was 1 percent of earnings up to $3,000. By 1963, the percentage rate had increased to 3.625 percent, and the rate has continued to climb. In 1994, a 7.65 percent tax was imposed on each employee's wages up to a maximum of $60,600 to pay for Social Security and Medicare. All earned income over $60,600 was taxed at a 1.45 percent rate for Medicare. By 2002, the maximum earnings subject to Social Security withholding had increased to $84,900. Figure 18.4 on the following page shows the growth in employees' Social Security withholding rates. Employers contribute matching amounts to the amounts paid by employees. (You should realize, though, that employees pay more for the program, because their wages could be higher if the employer did not have to contribute.) Self-employed individuals, since they are effectively

**Figure 18.4** Changes in Social Security Tax Rates for Employers and Employees, 1963–2002

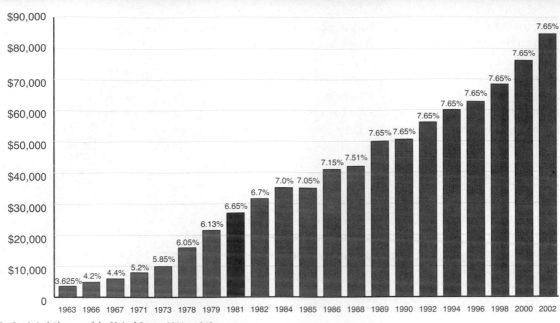

SOURCE: *Statistical Abstract of the United States, 2001*, p. 345.

both the employer and the employee, pay twice what other employees contribute. In 2002, self-employed persons paid 15.3 percent (2 × 7.65 percent) of their earned income to Social Security (but got a tax deduction for the employer's part).

You may recall from Chapter 8 that taxes that fall more heavily on people with lower incomes are regressive. Social Security taxes are regressive because once people's income exceeds the maximum taxable amount, they pay no more taxes. In 2003, a person who earned $50,000 paid $3,825, or 7.65 percent in Social Security and Medicare taxes; a person who earned $100,000 paid $6,840.00, or 6.84 percent; and a person who earned $150,000 paid $7,569.00, or 5.05 percent. This is clearly a regressive tax structure.

## Working While Receiving Social Security Benefits

Before 2000, people who chose to work beyond their retirement age were often penalized by having their Social Security benefits reduced. In 1999, for example, individuals between the ages of sixty-five and seventy-one could earn up to $17,000 without suffering any reduction in their benefits. For income earned beyond this threshold, however, they lost one dollar in Social Security benefits for every three they earned. At the same time, their extra earnings were taxed by the federal and state governments.

Although rules that discouraged older people from working might have made sense in 1935 when unemployment exceeded 18 percent, in the 1990s they came to be viewed as unfair and as contributing to labor shortages. In April 2000, Congress and President Bill Clinton finally acted to eliminate the earnings penalty. Now Americans aged sixty-five and over may earn any amount of income without suffering any reduction in their Social Security benefits. Another advantage of this decision is that the taxes paid on the extra income older Americans earn will increase government tax revenues and strengthen the Social Security system.

## Problems with Social Security

A number of respected authorities have reached some pretty depressing conclusions about Social Security. In the first place, you have to remember that Social Security is not really an insurance policy in the sense that you are guaranteed a certain amount of money. Your dependents get that amount of money if you die, just as they would with a regular insurance policy; but if you live, you get retirement payments, the amount of which is established by Congress and based on how much money you paid in when you were employed. Future legislators may not be as generous as past legislators, so you may find yourself with a very small retirement income if you rely only on Social Security.

## There Will Always Be a Problem with Social Security

No matter what Congress does, the Social Security system will soon face a major problem. Population trends will force the government to reduce benefits, or increase taxes, or both. With choices limited to these alternatives, some groups of U.S. citizens will be displeased no matter what is done. In 1950, a little more than fifty years ago, for each person receiving Social Security benefits, there were sixteen workers who, along with their employers, were paying taxes to provide the benefits. By the year 2000, this ratio had fallen to slightly more than six workers for each beneficiary. And, if government projections are correct, by 2030 the ratio will be just over two workers for each beneficiary. Demographers (experts who study population trends) believe that unless the Social Security system is changed, taxes will have to be increased dramatically over the next forty years. In this time many people who are young workers today will retire. You may want to receive the same benefits as your parents or grandparents, but the reality is that you won't. For this reason it is important for you to prepare for your future retirement by saving and investing your own funds.

# Confronting Consumer Issues: How to Meet Your Insurance Needs

Before you figure out how much life insurance you should buy, what type it should be, and where you should get it, first consider who should be insured in your family. You have to take into account the Social Security benefits you have coming, and that sometimes is not easy. You then have to look at the actual economic (or financial) dependency that anybody has on a particular member of a household. If you are a single college student, for example, it is usually not recommended that you have any insurance at all (unless you want to use it as a forced savings mechanism or as protection against becoming medically uninsurable later on in life). By the same token, it is usually pointless for a family to insure its children unless the children contribute a substantial amount to the family income. If one of them dies, the family's earning power generally will not fall. This is not necessarily true for a homemaker, however, who frequently contributes to the family earnings stream by employment outside the home, as well as implicitly through the value of services rendered to the family. In this case, the family may want to take out an insurance policy on the homemaker's life. The basic wage earner should, of course, be the one with the most insurance because, if she or he dies prematurely, the household will suffer the greatest loss.

## SOME INSURANCE BUYING RULES

Like a new bicycle, a new car, or a new house, insurance is competing for your consumer dollar. When you make an expenditure on life insurance, you obtain a certain amount of satisfaction in knowing that your dependents will be financially secure in the event of your premature death. Note, however, that there are other possible uses of these same funds that also yield satisfaction; thus, there is no pat answer or formula that will tell you exactly how much insurance is best for you.

In determining the type and amount of insurance coverage you need, the following rules should be considered:

1. Identify the major risks that you and your family reasonably face; insure against these risks according to the *potential* loss they can produce.

2. Insure against big losses, not small ones.

3. Never buy an insurance policy until you have compared policies from at least two companies (and perhaps more), not only on the costs but also on the terms of coverage. Use the **interest-adjusted cost (IAC)** figure for comparison.

4. Limit your losses and control your risks through preventive measures.

When you invest for your future remember that Social Security benefits will not pay for all of your needs when you retire.

5. *Buy* insurance; don't have it sold to you.

## ARE YOU UNDERINSURED?

There is a good chance that you are underinsured if anybody depends on you for even part of his or her livelihood. If, however, you live alone or are young and unmarried, or even are married but your spouse also contributes to the family kitty, then you may not need much (if any) life insurance. If you are married and have children, or a spouse who depends on you for at least part of his or her income, then you probably should have some form of life insurance. You should first realize that Social Security is the basis of all your protection needs, assuming you are covered by Social Security. You will have to find out from your local Social Security office exactly what kinds of benefits your dependents have coming in case of your death.

In this discussion, we make the same assumption that you should make when trying to figure out your insurance needs: Assuming that you drop dead tomorrow, how much would be left to your dependents, in what form, and over what period? This is not easy to figure out, so plan on spending some time at it. You may want to work it out with an insurance agent, but you can probably do it on your own.

## DETERMINING YOUR FUTURE SOCIAL SECURITY INCOME

Each year the Social Security Administration sends a statement of account to every American who has worked for at least ten years. Part of such a statement appears in Figure 18.5.

> ## Figure 18.5 A Social Security Estimated Benefits Statement

### Your Estimated Benefits

**Retirement** You have earned enough credits to qualify for benefits. At your current rate of earnings, if you stop working . . .

At age 62, your payment would be about . . . . . . . . . . . . . . . . . . . . . . . . . . . . . . . . . . . . .$   1,157 a month
At your full retirement age (66 years of age), your payment would be about . . . . . . . . . . . . . .$   1,584 a month
At age 70, your payment would be about . . . . . . . . . . . . . . . . . . . . . . . . . . . . . . . . . . . . . .$   2,146 a month

*Note:* Your benefit amount increases when you continue working because of your additional earnings and the special credits you will receive for delaying your retirement. This increased benefit can be important to you later in your life. It can also increase the future benefit amounts your family and survivors could receive.

**Disability** You have enough credits to qualify for benefits. If you become severely disabled right now . . .

Your payment would be about . . . . . . . . . . . . . . . . . . . . . . . . . . . . . . . . . . . . . . . . . . . . .$   1,464 a month

**Family** If you get retirement or disability benefits, your spouse and children may also qualify for benefits.

**Survivors** You have enough credits for your family to receive the following benefits if you die this year.

Total family benefits cannot be more than . . . . . . . . . . . . . . . . . . . . . . . . . . . . . . . . . . . . .$   2,557 a month
Your child  . . . . . . . . . . . . . . . . . . . . . . . . . . . . . . . . . . . . . . . . . . . . . . . . . . . . . . . . . . .$   1,096 a month
Your spouse who is caring for your child . . . . . . . . . . . . . . . . . . . . . . . . . . . . . . . . . . . . . .$   1,096 a month
Your spouse who reaches full retirement age . . . . . . . . . . . . . . . . . . . . . . . . . . . . . . . . . . .$   1,461 a month
Your spouse or minor child may be eligible for a special one-time death benefit of $255.

**Medicare** You have enough credits to qualify for Medicare at age 65.

### We based your benefit estimates on these facts

Your Name . . . . . . . . . . . . . . . . . . . . . . . . . . . . . . . . . . . . . .
Your Date of Birth . . . . . . . . . . . . . . . . . . . . . . . . . . . . . . .
Your estimated taxable earnings per year after 2003  . . . . . .
Your Social Security Number . . . . . . . . . . . . . . . . . . . . . . .

The statement includes estimates of what the person's monthly retirement benefits will be if he or she retires at 62, 66, or 70 years of age, estimates of disability benefits, and estimates of the amounts that would be paid to a surviving spouse or children. One of the most important parts of these statements is an earnings record that reports the amounts paid into the Social Security and Medicare systems each year. If you believe any of these amounts is incorrect, you should call (1-800-772-1213) or write the Social Security Administration at P.O. Box 33018, Baltimore, MD 21290-3018, to investigate the discrepancy.

Remember, unless you are now at retirement age, your statement will provide only a rough estimate. The calculations used to determine retirement benefits make certain assumptions that may or may not apply to you regarding your health, your average pay increases, how continuously you will work, and a fixed inflation rate. In general, experts say that if you earn at or above the maximum taxable

amount, you can count on Social Security to replace 22 percent of your earnings. That means that if you are currently earning $84,900 and plan to retire next year, you will receive approximately $18,678 in Social Security benefits, which would replace 22 percent of your earnings. If, however, you are currently receiving your highest income at a salary of $16,000 and plan to retire next year, you can expect to receive $6,400 in retirement benefits, or 40 percent of your current earnings.

Now that you have this information, you can figure out the financial condition of your family. To do this, look at Chapter 14, where we discuss how net-worth statements are calculated when you apply for a loan. Figuring out your net worth gives you a starting point.

You now have two major details of your financial situation in case you have dependents and die tomorrow—the Social

*(Continued on next page)*

## Confronting Consumer Issues (Continued)

Security payments your dependents would receive and the net worth you would leave them. Now you must figure out a monthly income goal for a spouse and children under age eighteen, a lump-sum education-fund goal for each child, a monthly retirement-income goal for a widow or widower starting at age sixty-two, and a monthly income goal, if any, for a widow or widower between child rearing and retirement. The latter is optional, depending on whether the family wants the widow or widower to have to work.

> ### Question for Thought & Discussion #9:
> How much life insurance should your family have on its adult members? Does your family have this amount of coverage? Why or why not?

## HOW MUCH LIFE INSURANCE SHOULD YOU BUY?

Neither you nor anyone else can calculate *exactly* how much life insurance you should buy. Some experts suggest that you should either own liquid assets (stocks, bonds, bank accounts, and the like) or purchase life insurance equal to at least seven times your annual income.

The appropriate amount for you, of course, depends on the factors already mentioned as well as many others. Your accumulated wealth has a good deal to do with your need for life insurance. A person who has inherited or accumulated several million dollars in liquid assets probably doesn't need life insurance. A single person with no dependents may not need much life insurance either. Your need to feel "safe" should also be considered when you decide how much life insurance to buy. After all, buying life insurance means that part of your income can no longer be used for other purchases. You have to decide how much consumption spending you want to give up to protect your beneficiaries if you suffer an untimely death.

Nonetheless, you can get a general idea of how much life insurance you should purchase by using the decision-making process to identify and evaluate your alternatives. Ultimately, your choice is likely to rely on fairly complicated mathematical calculations. In recent years many Web sites that can help you make these calculations have appeared on the Internet. To find such a Web site, complete a Web search using the key words *life, insurance, calculator*. In 2003, Insurance Quotes 4 Less placed an insurance calculator on its Web site at **http://www.insurancequotes4less.com/calculator.php**. When personal financial data and information about your family situation were entered, the calculator would estimate how much

life insurance you should purchase. As the Web site emphasizes, this is only an estimate. Still, by comparing your own calculations with the amounts suggested by this or a similar calculator, you may be better able to make a rational choice.

## WHAT TYPE OF INSURANCE SHOULD YOU BUY?

Suppose you decide you need $100,000 worth of life insurance. Which type should you purchase? Of the several life-insurance plans just discussed, the most important are term, whole life, and universal life. All but term insurance include some element of saving. Thus, you are not only buying pure insurance, you are also investing and getting a rate of return. Your decision whether to buy pure insurance or to buy savings will determine the payments you must make to the insurance company. The cheapest way to buy insurance is, of course, to buy term, because that way you buy only protection. If you already have a satisfactory saving program, you may not wish to save additional sums with an insurance company, since you can usually get a higher rate of return by going to other sources.

### Should You Invest through Life Insurance?
Consumers Union and several other research organizations point out that if purchasing whole life insurance is compared with buying term and investing the difference—that is, the difference between the whole life premium and the lower term premium—the combination of term and other investments will yield a larger sum of money at the end of any period.

Insurance salespersons, however, argue persuasively that you should buy whole life, not term, insurance. They say that whole life is a bargain, or even "free," because you eventually get back much or all of your money. Note, however, that if you die, your beneficiary will get only the face value on the policy, not the additional cash value. Salespersons use the cash-value aspect of whole life to tout its desirability over term insurance. Because term has no cash value, salespeople will tell you that buying it is "just throwing money down the drain." This "down the drain" argument ignores the fact that the term premiums are much lower than whole life premiums in the early years. For a person twenty-five years old, whole life premiums in the early years may cost three to four times more than term premiums.

### Life Insurance Sold on Campus
Insurance agents have become familiar figures on many campuses, where they sometimes contact students four to six times a year. If students say they cannot afford the premiums, the

insurance agent offers to finance the first annual premium on credit and even the second, with a loan to be paid off perhaps five years later. The student policyholder typically signs a policy assignment form, which makes the insurance company the first beneficiary if the student dies. Thus, the insurance company makes sure that it can collect the unpaid premium and interest. Generally, college students are advised not to buy life insurance because they usually have no dependents.

Additionally, the cost of campus life insurance is often high compared with policies available to the general public. If college students need to be insured, they should look at a standard life-insurance policy, either term or whole life.

> ## Question for Thought & Discussion #10:
> Have you been approached by someone who wanted to sell you life insurance? How did you react to this solicitation? Did you make a rational choice?

## THE FORCED-SAVINGS ASPECT OF INSURANCE

If you like the idea of having forced savings, then buying whole life insurance may be the way to do it. You'll feel compelled to pay the insurance premiums, and you know that part of the premium goes to a saving plan. If you would not otherwise have saved at all, having at least some savings compensates for the lower rate of return.

You do have other options. In some instances, you can have your employer take out a payroll deduction every month to put in your credit union account. You can also have your employer withhold a certain amount of money each month for U.S. savings bonds. In both instances, you will have more liquidity if you need it than if an insurance company had been doing your forced saving for you. (But if you don't trust yourself, you may prefer less liquidity.)

## SOME ADDITIONAL CONSIDERATIONS

A fact we have not yet mentioned about a permanent or whole life policy is that it is essentially a piece of property and has certain characteristics that are perhaps unique. Under current law, provided that the permanent insurance plan is set up properly, it can accumulate income tax free: interest on cash value is not taxable as current income. Essentially, then, you get a higher return than is actually shown in your life-insurance saving plan because you are not paying a tax on the savings you are accumulating. Remember, if you have a regular saving account, you have to pay federal and sometimes state income tax on the interest earnings of that account.

Remember, too, that death benefits on ordinary or straight life-insurance policies usually go to age one hundred; except in very rare cases, there is always going to be a death benefit.

## TAKE ADVANTAGE OF GROUP PLANS

Whenever you can take advantage of group term-insurance plans, you probably should do so to take care of at least part of your life-insurance needs. For reasons mentioned in the chapter, group insurance is generally cheaper than individually issued insurance (unless you happen to be significantly younger than the average age of the group).

## SHOPPING AROUND FOR INSURANCE

It generally is unwise to buy from the first insurance salesperson to knock on your door. Because large sums of money may be involved, look over several plans. Be aware, however, that life-insurance policies are incredibly complex. Seek out a knowledgeable insurance salesperson who represents a large number of companies and who can explain clearly the benefits of each program and the average annual costs per $1,000 of five-year–renewable term insurance. Similar information is available from many Internet insurance marketers.

To check the overall performance and financial stability of insurance companies, your most reliable resource is *Best's Insurance Reports*, published by A. M. Best Company and available in most libraries. *Best's* rates insurers from A+ all the way down to "omitted," and although the guide isn't infallible, it is one of the best available for consumers shopping for life insurance. The National Association of Insurance Commissioners (NAIC) evaluates life-insurance companies. These evaluations may be accessed from the NAIC Web site at **http://www.naic.org**. By checking these evaluations and *Best's Insurance Reports* before purchasing life insurance, you stand a greater chance of avoiding the plight that befalls some consumers when their insurance companies go bankrupt. All states have established guaranty funds for life insurance. If your insurance provider fails, however, there is no guarantee that benefits will be paid quickly or entirely by your state's fund.

*(Continued on next page)*

**Confronting Consumer Issues (Continued)**

## SOME WAYS TO CUT INSURANCE COSTS

The following suggestions can help you cut your insurance costs:

1. Don't carry insurance on children. Either save the premiums or use them to buy additional term insurance for yourself.

2. Consider term as opposed to whole life insurance.

3. Some insurance companies give premium discounts for nonsmokers, nondrinkers, people with homeowners' insurance, those who exercise regularly, and so on. Find out if your insurance company offers such discounts and take advantage of them.

4. Attempt to buy insurance via group plans through your employer or any organization of which you are a member.

5. Pay your premium annually instead of quarterly or monthly.

6. If you have a participating policy, don't let your dividends or refunds accumulate on deposit with the insurance company at a lower rate than the money could earn at a saving institution.

## Key Term:

**Interest-Adjusted Cost (IAC)**   An insurance cost index that takes into account dividends, interest, and earnings of the policy.

# Chapter Summary

1. The major hazards to financial security are illness, accident, unemployment, old age, premature death of the person providing financial support, desertion, divorce, and unexpected catastrophic expenses.

2. Life insurance can take many forms, the most popular being term and whole life.

3. Term insurance is generally for a five-year period, after which time a higher premium must be paid to obtain the same face value in insurance because the probability of death increases as the individual becomes older.

4. Whole, straight, or ordinary life insurance provides both pure protection and a saving plan whereby part of the premiums is put into investments that return interest to the policyholder. At any time, a whole life policyholder has a cash value in his or her policy.

5. Whole life insurance has living benefits. You can, for example, borrow on your cash value; you can get protection for the rest of your life at retirement; you can get a cash settlement; or you can convert your whole life policy to a stream of income called an annuity—over a certain period of time.

6. Death benefits can be paid to your beneficiaries in a lump sum equal to the face value, as interest on the face value of the insurance policy plus principal at the end of a specific period, or in installments until the face value has been paid.

7. There are several specialized life-insurance policies, such as a family plan and a family income plan.

8. Social Security is a form of social insurance in the United States. It provides for living and death benefits.

9. Social Security is not an insurance policy in the normal sense of the word. Basically, contributions to Social Security are merely transfers of income from those who work to those who do not work.

10. Social Security taxes are paid both by the employer and by the employee. In the economy as a whole, however, employees receive salaries that are lowered by the amount that employers must pay to Social Security. After all, that payment is a cost of hiring employees.

11. Since the year 2000, individuals aged sixty-five to seventy who work no longer lose benefits from Social Security. When they work, however, they continue to pay Social Security taxes.

12. In deciding whom to insure, consumers should base their decisions on who depends on whom and what financial stress would be undergone should an individual die prematurely.

13. Information about the basic benefits of Social Security is sent to most employees every year by the Social Security Administration.

14. Shopping for life insurance requires the same skills as shopping for any other consumer product. Information is the key. Consumers need to find out what their alternatives are and evaluate each alternative in terms of their own financial situations. They can consult *Best's Insurance Reports* and other publications to compare insurance coverage and costs as well as the financial soundness of various insurance companies. Computerized information-search services are also available to help consumers make decisions.

# Key Terms

beneficiary **442**
cash value **442**
income transfer **450**

interest-adjusted cost (IAC) **458**
living benefits **443**
OASDHI **450**
premium **438**
rider **441**

term insurance **439**
underwriter **438**
uniform decreasing term insurance **440**
whole life insurance **439**

# Questions for Thought & Discussion

1. Do you know a family that has suffered the premature death of a parent? How was the family affected by this event? Could having more life insurance have helped?

2. In your opinion, when should young people first purchase life insurance? Explain your reasoning.

3. Why do you believe the rate of growth in sales of whole life insurance has declined in recent years?

4. Why won't a death benefit of $100,000 be as important to a family twenty years from now as it would be today?

5. Would you be likely to choose to purchase universal life insurance coverage in the next five years? Why or why not?

6. What advantages are there to purchasing whole life insurance when you are young?

7. Do you believe it would be a good idea to allow people to invest part of their Social Security taxes in stocks, as President George W. Bush has suggested? Why or why not?

8. Do you know a retired person who is living on Social Security benefits alone? How is this person doing? What does this show you about your need to save for your own retirement?

9. How much life insurance should your family have on its adult members? Does your family have this amount of coverage? Why or why not?

10. Have you been approached by someone who wanted to sell you life insurance? How did you react to this solicitation? Did you make a rational choice?

## Things to Do

1. Investigate the cost of purchasing a $100,000 whole life policy from three different insurance companies by searching their Web sites on the Internet. How much difference is there in the prices charged? Can you find any reasons for these differences? If so, what are they?

2. Review the last pages of a tabloid newspaper at a local drugstore or supermarket. Find advertisements for mail-order life insurance. How much useful information do these advertisements provide? Why should consumers be suspicious of unsolicited offers for low-cost life insurance?

3. Search the Internet to find the lowest-cost term life insurance available. Check the financial strength of the firm that offers this insurance at Best's, Moody's, or Fitch's Web site (their Web addresses are listed on page 449). Is low-cost life insurance necessarily a good deal?

4. Select two nationally prominent politicians from different political parties and compare their stands on Social Security. In what ways do they agree, and in what ways do they disagree? Which politician do you feel has the better ideas? Be sure to credit your sources of information.

5. Imagine that you have reached retirement age. List the types and amounts of funds you believe you would need to have in order to maintain a reasonable standard of living for yourself and your spouse. Assume that Social Security will provide half of this amount. Describe the plans you should have made when you were young that would have allowed you to pay the other half of this amount from your own resources now that you are ready to retire.

## Internet Resources

### Finding Consumer Information on the Internet

The following Web sites have been selected for their relevance to topics discussed in this chapter. Search these sites to locate information that can add to your knowledge of online services that offer legal assistance to those who have been denied Social Security disability benefits. Remember, Web addresses change frequently. If any of these addresses no longer function, find similar sites to investigate using any of the search engines available to you.

1. Martindale Hubbell Lawyers.com maintains a Web site that lists lawyers throughout the nation according to their location and field of expertise. Use this Web site to find a lawyer in your area who specializes in Social

Security disability cases. It is located at **http://www.lawyers.com**.

2. Attorney Pages.com offers to link consumers to lawyers who specialize in Social Security disability litigation. Investigate its Web site at **http://attorneypages.com/attorneys/567**.

3. The Attorney Search Network offers to help California residents find a lawyer who is qualified to protect their Social Security disability rights. Review its Web site at **http://www.lawyerbureau.com**.

### Shopping on the Internet

The following Web sites have been selected because they offer consumers services similar to those described in this chapter. These are commercial sites that are designed to market products. They do not represent a comprehensive or balanced description of all firms that offer term life insurance online. How do the services at these Web sites compare with those that are available from local insurance salespeople? Remember, Web addresses change frequently. If any of these addresses no longer function, find similar sites to investigate using any of the search engines available to you.

1. Intelliquote.com offers to obtain quotes for term life insurance from as many as three hundred different insurance carriers. Investigate its Web site at **http://www.intelliquote.com**.

2. Wholesaleinsurance.net advertises that it can find the best price for term insurance by searching policies offered by more than 1,300 firms. Its Web site is located at **http://www.wholesaleinsurance.net**.

3. Financialone offers to let you use its "real-time system to search for the most competitive offers from top insur-ance companies in America." Review its Web site at **http://www.financialone.com**.

## InfoTrac Exercises

 Purchasers of new copies of this text are provided with access to the InfoTrac Web site. This Web site links students to thousands of recent articles published in hundreds of periodicals. Use the key words **Social Security reform** or other terms from this chapter to conduct a key-word search. Choose one article that is of particular interest to you and write a brief essay describing what you have learned from the article. Be sure to cite the author and title of the article and the name and date of the publication in which it appeared.

## Selected Readings

Gramlich, Edward M. "Social Security Reform in the 21st Century: The United States." *Journal of Aging and Social Policy,* Spring 2002, pp. 67–80.

Haggin, Leslie. "Make the Most of Cheaper Term Life Insurance." *Money,* December 1, 2002, p. 193.

Henderson, Sam. "Social Security Reform Said to Slack Poor's Cut." *Investment News,* April 15, 2002, p. 9.

Levy, Daniel S. "Your Family." *Time,* May 31, 1999, p. 103.

Orszag, Peter, and Peter Diamond. "Social Security: The Right Fix." *The American Prospect,* September 23, 2002, pp. 38–39.

"Retirement Age for Full Benefits." *Kiplinger's Retirement Report,* February 2003, pp. 6–7.

"Scope Out Your Benefits." *Business Week,* July 19, 1999, p. 120.

"Social Security Reform in Sight." *Accounting Today,* September 3, 2001, p. 45.

"Tips on Choosing Life Insurance." *Consumers' Research Magazine,* November 1996, pp. 20–24.

"Variable Life Insurance Variables." *Business Week,* April 8, 2002, p. 78.

Weiner, Leonard. "Enticing Prices on Terms of Endearment." *U.S. News & World Report,* October 25, 1999, p. 68.

# CHAPTER 19

# Looking to the Future

**After reading this chapter, you should be able to answer the following questions:**

- How will new technologies change the decisions consumers make in the future?

- What impact has the globalization of the world's economy had on U.S. consumers?

- How are consumers' rights to personal privacy being affected by new technologies?

- What ethical choices will consumers be forced to make because of the development of new medical technologies?

- Why is environmental protection a global issue?

- What steps can individual consumers take to help protect the environment?

- How can consumers prepare for a changing future?

ou are approaching the end of your class in consumer economics. Over the past weeks you have learned much and acquired important consumer skills. Does this mean that you know everything you will ever need to be a rational consumer throughout the rest of your life? Not by a long shot!

# What the Future Holds

No one can tell what the future holds with any degree of certainty. Consumers, however, can make reasonable predictions about changes that are likely to take place. By staying aware of the world around you, you will be able to identify alternatives you will need to evaluate to make the best possible consumer decisions for your life.

Thousands of developments that are taking place in the world right now will affect your life as a consumer for better or for worse. Rather than make a futile attempt to discuss them all, we will concentrate on five specific situations that every U.S. consumer will face in the coming years:

1. Changes in technology that will necessitate new decisions.
2. The globalization of the world's economy.
3. Assaults on individuals' rights to privacy.
4. Advances in medical technology that will force us to make ethical choices.
5. Our need to confront environmental issues.

The way we choose to deal with these situations, individually and as a society, will show much about the way we are likely to react to other important issues in our changing world.

# The Future of Technology

When your grandparents were children, they may have read Dick Tracy comic books that showed the famous crime fighter communicating on his wristwatch video-telephone. Although you may never have heard of Dick Tracy, you certainly have heard of portable devices that transmit and receive pictures as well as sounds. Who knows, you may even own one. As time passes, the number of new technologies that will become available for you to evaluate will continue to expand. Inevitably, you will need to learn about these advances to be able to make rational consumer choices in the future. There is no escaping this challenge. Unless you become a hermit in the Alaskan wilderness, you will be forced to deal with new technologies throughout your life.

## New Technologies—New Products

The world is changing at an accelerating pace. The development of another new technology seems to be announced almost every day. Many of the "facts" that you once accepted as true are now obsolete—others will become outdated in the future. Being a rational consumer requires you to commit yourself to a life of learning. Consider advances in automobile technology as just one example.

**Question for Thought & Discussion #1:**
What single new technology do you believe has changed the lives of U.S. consumers the most in the past year? Explain your choice.

> "The more complicated products become, the less able we will be to keep up with technological change."

**HYBRID VEHICLES**  In recent years manufacturers have developed and marketed hybrid vehicles that have small gasoline engines and electric motors. When these cars accelerate, they use both sources of power, but when they cruise at highway speeds, the electric motor becomes a generator and is turned by the gasoline engine to charge the car's batteries. Such a system requires an onboard computer to measure the car's need for power and complex mechanisms to transfer power to and from the electric motor. Small hybrid cars are able to travel more than fifty miles per gallon of gasoline. Larger hybrid cars and trucks are promised soon. In December 2002, General Motors and Toyota signed an agreement to jointly develop and market a hybrid sport utility vehicle by 2004. These hybrid vehicles represent a major step toward reaching the Environmental Protection Agency's goal of developing vehicles that can achieve seventy or more miles per gallon of gasoline by 2010.

**TRADE-OFFS OF NEW TECHNOLOGIES**  Advanced technologies offer consumers the promise of greater efficiency, a wider variety of products to choose from, and an improved standard of living, but at a cost. The more complicated products become, the less able we will be to keep up with technological change. Consider the goal of building a car that can get seventy miles per gallon of gasoline. It is already apparent that different manufacturers will develop and market different types of these vehicles. They will not all be exactly the same. Consumers who consider purchasing such a hybrid vehicle will need to learn about these different products and the technologies they use to rationally choose which to buy. Furthermore, consumers who purchase these vehicles will not be able to have them serviced by their local amateur mechanic in a backyard shop. Advanced technologies will force consumers to be more reliant on the expertise of technicians who have been trained in skills that most people will only vaguely understand.

## New Technologies—New Laws

The laws that govern transactions between consumers and businesses are also certain to change. Federal, state, and local governments are in a constant state of flux. No less than consumers, they must deal with new technologies.

**TAX RECEIPTS AND THE INTERNET**  Consider the growth of Internet marketing as an example. An important political issue at the start of the twenty-first century was whether transactions carried out on the Internet should be taxed. State and local governments anticipated a rapid decline in the sales tax revenues they receive because many consumers order products over the Internet and do not pay sales taxes on these purchases. E-commerce has put local stores that are required to charge sales taxes at a competitive disadvantage because consumers have to pay more to buy their products. Efforts to impose taxes on Internet transactions have met with considerable opposition, however. States that ask consumers to report their online purchases and pay taxes on them have had little success.

For years Internet marketers argued that it would be difficult or impossible for them to record the addresses of their customers, charge the appropriate amount of tax, and send these funds to thousands of state and local governments across the nation—to say nothing of keeping track of customers in foreign nations. They insisted that laws requiring tax collections for online purchases would effectively kill Internet marketing and the advantages it offers consumers. Regardless of these arguments, several large Internet marketers, including Wal-Mart, Target, and Toys 'Я' Us, agreed to begin charging sales tax for online sales at the end of 2002. It remains to be seen how successful they will be, or how much collecting sales taxes will cost them. If it proves to be expensive, you may be sure that these costs, along with the sales tax, will become part of the price consumers pay to shop the Internet.

**GOOD DEALS FOR CONSUMERS MEAN BAD DEALS FOR GOVERN-MENTS**   You might consider buying a new winter coat online with no sales tax as a "good thing," at least for you. But if millions of similar transactions go untaxed, governments that depend on sales tax revenue have only two choices: increase other taxes to make up for the lost sales tax revenue, or reduce their spending on programs that benefit consumers. Sales taxes, for example, are a source of funding for public education in most states. Few people would want school funding to be reduced because of technological advances.

How this issue will finally be resolved is not clear. It is certain, however, that rules that governed consumer transactions in the past will be changed to accommodate the technological transformation of our economy. In fact, the only thing we can be sure of when it comes to the government (in addition to taxes) is that whatever the laws are today, they will be different next year.

**Question for Thought & Discussion #2:**
Do you feel any guilt when you buy clothes online without paying sales tax? Should you? Explain your answer.

# Globalization of the World Economy

In 1945, the last year of World War II, less than 2 percent of the products sold in the United States were imported from other nations. At that time, U.S. consumers had little choice about where the products they bought had been made. Most industrialized nations were involved in the war, and the productive capacity of many had been largely destroyed. Other nations were in no position to export goods or services to the United States.

How would your life change if you could not purchase goods imported from other countries?

## The Growth of U.S. Trade

Now, nearly sixty years later, the situation has changed dramatically. In 2001, Americans purchased more than $1,350 billion worth of goods and services from other nations. This was almost 14 percent of the value of all products sold in the United States in that year. The trend toward increased purchases of foreign-made goods is even more apparent when specific types of products are considered. Roughly 22 percent of the automobiles sold in the United States in 2000 were produced in other countries. Almost all of the televisions, VCRs, DVD players, and tennis shoes we purchase are imported, as are most microwave ovens. Significantly, more than half the oil consumed in the United States (about 55 percent) is extracted from the ground in other nations. The list of foreign-made products we buy goes on and on and is likely to get even longer in the future.

At the same time, many U.S. workers earn their livings by producing products that are exported to other nations. In 2001, for example, foreign sales accounted for just over $1 trillion worth of American-made goods and services and helped to create roughly 18 million U.S. jobs. By examining Figure 19.1 on the next page,

**Figure 19.1** Value of U.S. Exports and Imports (in Billions of Dollars), 1992–2001

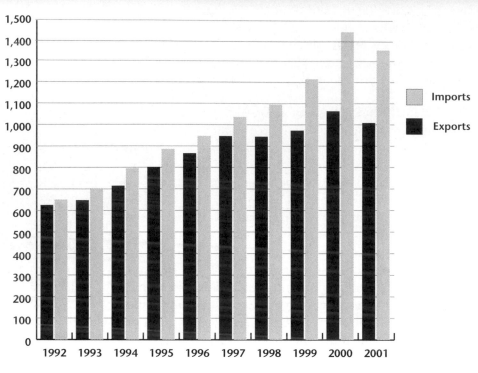

SOURCE: *Economic Indicators*, December 2002, p. 35.

you can see the growing importance of international trade in the U.S. economy. There is every indication that the trend toward increased trade will continue into the future despite the temporary downturn caused by the recession of 2001–2002.

## Benefits of Trade

Not all nations are equally well suited to produce all types of goods and services. Because countries have different types and qualities of raw materials, climates, workers, transportation systems, and technology, they can benefit when they **specialize** in producing goods for which they are best suited and then trade for other goods that they cannot produce as efficiently. By specializing and trading, countries can improve the standard of living their people enjoy.

**PRODUCTS WE EXPORT**   The United States, for example, is well suited to the efficient production of many technology-based goods and services and agricultural products. We have a well-trained work force, modern factories, and institutions that carry out many types of research and development. We also possess vast areas of fertile land with sufficient rainfall to grow far more food than we need for our own consumption in most years. Our extensive highway system, navigable rivers, seaports, railroads, airports, and advanced communications systems allow U.S. businesses to manufacture goods and

**Specialization**  The concentration of efforts on one area of production with the aim of having an advantage in the marketplace. Students specialize when they major in a certain subject at college, thus allowing them to have a comparative advantage in the job market later—assuming there is a demand for their specialized knowledge.

services and transport them quickly at low cost to almost any location in the world. It makes sense for consumers in other nations to buy American-made products that are less expensive or of better quality than their own.

**PRODUCTS WE IMPORT**    Similarly, it makes sense for U.S. consumers to purchase imported goods that are marketed here at lower prices or are superior in quality to American-made goods. In addition, some products that Americans want to own are not produced in the United States. No television picture tubes, for example, are manufactured in this country. It is impossible for a U.S. consumer to buy a television set that is totally American made, regardless of the label or trademark under which it is sold.

## Costs of Trade

Although an overwhelming majority of Americans benefit from international trade, clearly some others do not. When U.S. consumers choose to buy imported goods, they provide income to foreign firms and their employees. Competition from imported products has forced many businesses in this country to lay off workers and close. It has been estimated, for example, that over 80 percent of the jobs that once existed in the U.S. textile industry have been lost to foreign competition. Certainly, the individual workers who lost their jobs and the owners of the firms that were forced out of business are not better off because of international trade.

**DANGERS OF A NEGATIVE BALANCE OF TRADE**    When a country imports a greater value of goods than it exports, it has a **negative balance of trade.** Countries that have a negative balance of trade often finance their trade deficit by borrowing from lenders in other nations. This practice is common for many developing nations.

Loans to developing nations are most often made in the currencies of developed nations—U.S. banks loan dollars, British banks loan pounds, German banks loan euros, and so on. When these loans come due, they must be repaid in the same currency in which they were originally borrowed. Debtor nations, therefore, must have a way to earn these foreign currencies to repay their loans. When the prices of their export goods fall, it can become difficult or impossible for them to make their payments. On several occasions, debtor nations have experienced such economic problems that they were unable to repay their loans without assistance from developed nations.

*The Case of Thailand*    In 1997, for example, Thailand was unable to make payments on the billions of dollars it had borrowed. The problem started when the Thai stock market fell by more than 50 percent in just a few months. As foreigners withdrew their funds from Thai investments, the value of the Thai currency (the baht) declined rapidly. When Thai manufacturers sold products to other nations, they received fewer dollars, pounds, and the like for the products they sold. It became progressively more and more difficult for Thailand's businesses to earn the funds they needed to pay their debts. Ultimately, the Thai economy was rescued by a multibillion-dollar loan from the International Monetary Fund (IMF), which required the Thai government to impose economic restrictions that created hardships for most of the country's population. Similar events have taken place in Brazil, Argentina, and Indonesia in recent years.

**Negative Balance of Trade**    The condition when the value of the goods and services a country imports exceeds the value of its exports.

**Question for Thought & Discussion #3:**
Do you know anyone who believes he or she lost a job because of foreign competition? Would you be willing to pay higher prices to help this person regain employment?

*Why Should Americans Care?*   Economic difficulties in other nations are important to U.S. consumers because these nations export many products to the United States. They are also major customers for products we produce. Furthermore, many U.S. banks could be seriously harmed if these nations fail to repay their loans.

**DANGERS OF INTERNATIONAL DEPENDENCY**   The growth of international trade has created a form of international dependency among nations. Countries that have become dependent on imports from other nations are exposed to the possibility of economic difficulties if trade is interrupted for any reason. A situation that developed in Venezuela at the end of 2002 clearly illustrates this danger.

*Venezuela's Problems*   In November 2002, Venezuela exported 2.7 million barrels of oil each day. A large part of these exports flowed to the United States, accounting for nearly 15 percent of our petroleum imports. On December 2, 2002, managers and workers of the government-owned Venezuelan oil company went on strike to demonstrate their anger over restrictions imposed on the firm by the government of President Hugo Chavez. This strike, which lasted well into 2003, reduced the flow of oil from Venezuela by more than 80 percent. Reduced oil exports resulted in a corresponding reduction in Venezuela's earnings of foreign currencies that it needed to make payments on its $16 billion foreign debt. Furthermore, the political unrest in Venezuela caused banks in developed nations to withhold new loans that the country needed to maintain its services and prevent economic chaos. U.S. firms that had intended to sell products to people and businesses in Venezuela found that their customers were unable to pay for the goods and services the firms had expected to sell.

*Venezuela's Problems Are American Problems Too*   For U.S. consumers, Venezuela's political and economic problems were one contributing factor to an increase in the price of gasoline of roughly $.40 per gallon during late 2002 and early 2003. The low prices of many imported goods benefit U.S. consumers but leave them vulnerable to interruptions in trade that can increase the prices they pay, cause shortages, and harm our economy. The more imported goods we purchase, the greater our dependency on a steady flow of goods and services among nations. International trade is not an unqualified boon for U.S. consumers.

## Being a Responsible Consumer in the Global Economy

Would you be willing to buy products made in the United States to help strengthen our economy and provide more jobs for Americans? That's the basic message communicated by labels that implore you to "Buy American." When you see these messages on products, do they make any difference in your purchase decision? Millions of Americans apparently do not heed these labels—they continue to buy imported clothing, electronic products, cameras, and other goods or services in seemingly ever-increasing volumes.

But suppose you really do want to buy American. How easy is it for you to act on your convictions? Buying American is not a simple task or even possible in many situations. The globalization of our economy has made it difficult to determine whether a product is an import or a domestic good or service.

**Question for Thought & Discussion #4:**
Would you support legislation that would force Americans to buy more fuel-efficient cars so we would be less dependent on imported oil? Explain your answer.

## Trade with Chile

Will Chile be the next nation to join the United States, Canada, and Mexico in the North American Free Trade Agreement (NAFTA)? And if it is, how will this development affect consumers in Chile and the United States?

In December 2002, President George W. Bush signed a free trade deal with Chile that established common standards in banking and copyright law. In addition, if ratified, the agreement will eliminate 85 percent of tariffs between the two nations as it is phased in over the next twelve years. This means that the prices of imported Chilean grapes, fish, wine, and copper products for U.S. consumers might eventually fall by as much as 20 percent. At the same time, consumers in Chile will be able to purchase American-made machinery, cars, and computers for significantly lower prices. Some observers see this agreement as the first step toward including Chile and eventually other South American nations in an expanded NAFTA

that could ultimately include all nations in the Western Hemisphere.

Not all Americans or Chileans support the agreement. Workers in Chilean automobile factories worry that they will have to compete with workers from the United States. And, if the agreement is extended to Mexico, they will have to compete for jobs with Mexican workers as well. At the same time, fruit growers in the United States have expressed opposition to the agreement. They point out that farm workers in Chile are paid much lower wages than farm workers in the United States. Chilean grapes already sell for relatively low prices in this country. If the tariff on Chilean grapes is eliminated, the growers argue, they may be forced out of business.

**If you were a U.S. senator, would you vote to approve this agreement? What factors would be most important in helping you make your decision?**

**WHAT IS AN AMERICAN CAR?**   Suppose you have decided to buy a new Ford because you believe it is an American product. Well, maybe you're right, but, then again, maybe you're not. Your Ford might have been assembled in the United States (more than half of Ford automobiles are), but it is guaranteed to contain parts that were manufactured in other countries that include Mexico, Canada, Japan, and Taiwan, to name only a few. Further, a significant proportion of the Ford vehicles sold in the United States are assembled in Canada or Mexico, in part due to the removal of trade restrictions by the North American Free Trade Agreement (NAFTA), which was signed in 1993.

Maybe you decide that your Ford is an American product because Ford is an American corporation. Again, the accuracy of your belief is not clear. Although the Ford Motor Company was originally organized in the United States, it has become a *multinational corporation*. Ford factories can be found in nations all over the world. How accurate is it to state that Ford—or almost any other large corporation—belongs to any single nation? It's hard to say.

**GLOBALIZATION AFFECTS ALL CONSUMERS**   American consumers sometimes forget that the globalization of the world's economy affects consumers in all nations. Recent trade agreements between the United States and Cuba have allowed Cuban citizens to purchase apples grown in western New York. Japanese consumers cook rice that was grown in California. Mexican graphic artists use IBM computers. European power producers use gas turbines manufactured in Pennsylvania. The list goes on and on. If it's hard for American consumers to buy American, it is no less difficult for foreign consumers to buy products made exclusively in their own nations.

Many consumers across the world have apparently given up shopping for their own domestic goods and have chosen to look for products that offer the best quality at the lowest price regardless of where they were manufactured. As globalization of the world's economy progresses, this perspective toward spending decisions is likely to become more the rule than the exception.

**Question for Thought & Discussion #5:**
How might future consumers benefit from a total globalization of the world economy? What costs might they pay?

**WILL THERE BE A SINGLE WORLD ECONOMY?** Some people believe that at some time in the future, the economic boundaries that separate nations may disappear entirely. You can see the beginnings of this trend in the signing of NAFTA and the creation of the European Union (EU). These agreements have brought about reduced tariffs, the standardization of some products, and an increased flow of goods and services among nations. The creation and implementation of NAFTA and the EU have not taken place without discord. Many people oppose them for economic, political, or nationalistic reasons. But for better or worse, these agreements have been made, and change is taking place. Some people believe that by the turn of the next century the notion of individual national economies may have become obsolete. If all businesses and consumers come to think only in terms of a world economy, if barriers to trade disappear, and if national currencies are replaced by a single currency (this has already taken place within most of the EU countries), globalization of the economy will be complete.

# The Technological Assault on Privacy

What personal information do you provide when you shop the Internet? Do you worry about how this information may be used?

New technologies have brought consumers a long list of benefits, but these developments also have a darker side. Take your right to personal privacy as an example. Although the U.S. Constitution does not specifically guarantee a right to privacy, the United States Supreme Court has ruled that such a right is implied under various constitutional provisions, such as the Fourth Amendment, which guards against unreasonable searches and seizures. The courts have ruled, however, that it is legal for the government and private organizations to gather and "use appropriately" any information about individuals in which they have a "legitimate interest." The problem is that we have not been able to determine exactly what the terms *legitimate interest* and *appropriate use* mean. Furthermore, more advanced technologies are continually enhancing our ability to gather and use information about individuals. Consider the following situations that already exist.

## Privacy in the Workplace

In recent years many employers have joined Electronic Banking Systems, Inc. (EBS), in monitoring the work of their employees. EBS is a firm that processes donations from private individuals to charitable causes. The firm's workers sit at computer terminals and enter data about contributors and their contributions. Names, addresses, phone numbers, and amounts given are keyed into a massive database that can be accessed, reconfigured, and sold to other organizations for their own uses. At EBS row after row of workers occupy small desks in a room with no windows. They are not allowed to have anything on their desks except their work—no pictures, no flowers, no signs, nothing to indicate their individuality or even humanity. Each worker is given a quota of work to meet. Workers must open and process an average of three pieces of mail each minute, making 8,500 keystrokes per hour. The computers they use count their keystrokes and are programmed to allow managers to replay a copy of every entry they make. This makes it impossible for them to make mistakes without risking discovery. And they are watched through closed-circuit television while they work.

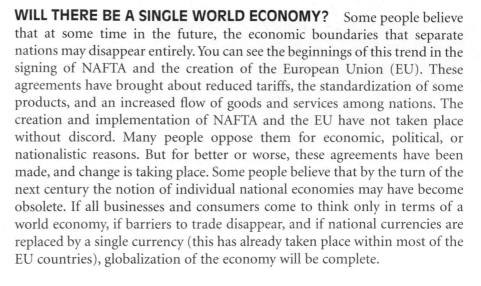

EBS is not alone. Have you ever called a business or help line only to hear a message that sounds something like "Calls may be monitored to assure the quality of customer service" before you are allowed to talk to a person? If you have, you can be sure that the business is keeping a close watch over its employees.

**EMPLOYER VERSUS EMPLOYEE RIGHTS**   At least two larger issues should concern you here. First, should employees have a right to privacy while they are working? Do you believe your employer should be allowed to watch your every move while you work? At what point does the employer's right to expect an honest day's work for a day's pay leave off and the right of individuals to have a degree of personal privacy begin? How much should a $10-per-hour wage buy an employer?

**CUSTOMER AND CONTRIBUTOR RIGHTS**   The other larger question concerns the privacy rights of customers or, in the case of EBS, contributors to charitable causes. It is one thing to send $10 or $20 to your favorite charity—it is something else entirely to realize that personal information about you and the gift you make will become available to many other organizations, some of them charitable, some of them not. In reaction to contributor protests, many charitable organizations have stopped selling information about the people who make donations. Unfortunately, quite a few organizations and businesses still gather and sell this type of information.

Federal law now requires some businesses, such as financial intermediaries, to publish their privacy policies and to make them available to all customers. Many other firms do this voluntarily. The next time you shop online, click on the privacy policy link at the business's Web site. It should explain how information about you and your purchases will be used. It is probably a good idea to avoid shopping at any Web site that does not include its privacy policy. You might also consider steering clear of businesses that have privacy policies you don't like.

> **Question for Thought & Discussion #6:**
> What costs and benefits do you experience because organizations gather, store, and distribute information about you?

## Privacy and Your Health

Obviously, information about your health needs to be gathered, stored, and made available to your health-care providers and other parties who have a legitimate reason to know about your medical conditions. You would not want to have to provide your complete medical history every time you visited a doctor's office or hospital. Insurance companies that provide you with medical coverage need to know what treatments you have received to keep you healthy. If they don't, they can't pay the medical professionals who furnish your care. And you want these professionals to share their medical information about you. Without sharing, one of your doctors could prescribe a drug that would cause a dangerous interaction with another drug you take that was prescribed by a different doctor. Such an interaction could harm your health or even endanger your life.

**HOW MUCH KNOWLEDGE IS TOO MUCH KNOWLEDGE?**   The question is, at what point does a legitimate interest in your health leave off and an invasion of your privacy begin? Would you want your employer, bank, or even your neighbors to know that you had a serious illness? What might this knowledge do to your chance of being promoted at work or your ability to

## CONSUMER CLOSE-UP
# When Your Boss Reads Your E-Mail

Employees have probably been telling jokes of questionable taste since humans moved out of a cave in prehistoric times and started to work together. Until quite recently, most "dirty" jokes were probably told in changing rooms, during coffee breaks, or around the water cooler. No records were kept, only those present heard the jokes, and although they may have been crude, probably few people were offended. Then came office e-mail. Some employees started to use their computers to send examples of doubtful humor to their friends and to other office workers who had no interest in their jokes.

A few years ago, for example, workers at the New York Times Company's office in Norfolk, Virginia, circulated jokes in an effort to develop a list of the ten jokes demonstrating the worst taste. Managers discovered the list by accident when they were reviewing employee e-mails as part of an investigation of a false claim for unemployment insurance. The Times had no consistent policy of reviewing employee e-mails, but the capability was there whenever management chose to look. Thus, the company had an electronic record of all employees who had participated in the "fun." On discovering the e-mails, the newspaper fired twenty-two employees for misuse of company equipment. Most of these workers had previously been in good standing with their employer. One had just received the "Employee of the Quarter" award.

**Do you believe the Times was right to fire these workers? Or did the company invade their privacy? Is telling jokes of questionable taste worse on e-mail than when standing around a water cooler? What does this example show about workers' need to be aware of the capabilities of electronic communications and how they may diminish individual privacy?**

obtain a mortgage to buy a new home? Would a bank's employees be willing to lend you funds if they knew you were ill? What if you had a communicable disease? Might the parents in your neighborhood refuse to allow their children to play with your son or daughter because of your illness? And, even if it made no difference to other people, is the status of your health anyone's business but your own?

**SHOULD I BE TESTED?**   Breakthroughs in medical research have made amazing strides in recent years. Tests are now available that can detect which people are likely to fall victim to a variety of diseases in their later years. Such tests can be used for breast and uterine cancer, Alzheimer's disease, diabetes, and a number of other serious illnesses. You might think, "How wonderful. Now people can be tested for these diseases when they are young and be treated early to either prevent or slow the progress of these illnesses in their later lives." But it's not that simple.

Suppose that your grandmother suffered from Alzheimer's disease in her old age. You worry that you might eventually suffer from it too, so you have yourself tested. When the results come back, they show that there is a high probability that you too will be afflicted with this dread disease. It is bad enough for you to learn about your future, but if your employer or medical insurer finds out, things could become much worse. Maybe you're thirty years old and a successful middle manager. You expect to be promoted regularly and earn a good income. Would your employer promote you or give you important responsibilities if she found out about your medical situation? Many people who might benefit from being tested have chosen not to because they fear the consequences of a negative report. What rights to privacy should people who are tested be given?

# Privacy and the Government

The federal government gathers information about consumers, but it has also attempted to place some limitations on who can see this information and how it may be used. Although laws have been passed that are intended to protect individual privacy (see Table 19.1), in many ways these efforts have not been particularly successful. These laws give people access to federal government records about them and enable them to find out how this information has been used. The laws, however, do not give individuals control over what is done with the information, the ability to change data they believe are wrong or misleading, or the right to delete information they feel is incomplete or no one's business but their own. Table 19.2 on the following page lists some of the many government agencies that collect data about people and provides some indication of how this information is used.

## OBTAINING INFORMATION FROM THE FEDERAL GOVERNMENT

Under the Freedom of Information Act of 1966 and the Privacy Act of 1974, you have a right to gain access to any information about you that has been collected by the federal government, with the exception of information that concerns national security. To obtain a copy of such information, all you need to do is write to the government agency and state, "Under provisions of the Privacy

| TABLE 19.1 | Federal Legislation Relating to Privacy |
|---|---|
| Freedom of Information Act (1900) | Provides that individuals have a right to obtain access to information about them collected in government files. |
| Fair Credit Reporting Act (1970) | Provides that consumers have the right to be informed of the nature and scope of a credit investigation, the kind of information that is being compiled, and the names of the firms or individuals who will be receiving the report. |
| Crime Control Act (1973) | Safeguards the confidentiality of information amassed for certain state criminal systems. |
| Family Educational Rights and Privacy Act (1974) | Limits access to computer-stored records of education-related evaluations and grades in private and public colleges and universities. |
| Privacy Act (1974) | Protects the privacy of individuals about whom the federal government has information. Specifically, the act provides that: 1. Agencies originating, using, disclosing, or otherwise manipulating personal information must ensure the reliability of the information and provide safeguards against its misuse. 2. Information compiled for one purpose cannot be used for another without the concerned individual's permission. 3. Individuals must be able to find out what data concerning them are being compiled and how the data will be used. 4. Individuals must be given a means through which to correct inaccurate data. |
| Tax Reform Act (1976) | Preserves the privacy of personal financial information. |
| Right to Financial Privacy Act (1978) | Prohibits financial institutions from providing the federal government with access to a customer's records unless the customer authorizes the disclosure. |
| Electronic Fund Transfer Act (1978) | Requires financial institutions to notify an individual if a third party gains access to the individual's account. |
| Counterfeit Access Device and Computer Fraud and Abuse Act (1984) | Prohibits use of a computer without authorization to retrieve data in a financial institution's or credit reporting agency's files. |
| Cable Communications Policy Act (1984) | Regulates access to information collected by cable service operators on subscribers to cable services. |
| Electronic Communications Privacy Act (1986) | Prohibits the interception of information communicated by electronic means. |

## TABLE 19.2    Personal Data Collected by Government Agencies

**Census Bureau**

The Census Bureau collects data every ten years on the U.S. population. Information collected includes the age, address, gender, ethnic origin or race, marital status, income, and place of employment of every person living in every household.

**Federal Bureau of Investigation (FBI)**

The FBI has a file on anyone who has ever been employed by the federal government, been in the military, or attempted to obtain a security clearance. The FBI also keeps files on some other individuals—for example, those who subscribe to publications by, contribute to, or are a member of certain organizations and causes.

**Social Security Administration (SSA)**

If you have ever been an employee and received wages, the Social Security Administration will have a record of it. In other words, your entire employment record—where you have worked and when—is contained in these records. Additionally, the Social Security Administration has information concerning the income, indebtedness, marital status, medical history, and household arrangements of anyone who has applied for Social Security benefits, Medicare or Medicaid, or Temporary Assistance for Needy Families (TANF).

**Internal Revenue Service (IRS)**

The IRS has information concerning age, gender, marital status, and all sources of income—jobs, checking and saving accounts, investments, and so on.

**Veterans Administration (VA)**

The VA files contain information on all persons who have served in the military. This includes not only military records but also information concerning jobs held, medical histories, education, and residences prior to entering the service. Additional information will be entered if a veteran is receiving service-related benefits, such as a VA mortgage loan, educational assistance, or medical care.

**Federal Housing Administration (FHA)**

The FHA has information on all persons applying for federal housing assistance.

**Selective Service System**

This agency has information on draft registration.

**Defense Department**

The Defense Department has information on anyone who has sought a security clearance.

**Securities and Exchange Commission (SEC)**

The SEC has information on anyone involved in the sale of securities.

**Department of Homeland Security (DHS)**

The DHS has information about people who are suspected of being associated with terrorist activities.

---

Act of 1974, 5 U.S.C. Section 522a, I hereby request all information concerning myself in your files." Send your letter to the relevant agency and address it to the attention of the "Freedom of Information/Privacy Acts Officer"—every federal agency has one. Be sure to include any information that will help the government personnel locate your file and include the address to which you wish it to be sent.

**INFORMATION GATHERED BY STATES**    You may have noticed that all of the laws in Table 19.1 were passed by the federal government. Unfortunately, many state and local governments also collect information about individual consumers and are not restrained in what they do with that information by these or other laws. Some states, for example, sell information they gather or provide it free of charge to private businesses. Some of this provision of information is legitimate. State court systems, for example, generally provide information concerning driving violation convictions to automobile insurance companies. These insurance companies have a legitimate need for this type of information about individual drivers. After all, most people agree that drivers who break traffic laws should pay higher insurance premiums than those who don't.

A number of other situations also appear to be reasonable examples of states providing information about individuals. People who are delinquent in paying their state or local taxes should expect a record of this fact to appear in their credit histories. Purchases of real estate, property taxes levied, and marriages or divorces are all considered matters of public record. But at some point, state and local governments cross the line and invade what many people regard as private information. Some states have sold information about automobile registrations collected by their Bureaus of Motor Vehicles to private marketing firms. Others have sold lists of people who have purchased hunting or fishing licenses. This type of information is valuable to businesses that want to market outdoors equipment. The question remains, should governments distribute this information even if they are paid to do so?

## Privacy in Your Home

Have you ever been called by a television rating service that asked about the shows you watch? Maybe you agreed to answer the questions, and maybe you didn't. But, whatever you did, it was your choice. This type of personal privacy is no longer guaranteed. People who subscribe to cable television services are often "surveyed" without their knowledge. Technology is being developed that will allow these firms to know exactly what programs you watch and when you watch them. Again, this is very important information to marketers. If you spend ten hours every week watching either boxing or wrestling, you are likely to receive different advertising solicitations in the mail than someone who spends every afternoon watching the "soaps." Similarly, the Web sites you visit on the Internet are tracked very closely. Many Web sites gather information about customers and sell it to other firms. For some sites, this is their primary source of income.

At first glance the situations described here may not seem to represent an important problem. But they have contributed to a gradual erosion of our personal right to privacy. As time goes by and technology becomes more advanced, we are likely to see this trend continue. Consumers need to stay aware of how information about them is gathered and distributed. You may not be able to do much to prevent this from happening, but you should at least know that it is going on.

**Question for Thought & Discussion #7:**
If you had the power, what restrictions would you place on how information about individuals could be collected and used? How would you enforce these restrictions?

# Ethical Choices and the Cost of Medical Technology

In Chapter 16 you learned that Americans spend more than $1.4 trillion every year for medical care. A large part of this spending is used to support research that increases our ability to treat and cure diseases or to improve and extend the lives of those who cannot be cured. The accomplishments of recent years are truly astounding, and there is every indication that even greater advances in medical technology will be made in the future. Just stop to think about the medical procedures we now take for granted.

## The Growing List of Medical Procedures and Treatments

It was only thirty-five years ago that Dr. Michael DeBakey performed the first human heart transplant surgery in Houston, Texas. In 1999, more than 2,200 of these procedures were completed in the United States. In the same year, 4,586 liver transplants and 12,400 kidney transplants were carried out. In fact, in 1999 nearly 24 million surgical procedures that required the patient

> ❝The more than $1.4 trillion that we spend for medical care each year works out to about $4,600 for every man, woman, and child in the United States.❞

to stay for at least one night in a hospital or other health-care facility were completed in the United States.

Not only have procedures that were once considered remarkable become commonplace, but new procedures and treatments are developed almost every day. In 2001, the cutting edge of medical technology saw the development of an artificial heart that kept a terminally ill patient alive for more than a year. Progress was made toward growing new organs or "spare parts" from human cells that are genetically identical to an individual's own DNA. When this process is perfected, patients who receive these organs will not need to use antirejection drugs because their bodies will not recognize that the new organs are different from their own. In 1999, a video camera was wired into the brain of a blind man that allowed him to see shapes for the first time in his life. Rapid advances in the field of neurology are giving people who have suffered spinal cord injuries, such as actor Christopher Reeve, hope that they may someday walk again. Researchers have found that interferon treatments are able to arrest the progress of multiple sclerosis. The list goes on and on. You may wonder why such advances would involve any degree of consumer choice. Many Americans would say that if we have the technological capability to cure or control a disease, we should do so. But the issue is not that simple.

## The Costs of Medical Technology

The minimum cost of a heart transplant today exceeds $120,000. Liver, lung, or kidney transplants can be even more expensive. Under the best of circumstances, a bone marrow transplant will cost $100,000, and if it takes a long time to find a suitable donor, the cost can be much greater. In 2003, the cost of hospitalization for U.S. consumers exceeded $400 billion without including the associated costs of doctors' fees or prescription drugs.

**WHO PAYS FOR MEDICAL CARE?**   The more than $1.4 trillion that we spend for medical care each year works out to about $4,600 for every man, woman, and child in the United States. Where does this money come from, or more specifically, who pays the bill? On the surface, the federal and state governments pay the largest portion of our medical costs. In 2000, total government spending for medical care reached $589 billion. Private insurance was a close second, contributing $438 billion to the bill. Private individuals and charitable organizations picked up the remaining $283 billion.

But where do the government, medical insurers, and charitable organizations get the money they use to pay medical bills? They get it from the people through taxes, premiums, or contributions. And even when employers purchase medical coverage for employees, it can be argued that the employers pay these costs by reducing the wages that they otherwise would have paid their workers. In the end, it appears that consumers pay for all medical care provided in the United States.

**MEDICAL-CARE SPENDING HAS NOWHERE TO GO BUT UP**   The U.S. population is rapidly aging. Remember that there was a "baby boom" in the years that followed World War II. These people will be in their sixties and seventies soon after the year 2010. Examine Table 19.3 to see how the population of the United States is aging. It is reasonable to believe that as people age, they will want to stay as healthy as possible and will demand as much medical care as the nation or they can afford.

| TABLE 19.3 | Population Age Distribution, 1980, 2000, and 2025 | | |
|---|---|---|---|
| Age Group | Percentage in 1980 | Percentage in 2000 | Percentage in 2025* |
| 0–17 | 24.7% | 19.0% | 17.8% |
| 18–34 | 32.1 | 16.3 | 14.4 |
| 35–54 | 21.5 | 29.9 | 25.6 |
| 55–74 | 17.5 | 28.6 | 34.4 |
| 75+ | 4.4 | 6.1 | 7.9 |

*Projected

SOURCE: *Statistical Abstract of the United States, 2000*, pp. 21, 25.

# Who Should Receive Medical Care?

In 2003, the United States devoted nearly 14 percent of its GDP (total production) to providing medical care or funding medical research. If current trends in medical spending continue into the future, this percentage will reach nearly 16 percent in 2010 and 18 percent in 2020. At some point in time, we will be forced to answer a basic question: How much of our scarce resources are we willing or able to allocate to providing medical care? The corollary to this question is: If we cannot provide all people with every possible medical treatment, how should we decide who receives which treatments? These questions present an unavoidable ethical dilemma for all U.S. consumers.

**MONEY ISN'T THE ONLY PROBLEM**  Many people wonder where the funds to provide medical care will be found. But it isn't just money that we must consider. Our knowledge of scarcity tells us that resources used to develop and provide medical treatments are resources that cannot be used in other ways. Where will we find the nurses, doctors, and other skilled caregivers we need to provide adequate levels of medical care?

More than 40 million Americans have no medical insurance. Tens of thousands of our nation's children have not had appropriate vaccinations. Many rural communities lack doctors and/or hospital facilities. The urban poor often do not receive regular medical care but must visit overcrowded emergency rooms when they become ill. As bad as this situation may seem, it is likely to become much worse in the future.

**THE ETHICS OF ALLOCATING MEDICAL CARE**  Imagine that you have been put in charge of our nation's medical-care delivery system. More than 285 million Americans demand the services that you provide. Some are young, some are old, and some are terminally ill. Many can have long and productive lives if they receive the proper treatments. A few are wealthy and can pay for any medical care they might want. Many more can barely purchase a bottle of aspirin. Every morning you read about new medical treatments that have been developed to improve people's health (if they can pay for them). You have a limited budget. You can't provide every possible treatment to every potential patient. It's up to you! How will you decide?

Now that's a pretty frightening situation to imagine, isn't it? You might think, "Yeah, but it's never going to happen." In a way you would be right, but in a more important way you would be wrong. Individually, no single

**Question for Thought & Discussion #8:**
What part of your income would you be willing to allocate to paying for medical care? For you, how much is too much?

American will ever be totally in charge of our nation's medical-care delivery system. In a broader sense, however, we all face this dilemma, and there is nothing we can do to avoid it. As a society, we will decide how to allocate our scarce medical-care resources. The more you learn about making rational choices, the better equipped you will be to participate in making these decisions.

## New Technologies—New Ethical Dilemmas

On a more personal level, medical procedures are being developed that may one day allow parents (if they choose to do so) to select the gender, height, complexion, or personality traits of their children before they are conceived. What knowledge and values would you need to call on to make such a choice? And what about cloning children? The technology to accomplish this exists and has been used to reproduce animals with specific desired traits. Should humans be cloned under any circumstances? This question is sure to surface in many heated debates throughout the world in the coming years. What are the moral and ethical issues in replicating life by cloning humans? What authority should make this decision? The U.S. Congress has outlawed this procedure in our country, but it may well be accomplished in other nations. Unconfirmed announcements that the first cloned humans had been born were made early in 2003. No one can say for sure how this issue will finally be resolved.

It is clear that medical technology is likely to advance more rapidly in the future than our ability to evaluate its implications for society. Furthermore, for every change we can anticipate, there are probably hundreds more that we cannot predict but that will affect our lives.

# Protecting the Environment Is Everyone's Responsibility

In recent years, consumers have become increasingly aware of the threat human activity presents to the world's environment. Your environment extends beyond the neighborhood or community where you live, work, and play. It includes the atmosphere that covers the earth, the oceans' plankton that produces most of the oxygen we breathe, the water in the Colorado River that is used to irrigate farms in California, and the soil in brown fields that once were the home of U.S. industries. Damage to the environment can be detrimental to your health and, according to some, may endanger human existence.

You have a voice in environmental decisions. Your consumer choices can either encourage or discourage a business's efforts to protect the environment. The same may be said for the way you vote or communicate with political leaders. Ultimately, our government is responsive to the wishes of the people. If enough people demand legislation to protect the environment, it will be passed into law. In addition, your personal lifestyle can contribute to environmental problems or diminish them. The future of environmental protection depends on the choices we all make.

What happens to the trash you discard each day? If you don't know, perhaps you ought to find out.

## The Scope of the Problem

If you are a typical U.S. consumer, you will create 4.6 pounds of trash today. Do you wonder where your trash goes? The problem of disposing of household waste is, of course, only the tip of the iceberg, so to speak. It is easy to recognize that space for trash in our landfills is limited, at least in areas with high population concentrations. In 2002, the city of New York began to ship much of its trash to landfills in western New York where more space was available. These shipments were not met with enthusiasm by local residents, and they almost doubled New York City's cost of garbage disposal.

Other environmental challenges are less obvious but no less important. Consider the following facts and how they affect consumers:

- In 1999, Americans produced over 1,200 million pounds of trash each day. Of this amount, 58 percent was discarded in landfills, 13 percent was incincrated, and 29 percent was recycled.
- In 1998, 96 percent of the Great Lakes shoreline was officially designated as polluted by the federal government.
- In 2000, the federal government designated 1,292 locations as hazardous waste sites that had not yet been cleaned. Many other sites existed but had not received federal designation.
- In 1999, Americans released 1.6 billion tons of carbon dioxide and 29 million tons of methane gas into the atmosphere. These gases are said to contribute to global warming.
- About 16 tons of sewage are dumped into our nation's waterways every minute of every day.
- In 2001, the federal government identified 1,488 plant and animal species as endangered.

## The Economics of Pollution

Much of the damage done to the world's environment takes place during the production or consumption of goods and services that satisfy human wants. Indeed, *pollution, production,* and *profit* are three concepts that are often linked together. In the United States and most other nations, businesses exist to earn profits for their owners. Even in nations that have government-owned enterprises, there is still an incentive to produce goods and services at the lowest possible cost. Unfortunately, low costs to producers often translate into increased costs and pollution to the environment.

## Who Pays the Costs of Pollution Control?

It is not easy to determine who pays for pollution abatement. At first glance it might appear that these costs are borne by the businesses that are required to install expensive equipment to reduce their emissions of noxious substances. After consideration, you might change your mind and conclude that the firms' customers would pay because the businesses simply pass the costs of pollution abatement on in higher prices. But there are other important considerations. All people, including a firm's customers, suffer from pollution of our environment. If pollution abatement eliminates or reduces these costs of pollution, possibly there is no added cost to society as a whole.

> "Much of the damage done to the world's environment takes place during the production or consumption of goods and services that satisfy human wants."

**A SPECIFIC EXAMPLE**    Consider the paper industry as an example. Manufacturing paper requires the use of many chemicals, including acids and bleach. After these chemicals are used, they must be eliminated in some way. At one time paper manufacturers typically dumped these chemicals into the nearest body of water, preferably a river or stream that would carry them a long distance from the manufacturing facility. This was by far the least expensive and most profitable way to dispose of the unwanted chemicals.

*Paper Producers Clean Up Their Acts*    Starting in the 1950s, states and then the federal government passed laws that required paper manufacturers to treat these chemicals to reduce their toxicity before they were discarded into the environment. Paper manufacturers were essentially left with two alternatives: install pollution-abatement equipment at a cost of millions of dollars per plant or close down. Many smaller paper mills did close. Others merged with larger firms that could better afford the equipment. By the mid-1960s the level of pollutants being discharged from paper mills in the United States had been reduced by roughly 80 percent. Economists who studied this situation concluded that almost all of the cost of pollution abatement was passed on to consumers in higher paper prices.

*Pollution Abatement Brings Many Benefits*    The higher price of paper products, however, is not the only result of these pollution-abatement laws. After a few years, the waterways that had been used as sewers began to recover. Fish and other wildlife returned to previously dead and toxic waters. The property value of land along these rivers and streams increased, and the incidence of illnesses among people living in communities downstream from paper mills declined. In the end it is difficult to say whether the costs or the benefits of these laws were greater. It is certain that the environment of many communities has benefited from their passage and enforcement.

**SHOULD THE GOVERNMENT PAY?**    Consider the environmental disaster that took place at Love Canal in 1978. From the 1940s though the 1960s, the Department of Defense and many private chemical manufacturers dumped dioxin and other dangerous chemicals into a giant clay-lined trench in upstate New York, less than one-quarter of a mile from the Niagara River. When it was discovered that these chemicals were leaking from the trench into the nearby river and endangering the health of millions of people, the U.S. government decided to act. It passed the Comprehensive Environmental Response Compensation and Liability Act of 1980, better known as the Superfund Act. Funds provided through this law were used to clean up Love Canal.

*Should Businesses Pay?*    You might wonder why the government didn't just force the businesses that had discarded chemicals into Love Canal to pay. This was impossible for several reasons. In the first place, many of the firms had failed or had no assets that could be used to pay for the clean-up. Firms that were still in business resisted pressures to pay. They argued in court that the disposal of chemicals into Love Canal had been totally legal at the time it took place and that much of this work had been completed under government contract. Why should they be held responsible for a failure of the government? Ultimately, they did agree to pay a portion of the cost, but the case took almost ten years to work its way through the court system.

*A Need for Quick Action*    If the government had not been willing to step in quickly, the chemicals would have continued to seep into the environment for many additional years. Whether the government (and therefore the taxpayers) should pay for environmental protection may be beside the point. In many cases, this appears to be the only way for environmental protection to be carried out.

## Environmental Legislation

Our experiences from the first half of the twentieth century clearly demonstrate that businesses are not likely to undertake environmental protection in any serious way unless they receive encouragement and leadership from the government. Since the 1950s, a series of federal and state laws have been passed that have gone a long way toward protecting our environment. Many people believe that the government could do much more.

**THE ENVIRONMENTAL PROTECTION AGENCY**    The Environmental Protection Agency (EPA) was created by an act of Congress in 1970. The agency was charged with administering the laws that protect the nation's land, air, and water resources. Included in its mandate are standards for water quality, solid and hazardous waste disposal, radiation, and air pollution. The EPA also regulates the use of pesticides and fertilizers for home use, landscaping, and farming.

In addition to enforcing regulations, the EPA provides technical assistance to firms that need help in meeting pollution-control standards. The EPA has offices in most cities across the United States. Its objective is to strike a balance between our need to protect the environment and our desire to produce goods and services that provide employment and satisfy our consumer needs.

**THE CLEAN AIR ACT OF 1990**    In 1990, Congress passed a comprehensive revision of the original Clean Air Act of 1970. This legislation established standards for car emissions that significantly exceeded those set in the 1970 law and required the use of cleaner-burning fuels in many urban areas. It attacked the problem of acid rain by forcing utilities to reduce smokestack emissions and required other industries to reduce the amount of cancer-causing compounds they release into the atmosphere. The law also gave the EPA the power to periodically reassess and change emission standards over time.

In 1999, the EPA announced that it would strengthen motor vehicle emission-control standards beginning in 2004. Under this plan the amount of sulfur compounds allowable in gasoline will be reduced by 70 percent from the 1999 standards by the year 2007. Furthermore, small trucks and sport utility vehicles (SUVs) would be included in mileage standards. The EPA estimated that the cost of the higher standards would include an increase in the cost of gasoline of roughly two cents per gallon and an increase in the price of new vehicles that would range from $100 to $200 each. Industry spokespeople dispute these estimates, arguing that the costs will be much greater.

In 1999, some business and political leaders expressed doubt that the benefits of the new standards would exceed their costs and suggested lowering the standards. Pointing out that it is impossible to eliminate all pollution, they insisted that the higher standards set by the EPA would increase the price of many consumer goods. Consumers were asked to consider whether the

**Question for Thought & Discussion #9:**
What role do you believe the government should play in environmental protection?

Would you be willing to pay a higher price for energy that is generated in an environmentally responsible way?

improved environmental protection was worth the costs they would have to pay. Recent events have given the opponents of the new standards most of what they wanted.

**RECENT CHANGES IN EPA ENFORCEMENT**  Although Congress writes the laws, the administrative agencies are in charge of enforcing them. In the case of the Clean Air Act, Congress gave the EPA a wide range of enforcement powers that it can use without further congressional approval. In November 2002, the EPA made sweeping changes in its clean-air rules that some have called an "assault on the air we breathe."

*New Rules for Power Companies*  Prior to the rule changes announced in 2002, power generators were required to measure their reduction in emissions against the amounts of pollutants they released into the atmosphere in the most recent two-year period. The new rules allow them to compare their current emissions with the worst two-year period within the past ten years. Effectively, this change allows power generators to count any improvements made in the past ten years as new improvements.

In the future, pollution rates will be averaged for an entire electricity-generating facility rather than for each individual boiler and generator as in the past. This means that older, highly polluting boilers may remain online if there are newer, low-polluting boilers at the same facility. Finally, power generators are allowed a ten-year exemption from the EPA rules if they can show that they have installed pollution-control equipment within the past fifteen years that was state-of-the-art equipment when it was installed. Thus, equipment that was placed in service in 1988 could prevent a firm from being required to install new equipment in 2003.

*Weighing the Costs and Benefits of the New Rules*  These rule changes clearly benefit U.S. businesses and power generators by lowering the cost of electric power. Consumers will benefit from these lower expenses as well. There is a question, however, as to whether the costs of these decisions to the environment will outweigh their expected benefits.

## State and Local Government Efforts

Environmental protection is an issue for state and local governments as much as for the federal government. In fact, state and local governments have often led the way in passing and enforcing laws designed to protect our environment.

**CALIFORNIA MOVES BEYOND THE CLEAN AIR ACT**  In 2000, the California legislature moved to make its automobile emission standards much stricter than the federal regulations. Under a 1991 California law, cars sold in that state were required by 2003 to meet hydrocarbon and nitrogen oxide emission standards that were 50 percent lower than the federal regulations. The 2000 California law lowered these standards another 40 percent.

As another example of state activism, after the EPA suggested expanding emission standards to include small trucks and SUVs in 1999, California took the rules one step further. In May 2002, California passed legislation designed to reduce the emissions from trucks waiting to be loaded or unloaded at California's docks and distribution centers. These emissions have been a growing problem in California and other states as hundreds of trucks idle for hours before they can be loaded and start their deliveries. The California law took effect on January 1, 2003. It imposes a fine of $250 for each truck that sits with its engine running for more than thirty minutes. It is expected that many other states will follow California's lead and pass similar laws in the future.

**STATES PROTECT WATER QUALITY**   Although there is federal clean-water legislation, many states have acted to protect the quality of their water resources by passing special legislation to meet their particular needs. The state of Maryland, for example, passed a law in 2000 designed to protect ground-water, watersheds, and the Chesapeake Bay from pollution from private septic systems. At that time many homes and businesses in rural Maryland were not tied into community sewage systems but relied on septic tanks and tile fields to dispose of human waste. Many of these private systems leaked pollutants into surrounding waterways. This caused a rapid growth of algae and other plants that clogged waterways and endangered fish and marine animals. The cost of retrofitting existing septic systems was estimated to be from $3,000 to $10,000 per home. Many of Maryland's residents and builders argued that they could not afford this expense. In response, Maryland's legislature decided to allow these costs to be applied as credits on state income tax. Similar legislation has been passed in Pennsylvania and Indiana.

**NEW YORK'S TOUGH REGULATIONS FOR SOLID WASTE DISPOSAL AND RECYCLING**   In 1988, the New York State Commissioner of Environmental Conservation announced a new set of regulations for the disposal of solid waste that were regarded as the toughest in the nation at that time. These regulations included rules for the disposal of toxic ash that results from burning trash, operating landfills, and maintaining hazardous waste–disposal sites. The commissioner also announced that every community in New York State would be required to institute recycling programs or face stiff fines from the state government. In the past fifteen years, these regulations have transformed New York from a state with one of the worst environmental protection records to a state with one of the best.

# Environmental Protection Is a Global Concern

Although individual nations, states, and communities can take steps to protect the environment, many problems require global action. We do not live in isolation. Acts that are destructive to the environment hurt us all, no matter where they take place or where we live.

**DAMAGE TO THE OZONE LAYER**   In 1991, the EPA announced that studies indicated that the level of ozone over the United States had fallen by an average of 4 to 5 percent in the years between 1978 and 1990. At that time, it was predicted that the total loss could reach as high as 20 percent by 2000.

There is a layer of ozone in the atmosphere at a height of thirty to sixty miles. This layer shields the earth from most of the sun's harmful ultraviolet rays. Experiments have shown that agricultural production drops by as much as 20 percent when plants are exposed to elevated levels of ultraviolet light. Greater concentrations of smog occur, and a higher incidence of cancer is likely to result.

Some scientists blame the destruction of the ozone layer on chemicals that have been released into the atmosphere by human-made products. The most

> **"Tougher regulations have transformed New York from a state with one of the worst environmental protection records to a state with one of the best."**

**Chlorofluorocarbons (CFCs)** A family of chemicals associated with the depletion of ozone in the earth's upper atmosphere.

**Greenhouse Effect** The gradual warming of the earth's atmosphere, primarily as a result of the release of carbon dioxide from the burning of fossil fuels.

important of these is a family of chemicals called **chlorofluorocarbons (CFCs)** that have been used in aerosol sprays and plastic foam and as refrigerants. Although many nations were slow to recognize the danger of CFCs, most countries had stopped production of these chemicals by 1995. The problem is that some CFCs are still produced in developing nations and continue to escape into the atmosphere. Although these chemicals may enter the air in only a few locations, they have the potential to destroy the ozone layer over the entire globe. In recent years developed nations have helped developing nations find alternative chemicals to use in place of CFCs. In 2000, scientists detected the first signs of regeneration in the ozone layer. It is too early to be sure that the problem has been overcome, but the initial signs are promising.

**THE GREENHOUSE EFFECT** It has been suggested that global warming, also known as the **greenhouse effect,** may represent as great a problem for the world's environment as damage to the ozone layer. Furthermore, the greenhouse effect is likely to be much more difficult to overcome. Measurements indicate that the world's average temperature increased by roughly .1 degree Fahrenheit per year during the 1980s and 1990s. That might not seem like much, but if this trend continues, by 2050 the world's average temperature will be five degrees higher than it is now. Such a temperature increase would melt the polar ice caps and raise the sea level by as much as forty feet. Many coastal regions would be flooded, and there would be a worldwide change in climate. Large numbers of plant and animal species could become extinct, including many varieties of fish that are important sources of food. Possibly the greatest potential danger is that the warming of the oceans might cause the death of marine plankton that generate most of the oxygen we breathe.

Many causes of global warming have been identified. One important cause is the release of carbon dioxide ($CO_2$) into the atmosphere from burning fossil fuel. In 1997, representatives of most nations gathered in Kyoto, Japan, to negotiate an agreement to reduce $CO_2$ emissions. After many days, an agreement was reached that called for an average 5 percent reduction in $CO_2$ emissions by participating nations. Although most nations have ratified this agreement, the United States has not. According to our government leaders, including President George W. Bush, abiding by this agreement would cause unacceptable harm to the U.S. economy and U.S. consumers. It was suggested that the reductions called for in the Kyoto Protocol would not stop the progress of global warming in any case because many developing nations did not sign the agreement. President George W. Bush has stated that the United States will do what it can to reduce $CO_2$ emissions in ways that will not harm the U.S. economy.

It is almost certain that environmental issues will grow in importance in the coming years. As an educated consumer, you will be better equipped to help sort out the hard choices that we will be forced to make in the future.

**Question for Thought & Discussion #10:**
What changes in lifestyle would you be willing to make to reduce the amount of $CO_2$ released into the earth's atmosphere? Would you drive a smaller car, invest in solar power, or use public transportation?

# Confronting Consumer Issues: Living a Green Life

Although numerous consumers believe they should do what they can to protect the environment, many feel powerless to do anything meaningful. Those who feel this way should realize that although their individual choices may have only a limited impact, collectively consumers have the power to make important differences through the decisions they make.

The loss of a single customer to a firm that irresponsibly pollutes the environment will make little difference. The loss of thousands of customers who refuse to patronize such a firm can force the business to close or change its practices. Consumers should also remember that they have a political voice in this nation. History shows that our government does respond to the will of the people, even though it is sometimes slow to do so. When consumers evaluate their alternatives, they should consider the effect their decisions will have on the environment both at the present and in the future. If we intend to leave a healthy world for our children, we need to make environmentally responsible choices now.

## THE GREEN CONSUMER

Consumers who make choices that protect the environment are sometimes called "green consumers." These people attempt to encourage businesses to produce products in environmentally responsible ways through their buying decisions.

**Green Shopping**   Businesses in the United States operate to earn a profit. They are, therefore, responsive to changes in consumer demand or public pressure that may influence the "bottom line." In the 1980s, for example, McDonald's restaurants sold hamburgers and fish sandwiches packaged in Styrofoam containers. Many people believed these containers were bad for the environment. They argued that the manufacture of these products released chemicals into the atmosphere that damaged the ozone layer and that when discarded in landfills, the containers would not degrade for thousands of years. Activists organized boycotts of McDonald's restaurants and carried out public information campaigns. Eventually, McDonald's replaced the plastic containers with cardboard ones. Although the firm never admitted that this decision had been influenced by consumer pressure, there seems to be little doubt that it was.

**Sources of Green Information**   Many publications rate businesses and the products they produce according to their records of being environmentally responsible. One of the best known, *Shopping for a Better World*, was first published by the Council on Economic

Is this your future home? How much style and beauty would you be willing to trade for energy efficiency?

Priorities in the early 1990s. This book rates firms by their environmental policies in a way that is easy to understand and use. Updated versions of this publication can be found online at Amazon.com or through many other booksellers.

A series of editions of *50 Simple Things You Can Do to Save the Earth,* published by the Earth Works Group, offers ideas for conservation and recycling. Probably the most convenient source of environmental information to use is the *National Green Pages,* prepared annually by Co-op America. This publication can be accessed online at **http://www.co-opamerica.org**. It identifies thousands of businesses that offer a wide variety of environmentally responsible products. A small selection of the "green" products offered through this publication is listed in Table 19.4 on the next page.

These examples are only a few of the thousands of sources individuals may use to find information about being environmentally responsible consumers. A simple Web search using any popular search engine will identify hundreds of organizations dedicated to protecting the environment. You might choose to join one or more of the organizations that have goals you support, or you could just use them as sources of information to help you make more responsible consumer decisions.

## RECYCLING, REUSE, AND CONSERVATION

Our environmental resources at any point in time are finite. Although some resources, such as fresh water, soil nutrients, and wildlife, are renewable, many are not. Oil or coal that is

*(Continued on next page)*

**Confronting Consumer Issues (Continued)**

| TABLE 19.4 | Green Products You Can Buy | |
|---|---|---|
| **Product** | **Firm's Name** | **Web Address** |
| Organic foods | Horizon Organic | http://www.horizonorganic.com |
| Children's books | Childsake | http://www.childsake.com |
| Energy-efficient appliances | Real Goods Renewables | http://www.gaiam.com |
| Solar power systems | Sierra Solar Systems | http://www.sierrasolar.com |
| Recycled paper products | Acorn Designs | http://www.acorndesigns.org |
| Pool purification systems | Carefree Cleanwater | http://www.carefreecleanwater.com |

burned today will not be available for future generations. Land that is used to store toxic waste may not be safe for other uses for hundreds or thousands of years. Groundwater that becomes polluted by either chemical or biological agents can take hundreds of years to regenerate. To assure adequate supplies of resources for our children and future generations, many consumers support efforts to recycle, reuse, or conserve the resources we consume today.

### The Potential of Recycling

Recycling is probably the most widely understood and supported aspect of the consumer environmental movement. Probably every U.S. family participates in recycling, if only by returning a soft-drink can or bottle for its deposit. Nearly 30 percent of household waste is recycled. Many states have set goals for trash recycling that far exceed current standards. California, Oregon, and Rhode Island, for example, have set 70 percent recycling goals for the future. Although recycling has become a major part of the consumer movement, there is a nagging question as to whether its benefits are greater than its costs, in both environmental and financial terms.

The cost of collecting, sorting, and transporting recyclable materials is substantial. Some localities are charged as much as $250 per ton for this service. Even when recycled materials are available, there is no guarantee that they will be used. The price of recycled paper fell so low in 2000 that many firms chose to store giant bales of used paper rather than sell them. Some recycling programs were stopped entirely because the available storage space had been filled.

Using recycled materials to produce products can be more expensive than producing brand-new products. Recycled paper, for example, must be treated with bleach and other chemicals to remove ink and dyes. Special materials are often added to recycled paper products to improve their texture and durability. All of this costs money and uses resources. Recycling may be the largest, most successful way to limit our use of resources, but it is not a perfect solution.

**Question for Thought & Discussion #11:**
Do you make a conscious effort to recycle, reuse, or conserve resources? Why do you do what you do?

**What Recycling Can't Do**   Although increased recycling may have positive effects, many environmental problems associated with trash disposal cannot be solved by recycling. For example, Americans use nearly 20 billion disposable diapers each year. These products cannot easily be recycled. When burned, they may create energy, but their incineration may cause atmospheric pollution and will leave a toxic residue that must be disposed of in specially sealed sites.

Many household products are made from composite materials that include a variety of metals, plastics, glass, and organic materials. Recycling these products is very expensive or impossible. Outdated home computers are a prime example. PCs contain lead and mercury that make them too dangerous to discard in ordinary landfills. They cannot be safely burned because these materials would be released into the atmosphere. Disassembling computers is very costly and still leaves the question of what to do with these dangerous materials. An estimated 50 million unused PCs are sitting in consumer basements, attics, and garages. This is a growing problem that recycling is unlikely to help resolve.

**Product Reuse**   For consumers who are serious about protecting the environment, reuse and conservation are better alternatives than recycling. Reusing products requires fewer resources than recycling. Consider soft-drink bottles as an example. When a plastic soft-drink bottle is returned for deposit, it will be ground up into plastic scraps that may be recycled into new products. This process requires additional resources that include heat and chemicals. A glass soft-drink bottle, however, may simply be washed, sterilized, and refilled. This also requires energy and water, but much less than recycling the plastic bottle.

As another example of reuse, consider the number of businesses that use discarded automobile tires to make other products. Playground equipment, boat bumpers on docks, and even soles for shoes and sandals are produced without the need to remanufacture the used tires in any significant way. Consumers who buy these products encourage reuse that is clearly environmentally responsible behavior.

**Conservation**   The best way to protect our scarce supplies of resources is to not use them at all. Conservation can be helpful to your wallet as well as beneficial to the environment. Consider the SUVs with which so many Americans seem to be in love. These vehicles typically receive mileage ratings that range from twelve to twenty miles per gallon. For every Lincoln Navigator or Cadillac Escalade on the road, four Toyota Prius hybrids could be driven using the same amount of gasoline. It has been estimated that the proliferation of SUVs has increased the consumption of gasoline in this country by as much as 15 percent over what it would otherwise have been. Certainly, SUVs offer some advantages. They are safer in accidents than smaller cars. They are better able to travel in snow or off-road conditions. And they can carry more people and luggage for large families. Ultimately, the type of vehicle consumers decide to drive is an individual choice. Still, for environmentally responsible consumers, mileage ratings are worth consideration.

Conservation does not have to involve large purchases such as automobiles. You can conserve by doing something as simple as taking a canvas bag with you when you go grocery shopping. Every grocery bag you don't use translates into reduced demand for scarce resources. In Europe, most stores expect customers to bring a canvas bag. They don't even offer plastic or paper bags to their customers.

## THE GREEN VOTER

The United States has a representative form of government. Political leaders are elected and must receive voter support to gain or retain office. This is true on the federal, state, and local levels. Our history is filled with examples of government policies that have been created or changed as a result of public pressure on politicians.

### Oil Drilling in the Alaskan National Wildlife Preserve   In the months following his inauguration in January 2001, President George W. Bush suggested opening up the Alaskan National Wildlife Preserve (ANWR) to oil drilling. This suggested change in government rules was part of a larger energy policy designed

to make the United States less dependent on sources of imported oil. There are an estimated 17 billion barrels of oil in the ground in the Alaskan Preserve as well as large deposits of natural gas. Unfortunately, extracting these resources would be difficult, and many believe the effort would result in an environmental disaster that could destroy the fragile environment of Alaska's North Slope. President Bush and his supporters have argued that oil-drilling technology has advanced to the point that there is no significant chance that an oil spill could take place.

The effort to approve drilling in the Alaskan Preserve has been blocked by U.S. senators who may have been influenced by political pressure from environmentalists, conservation groups, and individual voters. "I don't think that the American people support drilling in ANWR. . . . I think they would support a Democratic filibuster," then–Senate majority leader Tom Daschle said on NBC's *Meet the Press* news program in early 2002. Even after the Republican victory in November 2002, the Democratic Senate minority retained the ability to block new energy legislation by filibuster— ongoing debate that can be ended only by a two-thirds majority vote. In early 2003, several Republicans joined the Democrats to vote against the legislation.

This does not mean that all Americans oppose the Bush administration's energy policies. It does demonstrate the power U.S. citizens have to influence political decisions, including those that could affect the environment.

### Drilling in Florida   At the same time that the Bush administration was working to expand oil drilling in Alaska, political pressure appears to have caused the administration to prevent similar drilling in Florida. On May 19, 2002, the federal government announced that it would spend $235 million to purchase leases for more than 765,000 acres of land surrounding the Florida Everglades. These lands hold large oil and natural gas reserves, but surveys of Florida voters showed that 75 percent of them opposed further drilling. This purchase was seen as a boon to the reelection campaign of the president's brother, Jeb Bush, the governor of Florida, who stated, "[This is] a convergence of good politics and good public policy. I don't think we should be ashamed about it." Environmentalist David Reiner seemed to agree with Governor Bush when he said, "This is important because the Everglades have a unique problem: An oil spill goes everywhere. It's really one gigantic river."

### More Money for Michigan's Lakes and Wildlife   In the summer of 2002, Michigan voters approved a proposal that allocated additional funds from the Michigan State Park Endowment Fund and Natural

*(Continued on next page)*

## Confronting Consumer Issues (Continued)

Resources Trust Fund to support an expanded effort to protect water quality and the environment. A coalition of citizen groups had circulated petitions and lobbied state legislators to force the issue onto the state ballot. The Michigan director of the Nature Conservancy stated, "Passing [this proposal] means more money for local parks, lakes, and wildlife. We thank the voters for their continued support for preserving Michigan's natural heritage."

**Making a Difference**  It is clear that individual consumers can make a difference when it comes to protecting the world's environment. As resources become scarcer, and the costs of environmental damage increase, consumers, producers, and the government are likely to make choices that are more responsible toward the environment. The beginnings of this trend can already be seen. When the dangers of CFCs became apparent, the production of these chemicals was almost eliminated in just a few years. Appliances, air conditioners, and furnaces are much more efficient today than they were only a few years ago. Governments are providing grants and tax credits to businesses and consumers who use wind power or the sun to generate electricity. Many environmental challenges remain, but it is within our ability to overcome them, if we make rational choices.

**Question for Thought & Discussion #12:**
When you decide which candidate to vote for, how much consideration do you give to that person's stance on environmental issues? How much difference do you believe your opinion makes to politicians? Could you make them care more about what you think?

## THE TRUTH WILL CHANGE

The picture of reality that you hold today is not the same as what you will believe is true in the future. This does not mean you are wrong now—it means that tomorrow's reality will be different. There is no way to stop progress. It will take place whether we like it or not. The best we can hope for is to be able to keep up with change so that we can make rational decisions that are good for ourselves, our families, our communities, our country, and the world. The knowledge and skills you have gained in this course will help you achieve this goal.

# Chapter Summary

1. The world is changing at an accelerating pace. Consumers will be forced to deal with new technologies that are being developed at an increasing rate. Being a rational consumer requires consumers to commit themselves to a life of learning. Hybrid vehicles are only one of the many classes of new products that are becoming available to consumers. These products hold the promise of much-improved gas mileage but may result in increased maintenance costs.

2. New technologies also affect governments at all levels. Internet marketing has reduced sales tax receipts for state and local governments. Reduced tax receipts force governments to find other sources of revenue or reduce their spending for programs that benefit consumers. This situation has also given online retailers an advantage over traditional stores that are required to charge and collect sales tax.

3. The globalization of the world's economy affects consumers in every nation. In 2001, Americans purchased more than $1,350 billion worth of goods and services from other nations. This was almost 14 percent of the value of all products sold in the United States in that year. In the same year, foreign sales accounted for just over $1 trillion worth of American-made goods and services and helped to create roughly 18 million U.S. jobs.

4. Although many individual Americans benefit from international trade, clearly some others do not. When U.S. consumers buy imported goods, they provide income to foreign firms and their employees. Competition from imported products has forced many businesses in this country to lay off workers and close.

5. When a country imports a greater value of goods than it exports, it has a negative balance of trade. Countries that have a negative balance of trade often finance their trade deficit by borrowing from lenders in other nations. Many developing nations engage in this practice. On several occasions, debtor nations have experienced economic problems that left them unable to repay their loans without assistance from developed nations. Interruptions in the flow of trade among countries can be damaging to both importing and exporting nations.

6. The globalization of the world's economy has made it difficult to determine whether goods and services are imported or are domestic products. Multinational corporations are owned and operated by people from many nations. Many consumers now choose to buy products that have the best quality and lowest prices regardless of where they were manufactured.

7. Although the U.S. Constitution does not specifically guarantee a right to privacy, the United States Supreme Court has ruled that such a right is implied under various constitutional provisions. New technologies make it possible to gather and distribute vast amounts of information about individuals that many people believe should be private.

8. Information about individuals is gathered by employers, marketers, and the federal, state, and local governments. The possible distribution of medical information in inappropriate ways is particularly worrisome to many consumers. The Freedom of Information Act guarantees individuals the right to receive copies of information held by the federal government that concerns them. No such right is extended by state or local governments.

9. Advances in medical technology have enabled Americans to live longer, healthier lives but have also resulted in an ethical dilemma. It is impossible to provide all patients with every treatment that could help them. As a society, we must decide how many of our scarce resources should be allocated to providing medical care and how our medical-care spending should be allocated.

10. The need to protect the world's environment has become more pressing in recent years as human populations and demand for scarce resources have grown. Much of the damage done to the world's environment takes place during the production or consumption of goods and services that satisfy human wants. Most businesses exist to earn profits for their owners. Unfortunately, low costs to producers often translate into increased costs and pollution to the environment.

11. Pollution-abatement efforts are expensive. Although the direct costs of abatement programs are borne by businesses and the government, ultimately consumers pay through increased prices or taxes. Reducing pollution yields nonmonetary benefits that include improved public health and increased property values. These benefits are real but can be difficult to measure.

12. The Environmental Protection Agency is charged with overseeing the enforcement of federal environmental legislation. This agency has the power to change pollution-abatement rules without the approval of Congress. In 1999, many pollution standards were made stricter, only to be eased in 2002. Many state and local governments have passed environmental protection laws that frequently have gone beyond the standards set by the federal government.

13. Threats to the environment are a global problem. In the 1990s, most nations joined together to reduce the production of chlorofluorocarbons (CFCs) that damage the earth's ozone layer. There has been less success in reducing emissions of gases, particularly carbon dioxide, that contribute to global warming. Although an international agreement was reached in 1997 to reduce these gases, the

United States and a few other nations have not ratified the agreement, and there is doubt as to its efficacy even if it were followed.

14. Consumers are able to influence environmental conditions by leading a "green life." This happens when people make choices that protect the environment. There are many listings of businesses that work to protect the environment. The *National Green Pages* is probably the most widely used. By choosing to make purchases from these businesses, consumers can safeguard the environment.

15. Individual consumers can help protect the environment by making efforts to recycle, reuse, and conserve scarce resources. Although the actions of an individual consumer will have little impact on the situation, collectively consumers can make an important difference. Choosing more fuel-efficient vehicles, for example, could reduce our consumption of gasoline by 10 percent or more.

16. Consumers may also pressure the government to pass legislation or follow policies that are intended to protect the environment. Examples of this type of pressure can be seen in the decision not to drill for oil in the Alaskan National Wildlife Preserve or near the Florida Everglades.

17. There is no way to stop progress. It will take place whether we like it or not. The best we can hope for is to be able to keep up with change so that we can make rational decisions that are good for ourselves, our families, our communities, our country, and the world.

## Key Terms

chlorofluorocarbons (CFCs) **484**

greenhouse effect **484**

negative balance of trade **467**

specialization **466**

## Questions for Thought & Discussion

1. What single new technology do you believe has changed the lives of U.S. consumers the most in the past year? Explain your choice.

2. Do you feel any guilt when you buy clothes online without paying sales tax? Should you? Explain your answer.

3. Do you know anyone who believes he or she lost a job because of foreign competition? Would you be willing to pay higher prices to help this person regain employment?

4. Would you support legislation that would force Americans to buy more fuel-efficient cars so we would be less dependent on imported oil? Explain your answer.

5. How might future consumers benefit from a total globalization of the world economy? What costs might they pay?

6. What costs and benefits do you experience because organizations gather, store, and distribute information about you?

7. If you had the power, what restrictions would you place on how information about individuals could be collected and used? How would you enforce these restrictions?

8. What part of your income would you be willing to allocate to paying for medical care? For you, how much is too much?

9. What role do you believe the government should play in environmental protection?

10. What changes in lifestyle would you be willing to make to reduce the amount of $CO_2$ released into the earth's atmosphere? Would you drive a smaller car, invest in solar power, or use public transportation?

11. Do you make a conscious effort to recycle, reuse, or conserve resources? Why do you do what you do?

12. When you decide which candidate to vote for, how much consideration do you give to that person's stance on environmental issues? How much difference do you believe your opinion makes to politicians? Could you make them care more about what you think?

## Things to Do

1. Assume that you have decided to purchase an "American-made" digital camera. Gather information to determine if this is possible. If it is, compare the price and quality of the camera you identify with that of a competing imported product.

2. Prepare a list of organizations that you believe have files of information about you. Write an essay describing your feelings about this situation and explaining what, if anything, you believe the government should do to further protect your right to privacy.

3. Identify an advance in medical technology that has been beneficial to a member of your family or to someone you know personally. Investigate the development of this technology, including the funds spent on the effort. Compare the benefits and costs of this technology. Do you believe the benefits warranted the costs? Why or why not?

4. Identify a national environmental problem that you believe represents a significant danger to future generations. Describe the scope of this problem. Write and explain a proposal for a law that would reduce or eliminate the problem.

# Internet Resources

## Finding Consumer Information on the Internet

 The following Web sites have been selected for their relevance to topics discussed in this chapter. Search these sites to locate information that can add to your knowledge of developments that are likely to change consumers' lives in the future. Remember, Web addresses change frequently. If any of these addresses no longer function, find similar sites to investigate using any of the search engines available to you.

1. The NAFTA Secretariat is responsible for the administration of the dispute-settlement provisions of the agreement. Investigate its Web site at **http://www.nafta-sec-alena.org**.

2. The *Journal of the American Medical Association* publishes articles about new medical technologies. Its Web site can be found at **http://www.jama.ama-assn.org**.

3. The mission of the Environmental Protection Agency is to protect human health and safeguard the natural environment. Find out what is new at the EPA by searching its Web site at **http://www.epa.gov**.

## Shopping on the Internet

 The following Web sites have been selected because they offer consumers services similar to those described in this chapter. These are commercial sites that are designed to market products. They do not represent a comprehensive or balanced description of all firms that offer new types of products online. How do the services at these Web sites compare with those that are available from local businesses? Remember, Web addresses change frequently. If any of these addresses no longer function, find similar sites to investigate using any of the search engines available to you.

1. HybridCars.com is an online magazine covering environmental innovation in the automotive industry. Investigate recent developments in the auto industry by visiting its Web site at **http://www.hybridcars.com**.

2. Automotive Japanese Imports & Internet Car offers quality car imports and automotive services to end users and professionals, from stock or auctions. Find out what products this organization is marketing at its Web site located at **http://www.bestjapancar.com**.

3. Environmentally-Friendly Cleaning Products is an Internet marketer of nontoxic laundry detergents. Search this firm's Web site to find products you might be able to use. It is located at **http://www.ecomall.com/biz/cleaning.htm**.

## InfoTrac Exercises

 Purchasers of new copies of this text are provided with access to the InfoTrac Web site. This Web site links students to thousands of recent articles published in hundreds of periodicals. Use the key words **environmental protection, medical technology,** or other terms from this chapter to conduct a key-word search. Choose one article that is of particular interest to you and write a brief essay describing what you have learned from the article. Be sure to cite the author and title of the article and the name and date of the publication in which it appeared.

# Selected Readings

Baskin, Kathy. "Do You Have a Right to Privacy?" *PC World,* May 1999, p. 15.

Brenner, Robert. *The Boom and the Bubble: The U.S. in the World Economy.* New York: Verso Books, 2002.

Bunyard, Peter. "How Ozone Depletion Increases Global Warming." *The Ecologist,* March–April 1999, p. 85.

Colker, David. "Web Sales Tax Plan Pushed in California." *Los Angeles Times,* February 23, 2003, p. F1.

Elington, John, Julia Hailes, and Joel Makowen. *The Green Consumer.* New York: Penguin, 1990.

Epstein, Eve. "Converging Futures." *InfoWorld,* January 10, 2000, p. 27.

Glenn, Jim. "The State of Garbage in America." *BioCycle,* April 1999, p. 60.

Heinkel, Robert, Alan Kraus, and Joseph Zechner. "The Effect of Green Investment on Corporate Behavior." *Journal of Financial and Quantitative Analysis,* December 2001, pp. 431–450.

Monastersky, R. "A Sign of Healing Appears in the Stratosphere." *Science News,* December 18, 1999, p. 391.

Rothfeder, Jeffery. *Every Drop for Sale: Our Desperate Battle over Water.* East Rutherford, NJ: Tarcher, 2001.

Ward, Janet. "2001: A Trash Odyssey." *American City & County,* May 1999, p. 74.

Weiss, Stefan C. "Economics, Ethics, and End-of-Life Care." *JAMA,* December 1, 1999, p. 2076.

Young, Deborah. "Privacy vs. Data Collection." *Wireless Review,* April 1, 2001, p. 10.

# Glossary

**Abstract** A short history of title to land; a document listing all records relating to a given parcel of land.

**Acceleration Clause** A clause contained in numerous credit agreements whereby, if one payment is missed, the entire unpaid balance becomes due, or the due date is accelerated to the immediate future.

**Add-On Clause** A clause in an installment contract that makes your earlier purchases from that source security for the new purchase.

**Age/Earnings Profile** The profile of how earnings change with your age. When you're young and just starting out, your earnings are low; as you get older, your earnings increase because you become more productive and work longer hours; finally, your earnings start to decrease.

**Amortization Schedule** A table showing the amount of monthly payments due on a long-term loan, such as a mortgage; it indicates the exact amounts going toward interest and toward principal.

**Anesthesiologist** A doctor who specializes in administering anesthesia to patients before surgery.

**Annual Percentage Rate (APR)** The annual interest cost of consumer credit.

**Annual Percentage Yield (APY)** The standard annualized return on a saving deposit that all savings institutions must provide to depositors under the Truth in Savings Act.

**Antitrust Laws** Laws designed to prevent business monopolies. Antitrust laws are part of government antitrust policies that are aimed at establishing and maintaining competition in the business world to assure consumers of fair prices and goods of adequate quality.

**Asset** Something of value that is owned by an individual, a business, or the government.

**Assigned Risk** A person seeking automobile insurance who has been refused coverage. That person is assigned to an insurance company that is a member of the assigned-risk pool in the person's state.

**Automated Teller Machine (ATM)** An electronic customer-bank communication terminal that, when activated by an access card and personal identification number, can conduct routine banking transactions.

**Bait and Switch** A selling technique that involves advertising a product at a very attractive price (the "bait"); then informing the consumer, once he or she is in the store, that the advertised product either is not available, is of poor quality, or is not what the consumer "really wants"; and finally, promoting a more expensive item (the "switch").

**Bankruptcy** The state of having come under the provisions of the law that entitles a person's creditors to have that person's assets administered for their benefit.

**Basic Form Policy** A homeowners' insurance policy that covers eleven risks.

**Bed-and-Breakfast** A business run by an individual who rents rooms in his or her home and provides breakfasts to travelers.

**Beneficiary** The designated person or persons for any insurance policy. In a life-insurance policy, the beneficiary is the person who receives the benefits when the insured dies.

**Broad Form Policy** A homeowners' policy that covers eighteen risks.

**Bushing** Adding unordered accessories to a product to increase its price.

**Capital Gain** An increase in the value of something you own. Generally, you experience a capital gain when you sell something you own, such as a house or a stock. You compute your capital gain by subtracting the price you paid for whatever you are selling from the price you receive when you sell it.

**Capital Loss** The difference between the buying price and the selling price of something you own when the selling price is lower than the buying price.

**Capitalism** An economic system based on private ownership of the means of production and on a demand-and-supply market. This system emphasizes the absence of government restraints on ownership, production, and trade.

**Capitation** A plan for medical care in which providers are given a specific amount of money for each patient regardless of what treatment is provided.

**Carcinogenic** Cancer causing.

**Cash Value** Applied to whole life policies only, it represents the amount of "savings" built up in the policy and available to the living policyholder, either to borrow against or to receive if the policy is canceled.

**Cashier's Check** A check drawn against the funds of a bank to a designated person or institution. A cashier's check is paid for before it is obtained.

**Cease-and-Desist Order** An administrative or judicial order commanding a business firm to cease conducting the activities that the agency or a court has deemed to be "unfair or deceptive acts or practices."

**Certificate of Deposit (CD)** A deposit that cannot be cashed in before a specified time without paying a penalty in the form of reduced interest.

**Certified Check** A check that a bank has certified, indicating that sufficient funds are available to cover it when it is cashed.

**Chlorofluorocarbons (CFCs)** A family of chemicals associated with the depletion of ozone in the earth's upper atmosphere.

**Collateral** The backing that people often must put up to obtain a loan. Whatever is used as collateral for a loan can be sold to repay that loan if the debtor cannot pay it off as specified in the loan agreement. For example, the collateral for a new-car loan is generally the new car itself. If the finance company does not get paid for its car loan, it can then repossess the car and sell it to recover the amount of the loan.

**Common Law**   The unwritten system of law governing people's rights and duties, based on custom and fixed principles of justice. Common law is the foundation of both the English and the U.S. legal systems (excluding Louisiana, where law is based on the Napoleonic Code).

**Common Stock**   A unit of ownership that has a legal claim to the profits of a company. For each share owned, the common-stock owner generally has the right to one vote on such questions as merging with another company or electing a new board of directors.

**Comparative Advertising**   Advertising that makes comparisons between a product and specific competing products.

**Comparison Shopping**   Acquiring and comparing information about different sellers and different products in order to find the best price for products of substantially the same quality.

**Comprehensive Form Policy**   A homeowners' policy that covers all risks except, usually, flood, war, and nuclear attack.

**Condominium**   An apartment house or complex in which each living unit is individually owned. Each owner receives a deed allowing her or him to sell, mortgage, or exchange the unit independent of the owners of the other units in the building. Title to a condominium also gives the purchaser shared ownership rights in common areas.

**Conspicuous Consumption**   Consumption of goods more for their ability to impress others than for the inherent satisfaction they yield.

**Consumer Price Index (CPI)**   A price index based on a fixed representative market basket of about four hundred goods and services purchased in eighty-five urban areas.

**Consumer Redress**   The right of consumers to seek and obtain satisfaction for damages suffered from the use of a product or a service; protection after the fact.

**Consumers**   Individuals who purchase (or are given), use, maintain, and dispose of products and services in their final form in an attempt to achieve the highest level of satisfaction possible with their income limitations.

**Consumer Sovereignty**   A situation in which consumers ultimately decide which products and styles will survive in the marketplace; that is, producers do not dictate consumer tastes.

**Cooling-Off Period**   A specific amount of time in which a consumer has the right to reconsider and back out of a transaction.

**Cooperative**   An apartment building or complex in which each owner owns a proportionate share of a nonprofit corporation that holds title or a legal right to use the building.

**Cost-Benefit Analysis**   A way to reach decisions in which all the costs are added up, as well as all the benefits. If benefits minus costs are greater than zero, then a *net benefit* exists and the decision should be positive. Alternatively, if benefits minus costs are less than zero, then a *net cost* exists and the decision should be negative.

**Counteradvertising**   New advertising that is undertaken pursuant to a Federal Trade Commission order for the purpose of correcting earlier false claims made about a product.

**Debit Card**   A card similar to a credit card that allows a consumer to transfer funds from accounts by using a computerized banking system.

**Deductions**   Different types of expenses taxpayers may subtract from their incomes before figuring their tax liability.

**Defensive Advertising**   Advertising intended to rebut claims made by competing firms about a firm's product or business practices.

**Demand**   The quantity of a product that will be purchased at each possible price.

**Depository Institution**   Any organization that accepts deposits and assures depositors that they will be able to withdraw their funds when they need them.

**Dietary Reference Intakes (DRIs)**   Nutrient standards established by the National Academy of Sciences and expressed as recommended dietary allowances (RDAs) or estimated average requirements (EARs).

**Dietetic Foods**   Low-calorie or reduced-calorie food, or food intended for a special dietary purpose—for example, for low-sodium diets.

**Discount Points**   Additional charges added to a mortgage that effectively raise the rate of interest you pay.

**Dividend**   A share of profit paid by a corporation to its stockholders.

**Drug Interactions**   Reactions among drugs that can sometimes be life threatening.

**Dual Agency**   When an agent represents both the buyer and the seller in a transaction.

**Durable Goods**   Goods that typically last three or more years.

**Earnest Money**   Sometimes called a *deposit on a contract* or an *offer* to purchase a house. It is the amount of money you put up to show that you are serious about the offer you are making to buy a house. Generally, you sign an earnest agreement, or a contract that specifies the purchase price you are willing to pay for the house in question. If the owner selling the house signs, then you are committed to purchase the house; if you back down, you could lose the entire earnest money.

**Economic System**   A set of understandings that governs the production and distribution of goods and services that satisfy human wants.

**Electronic Fund Transfer Systems (EFTS)**   Systems for transferring funds electronically.

**Electronic Fund Transfer**   A transfer of funds via an electronic terminal, telephone, computer, or magnetic tape.

**Energy Guide Label**   A label that shows the expected energy costs of operating an appliance for a year; required by law for many consumer durables.

**Energy Star Program**   A voluntary labeling program initiated by the Environmental Protection Agency to identify and promote energy-efficient products and thereby reduce greenhouse gas emissions.

**Engel's Law**   A proposition, first enunciated by Ernst Engel, that states that as a family's income rises, the proportion spent on food falls.

**Equilibrium Price**   A price at which the quantity of a good or service demanded is exactly equal to the quantity that is supplied.

**Equity**   A legal claim to the profits of a company. This is another name for stock, generally called common stock.

**Ethical Behavior**   Behavior that is directed by moral principles and values; determining what is "right" in a given situation and acting in accordance with that determination.

**Ethical Investing**   Investing in corporations that are deemed to be socially responsible according to a given set of ethical criteria.

**Ethical Shopping**   Purchasing products manufactured by socially responsible business firms and refusing to purchase products manufactured by firms whose ethical behavior is perceived to be reprehensible.

**Excise Tax**   A tax that is collected from the manufacturer of a product.

**Exemption**   An amount of money that may be subtracted from income for each person a taxpayer supports.

**Filing Status**   The family situation under which taxes are filed: single, married filing jointly, head of household, and so on.

**Finance Charge**   The total costs you pay for credit, including interest charges, possible credit-insurance premium costs and appraisal fees, and other service charges.

**Fixed Expenses**   Expenses that occur at specific times and cannot be altered. Once a house is purchased or rented, a house payment is considered a fixed expense; so is a car payment.

**Flexible Expenses**   Expenses that can be changed in the short run. The amount of money you spend on food can be considered a flexible expense because you can buy higher- or lower-quality food than you now are buying. These are also known as variable expenses.

**401(k) Plan**   A form of retirement plan that can be used by corporations to set aside part of employee earnings for retirement savings before taxes are withheld.

**Fractional Reserve System**   A banking system in which depository institutions are required to keep a minimum share (the *reserve requirement*) of deposits they receive on reserve.

**Fraud**   Making a false statement with knowledge of its falsity, or with reckless disregard as to its truth, with the intent to cause someone to rely on the statement and therefore give up something of value.

**Free-Rider Problem**   When individuals attempt to receive benefits from a good or service without paying their appropriate share.

**Garnishment**   A court-ordered withholding of part of wages, the proceeds of which are used to satisfy unpaid debts. Also called *wage attachment.*

**Generalist**   A lawyer who does not practice in a specialized field of law but is qualified to provide legal advice in most normal situations.

**General Practitioner (GP)**   A doctor who has a family practice rather than a specialized practice.

**Generic Drugs**   Non–brand-name medicines that are often sold at much lower prices than the same brand-name products.

**Goods**   Tangible objects that have the ability to satisfy human wants.

**Greenhouse Effect**   The gradual warming of the earth's atmosphere, primarily as a result of the release of carbon dioxide from the burning of fossil fuels.

**Gross Capitalized Cost (GCC)**   The stated value of a new car that is leased.

**Health Maintenance Organization (HMO)**   A type of insurance plan in which members pay a flat fee in return for all medical services, provided they are administered by participating doctors and hospitals.

**High-Balling**   Offering an artificially high value for a product that is traded in and then inflating the price of the product that is sold.

**Home Warranty**   A form of insurance that protects a home buyer from the cost of major repairs that result from defects in construction.

**Hospice**   An organization that helps the families of terminally ill patients.

**Human Capital**   The skills and abilities humans have that allow them to produce goods and services from other productive resources.

**Identity Theft**   A crime in which a thief uses a person's Social Security number and other information to assume the person's identity as a consumer, borrower, saver, and investor.

**Imperfect Competition**   A market condition in which individual businesses have some power to set the price and quality of their products.

**Implied Warranty of Fitness**   An implicit warranty of fitness for a particular purpose, meaning that a seller guarantees the product for the specific purpose for which a buyer will use the goods, when the seller is offering his or her skill and judgment as to suitable selection of the right products.

**Implied Warranty of Merchantability**   An implicit promise by a seller that an item is reasonably fit for the general purpose for which it is sold.

**Income Transfer**   A transfer of income from some individuals in the economy to other individuals. This is generally done by the government. It is a transfer in the sense that no current services are rendered by the recipients. Unemployment insurance, for example, is an income transfer to unemployed individuals.

**Individual Retirement Account (IRA)**   An investment account on which the earnings are not taxed until funds are withdrawn from the account, usually at retirement. IRA contributions may also be tax deductible, depending on one's income level.

**Inflation**   A sustained rise in the weighted average of all prices.

**Informative Advertising**   Advertising that simply informs.

**Injunction**   A legal order requiring that an activity be stopped, corrected, or undertaken.

**Inpatient Services**   Services rendered to an individual by doctors and/or a hospital while the patient remains in the hospital for at least one night.

**Inside Information**   Information about a company's financial situation that is obtained before the public obtains it. True inside information is usually known only by corporate officials or other insiders.

**Insurable Interest**   Something of value that is to be insured.

**Insurable Risk**   An insurer's understanding of the risk of insuring a particular object or condition.

**Interest**   The cost of using someone else's money.

**Interest-Adjusted Cost (IAC)**   An insurance cost index that takes account of dividends, interest, and earnings of the policy.

**Internet Service Provider (ISP)**   A business that provides subscribers with access to the Internet through local telephone or cable lines.

**Investment**   The act of giving up something of value at present to be able to receive something else of greater value in the future.

**Keogh Plan**   A retirement program designed for self-employed persons by which a certain percentage of their income can be sheltered from taxation. As with IRAs, interest earnings are not taxed until withdrawal.

**Law of Demand**   A basic economic principle that states that as the price of goods or services rises, the quantity of those goods and services demanded will fall. Conversely, as the price falls, the quantity demanded will rise.

**Law of Supply**   A basic economic principle that states that as the price of goods or services rises, the quantity of those goods and services supplied will increase. Conversely, as the price falls, the quantity supplied will also decline.

**Lease**   A contract by which one conveys real estate for a specified period of time and usually for a specified rent; and the act of such conveyance or the term for which it is made.

**Liability**   Something for which you are liable or responsible according to law or equity, especially pecuniary debts or obligations.

**Lien**   A claim placed on the property of another as security for some debt or charge.

**Life-Span Goal**   A central goal that you wish to achieve within your life span.

**Litigants**   Those people involved in a lawsuit; that is, in the process of litigation.

**Living Benefits**   Benefits paid on a whole life type of insurance policy while the person is living. Living benefits include fixed and variable annuities.

**Low-Balling**   Offering to sell a product at a low price in a telephone conversation and then increasing the price when the consumer visits the firm to purchase the product.

**Luxury Good**   A good whose purchase increases more than in proportion to increases in income. Jewelry, gourmet foods, and sports cars usually fall into this category.

**Marginal Tax Rate**   The share of the next dollar earned that must be paid in taxes.

**Market Economy**   An economy that is characterized by exchanges in markets that are controlled by the forces of demand and supply.

**Market Power**   The ability of producers to change price and/or quality without substantially losing sales.

**Market**   The sum of all transactions that take place between buyers and sellers of a particular type of product.

**Money Income**   The total amount of actual dollars you receive per week, per month, or per year.

**Money Market Accounts**   Accounts that pay interest rates that fluctuate from day to day as prevailing interest rates in the economy change.

**Money Price**   The price that we observe today in terms of today's dollars. Also called the *absolute, nominal,* or *current price.*

**Monopoly**   The only producer of a product that has no substitutes.

**Mortgage**   A loan obtained for the purpose of purchasing land or buildings, in which the property is pledged as security.

**Mutual Fund**   A fund that purchases the stocks of other companies. If you buy a share in a mutual fund, you are, in essence, buying shares in all the companies in which the mutual fund invests. The only business of a mutual fund is buying other companies' stocks.

**Negative Balance of Trade**   The condition when the value of the goods and services a country imports exceeds the value of its exports.

**Net Worth**   The difference between the value of your assets and your liabilities— that is, what you are actually worth.

**No-Fault Auto Insurance**   A system of auto insurance whereby, no matter who is at fault, the individual is paid by his or her insurance company for a certain amount of medical costs and for the damage to the car.

**Nondurable Goods**   Products that have a useful life of less than three years.

**Nonlegal Professional**   A person who is not a lawyer but is qualified to provide legal advice about specific situations.

**OASDHI**   Old-Age, Survivors, Disability, and Hospital Insurance—the government name for Social Security insurance.

**Opportunity Cost**   The value of a second-best choice that is given up when a first choice is taken.

**Outpatient Services**   The services of doctors and/or hospitals that do not require the individual to remain as a registered patient in the hospital.

**Parkinson's Law**   Work will expand to fit the time allotted for it.

**Perfect Competition**   A market condition in which many businesses offer the same product for sale to many customers at the same price.

**Personal Identification Number (PIN)**   A number given to the holder of a debit card that allows the card to be used to transfer funds from accounts electronically. Typically, the card will not provide access to an electronic fund transfer system without the number.

**Persuasive Advertising**   Advertising intended to associate a specific product with a certain lifestyle or image in the minds of consumers.

**Plaintiff**   One who initiates a lawsuit.

**Point-of-Sale System**   An electronic communication system that can debit a customer's account when a debit card is used to cover a purchase from a merchant.

**Preferred Provider Organization (PPO)**   A type of insurance plan similar to an HMO but more flexible. In a PPO, members are allowed to choose the services of non–PPO medical providers in return for a higher copayment or deductible.

**Preferred Stock**   A unit of ownership in a corporation; each share entitles the owner to a fixed dividend that the corporation must pay before it pays any dividends to common stockholders; owners have a claim on the firm's assets before common stockholders if the firm fails, but have no vote in choosing the firm's board of directors.

**Premium**   The payment that must be made to the insurance company to keep an insurance policy in effect. Premiums usually are paid quarterly, semiannually, or annually.

**Prepayment Privileges**   With a mortgage loan with prepayment privileges, you can prepay your loan before the maturity date and not have to pay a penalty.

**Preventive Medicine**   Medical procedures carried out with the intention of preventing people from becoming ill.

**Private Costs**   The costs that are incurred by an individual and no one else. The private costs of driving a car, for example, include depreciation, gas, and insurance.

**Pro Rata**   Proportionately; that is, according to some exactly calculable factor.

**Productive Resources**   Raw materials, tools, and labor that may be used to produce other goods or services that have the ability to satisfy human wants.

**Profit**   The difference between the total amount of money income received from selling a good or a service and the total cost of providing that good or service.

**Progressive Tax**   A tax imposed in such a way that the greater your income is, the greater the share of that income you will pay in tax.

**Proportional Tax**   A tax imposed in such a way that all people pay the same share of their income in tax.

**Psychic Income**   The satisfaction derived from a work situation or occupation; nonmonetary rewards from doing a particular job.

**Pyramid Scheme**   An illegal sales plan through which people collect fees and a share of income earned from sales made by other individuals they recruit into the program.

**Rational Consumer Decision Making**
Making consumer decisions that maximize the satisfaction you can obtain from your time and money resources and that assist you in attaining lifelong, as well as short-term, goals.

**Real Values**   Dollar values that have been adjusted for inflation.

**Regressive Tax**   A tax imposed in such a way that the greater your income is, the smaller the percentage of that income you will pay in tax.

**Regulation E**   The rules issued by the Federal Reserve Board to protect users of electronic banking services.

**Relative Price**   The price of a commodity expressed in terms of the price of another commodity or the (weighted) average price of all other commodities.

**Residence Contents Broad Form Policy**
A renters' insurance policy that covers possessions against eighteen risks. It includes additional living expenses and liability coverage in case someone is injured in the apartment or house you are renting.

**Residual Value**   The predicted value a car will have at the end of a lease agreement.

**Résumé**   A brief summary of your education, training, and other achievements that you give to a prospective employer.

**Rider**   A written attachment to an insurance policy that alters the policy to meet certain conditions, such as convertibility, double indemnity, and so on.

**Right of Rescission**   The right to cancel a contract or an agreement that has been signed. For example, if you sign an agreement to buy a set of encyclopedias from a door-to-door salesperson, you have the right to cancel the agreement within a three-day period.

**Saving**   The act of not consuming or not spending your money income to obtain current satisfaction.

**Scarcity**   The condition in which we are unable to provide enough products to satisfy all people's needs and wants because of our limited resources.

**Service Flow**   The service that a product will yield over its expected life.

**Services**   Intangible actions that have the ability to satisfy human wants.

**Social Costs**   The costs that society bears for an action. For example, the social costs of driving a car include any pollution or congestion caused by that automobile.

**Socialist Economic System**   An economic system in which there is group (most often government) ownership of productive resources and control over the distribution of goods and services.

**Specialization**   The concentration of efforts on one area of production with the aim of having an advantage in the marketplace. Students specialize when they major in a certain subject at college, thus allowing them to have a comparative advantage in

the job market later—assuming there is a demand for their specialized knowledge.

**Standard Deduction**   An amount all taxpayers are allowed to subtract from their incomes before figuring their tax liability. The amount varies, depending on the taxpayer's filing status.

**Stock Market**   An organized market where shares of ownership in businesses are traded. These shares generally are called stocks. The largest centralized stock market in the United States is the New York Stock Exchange.

**Stop-Payment Order**   An order to one's bank not to honor a particular check when it is presented for payment.

**Supply**   The quantity of a product businesses are willing to offer for sale at each possible price.

**Tax Preferences**   A reduced tax rate applied to specific types of income; a legal method of reducing tax liabilities.

**Taxable Income**   A person's income that is subject to tax; computed by subtracting various deductions and exemptions from gross income.

**Technological Obsolescence**   When products lose value because they are technologically out of date rather than because they are worn out.

**Term Insurance**   Life insurance that is for a specified term (period of time) and provides only a death benefit; it is a form of pure insurance with no saving aspect.

**Time-Sharing**   An agreement through which consumers purchase the right to use a vacation facility for a specified period of time each year.

**Title Insurance**   Insurance that you pay for when you buy a house so you can be assured that the title, or legal ownership, to the house is free and clear.

**Title**   The physical representation of your legal ownership of a house. The title is sometimes called the *deed*.

**Townhouse**   A house that shares common side walls with other, similar houses.

**Trade-Off**   A term relating to opportunity cost. To get a desired economic good, it is necessary to trade off some other desired economic good whenever we are in a world of scarcity. A trade-off, then, involves a sacrifice that must be made to obtain something.

**Transaction Costs** All the costs associated with completing an exchange beyond the price of the product that is purchased.

**Traveler's Check** A guaranteed check, drawn against the funds of a financial institution, that is often used by consumers in place of cash when they travel. Like a cashier's check, it is purchased in advance. A traveler's check can be traced, and if stolen will be replaced by the issuing institution.

**U.S. Recommended Daily Allowances (U.S. RDAs)** A condensed system of nutrient standards that indicate the maximum amount of each nutrient needed for four broad categories of the population.

**Umbrella Policy** A type of supplemental insurance policy that can extend normal liability limits to $1 million or more for a relatively small premium.

**Underwriter** The company that stands behind the face value of any insurance policy. The underwriter signs its name to an insurance policy, thereby becoming answerable for a designated loss or damage on consideration of receiving a premium payment.

**Uniform Decreasing Term Insurance** A term insurance policy on which the premiums are uniform throughout the life of the policy but its face value declines.

**Values** Fundamental concepts or high-level preferences that regulate our behavior. High-level values determine lower-level tastes and preferences that affect our everyday lives.

**Voluntary Exchange** Transactions completed through the free will of those involved.

**W-2 Form** The form used by employers to report employee income and withholding to the employee and the government.

**Warranty of Habitability** An implied warranty made by a landlord to a tenant that leased or rented residential premises are in a condition that is safe and suitable for human habitation.

**Web Browser** A program that gives users the ability to search the Internet for specific types of information.

**Whole Life Insurance** Insurance that provides both death and living benefits; that is, part of the premium is put into a type of saving account.

**World Wide Web (WWW)** An information retrieval system that organizes the Internet's resources in a graphic fashion to facilitate the transfer of information between computers.

**Zero Deductible** In the collision part of an automobile-insurance policy, the provision that the insured pays nothing for any repair to damage on the car due to an accident that is the fault of the insured. Zero deductible is, of course, more expensive than a $250 or $500 deductible policy.

# Index